HANDBOOK OF THE NORMAL DISTRIBUTION

STATISTICS: Textbooks and Monographs

A Series Edited by

D. B. Owen, Founding Editor, 1972–1991

W. R. Schucany, Coordinating Editor
Department of Statistics
Southern Methodist University
Dallas, Texas

R. G. Cornell, Associate Editor
for Biostatistics
University of Michigan

W. J. Kennedy, Associate Editor
for Statistical Computing
Iowa State University

A. M. Kshirsagar, Associate Editor
for Multivariate Analysis and
Experimental Design
University of Michigan

E. G. Schilling, Associate Editor
for Statistical Quality Control
Rochester Institute of Technology

1. The Generalized Jackknife Statistic, *H. L. Gray and W. R. Schucany*
2. Multivariate Analysis, *Anant M. Kshirsagar*
3. Statistics and Society, *Walter T. Federer*
4. Multivariate Analysis: A Selected and Abstracted Bibliography, 1957–1972, *Kocherlakota Subrahmaniam and Kathleen Subrahmaniam*
5. Design of Experiments: A Realistic Approach, *Virgil L. Anderson and Robert A. McLean*
6. Statistical and Mathematical Aspects of Pollution Problems, *John W. Pratt*
7. Introduction to Probability and Statistics (in two parts), Part I: Probability; Part II: Statistics, *Narayan C. Giri*
8. Statistical Theory of the Analysis of Experimental Designs, *J. Ogawa*
9. Statistical Techniques in Simulation (in two parts), *Jack P. C. Kleijnen*
10. Data Quality Control and Editing, *Joseph I. Naus*
11. Cost of Living Index Numbers: Practice, Precision, and Theory, *Kali S. Banerjee*
12. Weighing Designs: For Chemistry, Medicine, Economics, Operations Research, Statistics, *Kali S. Banerjee*
13. The Search for Oil: Some Statistical Methods and Techniques, *edited by D. B. Owen*
14. Sample Size Choice: Charts for Experiments with Linear Models, *Robert E. Odeh and Martin Fox*
15. Statistical Methods for Engineers and Scientists, *Robert M. Bethea, Benjamin S. Duran, and Thomas L. Boullion*
16. Statistical Quality Control Methods, *Irving W. Burr*
17. On the History of Statistics and Probability, *edited by D. B. Owen*
18. Econometrics, *Peter Schmidt*
19. Sufficient Statistics: Selected Contributions, *Vasant S. Huzurbazar (edited by Anant M. Kshirsagar)*
20. Handbook of Statistical Distributions, *Jagdish K. Patel, C. H. Kapadia, and D. B. Owen*

21. Case Studies in Sample Design, *A. C. Rosander*
22. Pocket Book of Statistical Tables, *compiled by R. E. Odeh, D. B. Owen, Z. W. Birnbaum, and L. Fisher*
23. The Information in Contingency Tables, *D. V. Gokhale and Solomon Kullback*
24. Statistical Analysis of Reliability and Life-Testing Models: Theory and Methods, *Lee J. Bain*
25. Elementary Statistical Quality Control, *Irving W. Burr*
26. An Introduction to Probability and Statistics Using BASIC, *Richard A. Groeneveld*
27. Basic Applied Statistics, *B. L. Raktoe and J. J. Hubert*
28. A Primer in Probability, *Kathleen Subrahmaniam*
29. Random Processes: A First Look, *R. Syski*
30. Regression Methods: A Tool for Data Analysis, *Rudolf J. Freund and Paul D. Minton*
31. Randomization Tests, *Eugene S. Edgington*
32. Tables for Normal Tolerance Limits, Sampling Plans and Screening, *Robert E. Odeh and D. B. Owen*
33. Statistical Computing, *William J. Kennedy, Jr., and James E. Gentle*
34. Regression Analysis and Its Application: A Data-Oriented Approach, *Richard F. Gunst and Robert L. Mason*
35. Scientific Strategies to Save Your Life, *I. D. J. Bross*
36. Statistics in the Pharmaceutical Industry, *edited by C. Ralph Buncher and Jia-Yeong Tsay*
37. Sampling from a Finite Population, *J. Hajek*
38. Statistical Modeling Techniques, *S. S. Shapiro and A. J. Gross*
39. Statistical Theory and Inference in Research, *T. A. Bancroft and C.-P. Han*
40. Handbook of the Normal Distribution, *Jagdish K. Patel and Campbell B. Read*
41. Recent Advances in Regression Methods, *Hrishikesh D. Vinod and Aman Ullah*
42. Acceptance Sampling in Quality Control, *Edward G. Schilling*
43. The Randomized Clinical Trial and Therapeutic Decisions, *edited by Niels Tygstrup, John M Lachin, and Erik Juhl*
44. Regression Analysis of Survival Data in Cancer Chemotherapy, *Walter H. Carter, Jr., Galen L. Wampler, and Donald M. Stablein*
45. A Course in Linear Models, *Anant M. Kshirsagar*
46. Clinical Trials: Issues and Approaches, *edited by Stanley H. Shapiro and Thomas H. Louis*
47. Statistical Analysis of DNA Sequence Data, *edited by B. S. Weir*
48. Nonlinear Regression Modeling: A Unified Practical Approach, *David A. Ratkowsky*
49. Attribute Sampling Plans, Tables of Tests and Confidence Limits for Proportions, *Robert E. Odeh and D. B. Owen*
50. Experimental Design, Statistical Models, and Genetic Statistics, *edited by Klaus Hinkelmann*
51. Statistical Methods for Cancer Studies, *edited by Richard G. Cornell*
52. Practical Statistical Sampling for Auditors, *Arthur J. Wilburn*
53. Statistical Methods for Cancer Studies, *edited by Edward J. Wegman and James G. Smith*
54. Self-Organizing Methods in Modeling: GMDH Type Algorithms, *edited by Stanley J. Farlow*
55. Applied Factorial and Fractional Designs, *Robert A. McLean and Virgil L. Anderson*
56. Design of Experiments: Ranking and Selection, *edited by Thomas J. Santner and Ajit C. Tamhane*
57. Statistical Methods for Engineers and Scientists: Second Edition, Revised and Expanded, *Robert M. Bethea, Benjamin S. Duran, and Thomas L. Boullion*
58. Ensemble Modeling: Inference from Small-Scale Properties to Large-Scale Systems, *Alan E. Gelfand and Crayton C. Walker*

59. Computer Modeling for Business and Industry, *Bruce L. Bowerman and Richard T. O'Connell*
60. Bayesian Analysis of Linear Models, *Lyle D. Broemeling*
61. Methodological Issues for Health Care Surveys, *Brenda Cox and Steven Cohen*
62. Applied Regression Analysis and Experimental Design, *Richard J. Brook and Gregory C. Arnold*
63. Statpal: A Statistical Package for Microcomputers—PC-DOS Version for the IBM PC and Compatibles, *Bruce J. Chalmer and David G. Whitmore*
64. Statpal: A Statistical Package for Microcomputers—Apple Version for the II, II+, and IIe, *David G. Whitmore and Bruce J. Chalmer*
65. Nonparametric Statistical Inference: Second Edition, Revised and Expanded, *Jean Dickinson Gibbons*
66. Design and Analysis of Experiments, *Roger G. Petersen*
67. Statistical Methods for Pharmaceutical Research Planning, *Sten W. Bergman and John C. Gittins*
68. Goodness-of-Fit Techniques, *edited by Ralph B. D'Agostino and Michael A. Stephens*
69. Statistical Methods in Discrimination Litigation, *edited by D. H. Kaye and Mikel Aickin*
70. Truncated and Censored Samples from Normal Populations, *Helmut Schneider*
71. Robust Inference, *M. L. Tiku, W. Y. Tan, and N. Balakrishnan*
72. Statistical Image Processing and Graphics, *edited by Edward J. Wegman and Douglas J. DePriest*
73. Assignment Methods in Combinatorial Data Analysis, *Lawrence J. Hubert*
74. Econometrics and Structural Change, *Lyle D. Broemeling and Hiroki Tsurumi*
75. Multivariate Interpretation of Clinical Laboratory Data, *Adelin Albert and Eugene K. Harris*
76. Statistical Tools for Simulation Practitioners, *Jack P. C. Kleijnen*
77. Randomization Tests: Second Edition, *Eugene S. Edgington*
78. A Folio of Distributions: A Collection of Theoretical Quantile-Quantile Plots, *Edward B. Fowlkes*
79. Applied Categorical Data Analysis, *Daniel H. Freeman, Jr.*
80. Seemingly Unrelated Regression Equations Models: Estimation and Inference, *Virendra K. Srivastava and David E. A. Giles*
81. Response Surfaces: Designs and Analyses, *Andre I. Khuri and John A. Cornell*
82. Nonlinear Parameter Estimation: An Integrated System in BASIC, *John C. Nash and Mary Walker-Smith*
83. Cancer Modeling, *edited by James R. Thompson and Barry W. Brown*
84. Mixture Models: Inference and Applications to Clustering, *Geoffrey J. McLachlan and Kaye E. Basford*
85. Randomized Response: Theory and Techniques, *Arijit Chaudhuri and Rahul Mukerjee*
86. Biopharmaceutical Statistics for Drug Development, *edited by Karl E. Peace*
87. Parts per Million Values for Estimating Quality Levels, *Robert E. Odeh and D. B. Owen*
88. Lognormal Distributions: Theory and Applications, *edited by Edwin L. Crow and Kunio Shimizu*
89. Properties of Estimators for the Gamma Distribution, *K. O. Bowman and L. R. Shenton*
90. Spline Smoothing and Nonparametric Regression, *Randall L. Eubank*
91. Linear Least Squares Computations, *R. W. Farebrother*
92. Exploring Statistics, *Damaraju Raghavarao*
93. Applied Time Series Analysis for Business and Economic Forecasting, *Sufi M. Nazem*
94. Bayesian Analysis of Time Series and Dynamic Models, *edited by James C. Spall*

95. The Inverse Gaussian Distribution: Theory, Methodology, and Applications, *Raj S. Chhikara and J. Leroy Folks*
96. Parameter Estimation in Reliability and Life Span Models, *A. Clifford Cohen and Betty Jones Whitten*
97. Pooled Cross-Sectional and Time Series Data Analysis, *Terry E. Dielman*
98. Random Processes: A First Look, Second Edition, Revised and Expanded, *R. Syski*
99. Generalized Poisson Distributions: Properties and Applications, *P. C. Consul*
100. Nonlinear L_p-Norm Estimation, *Rene Gonin and Arthur H. Money*
101. Model Discrimination for Nonlinear Regression Models, *Dale S. Borowiak*
102. Applied Regression Analysis in Econometrics, *Howard E. Doran*
103. Continued Fractions in Statistical Applications, *K. O. Bowman and L. R. Shenton*
104. Statistical Methodology in the Pharmaceutical Sciences, *Donald A. Berry*
105. Experimental Design in Biotechnology, *Perry D. Haaland*
106. Statistical Issues in Drug Research and Development, *edited by Karl E. Peace*
107. Handbook of Nonlinear Regression Models, *David A. Ratkowsky*
108. Robust Regression: Analysis and Applications, *edited by Kenneth D. Lawrence and Jeffrey L. Arthur*
109. Statistical Design and Analysis of Industrial Experiments, *edited by Subir Ghosh*
110. U-Statistics: Theory and Practice, *A. J. Lee*
111. A Primer in Probability: Second Edition, Revised and Expanded, *Kathleen Subrahmaniam*
112. Data Quality Control: Theory and Pragmatics, *edited by Gunar E. Liepins and V. R. R. Uppuluri*
113. Engineering Quality by Design: Interpreting the Taguchi Approach, *Thomas B. Barker*
114. Survivorship Analysis for Clinical Studies, *Eugene K. Harris and Adelin Albert*
115. Statistical Analysis of Reliability and Life-Testing Models: Second Edition, *Lee J. Bain and Max Engelhardt*
116. Stochastic Models of Carcinogenesis, *Wai-Yuan Tan*
117. Statistics and Society: Data Collection and Interpretation: Second Edition, Revised and Expanded, *Walter T. Federer*
118. Handbook of Sequential Analysis, *B. K. Ghosh and P. K. Sen*
119. Truncated and Censored Samples: Theory and Applications, *A. Clifford Cohen*
120. Survey Sampling Principles, *E. K. Foreman*
121. Applied Engineering Statistics, *Robert M. Bethea and R. Russell Rhinehart*
122. Sample Size Choice: Charts for Experiments with Linear Models: Second Edition, *Robert E. Odeh and Martin Fox*
123. Handbook of the Logistic Distribution, *edited by N. Balakrishnan*
124. Fundamentals of Biostatistical Inference, *Chap T. Le*
125. Correspondence Analysis Handbook, *J.-P. Benzécri*
126. Quadratic Forms in Random Variables: Theory and Applications, *A. M. Mathai and Serge B. Provost*
127. Confidence Intervals on Variance Components, *Richard K. Burdick and Franklin A. Graybill*
128. Biopharmaceutical Sequential Statistical Applications, *edited by Karl E. Peace*
129. Item Response Theory: Parameter Estimation Techniques, *Frank B. Baker*
130. Survey Sampling: Theory and Methods, *Arijit Chaudhuri and Horst Stenger*
131. Nonparametric Statistical Inference: Third Edition, Revised and Expanded, *Jean Dickinson Gibbons and Subhabrata Chakraborti*
132. Bivariate Discrete Distribution, *Subrahmaniam Kocherlakota and Kathleen Kocherlakota*
133. Design and Analysis of Bioavailability and Bioequivalence Studies, *Shein-Chung Chow and Jen-pei Liu*
134. Multiple Comparisons, Selection, and Applications in Biometry, *edited by Fred M. Hoppe*

135. Cross-Over Experiments: Design, Analysis, and Application, *David A. Ratkowsky, Marc A. Evans, and J. Richard Alldredge*
136. Introduction to Probability and Statistics: Second Edition, Revised and Expanded, *Narayan C. Giri*
137. Applied Analysis of Variance in Behavioral Science, *edited by Lynne K. Edwards*
138. Drug Safety Assessment in Clinical Trials, *edited by Gene S. Gilbert*
139. Design of Experiments: A No-Name Approach, *Thomas J. Lorenzen and Virgil L. Anderson*
140. Statistics in the Pharmaceutical Industry: Second Edition, Revised and Expanded, *edited by C. Ralph Buncher and Jia-Yeong Tsay*
141. Advanced Linear Models: Theory and Applications, *Song-Gui Wang and Shein-Chung Chow*
142. Multistage Selection and Ranking Procedures: Second-Order Asymptotics, *Nitis Mukhopadhyay and Tumulesh K. S. Solanky*
143. Statistical Design and Analysis in Pharmaceutical Science: Validation, Process Controls, and Stability, *Shein-Chung Chow and Jen-pei Liu*
144. Statistical Methods for Engineers and Scientists: Third Edition, Revised and Expanded, *Robert M. Bethea, Benjamin S. Duran, and Thomas L. Boullion*
145. Growth Curves, *Anant M. Kshirsagar and William Boyce Smith*
146. Statistical Bases of Reference Values in Laboratory Medicine, *Eugene K. Harris and James C. Boyd*
147. Randomization Tests: Third Edition, Revised and Expanded, *Eugene S. Edgington*
148. Practical Sampling Techniques: Second Edition, Revised and Expanded, *Ranjan K. Som*
149. Multivariate Statistical Analysis, *Narayan C. Giri*
150. Handbook of the Normal Distribution: Second Edition, Revised and Expanded, *Jagdish K. Patel and Campbell B. Read*

Additional Volumes in Preparation

Bayesian Biostatistics, *edited by Donald A. Berry and Dalene K. Stangl*

HANDBOOK OF THE NORMAL DISTRIBUTION

Second Edition, Revised and Expanded

JAGDISH K. PATEL
University of Missouri—Rolla
Rolla, Missouri

CAMPBELL B. READ
Southern Methodist University
Dallas, Texas

Marcel Dekker, Inc.   New York•Basel•Hong Kong

ISBN: 0-8247-9342-0

The publisher offers discounts on this book when ordered in bulk quantities. For more information, write to Special Sales/Professional Marketing at the address below.

This book is printed on acid-free paper.

Copyright © 1996 by MARCEL DEKKER, INC. All Rights Reserved.

Neither this book nor any part may be reproduced or transmitted in any form or by any means, electronic or mechanical, including photocopying, microfilming, and recording, or by any information storage and retrieval system, without permission in writing from the publisher.

MARCEL DEKKER, INC.
270 Madison Avenue, New York, New York 10016

Current printing (last digit):
10 9 8 7 6 5 4 3 2 1

PRINTED IN THE UNITED STATES OF AMERICA

PREFACE TO THE SECOND EDITION

Since the First Edition of the Handbook appeared in 1982 much further research has been published, and we felt that there was a need to update the material.

All chapters in the First Edition (except Chapter 9, now deleted) have been brought up to date with new material. Several First Edition chapters have been reorganized, notably Chapters 2, 3, 6 and 8. The bivariate normal distribution is more thoroughly covered, with expanded material which has been divided into Chapters 9 (distributional properties) and 10 (sampling distributions). This Second Edition also introduces material covering estimation procedures for normally distributed samples, Chapters 11 (on point estimation) and 12 (on statistical confidence, tolerance and prediction intervals).

First and foremost we would like to thank Sheila Crain, Administrative Assistant for the project, who spent many patient hours setting up the finished text; also Tina Martin, and Walter Brownfield and the staff at Marcel Dekker, Inc.

Jagdish K. Patel
Campbell B. Read

PREFACE TO THE FIRST EDITION

This book contains a collection of results relating to the normal distribution. It is a compendium of properties, and problems of analysis and proof are not covered. The aim of the authors has been to list results that will be useful to theoretical and applied researchers in statistics as well as to students.

Distributional properties are emphasized, both for the normal law itself and for statistics based on samples from normal populations. The book covers the early historical development of the normal law (Chapter 1); basic distributional properties, including references to tables and to algorithms suitable for computers (Chapters 2 and 3); properties of sampling distributions, including order statistics (Chapters 5 and 8), Wiener and Gaussian processes (Chapter 9); and the bivariate normal distribution (Chapter 10). Chapters 4 and 6 cover characterizations of the normal law and central limit theorems, respectively; these chapters may be more useful to theoretical statisticians. A collection of results showing how other distributions may be approximated by the normal law completes the coverage in the book (Chapter 7).

Several important subjects are not covered. There are no tables of distributions in this book, because excellent tables are available elsewhere; these are listed, however, with the accuracy and coverage in the sources. The multivariate normal distribution other than the bivariate case is not discussed; the general linear model and regression models based on normality have been amply documented elsewhere; and the applications of normality in the methodology of statistical inference and decision theory would provide material for another volume on their own.

In citing references, the authors have tried to balance the aim of giving historical credit where it is due with the desirability of citing easily obtainable sources that may be consulted for further detail. In the latter case, we do not aim to cite every such work, but only enough to give the researcher or student a readily available source to which to turn.

We would like to thank the following persons for reviewing parts of the manuscript and giving helpful suggestions: Lee J. Bain, Herbert A. David, Maxwell E. Engelhardt, C. H. Kapadia, C. G. Khatri, Samuel Kotz, Lloyd S. Nelson, Donald B. Owen--who also gave editorial guidance, Stephen M. Stigler, Farroll T. Wright, and a referee. For assistance in typing the manuscript and for their infinite patience, we thank Connie Brewster, Sheila Crain, Millie Manley, and Dee Patterson; we would like to thank Dr. Maurits Dekker and the staff at MDI for their work in taking the manuscript through production. We would also like to thank Southern Methodist University for giving one of us (C.B.R.) leave for a semester in order to do research for the manuscript.

Jagdish K. Patel
Campbell B. Read

CONTENTS

PREFACE TO THE SECOND EDITION		iii
PREFACE TO THE FIRST EDITION		v

1. GENESIS: A HISTORICAL BACKGROUND — 1

2. BASIC PROPERTIES — 19

    2.1 Definitions and Properties — 19
    2.2 Moments, Cumulants, and Generating Functions — 24
    2.3 Member of Some Families of Distributions — 26
    2.4 Some Basic Distributions Derived from the Normal — 31
    2.5 Normal Integrals — 34
    2.6 Tables — 37
          References — 40

3. EXPANSIONS AND ALGORITHMS — 45

    3.1 Nomograms and Computing Algorithms — 45
    3.2 The Standard Normal Density Function — 47
    3.3 Expressions Relating to the Distribution Function – I — 48
    3.4 The Distribution Function – II — 51
    3.5 The Distribution Function – III — 53
    3.6 Approximations Relating to Mills' Ratio: Expansions — 55
    3.7 Further Expressions Relating to Mills' Ratio — 62
    3.8 From Mills' Ratio to the Distribution Function — 63
    3.9 Quantiles — 66
    3.10 Approximating the Normal by Other Distributions — 70
          References — 73

4. CHARACTERIZATIONS — 81

    4.1 Characterizations by Linear Statistics — 82
    4.2 Linear and Quadratic Characterizations — 85
    4.3 Characterizations by Conditional Distributions and
          Regression Properties — 88
    4.4 Independence of Some Statistics — 92
    4.5 Characteristic Functions and Moments — 94
    4.6 Characterizations from Properties of Transformations — 96
    4.7 Sufficiency, Estimation, and Testing — 99
    4.8 Miscellaneous Characterizations — 103
    4.9 Near-characterizations — 105
          References — 106

5. SAMPLING DISTRIBUTIONS — 113

- 5.1 Samples not Larger than Four — 113
- 5.2 The Sample Mean: Independence — 115
- 5.3 Sampling Distributions Related to Chi-Square — 116
- 5.4 Sampling Distributions Related to t — 121
- 5.5 Distributions Related to F — 126
- 5.6 The Sample Mean Deviation — 129
- 5.7 The Moment Ratios $\sqrt{b_1}$ and b_2 — 132
- 5.8 Miscellaneous Results — 136
- References — 139

6. LIMIT THEOREMS AND EXPANSIONS — 145

- 6.1 Classical Central Limit Theorems — 146
- 6.2 Further Central Limit Theorems — 151
- 6.3 Asymptotic Normality — 153
- 6.4 Rapidity of Convergence to Normality — 159
- 6.5 Limit Theorems for Sample Fractiles — 165
- 6.6 Expansions — 167
- References — 174

7. NORMAL APPROXIMATIONS TO DISTRIBUTIONS — 179

- 7.1 The Binomial Distribution — 180
- 7.2 The Poisson Distribution — 188
- 7.3 The Negative Binomial Distribution — 191
- 7.4 The Hypergeometric Distribution — 195
- 7.5 Miscellaneous Discrete Distributions — 198
- 7.6 The Beta Distribution — 199
- 7.7 The von Mises Distribution — 202
- 7.8 The Chi-Squared and Gamma Distributions — 203
- 7.9 Noncentral Chi-Square — 210
- 7.10 Student's t Distribution — 212
- 7.11 Noncentral t — 216
- 7.12 The F Distribution — 218
- 7.13 Noncentral F — 222
- 7.14 Miscellaneous Continuous Distributions — 223
- 7.15 Normalizing Transformations — 226
- References — 232

8. ORDER STATISTICS FROM NORMAL SAMPLES — 241

- 8.1 Order Statistics: Basic Results — 242
- 8.2 Moments — 250
- 8.3 Ordered Deviates from the Sample Mean — 258
- 8.4 The Sample Range — 263
- 8.5 Quasi-Ranges — 268
- 8.6 Median and Midrange — 269
- 8.7 Asymptotic Properties — 273
- 8.8 Quantiles — 278
- 8.9 Miscellaneous Results — 280
- References — 287

9. THE BIVARIATE NORMAL DISTRIBUTION — 295

- 9.1 Definitions and Basic Properties — 295
- 9.2 Probability Functions and Tables — 301
- 9.3 Algorithms and Approximations — 307
- 9.4 Characterizations — 313
- 9.5 Associated Distributions — 316
- 9.6 Offset Circles and Ellipses — 324
- References — 332

10. BIVARIATE NORMAL SAMPLING DISTRIBUTIONS — 341

- 10.1 BVN Statistics–General — 341
- 10.2 The Sample Correlation Coefficient R — 348
- 10.3 Approximations to the Distribution of R — 353
- References — 359

11. POINT ESTIMATION — 365

- 11.1 Sufficiency and Completeness — 365
- 11.2 Estimators of μ and Its Functions — 367
- 11.3 Estimators of σ^2 and Its Functions — 371
- 11.4 Joint Estimators of μ and σ^2, and Their Functions — 374
- 11.5 Estimators of Parametric Functions: Two Samples — 384
- References — 386

12. STATISTICAL INTERVALS — 391

- 12.1 Confidence Intervals — 391
- 12.2 Tolerance Intervals — 402
- 12.3 Prediction Intervals — 413
- References — 419

INDEX — 427

Chapter 1
GENESIS: A HISTORICAL BACKGROUND

> I know of scarcely anything so apt to impress the imagination as the wonderful form of cosmic order expressed by the "Law of Frequency of Error". The law would have been personified by the Greeks and deified, if they had known of it. It reigns with serenity and in complete self-effacement amidst the wildest confusion.

So wrote Sir Francis Galton (1889, p. 66) about the normal distribution, in an age when the pursuit of science was tinged with the romanticism of the nineteenth century. In this age of computers, it is hard to find enthusiasm expressed with the sense of wonder of these men of letters, so much do we take for granted from modern technology.

In the seventeenth century Galileo (trans. 1953; 1962, pp. 303-309), expressed his conclusions regarding the measurement of distances to the stars by astronomers (Maistrov, 1974, pp. 31-34). He reasoned that random errors are inevitable in instrumental observations, that small errors are more likely to occur than large ones, that measurements are equally prone to err in one direction (above) or the other (below), and that the majority of observations tend to cluster around the true value. Galileo revealed here many of the characteristics of the normal probability distribution law, and also asserted that (random) errors made in observation are distinct from (systematic) final errors arising out of computation.

Although the study of probability began much earlier, modern statistics made its first great stride with the publication in 1713 of Jacob Bernoulli's *Ars Conjectandi*, in which Bernoulli proved the Weak Law of Large Numbers. The normal distribution first appeared in 1733 as an approximation to the probability for sums of binomially distributed quantities to lie between two values, when Abraham De Moivre communicated it to some of his contemporaries. A search by Daw and Pearson (1972) confirmed that several copies of this note had been bound up with library copies of De Moivre's *Miscellanea Analytica*, printed in 1733 or later. The theorem appeared again in his book *The Doctrine of Chances* (1738; 1756; 1967); see also David (1962) where it appears as an Appendix. Although the main result is commonly termed "the de Moivre-Laplace limit theorem" (see [3.4.8]), the same approximation to binomial probabilities was obtained by Daniel Bernoulli in 1770-1771, but because he published his work through the Imperial Academy of Sciences in St. Petersburg, it remained there largely unnoticed until recently (Sheynin, 1970). Bernoulli also compiled the earliest known table of the curve $y = \exp(-\mu^2/100)$; see Table 1.1.

The natural development of probability theory into mathematical statistics took place with Pierre Simon de Laplace, who "was more responsible for the early development of mathematical statistics than any other man" [Stigler (1975a), p. 503]. Laplace (1810; 1811; 1878-1912) developed the characteristic function as a tool for large sample theory and proved the first general Central Limit Theorem; broadly speaking, central limit theorems show how sums of random variables, when standardized to have mean 'zero' and unit variance tend to behave like standard normal variables as the sample size becomes large; this happens, for instance, when they are drawn as random samples from "well-behaved" distributions; see [6.1].

Laplace showed that a class of linear unbiased estimators of linear regression coefficients is approximately normally distributed if the sample size is large; he proved (Laplace, 1812, Chap. VIII) that the probability distribution of the expectation of life at any specified age tends to the normal (Seal, 1967, p. 207). He derived the asymptotic distribution of a single order statistic in a linear regression problem as normal, when the parent distribution is symmetric about zero and well-behaved. In 1818 he showed that when the parent distribution is normal, the least squares estimator (LSE) has smaller variance than any linear combination of observations (Stigler, 1973). In the course of deriving this result, Laplace showed that the asymptotic joint distribution of the LSE and his order statistic estimator is bivariate normal (see Chapter 10), and obtained the minimum variance property of the LSE under normality while trying to combine the two estimators to reduce the variance.

Laplace published his seminal work on the central limit theorem in two memoirs (1810 and 1811) and a book (1812). The 1810 memoir was broader than de Moivre's normal approximation to binomial probabilities, because it covered sums and means of variables (errors) based on other distributions. Laplace must have come across another major work, that of Gauss (1809), causing him hastily to write a supplement to his 1810 memoir and leading him to relate the limit theorem to linear estimation and least squares at greater length in 1811 and 1812 (Stigler, 1986, pp. 143, 147-148).

Problems arising from the collection of observations in astronomy led Legendre in 1805 to state the Least Squares principle, that of minimizing the sum of squares of "errors" of observations about what we would call in modern terms a regression plane; Legendre also obtained the normal equations. In 1800, Carl Friedrich Gauss published his *Theoria Motus Corporum Coelestium*, stating that he had used the least squares principle since 1795. This led to some controversy as to priority, involving Gauss, Laplace, Legendre and several colleagues of Gauss (Plackett, 1972), but it all hinged upon whether publication should be the criterion for settling the issue or not. In the nineteenth century, researchers often worked independently, without knowledge of the achievements of others, as we shall see later. It comes as no surprise, then, that Gauss may have known nothing of Legendre's earlier work when he published his *Theoria Motus*.

In this work, Gauss showed that the distribution of errors, assumed continuous, must be normal if the location parameter (again in modern terminology) has a uniform prior, so that the arithmetic mean is the mode of the posterior distribution (Seal, 1967). Gauss's linear least squares model was thus appropriate when the "errors" come from a normal distribution. An American mathematician, Robert Adrain (1808), who knew nothing of Gauss's work but who may have seen Legendre's book, derived the univariate and bivariate normal distributions as distributions of errors, and hence the method of least squares (Stigler, 1977), but his work did not influence the development of the subject.

The study of least squares, or the Theory of Errors, was to proceed for several decades without much further interaction with developing statistical theory. The normal distribution had not yet found its place in either theoretical or applied branches of the subject, and Gauss gave little further consideration to it (Seal, 1967). However, he pointed out (see Maistrov, 1974, pp. 155-156) that under the normal law, errors of any magnitude are possible. Once the universality of the normal law was accepted and then assumed, as it was to be for most of the nineteenth century, scientists also assumed that all observations should therefore be retained, resulting in a delay in developing methods for identifying and discarding outliers. For further details about Gauss's contributions to statistics and the theory of least squares, see Eisenhart (1983, pp. 547-549), Sprott (1978; 1983), Stigler (1986, pp. 140-148) and Whittaker and Robinson (1924; 1926).

One of the first to fit a normal curve to data in the social sciences was the Belgian scientist Adolphe Quetelet (1846, pp. 400-402), but he did so indirectly. He was familiar with Laplace's central limit theorem, and figured that he could approximate the normal curve via a symmetrical binomial distribution based on 1000 trials. He fitted this to the height and chest measurements of 5,732 Scottish soldiers and, separately, to the heights of 100,000 French conscripts. This indirect approach enabled Quetelet to avoid the need for calculus or normal probability tables. Stigler (1986, Chapter 5) gives a full explanation and discussion. Earlier, the astronomer Friedrich Wilhelm Bessel (1818) had published a comparison of the observed residuals and those expected from Gauss's normal law of errors for three sets of 300 or more measurements of angular coordinates of stars, and found a remarkably close agreement.

Not only was Quetelet one of the first to introduce statistical ideas outside the field of physics, but he was also instrumental in popularizing the idea that Gauss and Laplace's Law of Errors was universal. In essence, laws of error are distributions of the "errors" that occur in measuring a fixed quantity repeatedly according to some procedure that is carried out under unchanging conditions. The quotation from Galton with which this essay opens is a colorful affirmation of the universality of the normal law. The notion of laws of error goes back to the eighteenth century (Eisenhart 1983), but the erroneous idea that all sets of data conform to the normal form held sway through much of the nineteenth.

Two factors contributed to the acceptance of this idea. One was the appeal of Gauss's derivation of the normal law as that which leads to least squares; the other was what became known as the Hypothesis of Elementary Errors. This was expressed by Laplace (1810) in the supplement to his memoir on the central limit theorem (Stigler, 1986, pp. 143-144, 202). It states essentially that, if each of the observable errors is itself the aggregate of a very large number of errors -- elementary errors -- that are very small but independently distributed, then the observable errors themselves will have the Gaussian distribution, a property that would follow from Laplace's central limit theorem. Laplace's proof of the theorem, of course, did not require the assumption of this hypothesis.

Other derivations of the normal law followed the preeminent contributions of Laplace and Gauss. Many of these were based on the Hypothesis of Elementary Errors, which became firmly established until the end of the nineteenth century. For example Hagen (1837) derived the law under the assumption that the elementary errors take discrete values on a lattice with values spaced apart by an arbitrarily small distance ϵ, positive and negative multiples of any size being equally likely. This result in turn was generalized by Bessel (1838). Astronomers like G. B. Airy (1861) incorrectly interpreted Laplace's central limit theorem as though the Hypothesis was an underlying assumption. Adams (1974, pp. 59-67) and Eisenhart (1983, pp. 550-559) provide more detail on these and other derivations.

In 1860 the Scottish mathematical physicist James Clerk Maxwell published the first of his two great papers on the kinetic theory of gases. Using geometrical considerations, he derived the normal distribution as the distribution of orthogonal velocity components of particles moving freely in a vacuum (Maxwell, 1860, pp. 22-23; 1952, pp. 380-381). His results lead through the work of Boltzmann to the modern theory of statistical mechanics and are notable as the first attempt to describe the motion of gases by a statistical function rather than a deterministic one. "The velocities are distributed among the particles" he wrote, "according to the same law as the errors are distributed among the observations in the theory of the method of least squares."

The first scientist in England to make use of earlier work on the continent was Francis Galton, who had a remarkable career in exploration, geography, the study of meteorology and, above all, anthropometry; the last-named is the study of anthropology through analysis of physical measurements. In his book *Natural Inheritance* (1889), Galton drew on the work of Quetelet, noting his application of normal curve-fitting to human measurements, and developing it to fit a model to describe the dependence of such measurements on those of an offspring's parents. Galton plotted his data in two dimensions and noted that data points of equal intensity appeared to lie on elliptical curves. From all of this he developed the linear regression model, the concept of correlation (1888), and the equation of the bivariate normal distribution (1886), with the help of a mathematician, Hamilton Dickson (see Chapter 10). The bivariate and trivariate normal distributions had already been developed independently by Bravais (1846), with application to data by Schols (1875), but although aware of the notion of correlation, these workers did not find it to have the degree of importance which Galton gave it.

A good account of how Galton arrived at the elliptical curves and bivariate normal distribution is given by Pearson (1920; 1970, p. 196) including the diagram from which he discovered observationally the form of the bivariate normal surface.

The publication of *Natural Inheritance* was the catalyst for the English school of biometry to begin to take major strides. The zoologist W.F.R. Weldon sought the help of an applied mathematician, Karl Pearson, realizing that statistical methods might establish evidence to support Darwin's theory of natural selection. The "law of error", as Galton termed it, prompted Weldon to assume initially that all physical characters in homogeneous animal populations would be normally distributed (Pearson, 1965; 1970, p. 328).

Galton realized, however, that sets of data might well follow some other frequency law. The geometric mean of a set of observation, he wrote, might better represent the most probable value of a distribution, and if so, the logarithms of the observations might be assumed to follow the normal law. This led to the log-normal distribution, but it stimulated Karl Pearson to develop a system of frequency curves, depending on a set of parameters, which would cover all distributions occurring in nature, at least those which are continuous (Pearson, 1967; 1970, pp. 344-345).

Karl Pearson generalized and gave precision to Galton's discussion of correlation by developing a theory of multiple correlation and multiple regression. Such a theory, he realized, would be necessary to answer the kind of questions which were being posed by Weldon. Working from two, three and four variables, Francis Edgeworth (1892) provided the first statement of the multivariate normal distribution, and Pearson (1896) gave an explicit derivation of it (Seal, 1967).

The work of Legendre, Gauss and the least squares school was neglected and unnoticed during the early years of biometry in England (1885-1908), possibly because of Pearson's preoccupation with the multivariate normal distribution. Pearson's predictive regression equation (1896) was not seen to be identical in form and solution to Gauss's 1809 model. Duplication of research in least squares theory continued in the work of R. A. Fisher and others at Rothamsted until well after 1930 (Seal, 1967). In trying to fit data to his frequency curves, Pearson was faced with the need to test the goodness of fit. Ernst Abbe (1863; 1906) had derived the distribution of ΣX_i^2, where $X_1,...,X_n$ is a random sample from a normal distribution with mean zero (Sheynin, 1966; Kendall, 1971), and Helmert (1876) derived the distribution of $\Sigma(X_i - \bar{X})^2$, where $\bar{X}$ is the arithmetic mean of $X_1,...,X_n$. These yield the Chi-square distribution with n and n − 1 degrees of freedom, respectively. Abbe's work went unnoticed until 1966, credit prior to that time having been given to Helmert. The matrix transformation of Helmert, however, is still used as an instructional tool for deriving the distribution of $\Sigma(X_i - \bar{X})^2/(n-1)$ in a normal sample.

The problem is quite different if the parent population is not normal. Lancaster (1966) shows how Laplace provided the necessary techniques for

Bienaymé in 1838 and 1852 to obtain Chi-square as an asymptotic large-sample distribution without any assumption of normality. Bienaymé obtained something close to "$\Sigma(\text{observed-expected})^2/(\text{expected})$", which is Pearson's Chi-square statistic, summation being over classes.

Weldon had found a noticeable exception to the fit of the normal curve to a set of data for the relative frontal breadth of Naples crabs (Pearson, 1965; 1970, p. 328). Thinking it might be compounded from two sub-species, he fitted a compound normal distribution (see [1.6]), or "double-humped curve" as he called it; but the fit was done purely by trial and error, in 1892. Karl Pearson's first statistical paper in 1894 tackled this problem, introducing the method of moments as a technique for fitting a frequency distribution. But the question of whether the fitted distribution was reasonable led Pearson to the work which resulted in his paper of 1900 and established the Chi-square goodness-of-fit test firmly as a cornerstone of modern statistics. Ironically, the first sacred cow to fall with this new tool was the Law of Errors; Pearson soundly berated the astronomer G. B. Airy (1861), who had tried to illustrate the universality of the normal law with an unlucky set of data. Using the same data and his new Chi-Square test, Pearson showed that the normal law gave an unacceptable fit after all. "How healthy is the spirit of skepticism", he wrote, "in all inquiries concerning the accordance of theory and nature" (Pearson, 1900, p. 172).

Pearson and the English school of biometry preferred to work with large data sets for the statistical analysis of the various problems which they faced. But W. S. Gosset (Student) was compelled by circumstances arising in the Guinness brewery company in Dublin to solve problems for which small samples only could be taken. A year of work in the Biometric Laboratory under Pearson led to his famous paper "The Probable Error of a Mean" (Student, 1908) in which he deduced the t distribution, via the ratio of the sample mean to standard deviation in a normal sample. Pearson seemed to have little interest in Gosset's results, perhaps because he felt wary of letting any biologist or medical research worker believe that there was a simple method of drawing conclusions from scanty data.

Gosset was unaware of Helmert's derivation of the sampling distribution of $\Sigma(X_i - \bar{X})^2$, but he inferred its distribution and showed it to be uncorrelated with $\bar{X}$. For other work by Gosset, see Pearson (1970, pp. 348-351, 360-403), or Pearson and Wishart (1958).

The English school of biometry did not undertake a serious study of probability theory, unlike the Russian school of Pafnuti Lvovich Tchebyshev and his pupils Markov and Lyapunov. From the middle of the nineteenth century the St. Petersburg school applied mathematical rigor to laws of large numbers, dependent events and central limit properties; with the introduction of the concept of a random variable, it was able to establish sufficient conditions for standardized sums of dependent, as well as of independent random variables to converge to the normal law. The first clear statement of the problem, together with a proof which later required revisions and additional proof (Maistrov, 1974, pp. 202-208) was given by Tchebyshev in 1887 (reprinted 1890) using the method of moments. The importance of Tchebyshev's approach to Central Limit Theorems lay in the clearly defined mathematical character he ascribed to random variables. He did his by establishing restrictions on the applicability of results in probability theory, so that in every set of circumstances one might determine whether or not the limit theorems hold. It was left to Andrei Andreevich Markov (1898) to correct Tchebyshev's theorem, and to Alexander Mikhailovich Lyapunov (1901; 1954-1965, Vol. I, pp. 157-176) to produce a Central Limit Theorem of great generality, rigorously proved with the tools of classical analysis, including that of characteristic functions. Since this does not require the existence of moments of any order, it might appear that Tchebyshev's method of moments had been outlived; apparently challenged by this, Markov (1913; 1951, pp. 319-338) proved Lyapunov's theorem by the method of moments with the now well-known device of introducing truncated random variables. For a statement of these and other limit theorems as they developed historically, see [6.1], and Uspensky (1937, Appendix II). This work was to be continued and put on a firm axiomatic basis by Bernstein, Khinchine and Kolmogorov. For further accounts of the early history of the Russian school, see Maistrov (1974), Adams (1974), and Seneta (1982; 1983; 1985).

In the same paper, Tchebyshev (1890) developed a series expansion and an upper bound for the difference between the cumulative distribution function $F_n(x)$ of a standardized sum $(\Sigma X_i)/(\sqrt{n}\sigma)$ and that of a standard normal variable, where σ is the standard deviation in the population of interest. In 1905, Charlier introduced a series to improve the Central Limit Theorem approximation to the density function of $(\Sigma X_i)/(\sqrt{n}\sigma)$ in terms of the standard normal density ϕ: Edgeworth (1905) developed his series expansion for $F_n(x)$ with the same purpose in mind, although he had produced versions for symmetrical densities as early as 1883. See [3.5], Gnedenko and Kolmogorov (1968, Chapter 8), and Stigler (1978).

The early story of the normal distribution is largely the story of the beginnings of Statistics as a science. We leave it at this point; with R. A. Fisher modern statistics begins to branch out and accelerate. From 1915 onwards Fisher found the distribution of the correlation coefficient, of the absolute deviation $\Sigma |X_i - \bar{X}|/n$ in normal samples, of regression coefficients, correlation ratios, multiple regression and partial correlation coefficients, and of the ratio F of sample variances from two normal populations. At the same time he developed his ideas of estimation, sufficiency, likelihood, inference, the analysis of variance and experimental design (Kendall, 1970; Savage, 1976). Normality assumptions have played a key role in statistical analysis through the years, but since the 1960's considerably more attention has been given to questioning these assumptions, requiring estimators that are robust when such assumptions are violated, and to devising further enlightening tests of their validity.

A final word about nomenclature. The normal distribution has been named after various scientists, including Laplace and Gauss. Kac (1975, pp. 6-7), recalls that it has also been named after Quetelet and Maxwell. Stigler (1980) points out that although a few modern writers refer to it as the Laplace or Laplace-Gauss distribution, and engineers name it after Gauss, no modern writers name it after its originator Abraham de Moivre, as far as is known. Francis Galton called it by several names, indicating that before 1900 no one term had received common acceptance. Among these names were "law of frequency of error" and "the exponential law" (1875), "law of deviation from an average" and "law of errors of observation" (1869). Quetelet (1846, p. 386) called it "la courbe de possibilité."

In 1877, Galton first used the name "normal law" in the sense that it is commonly encountered in Statistics, but the earliest known use of the term, and that in the same sense, was in 1873, by the American Charles Sanders Pierce (Kruskal, 1978). The label "normal law" or "normal distribution" gained acceptance with the English school of biometry; Karl Pearson (1921-1933; 1978, p. 156), claimed to have coined it, but was apparently unaware of earlier uses.

REFERENCES

Abbe, E. (1863; 1906). Uber die Gesetzmässigkeit in der Vertheilung der Fehler bei Beobachtungsreihen, *Gesammelte Abhandlungen*, Vol. II, Jena: Gustav Fischer. English translation (1968), ref. P. B. 191, 92, T., from National Technical Information Service, Springfield, Virginia.

Adams, W. J. (1974). *The Life and Times of the Central Limit Theorem*, New York: Caedmon.

Adrain, R. (1808). Research concerning the probabilities of errors which happen in making observations, etc. *The Analyst; or Mathematical Museum* 1 (4), 93-109.

Airy, G. B. (1861). *On the Algebraical and Numerical Theory of Errors of Observations and the Combination of Observations*, London: MacMillan.

Bernoulli, D. (1770-1). Mensura sortis ad fortuitam successionem rerum naturaliter contingentium applicata, *Novi Commentarii Academiae Scientiarum Imperialis Petropolitanae* 14, 26-45; 15, 3-28.

Bernoulli, J. (1713). *Ars Conjectandi* (German version (1899), *Wahrscheinlichkeitsrechnung*, Leipzig: Engelmann).

Bessel, F. W. (1818). *Astronomiae pro anno MDCCLV deducta ex observationibus viri incomporabilis James Bradley specula Grenovicensi per annos 1750-1762 institutis*, Regiomonti.

Bessel, F. W. (1838). Untersuchungen über der Wahrscheinlichkeit der Beobachtungsfehler, *Astronomische Nachrichten* 15, 368-404.

Bienaymé, J. (1838). Sur la probabilité des resultats moyens des observations; demonstration directe de la règle de Laplace, *Mémoires des Savans Étrangers, Académie (Royale) des Sciences de l'Institut de*

France, Paris, 5, 513-558.

Bienaymé, J. (1852). Sur la probabilité des erreurs d'après la méthode des moindres carrés. (Reprinted (1858), *Mémoires des Savans Étrangers, Académie (Royale) des Sciences de l'Institut de France*, Paris, 5, 615-6634.

Bravais, A. (1846). Analyse mathematique sur les probabilitiés des erreurs de situation d'un point, *Mémoires des Savans Étrangers, Académie (Royale) des Sciences de l'Institut de France*, Paris, 9, 255-332.

Burgess, J. (1898). On the definite integral...[i.e....], with extended tables of values, *Transactions of the Royal Society of Edinburgh* 39, 257-321.

Charlier, C. V. L. (1905). Über die Darstellung willkürlicher Funktionen, *Arkiv för Matematik, Astronomi och Fysik* 2, No. 20, 1-35.

David, F. N. (1962). *Games, Gods and Gambling*, New York: Hafner.

Daw, R. H. and Pearson, E. S. (1972). Studies in the history of Probability and Statistics. XXX. Abraham de Moivre's 1733 derivation of the normal curve: A bibliographical note, *Biometrika* 59, 677-680.

De Moivre, A. (1733). *Approximatio ad Summam Ferminorum Binomii $(a + b)^n$ in Seriem expansi*.

De Moivre, A. (1738, 1756). *The Doctrine of Chances* [(1967) reprint, New York: Chelsea].

De Morgan, A. (1837). Theory of probabilities, in *Encyclopaedia Metropolitana* 2, 359-468.

De Morgan, A. (1838). *An essay on probabilities and on their application to life contingencies and insurance offices*, London: Longmans.

Edgeworth, F. Y. (1883). The law of error, *Philosophical Magazine* 16, 300-309.

Edgeworth, F. Y. (1892). Correlated averages, *Philosophical Magazine* Ser. 5, 34, 190-204.

Edgeworth, F. Y. (1905). The law of error, *Proceedings of the Cambridge Philosophical Society* 20, 36-65.

Eisenhart, C. (1983). Laws of error, *Encyclopedia of Statistical Sciences*, Vol. 4, Kotz, S., Johnson, N. L. and Read, C. B., eds., 530-566, New York: Wiley.

Galileo, G. (tr. 1953, 2nd edn. 1962). *Dialogue Concerning the Two Chief World Systems - Ptolemaic and Copernican* (transl. S. Drake), Berkeley: University of California Press.

Galton, F. (1869). *Hereditary Genius*, London: MacMillan.

Galton, F. (1875). Statistics by intercomparison, with remarks on the law of frequency of error, *Philosophical Magazine* 49, 33-46.

Galton, F. (1886). Family likeness in stature, *Proceedings of the Royal Society* 40, 42-73.

Galton, F. (1888). Co-relations and their measurement, chiefly from anthropometric data, *Proceedings of the Royal Society* 45, 135-145.

Galton, F. (1889). *Natural Inheritance*. London: MacMillan.

Gauss, C. F. (1809). *Theoria Motus Corporum Coelestium*, Lib. 2, Sec. III, 205-224, Hamburg: Perthes u. Besser.

Gauss, C. F. (1816). Bestimmung der Genauigkeit der Beobachtungen, *Zeitschrift Astronomi* 1, 185-197.

Glaisher, J. W. L. (1871). On a class of definite integrals - Part II, *Philosophical Magazine* 42, 421-436.

Gnedenko, B. V. and Kolmogorov, A. N. (1968). *Limit Distributions for Sums of Independent Random Variables* (trans. from Russian), Reading, Mass: Addison-Wesley.

Greenwood, J. A. and Hartley, H. O. (1962). *Guide to Tables in Mathematical Statistics*, Princeton, NJ: Princeton University Press.

Hagen, G. H. L. (1837). *Grundzüge der Wahrscheinlichkeitsrechnung*, Berlin: Ernst & Korn.

Helmert, F. R. (1876). Die Genauigkeit der Formel von Peters zur Berechnung des wahrscheinlichen Beobachtungsfehlers director Beobachtungen tungen gleicher Genauigkeit, *Astronomische Nachrichten* 88, 113-120.

Kac, M. (1975). Some reflections of a mathematician on the nature and the role of statistics, *Proceedings of the Conference on Directions for Mathematical Statistics*, 5-11. Applied Probability Trust.

Kendall, M. G. (1963). Ronald Aylmer Fisher, 1890-1962, *Biometrika* 50, 1-15. Reprinted (1970) in *Studies in the History of Probability and Statistics* (Pearson, E. S. and Kendall, M. G., eds.), 439-454. New York: Hafner.

Kendall, M. G. (1971). Studies in the History of Probability and Statistics. XXVI. The work of Ernst Abbe, *Biometrika* 58, 369-373; Corrigendum (1972), *Biometrika* 59, 498.

Kramp, C. (1799). *Analyse des réfractions astronomiques et terresires*, Leipsic: Schwikkert; Paris: Koenig.

Kruskal, W. (1978). Formulas, numbers, words: Statistics in prose, *The American Scholar* 47, 223-229.

Lancaster, H. O. (1966). Forerunners of the Pearson χ^2, *Australian Journal of Statistics* 8, 117-126.

Laplace, P. S. (1809-1810). Mémoire sur les approximations des formules qui sont fonctions de très grands nombres et sur leur application aux probabilités, *Mémoires de la Classe des Sciences mathématiques et physiques de l'Institut*, Paris, 353-415, 559-565, et passim (see *Oeuvres Complètes* 12, 301-353).

Laplace, P. S. (1811). Mémoire sur les intégrales définies et leur application aux probabilitiés, *Mémoires de la Classe des Sciences mathématiques et physiques de l'Institut*, Paris, 279-347, et passim (1810-1812: see *Oeuvres Complètes* 12, 357-412).

Laplace, P. S. (1812). *Théorie Analytique des Probabilités.* Paris. (*Oeuvres Complètes*, Vol. 7).

Laplace, P. S. (1878, 1912). *Oeuvres Complètes de Laplace*, 14 volumes, Paris: Gauthier-Villars.

Legendre, A. M. (1805). *Nouvelles Méthodes pour la Détermination des Orbites des Comètes*, Paris: Courcier.

Legendre, A. M. (1826). *Traite des fonctions elliptiques et des integrales Euleriennes, avec des tables pour en faciliter le calcul numerique*, Vol. 2, Paris: Huzard-Courcier.

Lyapunov, A. M. (1901). Nouvelle forme du théorème sur la limite de probabilité, *Mémoires de l'Académie Impériale des Sciences de St. Pétersbourg* 12, 1-24.

Lyapunov, A. M. (1954-1965). *Izbrannye Trudi* ("Selected Works"), Academy of Sciences, USSR.

Maistrov, L. E. (1967, 1974). *Probability Theory: A Historical Sketch*, (transl. by S. Kotz), New York: Academic Press.

Markov, A. A. (1888). Table des valeurs de l'integrale ..., St. Pétersbourg: Académie Impériale des Sciences.

Markov, A. A. (1899-1900). The law of large numbers and the method of least squares, *Izvestia Physiko-mathematicheskago Obschvestva pri Imperatorskom Kazanskom Universitet* 8, 110-128; 9, 41-43.

Markov, A. A. (1913). A probabilistic limit theorem for the cases of Academician A. M. Lyapounov, *Izvlyechenye iz knigi ischislyenye veroyatnostyei* \(Supplement to "Theory of Probability"), 4-e.

Markov, A. A. (1951). *Izbrannye Trudi* ("Selected Works"), Academy of Sciences, USSR.

Maxwell, J. C. (1860). Illustrations of the dynamical theory of gases, *Philosophical Magazine* 19, 19-32; 20, 21-37; (1952), *Scientific Papers of James Clerk Maxwell*, 377-409.

Pearson, E. S. (1965). Some incidents in the early history of Biometry and Statistics, *Biometrika* 52, 3-18. Reprinted (1970) in *Studies in the History of Statistics and Probability*, (Pearson, E. S. and Kendall, M. G., eds.), 323-338, New York: Hafner.

Pearson, E. S. (1967). Some reflexions on continuity in the development of mathematical statistics, 1885-1920, *Biometrika* 54, 341-355. Reprinted (1970), in *Studies in the History of Statistics and Probability* (Pearson, E. S. and Kendall, M. G., eds.), 339-353.

Pearson, E. S. (1970). William Sealy Gosset, 1876-1937; "Student" as a statistician, *Studies in the History of Statistics and Probability*, (Pearson E. S., and Kendall, M. G., eds.), 360-403, New York: Hafner.

Pearson, E. S. and Hartley, H. O. (1958). *Biometrika Tables for Statisticians*, 1 (2nd ed.), London: Cambridge University Press.

Pearson, K. (1896). Mathematical contributions to the theory of evolution. III. Regression, heredity and panmixia, *Philosophical Transactions of the Royal Society of London*, A 187, 253-318.

Pearson, K. (1900). On a criterion that a given system of deviations from the probable in the case of a correlated system of variables is such that it can be reasonably supposed to have arisen from random sampling. *Philosophical Magazine* (5)50, 157-175.

Pearson, K. (1920). Notes on the history of correlation, *Biometrika* 13, 25-45. Reprinted (1970) in *Studies in the History of Statistics and Probability*, (Pearson, E. S. and Kendall, M. G., eds.), 185-205, New York: Hafner.

Pearson, K. (1921-1933). *The History of Statistics in the 17th and 18th Centuries, against the Changing Background of Intellectual, Scientific and Religious Thought* (lectures, E. S. Pearson, ed., 1978), New York: MacMillan.

Plackett, R. L. (1972). Studies in the history of Probability and Statistics XXIX. The discovery of the method of least squares, *Biometrika* 59, 239-251.

Quetelet, L. A. J. (1846). *Lettres à S.A.R. Le Duc Régnant de Saxe-Cobourg et Gotha, sur la Théorie des Probabilitiés appliquée aux Sciences Morales et Politiques*, (English translation, 1849), Brussels: Hayez.

Savage, L. J. (1976). On rereading R. A. Fisher (with discussion), *Annals of Statistics* 4, 441-500.

Schols, C. M. (1875). Over de theorie des fouten in de ruimte en in het platte vlak, *Verhandelingen der koninklijke Akademie van Wetenschappen*, Amsterdam, 15, 1-75.

Seal, H. L. (1967). Studies in the history of Probability and Statistics. XV. The Historical development of the Gauss linear model, *Biometrika* 54, 1-24. Reprinted (1970) in Studies in the History of Probability and Statistics (Pearson, E. S. and Kendall, M. G., eds.), 207-230, New York; Hafner.

Seneta, E. (1982). Chebyshev (or Tchébichef), Pafnuty Lvovich, *Encyclopedia of Statistical Sciences*, Vol. 1, Kotz, S., Johnson, N. L. and Read, C. B., eds., 429-431, New York: Wiley.

Seneta, E. (1983). Liapunov, Alexander Mikhailovich, *Encyclopedia of Statistical Sciences*, Vol. 4, Kotz, S., Johnson, N. L. and Read, C. B., eds., 625-627, New York: Wiley.

Seneta, E. (1985). Markov, Andrei Andreevich, *Encyclopedia of Statistical Sciences*, Vol. 5, Kotz, S., Johnson, N. L. and Read, C. B., eds., 247-249, New York: Wiley.

Sheppard, W. F. (1898). On the application of the theory of error to cases of normal distribution and normal correlation, *Philosophical Transactions of the Royal Society of London* A 192, 101-167.

Sheppard, W. F. (1903). New tables of the probability integral, *Biometrika* 2, 174-190.

Sheppard, W. F. (1907). Table of deviates of the normal curve, *Biometrika* 5, 404-406.

Sheynin, O. B. (1966). Origin of the theory of errors, *Nature* 211, 1003-1004.

Sheynin, O. B. (1968). Studies in the history of probability and statistics. XXI. On the early history of the law of large numbers, *Biometrika* 55, 459-467. Reprinted (1970) in *Studies in the History of Statistics and Probability* (Pearson, E. S. and Kendall, M. G., eds.),

231-239, New York: Hafner.

Sheynin, O. B. (1970). Studies in the history of Probability and Statistics. XXIII. Daniel Bernoulli on the normal law, *Biometrika* 57, 199-202.

Sprott, D. A. (1978). Gauss's contributions to statistics, *Historia Mathematica* 5, 183-203.

Sprott, D. A. (1983). Gauss, Carl Friedrich, *Encyclopedia of Statistical Sciences*, Vol. 3, Kotz, S., Johnson, N. L. and Read, C. B., eds., 305-309, New York: Wiley.

Stigler, S. M. (1973). Laplace, Fisher, and the discovery of the concept of sufficiency, *Biometrika* 60, 439-445.

Stigler, S. M. (1975a). Studies in the history of probability and statistics. XXXIV. Napoleonic statistics: the work of Laplace, *Biometrika* 62, 503-517.

Stigler, S. M. (1975b). The transition from point to distribution estimation, *Bulletin of the International Statistical Institute* 46, (Proceedings of the 40th Session, Book 2), 332-340.

Stigler, S. M. (1977). An attack on Gauss, published by Legendre in 1820, *Historia Mathematica* 4, 31-35.

Stigler, S. M. (1978). Francis Ysidro Edgeworth, Statistician, *Journal of the Royal Statistical Society* A 141, 287-313. Followed by Discussion.

Stigler, S. M. (1980). Stigler's Law of Eponymy, *Transactions of the New York Academy of Sciences II* 39 (Science and Social Structure: a Festschrift for Robert K. Merton), 147-157.

Stigler, S. M. (1986). *The History of Statistics: The Measurement of Uncertainty before 1900*, Cambridge, Mass.: Harvard University.

'Student' (1908). On the probable error of the mean, *Biometrika* 6, 1-25.

Tchebyshev, P. L. (1890). Sur deux théorèmes relatifs aux probabilités, *Acta Mathematica* 14, 305-315. (Reprinted 1962 in Oeuvres, Vol. 2, New York: Chelsea).

Uspensky, J. V. (1937). *Introduction to Mathematical Probability*, New York: McGraw Hill.

Whittaker, E. T. and Robinson, G. (1926). *The Calculus of Observations*, London: Blackie.

Chapter 2
BASIC PROPERTIES

2.1 DEFINITIONS AND PROPERTIES

Some of the following results and others appear in Abramowitz and Stegun (1964, Sec. 26.2). Further properties may be found in Chapters 3 and 4; for distributions of sampling statistics from normal populations, see Chapters 5 and 7.

[2.1.1] The probability density function (pdf) of a normal random variable X is given by

$$f(x; \mu, \sigma^2) = \frac{1}{\sigma\sqrt{2\pi}} \exp\left[-\frac{(x-\mu)^2}{2\sigma^2}\right], \quad |x| < \infty, \ |\mu| < \infty, \ \sigma > 0.$$

The constants μ, σ, and σ^2 are, respectively, the mean, standard deviation, and variance of the normal distribution. Let $F(x; \mu, \sigma^2)$ denote the cumulative distribution function (cdf) of X.

[2.1.2] The pdf of a *standard* normal random variable Z is given by

$$\phi(z) = \frac{1}{\sqrt{2\pi}} \exp\left(-\frac{z^2}{2}\right), \quad -\infty < z < \infty.$$

Let $\Phi(z)$ denote the cdf of Z. Throughout the book the notation $N(\mu, \sigma^2)$ will denote the normal distribution with mean μ and variance σ^2. Also, the notations ϕ and Φ are used only in reference to normal distributions.

Z is termed a "standard" normal variable because if X has a $N(\mu, \sigma^2)$ distribution, then

$$Z = (X - \mu)/\sigma$$

"standardizes" X in the sense that Z has mean 0 and variance equal to 1.

Further, $F(x) = \Phi[(X-\mu)/\sigma]$, $\Phi(z) = F(\mu + \sigma z)$, and $\phi(0) = 1/\sqrt{2\pi}$. We shall denote the quantiles of Z by z_β, where

$$\Pr(Z \leq z_\beta) = \Phi(z_\beta) = 1 - \beta.$$

[2.1.3] The curves f(x) and $\phi(z)$ are symmetric about $x = \mu$ and $z = 0$, respectively. Hence, for all x and z, $f[(\mu - x)/\sigma] = f[(x - \mu)/\sigma]$,

$$F(-x) = 1 - F(x + 2\mu), \quad \phi(-z) = \phi(z), \quad \Phi(-z) = 1 - \Phi(z),$$

$$z_\beta = -z_{1-\beta}, \quad \Phi(z_\beta) - \Phi(z_{1-\beta}) = 1 - 2\beta, \quad 0 < \beta < 0.5.$$

[2.1.4] The pdf f(x) is unimodal with mean (expected value), median and mode at $x = \mu$; the corresponding central value of $\phi(z)$ is at $z = 0$. The curve f(x) has points of inflection at $\mu - \sigma$ and $\mu + \sigma$; for $\phi(z)$ these points are at -1 and 1. Further, f(x) is logconcave in x and $\phi(z)$ is logconcave in z (Tong, 1978, p. 661); this property is useful in deriving inequalities for normal probabilities.

[2.1.5] The incomplete gamma function ratio P(a, x) and $\phi(x)$ are related by $P(1/2, x) = 2\phi(\sqrt{2x}) - 1$, $x \geq 0$, where

$$P(a, x) = \left(\int_0^x t^{a-1} e^{-t} dt\right)/\Gamma(a).$$

[2.1.6] The distribution with pdf $\phi(\cdot)$ is not the only one to be treated as a standard, although it is by far the commonest. In the 18th and 19th centuries Laplace, Gauss and other adopted the *error function* erf(x) as a standard. This is defined by

$$\text{erf}(x) = 2\Phi(x\sqrt{2}) - 1, \quad x \geq 0,$$

and corresponds to a normal density with variance $\frac{1}{2}$,

$$f(x; 0, \tfrac{1}{2}) = (1/\sqrt{\pi}) e^{-x^2}.$$

Basic Properties

This version is still used by some engineers. Stigler (1982, pp. 137-138) presents an interesting argument for adopting

$$f(x; 0, 1/(2\pi)) = e^{-\pi x^2}$$

as a standard. There are pedagogical advantages, e.g., at $x = 0$, the density is equal to 1.0, and there is or are "no square root sign, no normalizing constant, no unneeded 2's."

Hefferman (1988, pp. 100-102) takes as a measure of spread

$$\delta = \{E[X_1 - X_2)^2]\}^{1/2} ,$$

based on pairwise differences, where X_1 and X_2 are independent observations from a population of interest. Then $\delta = \sqrt{2}\,\sigma$; if we take $\delta = 1$ as our standard, we get the normal density $f(x; 0, \frac{1}{2})$ corresponding to the error function above.

[2.1.7] Let X be a $N(\mu, \sigma^2)$ random variable and $Y = aX + b$, a and b are both real. Then Y has a $N(a\mu + b,\ a^2\sigma^2)$ distribution.

[2.1.8] *Repeated Derivatives of $\phi(x)$.* (a) Let

$$\left(-\frac{d}{dx}\right)^r \phi(x) = H_r(x)\phi(x)$$

where $H_r(x)$ is a Tchebyshev-Hermite polynomial of degree r in x. The polynomial $H_r(x)$ is the coefficient of $t^r/r!$ in $\exp(tx - t^2/2)$, so that

$$H_r(x) = x^r - \frac{r^{(2)}}{2 \cdot 1!} x^{r-2} + \frac{r^{(4)}}{2^2 \cdot 2!} x^{r-4} - \frac{r^{(6)}}{2^3 \cdot 3!} x^{r-6} + \cdots ,$$

where $r^{(a)} = r(r-1)\ldots(r-a+1)$. The first nine polynomials are given by

$H_0(x) = 1,$ $\qquad H_1(x) = x ,$
$H_2(x) = x^2 - 1,$ $\qquad H_3(x) = x^3 - 3x ,$
$H_4(x) = x^4 - 6x^2 + 3,$ $\qquad H_5(x) = x^5 - 10x^3 + 15 x,$
$H_6(x) = x^6 - 15x^4 + 45x^2 - 15 ,$
$H_7(x) = x^7 - 21x^5 + 105x^3 - 105x ,$
$H_8(x) = x^8 - 28x^6 + 210x^4 - 420x^2 + 105 .$

Expressions for $H_r(x)$ of higher degree can be obtained from the recurrence relation

$$H_r(x) = xH_{r-1}(x) - (r-1)H_{r-2}(x) .$$

See Stuart and Ord (1987, Secs. 6.14-15) and Draper and Tierney (1973, pp. 507-508), where expressions for $H_r(x)$ are given explicitly for $r = 0, 1, 2, ..., 27$.

(b) Let

$$\phi^{(m)}(x) = \frac{d^m}{dx^m} \phi(x) .$$

Then the differential equation

$$\phi^{(m+2)}(x) + x\phi^{(m+1)}(x) + (m+1)\phi^{(m)}(x) = 0$$

is satisfied. The value of $\phi^{(m)}(x)$ at $x = 0$ is

$$\phi^{(m)}(0) = \begin{cases} \dfrac{(-1)^{m/2} m!}{(\sqrt{2\pi}) 2^{m/2} (m/2)!} , & m = 2r, \quad r = 0, 1, 2, ... \\ 0 , & m \text{ odd}, \quad m > 0 \end{cases}$$

(Abramowitz and Stegun, 1964, Sec. 26.2.28). See [2.6] and Table 2.1 for listings of tables of these derivatives.

[2.1.9] *Repeated Integrals of $\phi(x)$.* (a) Let

$$I_n(x) = \int_x^\infty I_{n-1}(y) \, dy, \quad n \geq 0 ,$$

where $I_{-1}(x) = \phi(x)$.

(b) $I_n(x) = \left(-\dfrac{d}{dx}\right)^{n-1} \phi(x) = (-1)^{n-1} \phi^{(n-1)}(x), \quad n \geq 1$;

(c) $\left(\dfrac{d^2}{dx^2} + x\dfrac{d}{dx} - n\right) I_n(x) = 0$;

(d) $(n+1) I_{n+1}(x) + xI_n(x) - I_{n-1}(x) = 0, \quad n > -1$;

(e) $I_n(x) = \int_x^\infty \frac{(y-x)^n}{n!} \phi(y) \, dy, \quad n > -1$;

(f) $I_n(0) = L_n(0) = \left[\left(\frac{n}{2}\right)! 2^{1+n/2}\right]^{-1}$, n even

(Abramowitz and Stegun, 1964, Secs. 26.2.41-46).

For a presentation of various definite and indefinite integrals involving $\phi(x)$ and $\Phi(x)$, see Section [2.5].

[2.1.10] Sheppard (1898, pp. 104-106) gives an interesting geometrical property of the normal curve. Suppose that the mean is zero and the standard deviation σ. Let P be a point and let PM be the ordinate at P, so that OM = x. Let P move so that if the tangent PT to the locus of P intersects the x axis at T, then OM · MT = σ^2. The locus of points of P is then a normal curve, but the area under the curve will not necessarily be unity. The additional condition that the ordinate at 0 is of height $(\sigma\sqrt{2\pi})^{-1}$ yields the curve $y = f(x; 0, \sigma^2)$, the normal pdf.

[2.1.11] The rv X is *stochastically smaller* than the rv Y (written $X \stackrel{st}{<} Y$) if and only if, for all t,

$$\Pr(X \le t) \ge \Pr(Y \le t).$$

Suppose that $X \sim N(\mu, \sigma^2)$ and $Z \sim N(0, 1)$. Then $|X| \stackrel{st}{<} |Z|$ if and only if $f_X(0; \mu, \sigma) \ge \phi_z(0)$ (Horn, 1988, pp. 1327-1329).

[2.1.12] Two or more distributions are *equivalent* if their kth moments are equal for all k (k = 0,1,2,3,...). A random variable X is *determinate* if it is the only member of its equivalence class, and is *indeterminate* otherwise.

Berg (1988, pp. 910-913) showed that, if $X \sim N(\cdot, \cdot)$, then X^{2n+1} is indeterminate for $n \ge 1$. As an illustration, if $Z \sim N(0, 1)$, Z^3 has pdf d(x), where

$$d(x) = (3\sqrt{\pi})^{-1} |x|^{-2/3} \exp(-|x|^{2/3}), \quad |x| < \infty.$$

But the function

$$d(x)\{1 + a[\cos(\sqrt{3}\,|x|^{2/3}) - \sqrt{3}\sin(\sqrt{3}\,|x|^{2/3})]\}, \quad |x| < \infty,$$

is a pdf whenever $|a| \leq 0.5$, having the same moments as $d(x)$.

Berg also showed that $|X|^\alpha$ is determinate when $0 < \alpha \leq 4$ and indeterminate when $\alpha > 4$. "We are thus in the strange situation that X^3 and X^5 are indeterminate, whereas $|X^3|$ and X^4 are determinate."

2.2 MOMENTS, CUMULANTS, AND GENERATING FUNCTIONS

[2.2.1] Let X be a $N(\mu, \sigma^2)$ random variable (rv). For $r = 1, 2, 3, ...$, let $E(X^r) = \mu'_r$; $E(X - \mu'_1)^r = \mu_r$; $E|X - \mu'_1|^r = \nu_r$; and let κ_r be the cumulant of order r, defined by the equation

$$\exp\left[\kappa_1 t + \frac{\kappa_2 t^2}{2!} + \ldots + \frac{\kappa_r t^r}{r!} + \ldots\right] = 1 + \mu'_1 t + \frac{\mu'_2 t^2}{2!}$$
$$+ \ldots + \frac{\mu'_r t^r}{r!} + \ldots$$

for values of the dummy variable t in some interval. Other than κ_1, all cumulants remain unchanged if a rv has its origin shifted by some fixed quantity.

For a normal distribution, all absolute moments ν_r and hence moments μ'_r, μ_r, and cumulants κ_r of all orders exist. Carleman's condition, that a set of moments uniquely determines a distribution if $\Sigma_{i=0}^\infty (\mu_{2i})^{-1/(2i)}$ diverges, is satisfied by the normal distribution (Stuart and Ord, 1987, Sec. 4.23).

[2.2.2] The following results hold for $N(\mu, \sigma^2)$ variables:

Mean $= \mu'_1 = \mu$

Variance $= \mu_2 = \sigma^2$

Higher central moments:

$\mu_3 = 0$, $\mu_4 = 3\sigma^4$, $\mu_5 = 0$, $\mu_6 = 15\sigma^6$, $\mu_7 = 0$, $\mu_8 = 105\sigma^8$

Mean deviation $= \nu_1 = \sqrt{(2\sigma^2)/\pi}$, $\nu_2 = \sigma^2$, $\nu_3 = 2\sigma^3\sqrt{2/\pi}$,

$\nu_4 = 3\sigma^4$, $\nu_5 = 8\sigma^5\sqrt{2/\pi}$

Cumulants: $\kappa_1 = \mu$, $\kappa_2 = \sigma^2$, $\kappa_r = 0$ if $r \geq 3$

Coefficient of variation $= \sigma/\mu$

Coefficient of skewness: $\beta_1 = \mu_3^2/\mu_2^3 = 0$

Kurtosis: $\beta_2 = \mu_4/\mu_2^2 = 3$

Basic Properties

Coefficient of excess = $\beta_2 - 3 = 0$

[2.2.3] Further properties:

(a) $\mu'_{2r-1} = \sigma^{2r-1} \sum_{i=1}^{r} \dfrac{(2r-1)!}{(2i-1)!(r-i)!2^{r-i}} \left(\dfrac{\mu}{\sigma}\right)^{2i-1}$, $r = 1,2,3, \ldots$;

$\mu'_{2r} = \sigma^{2r} \sum_{i=0}^{r} \dfrac{(2r)!}{(2i)!(r-i)!2^{r-i}} \left(\dfrac{\mu}{\sigma}\right)^{2i}$, $r = 1,2,3, \ldots$;

$\mu'_{r+1} = r\sigma^2 \mu'_{r-1} + \mu \mu'_r$, $r = 1, 2, 3, \ldots$ (Bain, 1969, p. 34).

(b) $\mu_{2r+1} = 0$, $r = 0, 1, 2, 3, \ldots$

$\mu_{2r} = \nu_{2r} = \dfrac{(\sigma^2/2)^r (2r)!}{r!}$, $r = 0, 1, 2, \ldots$ (Bain, 1969, p. 34);

$\nu_{2r-1} = \dfrac{(r-1)! 2^r \sigma^{2r-1}}{\sqrt{2\pi}}$, $r = 1, 2, 3, \ldots$ (Lukacs, 1970, p. 13);

$\nu_r = \dfrac{2^{r/2} \Gamma\{(r+1)/2\} \sigma^r}{\sqrt{\pi}}$, $r = 1, 2, \ldots$ (Kamat, 1953, p. 21) .

[2.2.4] *Incomplete Moments.* Let $I_r(a) = \int_a^\infty y^r \phi(y)\, dy$; $r = 0, 1, 2, \ldots$ and $I_r(a) = 0$ if $r < 0$. Then $I_0(a) = 1 - \Phi(a)$ and

$$I_r(a) = a^{r-1} \phi(a) + (r-1) I_{r-2}(a),$$

$$I_r(a) = [a^{r-1} + (r-1)a^{r-3} + \ldots + (r-1)(r-3) \ldots 5 \cdot 3 \cdot a] \phi(a)$$

$$+ (r-1)(r-3) \ldots 5 \cdot 3 \cdot I_0(a), \quad r \text{ even},$$

$$I_r(a) = [a^{r-1} + (r-1)a^{r-3} + \ldots + (r-1)(r-3) \ldots 4 \cdot 2] \phi(a),$$

r odd (Elandt, 1961, p. 551) .

[2.2.5] *Generating functions of X*:
Moment-generating function = $E(e^{tX}) = \exp(t\mu + t^2 \sigma^2 / 2)$.
Characteristic function = $E(e^{itX}) = \exp(it\mu - t^2 \sigma^2 / 2)$.
Cumulant-generating function = $\log E(e^{itX}) = it\mu - t^2 \sigma^2 / 2$.

2.3 MEMBER OF SOME FAMILIES OF DISTRIBUTIONS

The normal distribution is a member of certain families of distributions such as the exponential, Pearson, stable, and infinitely divisible families. Hence the general properties of these families also hold for the normal distribution.

[2.3.1] *Exponential Families* (Lehmann, 1983, pp. 26-36; Ferguson, 1967, pp. 125-131; Mood et al., 1974, pp. 312-314; Bickel and Doksum, 1977, pp. 67-73).
(a) The pdf g(x; θ) of a *one-parameter exponential family* member, defined for θ in an interval I, is of the form

$$g(x;\theta) = \begin{cases} \exp[Q(\theta)T(x) + S(x) + h(\theta)], & a < x < b, \\ 0, & \text{elsewhere.} \end{cases}$$

If, further, neither a nor b depends upon θ, Q is a nontrivial continuous function of θ, T is differentiable with nonzero derivative almost surely and S is continuous in x when $a < x < b$, then g represents a *regular* case of the exponential family (Hogg and Craig, 1978, p. 357).

(i) The normal pdf with $\theta = \mu$ and σ^2 known is a regular member of this family with

$$Q(\theta) = \theta/\sigma^2, \quad T(x) = x, \quad S(x) = -x^2/(2\sigma^2) - \tfrac{1}{2}\log(2\pi\sigma^2),$$

$$h(\theta) = -\theta^2/(2\sigma^2), \quad a = -\infty, \quad b = +\infty.$$

(ii) The normal pdf with $\theta = \sigma^2$ and μ known is a regular member of this family with

$$Q(\theta) = -1/(2\theta), \quad T(x) = (x-\mu)^2, \quad S(x) = 0,$$

$$h(\theta) = -\tfrac{1}{2}\log(2\pi\theta), \quad a = -\infty, \quad b = +\infty.$$

Basic Properties

(b) The pdf $g(x; \theta_1, \theta_2)$ of a *two-parameter exponential family* member, defined for θ_j in an interval I_j (j = 1, 2), is of the form

$$g(x; \theta_1, \theta_2) = \exp\left[\sum_{i=1}^{2} Q_i(\theta_1, \theta_2) T_i(x) + S(x) + h(\theta_1, \theta_2)\right],$$

$$a < x < b.$$

Further, g represents a *regular case* if neither a nor b depends on θ_1 or on θ_2, the functions $Q_i(\theta_1, \theta_2)$ are nontrivial, functionally independent, and continuous in θ_1 and in θ_2 ($\theta_j \in I_j$; j = 1, 2), the functions $T_i(x)$ are differentiable in x with continuous derivatives $T_i'(x)$ for $a < x < b$, no one of the $T_i(x)$ is a linear combination of the others, and S is continuous in x ($a < x < b$) (Hogg and Craig, 1978, p. 366).

The normal pdf with $\theta_1 = \mu$ and $\theta_2 = \sigma^2$ is a regular member of this family, where

$$Q_1(\theta_1, \theta_2) = -1/(2\theta_2), \qquad Q_2(\theta_1, \theta_2) = \theta_1/\theta_2,$$

$$T_1(x) = x^2, \qquad T_2(x) = x, \qquad S(x) = 0,$$

$$h(\theta_1, \theta_2) = -\theta_1^2/(2\theta_2) - \tfrac{1}{2}\log(2\pi\theta_2), \qquad a = -\infty, \; b = +\infty.$$

[2.3.2]   *Linear Exponential-type Distributions*   (Blackwell and Girshick, 1954, pp. 179-180; Patil, 1963, p. 205; Patil and Shorrock, 1965, pp. 94-99). The pdf $g(x; \theta)$ of a *linear exponential type distribution* is of the form

$$g(x; \theta) = a(x) \exp(\theta x)/g(\theta),$$

where $a(x)$ is a nonnegative function which depends only on x, θ lies in an interval I, and $g(\theta)$ is finite and differentiable (Patil, 1963).

The normal pdf with $\theta = \mu/\sigma^2$, σ^2 known, is of linear exponential type with $a(x) = \exp\{-x^2/(2\sigma^2)\}$, $g(\theta) = (\sigma\sqrt{2\pi})^{-1} \exp(-\theta^2\sigma^2/2)$.

The normal pdf with $\theta = \sigma^2$ and μ known is not of linear exponential type.

[2.3.3] *The Pearson System* (Pearson, 1895; Johnson and Kotz, 1970, pp. 9-15; Ord, 1985, pp. 655-659; Stuart and Ord, 1987, Secs. 6.2-6.13). The pdf g(x) of a member of the Pearson system satisfies the differential equation

$$\frac{d}{dx}\{g(x)\} = \frac{(x-a)g(x)}{b_0 + b_1 x + b_2 x^2}.$$

The normal pdf is a member of this system with $a = \mu$, $b_0 = -\sigma^2$, $b_1 = 0$, $b_2 = 0$.

[2.3.4] *Monotone Likelihood Ratio (MLR) Distributions* (Ferguson, 1967, pp. 208-210; Mood et al., 1974, p. 423). In what follows, $L(x_1, x_2,...,x_n; \theta)$ denotes the likelihood function or joint pdf $\prod_{i=1}^{n} g(x_i; \theta)$ of a random sample of size n from a distribution with pdf $g(x; \theta)$.

The pdf $g(x; \theta)$ for θ in an interval I is said to have a MLR in a statistic $T(x_1, x_2,...,x_n)$ if the likelihood ratio

$$\frac{L(x_1, x_2,...,x_n; \theta_1)}{L(x_1, x_2,...,x_n; \theta_2)}$$

is either a nonincreasing function of T for every $\theta_1 < \theta_2$ or a nondecreasing function of T for every $\theta_1 < \theta_2$; $n = 1, 2,...$.

If $n = 1$, the normal pdf with $\theta = \mu$ and σ^2 known has MLR in $T(x) = x$. The normal pdf with $\theta = \sigma^2$ and μ known has MLR in $T(x) = (x-\mu)^2$, but not in $T(x) = x$.

[2.3.5] *Stable Distributions* (Feller, 1971, pp. 169-176, 574-583; Holt and Crow, 1973, pp. 143-198; Monrad and Stout, 1988, pp. 617-621). Let $X, X_1, ..., X_n$ be independently and identically distributed (iid) random variables and let $S_n = \sum_{i=1}^{n} X_i$. The distribution of X is *stable* if for each n there exist constants $c_n > 0$, and γ_n such that S_n has the same distribution as $c_n X + \gamma_n$ and the distribution of X is not concentrated at the origin. The distribution of X is *strictly stable* if $\gamma_n = 0$. Stable distributions are absolutely continuous and unimodal (Lukacs, 1970, pp. 138, 158).

The normal distribution is a stable distribution with $c_n = \sqrt{n}$ and γ_n

$= \mu\sqrt{n}(\sqrt{n}-1)$. It is strictly stable when $\mu = 0$.

[2.3.6] *Infinitely Divisible Distributions* (Billingsley, 1979, pp. 322-327; Feller, 1971, pp. 176-179, Chap. XVII; Lukacs, 1970, Chap. 5). A distribution G is an *infinitely divisible distribution* (IDD) if for each n it can be represented as the distribution of the sum $S_n \ \Sigma_{i=1}^n \ X_i$ of n iid random variables with common distribution G_n; all stable distributions are IDD (Feller, 1971, p. 176).

The $N(\mu, \sigma^2)$ distribution is IDD with G_n as the common $N(\mu/n, \sigma^2/n)$ distribution.

[2.3.7] *Unimodal Distributions* (Lukacs, 1970, p. 91-99). A distribution $G(x)$ is said to be *unimodal* if there exists at least one value a such that $G(x)$ is convex for $x < a$ and concave for $x > a$. The point a is called a *mode* or *vertex* of $G(x)$; all stable distributions are unimodal (Lukacs, 1970, p. 91). An equivalent definition of unimodality is that "the density function g (if it exists) is such that $\{x| \ g(x) \geq c\}$ is a convex set for all c ≥ 0" (Anderson, 1955, pp. 170-171).

The $N(\mu, \sigma^2)$ distribution is unimodal with mode at μ.

[2.3.8] *Polya-type Distributions* (Karlin, 1957, p. 282; Johnson and Kotz, 1970, p. 32). A family of distributions with pdfs $f(x_i, \theta)$ which are continuous in the real variables θ and x is said to be of *Polya type n* if for all $x_1 < x_2 < \ldots < x_m$, $\theta_1 < \theta_2 < \ldots < \theta_m$ and $1 \leq m \leq n$,

$$\begin{vmatrix} f(x_1; \theta_1) & f(x_1; \theta_2) & \ldots & f(x_1; \theta_m) \\ f(x_2; \theta_1) & f(x_2; \theta_2) & \ldots & f(x_2; \theta_m) \\ \ldots & & & \\ f(x_m; \theta_1) & f(x_m; \theta_2) & \ldots & f(x_m; \theta_m) \end{vmatrix} \geq 0 ,$$

and is *strictly of Polya type n* if strict inequality holds in the above determinant. If the family of distributions is of Polya type n for every n, then the family is said to be of *Polya type* ∞.

The normal distribution with $\theta = \mu$ or with $\theta = \sigma^2$ is strictly of Polya type ∞ (Karlin, 1957, p. 282).

[2.3.9] *Symmetric Power Distributions* (Box, 1953, p. 467; Box and Tiao, 1964, p. 170). A distribution is a *symmetric power distribution* if it has a pdf given by

$$g(x; \mu, \sigma, \beta) = k \exp\{-\tfrac{1}{2}|(x-\mu)/\sigma|^{2/(1+\beta)}\}, \quad -\infty < x < \infty ,$$

$$|\mu| < \infty, \; 0 < \sigma < \infty, \; |\beta| < 1 ,$$

$$k = 1/[\Gamma(\alpha) 2^\alpha \sigma], \quad \alpha = 1 + (1+\beta)/2 .$$

The $N(\mu, \sigma^2)$ distribution results when $\beta = 0$. If $\beta < 0$, the distribution is platykurtic ($\kappa_2 < 3$) with the uniform distribution in the limit as $\beta \to -1$. If $\beta > 0$, the distribution is leptokurtic ($\kappa_2 > 3$), with the double exponential distribution when $\beta = 1$. This family has been studied in connection with some tests of normality, and of robustness.

Vianelli (1983, pp. 3-10) has expressed these distributions via

$$g(x) = \{2r^{1/r} \Gamma(1 + \tfrac{1}{r}) S_r\}^{-1} \exp\{-|x-M|^r/(rS_r^r)\}, \quad |x| \leq \infty ,$$

where S_r^r is the rth absolute central moment of $g(\cdot)$. He calls S_r^r the "structural variability index" characterizing $g(\cdot)$, and refers to symmetric power distributions as "normal curves of order r". The case $r = 2$ yields the $N(\mu, \sigma^2)$ family.

Vianelli also considers the family

$$h(x) = \frac{r}{2bB(r^{-1}, q+1)} (1 - \frac{|x|^r}{b^r})^q, \quad |x| \leq b, \; q > 0, \; r \geq 1 ,$$

where $B(\cdot, \cdot)$ is the beta function. The normal distribution results when $b \to \infty$.

[2.3.10] (a) If a rv Y has pdf $g(y, \lambda)$ and cdf $G(y, \lambda)$, where

$$g(y, \lambda) = 2 \phi(y) \Phi(\lambda y), \quad |y| < \infty, \; |\lambda| < \infty ,$$

Basic Properties

then Y has a *skew-normal distribution with parameter* λ, denoted here as SN(λ). Azzalini (1985, pp. 171-178) exhibits several properties:

- The standard normal distribution is given by SN(0).
- If $Y \sim SN(\lambda)$ then $-Y \sim SN(-\lambda)$ and conversely.
- $g(\cdot, \lambda)$ is strongly unimodal, since log $g(y, \lambda)$ is a concave function of y.
- As $\lambda \to \infty$, $g(y, \lambda)$ converges to the pdf of the half-normal distribution, discussed in *[9.*.*]*.
- $G(y, \lambda) = 1 - G(-y, -\lambda)$.
- $G(y, 1) = \{\Phi(y)\}^2$
- If $Y \sim SN(\lambda)$, then $Y^2 \sim \chi_1^2$ (see *[5.3]*).

Henze (1986, pp. 271-275) proves that if Z_1 and Z_2 are independent N(0, 1) rvs, then

$$\frac{\lambda}{\sqrt{(1+\lambda^2)}} |Z_1| + \frac{1}{\sqrt{(1+\lambda^2)}} Z_2$$

has a SN(λ) distribution.

(b) One limitation of the family $g(y; \lambda)$ is that λ can only generate tails thinner than those of the standard normal. A class of densities that includes $\phi(\cdot)$ and allows thick tails is that of symmetric power distributions discussed in *[2.3.9]*:

$$h(y; w) = C_w \exp\{-|y|^w/w\}, \quad |y| < \infty, \quad w > 0,$$

$$C_w = 1/\{2w^{(1/w)-1} \Gamma(1/w)\}$$

(Box, 1953, pp. 465-468). Azzalini (1986, pp. 199-208) extends this family to a skew form via $2H(\lambda y) h(y; w)$, $H(\cdot)$ being the cdf of $h(\cdot; w)$.

2.4 SOME BASIC DISTRIBUTIONS DERIVED FROM THE NORMAL

Here we present three simple transformations and the truncated distributions of a normal variable; other, more complex related distributions and stochastic processes are discussed in Chapter 9.

[2.4.1] If Z has a standard normal distribution (see [2.1.2]), then Z^2 has a *chi-squared distribution* with one degree of freedom and pdf

$$(2\pi)^{-1/2} y^{-1/2} e^{-y/2}, \quad y > 0.$$

See also Sec. [5.3].

[2.4.2] Let the random variable Y satisfy $Y = \exp(X)$, where X has a $N(\mu, \sigma^2)$ distribution. Then Y has a *lognormal distribution* with pdf given by

$$g(y; \mu, \sigma^2) = \frac{1}{\sigma y \sqrt{2\pi}} \exp\left\{-\frac{(\log y - \mu)^2}{2\sigma^2}\right\}, \quad y > 0, \; -\infty < \mu < \infty,$$

$$\sigma > 0.$$

(i) Let $G(y; \mu, \sigma^2)$ be the cdf corresponding to the pdf $g(y; \mu, \sigma^2)$ and $F(x; \mu, \sigma^2)$ be the cdf of the rv X. Then

$$G(y; \mu, \sigma^2) = F(\log y; \mu, \sigma^2), \quad F(x; \mu, \sigma^2) = G(e^x; \mu, \sigma^2).$$

(ii) Let z_p and y_p be the quantiles of order p of a $N(0, 1)$ rv and $g(y; \mu, \sigma^2)$, respectively, so that $\Phi(z_p) = \Pr(Y \leq y_p) = 1 - p$; then (Aitchison and Brown, 1957, p. 9) $y_p = \exp(\mu + z_p \sigma)$.

(iii) The mean and variance of $g(y; \mu, \sigma^2)$ are $\exp(\mu + \sigma^2/2)$ and $\exp(2\mu + x^2)\{\exp(\sigma^2) - 1\}$, respectively.

For further information about the lognormal distribution, see Aitchison and Brown (1957) or Crow and Shimizu (1988, Chapter 1).

[2.4.3] *Uniform Distribution.* Let X be a normal random variable with cdf $F(x) = F(x; \mu, \sigma^2)$. Then the random variable defined by $Y = F(X)$ has a uniform distribution with pdf $g(y) = 1$, $0 < y < 1$.

[2.4.4] Let Z be a $N(0, 1)$ random variable. Let $g(\cdot)$ be an absolutely continuous function and let $g(Z)$ have a finite variance. Then

$$E\{[g'(Z)]^2\} \geq \text{Var}[g(Z)],$$

where g'(·) is the derivative of g, with equality if and only if g(Z) is linear in Z (Chernoff, 1981, p. 533).

[2.4.5] Let X be a $N(\mu, \sigma^2)$ random variable. The conditional distribution of X, given that $a < X < b$, is a *doubly truncated normal distribution* with pdf g(y), where

$$g(y) = \begin{cases} \frac{1}{\sigma}\phi\left(\frac{y-\mu}{\sigma}\right) / \left[\Phi\left(\frac{b-\mu}{\sigma}\right) - \Phi\left(\frac{a-\mu}{\sigma}\right)\right], & a < y < b \\ 0, & \text{otherwise} . \end{cases}$$

If $a = -\infty$, the distribution is *singly truncated from above*, and is *singly truncated from below* if $b = \infty$.

(a) When $a = 0$, $b = \infty$, and $\mu = 0$, the pdf g(·) is that of the half-normal distribution discussed in *[9.*.*]*.

(b) Let $h = (a-\mu)/\sigma$, $k = (b-\mu)/\sigma$. Then the mean and variance μ_t and σ_t^2 of the doubly truncated normal distribution are given by

$$\mu_t = \mu + \sigma\{\phi(h) - \phi(k)\}/[\Phi(k) - \Phi(h)] ,$$

$$\sigma_t^2 = \sigma^2[1 + \{h\phi(h) - k\phi(k)\}/\{\Phi(k) - \Phi(h)\}] - (\mu_t - \mu)^2.$$

(c) Let Y be a $N(\mu, \sigma^2)$ rv, singly truncated from below at a. If $u = (a-\mu)/\sigma$, then the mean and variance of Y are given by

$$\mu_t = \mu + \sigma/R(u), \quad \sigma_t^2 = \sigma^2\left[1 + h/R(h) - \{R(h)\}^{-2}\right],$$

where R(x) is Mills' ratio $[1 - \Phi(x)]/\phi(x)$, discussed in *[3.6]*–*[3.8]*. For details on tables see *[3.1.1(c)]*.

(d) Shah and Jaiswal (1966, pp. 107-111) give the pdf and moments of g(y) shifted so that the origin is at a. Johnson and Kotz (1970, p. 84) give a table of μ_t, σ_t, and the ratio (mean deviation)/σ_t for selected values of $\Phi(h)$ and of $1 - \Phi(k)$; see also *[2.6.3]*.

2.5 NORMAL INTEGRALS

A *normal integral* is any definite or indefinite integral in which $\Phi(\cdot)$ or $\phi(\cdot)$ appears in the integrand. Examples are the moments of a normal distribution. Owen (1980, 1981) has collected many of these in a list that includes results relating to bivariate and multivariate normal distributions, for some of which see Chapter 10. The list is extensive; the following selection illustrates the type of normal integrals tabulated; see also Chou (1981), and Section *[2.1.9]*.

[2.5.1] Indefinite Normal Integrals

$$\int x\phi(x)dx = -\phi(x),$$

$$\int x^2\phi(x)\,dx = \Phi(x) - x\phi(x),$$

$$\int x^n\phi(x)dx = \sum_{i=0}^{\frac{1}{2}(n-1)} 2^{\frac{1}{2}(n-2i-1)}\left(\frac{n-1}{2}\right)! \frac{x^{2i}\phi(x)}{i!}, \quad n \text{ odd},$$

$$\int x^n\phi(x)dx = (n-1)(n-3)\cdots 5\cdot 3\cdot 1 \left\{\Phi(x) - \sum_{i=0}^{\frac{1}{2}n-1} 2^i x^{2i+1} \frac{i!}{(2i+1)!}\phi(x)\right\}, \quad n \text{ even},$$

$$\int \{\phi(x)\}^2 dx = [2\sqrt{\pi}]^{-1}\Phi(x\sqrt{2}),$$

$$\int \phi(x)\phi(a+bx)\,dx = t^{-1}\phi(a/t)\Phi(tx+a/t), \quad t = (1+b^2)^{1/2},$$

$$\int x\phi(a+bx)dx = -\frac{1}{b^2}\phi(a+bx) - \frac{a}{b^2}\Phi(a+bx),$$

$$\int x^2\phi(a+bx)dx = \frac{a^2+1}{b^3}\Phi(a+bx) - \frac{bx-a}{b^3}\phi(a+bx),$$

$$\int \{\phi(a+bx)\}^n dx = \frac{1}{\sqrt{n}\,(2\pi)^{(n-1)/2}b}\Phi(\sqrt{n}(a+bx)),$$

$$\int \Phi(a+bx)dx = \frac{a+bx}{b}\Phi(a+bx) + \frac{1}{b}\phi(a+bx),$$

Basic Properties

$$2b^2 \int x\Phi(a+bx)dx = (b^2x^2 - a^2 - 1)\Phi(a+bx) + (bx-a)\phi(a+bx),$$

$$\int x^2\Phi(x)dz = \tfrac{1}{3}x^3\Phi(x) + \tfrac{1}{3}(x^2+2)\phi(x),$$

$$\int x^n\Phi(x)dx = \tfrac{1}{n+1}\left\{[x^{n+1} - nx^{n-1}]\Phi(x) + x^n\phi(x) \right.$$
$$\left. + n(n-1)\int x^{n-2}\Phi(x)dx\right\},$$

$$\int x\phi(x)\Phi(bx)dx = \frac{b}{(2\pi(1+b^2))^{1/2}}\Phi(x\sqrt{1+b^2})) - \phi(x)\Phi(bx),$$

$$\int \{\Phi(x)\}^2 dx = x\{\Phi(x)\}^2 + 2\Phi(x)\phi(x) - \pi^{-\tfrac{1}{2}}\Phi(x\sqrt{2}),$$

$$\int e^{cx}\{\phi(bx)\}^n dx = \frac{1}{b\{n(2\pi)^{n-1}\}^{1/2}} e^{c^2/(2nb^2)} \Phi(bx\sqrt{n} - \tfrac{c}{b\sqrt{n}}),$$

$$b \neq 0, n > 0.$$

[2.5.2] Definite Normal Integrals

$$\int_0^y x^n \phi(bx)dx = \frac{2^{(n-1)/2}\Gamma(\tfrac{1}{2}(n+1))}{\sqrt{2\pi}\, b^{n+1}} \Pr(Y \leq b^2 y^2),$$

where Y has a chi-squared distribution with $n+1$ degrees of freedom;

$$\int_{-\infty}^{\infty} x^2 \{\phi(x)\}^n dx = \left[n^{3/2}(2\pi)^{(n-1)/2}\right]^{-1},$$

$$\int_{-\infty}^{0} \phi(ax)\Phi(bx)dx = (2\pi a)^{-1}\arctan(a/b),$$

$$\int_{0}^{\infty} \phi(ax)\Phi(bx)dx = (2\pi a)^{-1}(\tfrac{\pi}{2} - \arctan\tfrac{b}{a}),$$

$$\int_0^\infty x\phi(x)\Phi(bx)dx = \frac{1}{2\sqrt{2\pi}}\left[1 + \frac{b}{(1+b^2)^{1/2}}\right],$$

$$\int_0^\infty x^2\phi(x)\,\Phi(bx)dx = \tfrac{1}{4} + \tfrac{1}{2\pi}\left(\tfrac{b}{1+b^2} + \arctan b\right),$$

$$\int_{-\infty}^\infty x\{\phi(x)\}^2\Phi(x)dx = 1/(4\pi\sqrt{3}),$$

$$\int_0^\infty \{\Phi(bx)\}^2 \phi(x)dx = (2\pi)^{-1}[\arctan b + \arctan\sqrt{1+2b^2}],$$

$$\int_{-\infty}^\infty \{\Phi(bx)\}^2\phi(x)dx = \pi^{-1}\arctan\sqrt{1+2b^2},$$

$$\int_{-\infty}^\infty x\phi(x)\Phi(bx)dx = b/[\sqrt{2\pi(1+b^2)}] = \int_{-\infty}^\infty x\phi(x)\{\Phi(bx)\}^2 dx,$$

$$\int_{-\infty}^\infty \Phi(a+bx)\phi(x)dx = \Phi(a/\sqrt{1+b^2}).$$

Further, let $t = \sqrt{1+b^2}$ and

$$T(h, a) = \int_0^a \frac{\phi(x)\phi(hx)}{1+x^2}dx, \quad |h| < \infty,\ 0 < a < \infty.$$

(For more on the latter integral see [10.2.2.2]). Then

$$\int_0^\infty x\Phi(a+bx)\phi(x)dx = (b/t)\phi(a/t)\Phi(-ab/t) + (2\pi)^{-1/2}\Phi(a),$$

$$\int_{-\infty}^\infty x\Phi(a+bx)\phi(x)dx = (b/t)\phi(a/t),$$

$$\int_{-\infty}^0 \phi(x)\Phi(a+bx)dx = \tfrac{1}{2}\Phi(a/t) - T(a/t, b),$$

Basic Properties

$$\int_0^\infty \phi(x)\Phi(a+bx)dx = \tfrac{1}{2}\Phi(a/t) + T(a/t, b),$$

$$\int_{-\infty}^\infty [\Phi(a+bx)]^2 \phi(x)dx = \tfrac{2b}{t} \phi(\tfrac{a}{t})\Phi\left(\tfrac{a}{t\sqrt{1+2b^2}}\right).$$

2.6 TABLES

A large choice of tables is available that give probabilities under the standard normal curve; the cdf $\Phi(x)$ is usually tabulated in textbooks dealing with introductory statistics and data analysis, both for general study and for applications in specific fields like agriculture, business, medicine and psychology. These tables commonly are accurate to four decimal places. There is also a choice of tables giving values of $\phi(x)$, quantiles z_α, derivatives of $\phi(x)$, Mills' ratio $R(x)$, and so on. Here we list some of these, including American, British, Russian, Indian and Japanese sources. Algorithms and expansions for generating these functions appear in Chapter 3, along with some useful approximations and inequalities.

[2.6.1] Greenwood and Hartley (1962) give a comprehensive list of tables published up to 1958, as do Fletcher et al. (1962a, pp. 318-356). Fletcher et al. (1962b, pp. 781-932) include a discussion of errors appearing in published tables cited, together with corrections. There is a comprehensive discussion in Johnson and Kotz (1970, pp. 41-45).

[2.6.2] Table 2.1 summarizes the information found in seven sources; some of them can be seen to be very detailed, particularly for the cdf $\Phi(x)$. The most accurate table that we could find of quantiles z_α [when $1-\alpha$, or $\Phi(z_\alpha)$, is given] was that of White (1970, pp. 636-637), giving z_α to 20 decimal places for $1-\alpha = 0.50$ (0.005) 0.995 and for $\alpha = 5 \times 10^{-k}$, 2.5×10^{-k}, and 10^{-k}, $k = 1(1)20$; see also Kelley (1948), with z_α to 8 places.

In order to use these tables, which relate to the upper 50 percent of the normal area only, results in *[2.1.4]* should be noted.

TABLE 2.1 Tables of the Unit Normal cdf Φ, pdf ϕ, Quantiles z_α^a, Derivatives of ϕ, and Mills' Ratio $R(x)^b$

Source	Function	Coverage	Decimal places	Significant figures
Abramowitz and Stegun (1964, pp. 966-977)[c]	$\Phi(x), \phi(x), \phi'(x)$	$x = 0.0(0.02)3.00$	15	
	$\Phi(x)$	$x = 3.0(0.05)5.0$	10	
	$\phi(x)$	$x = 3.0(0.05)5.0$	9	
	$\phi'(x)$	$x = 3.0(0.05)5.0$	7	
	$\phi^{(r)}(x), r=2,\ldots,6$	$x = 0.0(0.02)3.00(0.05)5.0$	7-10	
	$-\log_{10}[1 - \Phi(x)]$	$x = 5(1)50(10)100(50)500$	5	
	$\phi^{(r)}(x); r=7,\ldots,12$	$x = 0.0(0.1)5.0$	7	
	$\phi(z_\alpha), z_\alpha$	$1 - \alpha = 0.500(0.001)0.999$	5	
	z_α	$1 - \alpha = 0.975(0.0001)0.9999$	5	
	z_α	$\alpha = 10^{-r}; r = 4,\ldots,23$	5	
Owen (1962, pp. 3-13)	$\Phi(x), \phi(x)$	$x = 0.0(0.01)3.99$	6	
	$1 - \Phi(x)$	$x = 3.0(0.1)6.0(0.2)10(1)20, 25,30(10)100(25)200(50)500$		5
	$z_\alpha, \phi(z_\alpha)$	$1 - \alpha = 0.50(0.01)0.90(0.005)0.99 (0.001)0.999(0.0001)$ 0.9999, etc. to $1-10^9$	5	
	$\Phi(x)/\phi(x)$	$x = 0(0.01)3.99$	4	
	$R(x), \phi'(x), \phi''(x), \phi'''(x)$	$x = 0.0(0.01)3.99$	5	
	$R(x)$	$x = 3.0(0.1)6.0(0.2)10(1)20, 25,30(10)100(25)200(50)500$		5

[a] $\Phi(z_\alpha) = 1 - \alpha$
[b] $R(x) = \{1 - \Phi(x)\}/\phi(x)$
[c] Tables compiled by M. Zelen and N.C. Severo.

Basic Properties

Table 2.1 (continued)

Source	Function	Coverage	Decimal places	Significant figures
Pearson and Hartley (1966, tables 1-5)	$\Phi(x)$, $\phi(x)$	$x = 0.0(0.01)4.50$ $x = 4.50(0.01)6.00$	7 10	
	$-\log_{10}[1 - \Phi(x)]$	$x = 5(1)50(10)100(50)500$	5	
	z_α	$1 - \alpha = 0.50(0.001)0.999$ $0.98(0.0001)0.9999$ $1-10^{-r}$ (r=4,...,9)	4 4 4	
	$\phi(z_\alpha)$	$1 - \alpha = 0.50(0.001)0.999$	5	
Pearson and Hartley (1972, tables 1-2)	z_α, $\phi(z_\alpha)$	$1 - \alpha = 0.50(0.001)0.999$	10	
	z_α	$1 - \alpha = 0.999(0.0001)0.9999$	8	
	$\phi(z_\alpha)$	$1 - \alpha = 0.999(0.0001)0.9999$	9	
	$\Phi(x) - 0.5$, $\phi(x)$, $\phi'(x)$, $\phi''(x)$	$x = 0.0(0.02)6.20$	6	
	$\phi^{(r)}(x)$; $r = 3,...,9$	$x = 0.0(0.02)6.20$	5 to 1	
Rao et al. (1966, tables 3.1, 3.2)	$\Phi(x) - 0.5$	$x = 0.00(0.001)3.0(0.01)$ $4(0.1)(4.9)$	6	
	$\phi(x)$	$x = 0.0(0.01)3.0(0.1)4$	6	
	z_α	$2(1 - \alpha) = 0.01(0.01)0.99;$ $10^{-r}(r = 3,...,9)$	6 5	
	$2[1 - \Phi(x)]$	$x = 0.25, 0.5(0.5)5.0$	6	
Smirnov (1965, tables I, II, III)	$\Phi(x) - 0.5$, $\phi(x)$	$x = 0.00(0.001)2.50(0.002)$ $3.40(0.005)4.00(0.01)4.50$ $x = 4.50(0.01)6.00$	7 10	
	$-\log_{10}\{1 - \Phi(x)\}$	$x = 5(1)50(10)100(50)500$	5	
Yamauti (1972, tables A1, A2, A3)	$\phi(x)$	$x = 0.0(0.01)4.99$		5
	$1 - \Phi(x)$	$x = 0.0(0.01)4.99$		5
	$1 - \Phi(x)$	$x = 0.1(0.1)10.0$		35
	z_α	$\alpha = 0.0(0.001)0.499$	5	

[2.6.3] Clark (1957, p. 527-536) gives a table to four decimal places of the mean μ_{ab} and standard deviation σ_{ab} of a standard normal distribution truncated at a and b (a < b), for a = -3.0(0.25)0.50 and b = 0.0(0.25)3.0. $\mu_{-b,-a} = -\mu_{ab}$ and $\sigma_{-b,-a} = \sigma_{ab}$; see *[2.4.5]*.

[2.6.4] It is of historical interest to note the list of some of the earliest known tables relating to the normal distribution in Table 2.1. Although these were prone to errors, we must count it remarkable that such degrees of accuracy were obtained without the help of computers and, for the most part, without the use of any but the most primitive hand calculators.

[2.6.5] For tables of random normal deviates (having a N(0,1) distribution) see Wold (1948) or Sengupta and Bhattacharya (1958), pp. 250-286).

[2.6.6] Since $\Phi(x) = 1 - \phi(x)R(x)$, where $R(x)$ is Mills' ratio, the standard normal cdf may be obtained from $R(x)$. The most extensive tables of $R(x)$ are those of Sheppard (1939); these give 12 decimal places for x = 0.00(0.01)9.50 and 24 places for x = 0(0.1)10.

REFERENCES

Abramowitz, M. and Stegun, I. A. (eds.) (1970). *Handbook of Mathematical Functions*, Washington, D.C.: National Bureau of Standards. *[2.1; 2.1.8, 9; 2.6]*

Aitchison, J. and Brown, J. A. C. (1957). *The Lognormal Distribution*, London: Cambridge University Press. *[2.4.2]*

Anderson, T. W. (1955). The integral of a symmetric unimodal function over a symmetric convex set and some probability inequalities, *Proceedings of the American Mathematical Society* 6, 170-176. *[2.3.7]*

Azzalini, A. (1985). A class of distributions which includes the normal ones, *Scandinavian Journal of Statistics* 12, 171-178. *[2.3.10]*

Azzalini, A. (1986). Further results on a class of distributions which includes the normal ones, *Statistica* 46, 199-208. *[2.3.10]*

Bain, L. J. (1969). Moments of a non-central t and non-central F distribution, *The American Statistician* 23(4), 33-34. *[2.2.3]*

Berg, C. (1988). The cube of a normal distribution is indeterminate, *Annals of Probability 16*, 910-913. *[2.1.12]*

Bickel, P. J. and Doksum, K. A. (1977). *Mathematical Statistics: Basic Ideas and Selected Topics*, San Francisco: Holden-Day. *[2.3.1]*

Billingsley, P. (1979). *Probability and Measure*, New York: Wiley. *[2.3.6]*

Blackwell, D. and Girshick, M. A. (1954). *Theory of Games and Statistical Decisions*, New York: Wiley. *[2.3.2]*

Box, G. E. P. (1953). A note on regions for tests kurtosis, *Biometrika 40*, 465-568. *[2.3.9, 10]*

Box, G. E. P. and Tiao, G. C. (1964). A note on criterion robustness and inference robustness, *Biometrika 51*, 169-173. *[2.3.9]*

Chernoff, H. (1981). A note on an inequality involving the normal distribution, *Annals of Probability 9*, 533-535. *[2.4.4]*

Chou, Y-M. (1981). Additions to the table of normal integrals, *Communications in Statistics B10*, 537-538. *[2.5]*

Clark, F. E. (1957). Truncation to meet requirements on means, *Journal of the American Statistical Association 52*, 527-536. *[2.6.3]*

Crow, E. L. and Shimizu, K., eds. (1988). *Lognormal Distributions: Theory and Applications*, New York: Dekker. *[2.4.2]*

Draper, N. R. and Tierney, D. E. (1973). Exact formulas for additional terms in some important series expansions, *Communications in Statistics 1*, 495-524. *[2.1.8]*

Elandt, R. C. (1961). The folded normal distribution: Two methods of estimating parameters from moments, *Technometrics 3,*, 551-562. *[2.2.4]*

Feller, W. (1971). *An Introduction to Probability Theory and its Applications, Vol 2* (2nd edn.), New York: Wiley. *[2.3.5, 6]*

Ferguson, T. S. (1967). *Mathematical Statistics. A Decision Theoretic Approach*, New York: Academic. *[2.3.1, 4]*

Fletcher, A., Miller, J. C. P., Rosenhead, L. and Comrie, L. J. (1962a). *An Index of Mathematical Tables*, Vol. I (2nd edn.), Reading, Mass.: Addison-Wesley/Scientific Computing Service. *[2.6.1]*

Fletcher, A., Miller, J. C. P., Rosenhead, L. and Comrie, L. J. (1962b). *An Index of Mathematical Tables*, Vol. II (2nd edn.), Reading, Mass.: Addison-Wesley/Scientific Computing Service. *[2.6.1]*

Greenwood, J. A. and Hartley, H. O. (1962). *Guide to Tables in Mathematical Statistics*, Princeton, N. J.: Princeton University Press. *[2.6.1]*

Hefferman, P. M. (1988). New measures of spread and a simpler formula for the normal distribution, *American Statistician* 42, 100-102. *[2.1.6]*

Henze, N. (1986). A probabilistic representation of the skew-normal distribution, *Scandinavian Journal of Statistics* 13, 271-275. *[2.3.10]*

Hogg, R. V. and Craig, A. T. (1978). *Introduction to Mathematical Statistics* (4th edn.), New York: Macmillan. *[2.4.1]*

Holt, D. R. and Crow, E. L. (1973). Tables and graphs of stable probability density functions, *Journal of Research* (National Bureau of Standards) 77B, 143-198. *[2.3.5]*

Horn, P. S. (1988). On the stochastic ordering of absolute univariate Gaussian random variables, *Annals of Statistics* 16, 1327-1329. *[2.1.11]*

Johnson, N. L. and Kotz, S. (1970). *Distributions in Statistics: Continuous Univariate Distributions*, Vol. 1, New York: Wiley. *[2.3.3, 8; 2.4.5; 2.6.1]*

Kamat, A. R. (1953). Incomplete and absolute moments of the multivariate normal distribution with some applications, *Biometrika* 40, 20-34. *[2.2.3]*

Karlin, S. (1957). Polya type distributions, II, *Annals of Mathematical Statistics* 28, 281-308. *[2.3.8]*

Kelley, T. L. (1948). *The Kelley Statistical Tables*, Cambridge, Mass.: Harvard University Press. *[2.6.2]*

Lehmann, E. L. (1983). *Theory of Point Estimation*, New York: Wiley. *[2.3.1]*

Lukacs, E. (1970). *Characteristic Functions*, London: Griffin. *[2.2.3; 2.3.5, 6, 7]*

Mood, A. M., Graybill, F. A. and Boes, D. C. (1974). *Introduction to the Theory of Statistics* (3rd ed.), New York: McGraw-Hill. *[2.3.1, 4]*

Monrad, D. and Stout, W. (1988). Stable distributions, *Encyclopedia of Statistical Sciences*, Vol. 8 (Kotz, S., Johnson, N. L. and Read, C. B., eds.), 617-621. New York: Wiley. *[2.3.5]*

Ord, J. K. (1985). Pearson system of distributions, *Encyclopedia of Statistical Sciences*, Vol. 6 Kotz, S., Johnson, N. L. and Read, C. B., eds.), 655-659. New York: Wiley. *[2.3.3]*

Owen, D. B. (1962). *Handbook of Statistical Tables*, Reading, Mass: Addison-Wesley. *[2.6]*

Owen, D. B. (1980). A table of normal integrals, *Communications in Statistics B9*, 389-419. Errata (1981), ibid., *B10*, 541. *[2.5]*

Patil, G. P. (1963). A characterization of the exponential-type distribution, *Biometrika 50*, 205-207. *[2.3.2]*

Patil, G. P. and Shorrock, R. (1965). On certain properties of the exponential type families, *Journal of the Royal Statistical Society B27*, 94-99. *[2.3.2]*

Pearson, E. S. and Hartley, H. O. (1966). *Biometrika Tables for Statisticians*, Vol. 1 (3rd edn.), London: Cambridge University Press. *[2.6]*

Pearson, E. S. and Hartley, H. O. (1972). *Biometrika Tables for Statisticians*, Vol. 2, London: Cambridge University Press. *[2.6]*

Pearson, K. (1895). Contributions to the mathematical theory of evolution, II: Skew variations in homogeneous material, *Philosophical Transactions of the Royal Society of London A186*, 343-414. *[2.3.3]*

Rao, C. R., Mitra, S. K. and Matthai, A., eds. (1966). *Formulae and Tables for Statistical Work*, Calcutta: Statistical Publishing Society. *[2.6]*

Sengupta, J. M. and Bhattacharya, N. (1958). Tables of random normal deviates, *Sankhyā 20*, 250-286. *[2.6.5]*

Sheppard, W. F. (1898). On the application of the theory of error to cases of normal distribution and normal correlation, *Philosophical Transactions of the Royal Society 192*, 101-167.

Sheppard, W. F. (1939). *The Probability Integral*, British Association Mathematical Tables Vol. 7, London: Cambridge University Press. *[2.6.6]*

Smirnov, N. V. (1965). *Tables of the Normal Probability Integral, the Normal Density, and its Normalized Derivatives* (transl. from the 1960 Russian edn.), New York: Macmillan. *[2.6]*

Stigler, S. M. (1982). A modest proposal: a new standard for the normal, *American Statistician 36*, 137-138. *[2.1.6]*

Stuart, A. and Ord, J. K. (1987). *Kendall's Advanced Theory of Statistics*, Vol. 1 (5th edn.), New York: Oxford University Press. *[2.1.8; 2.2.1; 2.3.3]*

Tong, Y. L. (1978). An adaptive solution to ranking and selection problems, *Annals of Statistics 6*, 658-672. *[2.1.4]*

Vianelli, S. (1983). The family of normal and lognormal distributions of order r, *Metron 31*, 3-10. *[2.3.9]*

White, J. S. (1970). Tables of normal percentile points, *Journal of the American Statistical Association 65*, 635-368. *[2.6.2]*

Wold, H. (1948). *Random Normal Deviates*, Tracts for Computers, Vol. 25, London: Cambridge University Press. *[2.6.5]*

Yamauti, Z. (ed.) (1972). *Statistical Tables and Formulas with Computer Applications*, Tokyo: Japanese Standards Association. *[2.6]*

Chapter 3

EXPANSIONS AND ALGORITHMS

The standard normal cdf $\Phi(x)$ and related functions, such as the quantiles and Mills' ratio, can be expanded in series and in other forms such as continued fractions. Such expressions give rise to approximations to $\Phi(x)$ and other functions, by truncating an expansion at a point which provides a suitable degree of accuracy. The main use of such expansions is in providing suitable algorithms for computing numerical values of functions of interest. Some of the expansions were used in one form or another to compute the values appearing in published tables of the normal distribution (see *[2.6]*).

Where approximations are scrutinized for accuracy, or compared with one another, writers have generally examined the absolute error or absolute relative error; in approximating to the cdf $\Phi(x)$, for example, these would be $|G(x) - \Phi(x)|$ or $|G(x) - \Phi(x)|/\Phi(x)$, respectively, where $G(x)$ is the approximating function. In this chapter, the absolute error will be denoted briefly by |error|, and the absolute relative error by |relative error|.

Some sources included in this chapter contain errors that are corrected in later corrigenda. We do not claim to have detected all such errors, but those made between 1943 and 1969 may be cited in the *Cumulative Index to "Mathematics of Computation,"* vols. 1 to 23. Errors appearing in later sources are frequently reported in *Mathematics of Computation*.

3.1 NOMOGRAMS AND COMPUTING ALGORITHMS

A large body of computing algorithms for the normal cdf and quantiles is available, and for the generation of random normal deviates, that is, of simulated "samples" from normal populations. In this section, we shall identify some of these; many more will be given explicitly in Section *[3.2]* (the normal density function), Sections *[3.3]-[3.8]* (the normal cdf and Mills' ratio), and Section *[3.9]* (quantiles/percentage points).

[3.1.1] There is a nomogram for $\Phi(x)$ by Varnum (1950, pp. 32-34) and for the mean of a normal distribution singly truncated from below by Altman (1950, pp. 30-31) (see *[2.6.2]*).

[3.1.2] Some algorithms for computing $\Phi(x)$ use polynomial and rational approximations; these appear in *[3.4]*. Algorithms of Ibbetson (1963, p. 616) and Adams (1969, pp. 197-198) for computing the standard normal cdf $\Phi(x)$ give accuracy to seven and nine decimal places respectively; see also Hill (1969, p. 299) and Hill (1973, pp. 424-427). These algorithms are designed for everyday statistical use. Hill and Joyce (1967, pp. 374-375) give an algorithm providing accuracy within at least two decimal places of the machine capability. IMSL (1987, Sec. 17.2) has subroutines for computing normal percentiles, cdf values $\Phi(x)$, pdf values $\phi(x)$, and functions related to the distributions of chi-square, t, F, and bivariate normal distribution. There is a useful discussion from the viewpoint of statistical computing in Kennedy and Gentle (1980, Sec. 5.3.1).

[3.1.3] Algorithms for computing N(0, 1) quantiles z_α are those of Beasley and Springer (1977, pp. 118-121), Odeh and Evans (1974, pp. 96-97), and Marsaglia et al. (1964, pp. 4-10). See also the discussion in Kennedy and Gentle (1980, Sec. 5.3.2), and in *[3.9]*, where several algorithms and approximations are presented in detail.

[3.1.4] *Generation of random normal deviates.* Several methods, both exact and approximate, exist for generating sets of N(0, 1) variate values. Rather than list all of these, we will summarize the recommendations made in various books and surveys, based on a comparison of computing times. Sources of ready made tables of N(0, 1) variate values are given in *[2.6.6]*.

There is an extensive discussion of methodology in Kennedy and Gentle (1980, Sec. 6.5.1), in Bratley et al. (1987, Secs. 5.2.7, 5.2.9-10, 5.3.1), and in Knuth (1981, pp. 125-127, 552). Atkinson and Pearce (1976, pp. 431-461) found the "fast" method of Marsaglia et al. (1964, pp. 4-10), and the "convenient" method of Marsaglia and Bray (1964, pp. 260-264) to

Expansions and Algorithms 47

be fastest in Fortran. In subsequent discussion, however, the Box-Muller transformation was preferred for programming in APL. We discuss the latter in *[7.15.8]*; see Box and Muller (1958, pp. 610-611). This method has become well-established for generating standard normal deviates.

Dagpunar (1988, Chap. 4) favors the Marsaglia et al. "fast" method and the Marsaglia and Bray "convenient" method for good all-round performance. He also discussed the generation of values from the tail of a normal distribution. Ripley (1987, pp. 54, 60, 65, 82-87) tabulates computing times for several methods. "There is little to choose," he writes (p. 82), "between the Box-Muller, polar and ratio methods for sensitivity to pseudo-random number generators." By the polar method he means Marsaglia and Bray's adaptation of the Box-Muller method. The ratio or ratio-of-uniforms method dates back inter alia to Brent (1974, pp. 704-706), an algorithm designed for speed, but containing a large set of constants.

3.2 THE STANDARD NORMAL DENSITY FUNCTION

[3.2.1] The exponential series gives rise to the approximation

$$\phi(x) = \frac{1}{(2\pi)^{1/2}} \left\{ 1 + \sum_{i=1}^{n} \frac{(-x^2/2)^i}{i!} \right\} + \cdots,$$

which improves rapidly as n increases.

[3.2.2] *Rational Approximations* (Abramowitz and Stegun, 1964, Secs. 26.2.20 - 26.2.21; Hastings, 1955, pp. 151-153). For all values of x,

(a) $\phi(x) = [2.4908,95 + (1.4660,03)x^2 - (0.0243,93)x^4$
$+ (0.1782,57)x^6]^{-1} + \epsilon(x), \quad |\epsilon(x)| < 2.7 \times 10^{-3}$.

(b) $\phi(x) = [2.5112,61 + (1.1718,01)x^2 + (0.4946,18)x^4$
$- (0.0634,17)x^6 + (0.0294,61)x^8]^{-1} + \epsilon(x),$
$|\epsilon(x)| < 0.8 \times 10^{-3}$.

(c) $\phi(x) = [2.5052,367 + (1.2831,204)x^2 + (0.2264,718)x^4$
$+ (0.1306,469)x^6 - (0.0202,490)x^8$
$+ (0.0039,132)x^{10}]^{-1} + \epsilon(x)$, $|\epsilon(x)| < 2.3 \times 10^{-4}$.

[3.2.3] Shore (1982, pp. 108-111) presents two approximations to $\phi(x)$ in terms of the cdf $\Phi(x)$.

(i) $\hat{\phi}_1(x) = 1.4184[1 - \Phi(x)]^{0.8632}\Phi(x)$, $x \geq 0$;

$$|\hat{\phi}_1(x) - \phi(x)| = \begin{cases} 0.009, & 0 < x \leq 0.3, \\ 0.004, & 0.3 < x \leq 1.2, \\ 0.001, & 1.2 < x \leq 2.4, \\ 0.0002, & 2.4 < x \leq 2.6. \end{cases}$$

(ii) $\hat{\phi}_2(x) = 0.4115[1 - \Phi(x)]\{2 - \log[(1/\Phi(x)) - 1]\}$, $x \geq 0$;

this approximation is comparable in accuracy with $\hat{\phi}_1(x)$.

3.3 EXPRESSIONS RELATING TO THE DISTRIBUTION FUNCTION – I

There is a considerable literature on series expansions, continued fractions, rational functions and inequalities that lead to approximations to the standard normal cdf $\Phi(x)$. Many of these results are expressed in terms of Mills' ratio $R(x)$, viz.,

$$R(x) = [1 - \Phi(x)]/\phi(x).$$

Researchers are cautioned that discussions in the literature of convergence of series or continued fractions may refer to convergence to $\Phi(x)$, to the tail probability $1 - \Phi(x)$ when $x > 0$, to $R(x)$, or to $\overline{R}(x)$, where

$$\overline{R}(x) = \frac{\Phi(x) - 0.5}{\phi(x)}, \quad x > 0.$$

Expansions and Algorithms

Properties of an approximation to R(x), say, will have to be modified if the same expression is applied as an approximation to $\Phi(x)$. See also *[3.8.1.1]*.

In this section and the next two, we present some expressions for $\Phi(\cdot)$ not involving R(x). We begin in this section with some series expansions.

[3.3.1] *Laplace's 1785 Power Series.* If $x \geq 0$,

$$\Phi(x) = \tfrac{1}{2} + \frac{1}{\sqrt{2\pi}} \sum_{r=0}^{\infty} \frac{(-1)^r x^{2r+1}}{(2r+1)2^r r!}$$

$$= \tfrac{1}{2} + \frac{1}{\sqrt{2\pi}} \left(x - \frac{x^3}{6} + \frac{x^5}{40} - \frac{x^7}{336} + \frac{x^9}{3456} - \cdots \right).$$

This series "converges too slowly except when x is small" (Stuart and Ord, 1987, Sec. 5.37). Lee (1992, p. 114) suggests that when $x \leq 1$, the series provides "very good bounds of $\Phi(x)$"; it gives alternating upper and lower bounds of $\Phi(x)$ when summed from $r = 0$ to $r = n$ as n increases. If the same number of terms is used, the absolute error when $x \leq 2$ is less than that of Laplace's 1812 series, discussed in *[3.6.4]* (Lee, op.cit.). Good (1987, pp. 83-84) states that for a proportional accuracy of less than 1/1,000 (respectively, 1/10,000) in approximating $1 - \Phi(x)$ when $x < 1$, it suffices to take the expansion to the term in x^7 (respectively, x^9). See *[3.4.1]* for a polynomial approximation based on Laplace's 1785 series.

[3.3.2] An expansion of Kerridge and Cook (1976, p. 402) is applicable over the whole range of values of x, unlike many other expressions appearing in this chapter. The most computationally attractive form is given by

$$\Phi(x) = \tfrac{1}{2} + (\sqrt{2\pi})^{-1} \times \exp\!\left(\tfrac{-x^2}{8}\right) \sum_{n=0}^{\infty} \frac{\theta_{2n}(\tfrac{1}{2}x)}{2n+1}, \quad -\infty < x < \infty,$$

where $\theta_0(x) = 1$, $\theta_1(x) = x^2$, and $\theta_{2n}(x)$ is obtained from the recurrence relation

$$\theta_{r+1}(x) = x^2\{\theta_r(x) - \theta_{r-1}(x)\}/(r+1), \quad r = 1, 2, \ldots.$$

Alternatively, $\theta_r(x) = x^r H_r(x)/r!$, where $H_r(x)$ is the rth Hermite polynomial (see *[2.1.9]*). The number of terms required to achieve accuracy to 5, 10 or 15 decimal places are tabulated (for 10 decimal places 6, 13 and 24 terms when x = 1,3 and 6, respectively). Uniformly fewer such terms are required for $1 \leq x \leq 6$ than in either the Laplace series in *[3.3.1]* or in *[3.6.4]*, or than in Shenton's continued fraction discussed in *[3.6.2]* (Shenton, 1954, pp. 182-188). It requires fewer terms than Laplace's (1805) continued fraction (see *[3.6.1]*) if $x \leq 2$, but the latter begins to perform better if $x \geq 3$, as does Laplace's asymptotic series in $1/x$ of 1785 (see *[3.6.3]*).

[3.3.3] *Trigonometric Series.* Moran (1980, pp. 675-676) gives two series, which are accurate to nine decimal places if $|x| \leq 7$, when truncated as indicated:

(a) $\Phi(x) \simeq \frac{1}{2} + \frac{1}{\pi} \{ \frac{x}{3\sqrt{2}} + \sum_{n=1}^{12} \frac{\exp(-n^2/9)}{n} \sin\left(\frac{nx\sqrt{2}}{3}\right) \}$

(b) $\Phi(x) \simeq \frac{1}{2} + \frac{1}{\pi} \sum_{n=0}^{12} \frac{\exp(-n'^2/9)}{n'} \sin(\frac{n'x\sqrt{2}}{3})$, $n' = n + \frac{1}{2}$.

These approximations are generally better than that of Kerridge and Cook in *[3.3.2]*, but if $|x| > 7$ the accuracy decreases rapidly; see Stuart and Ord (1987, Sec. 5.38).

(c) The normal cdf $\Phi(\,\cdot\,)$ is related to the error function erf$(\,\cdot\,)$ via

$$\Phi(x) = \begin{cases} \frac{1}{2}[1 + \mathrm{erf}(x/\sqrt{2})], & x \geq 0, \\ \frac{1}{2}[1 - \mathrm{erf}(x/\sqrt{2})], & x < 0; \end{cases}$$

see *[2.1.7]*. The approximation

$$\mathrm{erf}(x) \approx \frac{2}{\pi}\left[\frac{x}{5} + \sum_{n=1}^{37} \frac{1}{n}\exp\left(-\frac{n^2}{25}\right)\sin\left(\frac{2nx}{5}\right)\right], \quad |x| \leq \frac{5\pi}{2},$$

is due to Strecok (1968, pp. 144-158). Based on this approximation, Craig (1984, pp. 232-236) provides subroutines in FORTRAN and BASIC that calculate $\Phi(\,\cdot\,)$ with 16 decimal places of accuracy. The programs "do not strive for extreme accuracy, but rather compromise between the needs for

Expansions and Algorithms 51

reasonable accuracy and speed of execution". The structure of this series is similar to that of Moran in (a).

[3.3.4] Allasia (1981, p. 327) applied the Central Limit Theorem (see *[6.1]*) to sums of independently and identically distributed variates having a common uniform distribution in the interval $(-\frac{1}{2}, \frac{1}{2})$, and showed that

$$\lim_{n\to\infty} F_n(z\sqrt{n/12}) = \Phi(z),$$

where

$$F_n(u) = V \ (1/n!) \sum_{k=0}^{[u+n/2]} (-1)^k \binom{n}{k}(u+\tfrac{1}{2}n-k)^n, \quad \begin{cases} 0, & u < -\tfrac{1}{2}n, \\ & |u| \leq \tfrac{1}{2}n, \\ 1, & u > \tfrac{1}{2}n. \end{cases}$$

Here [x] is the largest integer no greater than x. Application of the Berry-Esseen theorem (see *[6.3.2]*) yields a uniform bound on the error in approximating $\Phi(\cdot)$;

$$|F_n(z\sqrt{n/12}) - \Phi(z)| \leq 0.05(3/n)^{1/2}.$$

3.4 THE DISTRIBUTION FUNCTION – II

In this section we list some *rational function approximations*, requiring the use of several constants.

[3.4.1] Badhe (1976, pp. 173-176) derived an approximation from Laplace's 1785 power series (see *[3.3.1]*), when $0 < x \leq 2$:

$$\Phi(x) = 0.5 + x(a+y[b + y\{c + y(d + y[e + y\{f + y(g + hy)\}])\}]),$$
$$y = x^2/32,$$

a = 0.3989,4227,84	b = −2.1276,9007,9
c = 10.2125,6621,21	d = −38.8830,3149,09
e = 120.2836,3707,87	f = −303.2973,1534,19
g = 575.0731,3191,7	h = −603.9068,0920,58.

This approximation has maximum absolute error equal to 0.20×10^{-8} if $0 \le x \le 2$.

[3.4.2] The following three approximations are due to Hastings (1955, pp. 185-187); see also Abramowitz and Stegun (1964, Secs. 26.2.18-19):

(a) $2[1 - \Phi(x)] = [1 + (0.1968,54)x + (0.1151,94)x^2$
$\qquad + (0.0003,44)x^3 + (0.0195,27)x^4]^{-4}$
$\qquad + 2\epsilon(x), \quad x \ge 0,$

$|\epsilon(x)| < 2.5 \times 10^{-4}$.

(b) $2[1 - \Phi(x)] = [1 + (0.0997,9271)x + (0.0443,2014)x^2$
$\qquad + (0.0096,9920)x^3 - (0.0000,9862)x^4$
$\qquad + (0.0058,1551)x^5]^{-8} + 2\epsilon(x), \quad x \ge 0,$

$|\epsilon(x)| < 2 \times 10^{-5}$

(c) $2[1 - \Phi(x)] = [1 + (0.0498,6734,70)x + (0.0211,4100,61)x^2$
$\qquad + (0.0032,7762,63)x^3 + (0.0000,3800,36)x^4$
$\qquad + (0.0000,4889,06)x^5 + (0.0000,5838,30)x^6]^{-16}$
$\qquad + 2\epsilon(x), \quad x \ge 0,$

$|\epsilon(x)| < 1.5 \times 10^{-7}$

[3.4.3] Carta (1975, pp. 856-862) gives fractions of the form

$$1 - \Phi(x) \simeq \tfrac{1}{2}\left[\sum_{r=0}^{n} c_r x^{r-1}\right]^{-2^q}, \quad x > 0.$$

Values of the c_r for $n = 4,5,6$ and $q = 1(1)15$ are tabulated and the errors of approximation are diagrammed.

[3.4.4] Divgi (1979, pp. 903-906) approximated $R(x)$ by choosing coefficients $\{a_{nk}\}$ so as to minimize the integral

$$\int_0^\infty \left[1 - \Phi(x) - \phi(x) \sum_{k=0}^{n} a_{nk} x^k\right]^2 dx,$$

in which n is fixed. This leads to a least squares approximation

$$R(x) \simeq \sum_{j=0}^{n} a_{nj} x^j, \quad x \ge 0,$$

for certain choices of $a_{n0}, a_{n1}, \ldots, a_{nn}$. A series with $n = 10$ has a maximum absolute error of 3×10^{-7}, and to ten significant digits the

Expansions and Algorithms 53

coefficients (dropping 'n' from the subscripts) are:

$$a_0 = 1.25331\ 3402 \qquad a_1 = -0.99996\ 73043$$
$$a_2 = 0.62629\ 72801 \qquad a_3 = -0.33162\ 18430$$
$$a_4 = 0.15227\ 23563 \qquad a_5 = -5.98283\ 4993 \times 10^{-2}$$
$$a_6 = 1.91564\ 9350 \times 10^{-2} \qquad a_7 = -4.64496\ 0579 \times 10^{-3}$$
$$a_8 = 7.77108\ 8713 \times 10^{-4} \qquad a_9 = -7.83082\ 3677 \times 10^{-5}$$
$$a_{10} = 3.53424\ 4658 \times 10^{-6}$$

3.5 THE DISTRIBUTION FUNCTION – III

In this section we list a number of *simple approximations* to $\Phi(x)$, and some inequalities. These can be particularly useful if less accuracy is required than elsewhere.

[3.5.1] Shah (1985, p. 80) gives a simple approximation, for which the maximum absolute error is 0.0052:

$$\Phi(z) - 0.5 = \begin{cases} z(4.4 - z)/10, & 0 \leq z \leq 2.2, \\ 0.49, & 2.2 < z < 2.6, \\ 0.50, & z \geq 2.6. \end{cases}$$

[3.5.2] Lew (1981, p. 300) gives two simple approximations to $\Phi(\cdot)$;

$$1 - \Phi(z) \simeq q(z) \qquad \text{and} \ 1 - \Phi(z) \simeq q^*(z),$$

where

$$q(z) = \begin{cases} 0.5 - (2\pi)^{-\frac{1}{2}}(z - z^3/7), & 0 \leq z \leq 1, \\ (1 + z)\phi(z)/(1 + z + z^2), & z > 1; \end{cases}$$

$$q^*(z) = \begin{cases} 0.5 - 0.4z, & 0 \leq z \leq 0.5, \\ (1 + z)\phi(z)/(1 + z + z^2), & z > 0.5. \end{cases}$$

The maximum absolute relative percentage error of $q(\cdot)$ is 2.0, and that of $q^*(\cdot)$ is 3.0.

[3.5.3] $\Phi(x) \simeq 1/[1 + \exp(-2y)] = \frac{1}{2}(1 + \tanh y)$, $x \geq 0$,

$$y = (\sqrt{2/\pi})\, x\{1 + 0.044715)x^2\}.$$

This was developed by Page (1977, p. 75) as an improvement to an earlier approximation by Tocher (1963):

$$\Phi(x) \simeq 1/[1 + \exp(-2x\sqrt{2/\pi})]\,.$$

If $0 \leq x \leq 4.0$ in the latter, the largest absolute error is 0.17670, when $x = 1.5$. The largest such error for any $x > 0$ in Page's approximation is 0.000179.

[3.5.4] $\Phi(x) \simeq \frac{1}{2}[1 + \text{sgn}(x)\{1 - \exp(-t^2)\}^{1/2}]$, $|x| < \infty$,

$$t = 0.806|x|(1 - 0.018|x|)\,, \quad \text{sgn}(x) = |x|/x, \quad x \neq 0\,.$$

Hamaker (1978, p. 77) developed this to improve an approximation of Pólya (1949, p. 64), which is also an inequality:

$$\Phi(x) < \frac{1}{2}\{1 + [1 - \exp(-2x^2/\pi)]^{1/2}\}\,, \quad x > 0\,.$$

The largest relative error of the latter when $x > 0$ is of the order of 10^{-2}. See also Chu (1955, p. 263), and *[3.5.6]*.

Hamaker's formula, considered as an approximation to $1 - \Phi(x)$, has absolute relative error no greater than 0.01, if $0.0001 \leq 1 - \Phi(x) \leq 0.50$.

[3.5.5] $1 - \Phi(x) \simeq (1 + e^y)^{-1}$, $y = (4.2)\pi x/(9 - x)$, $0.5 \leq \Phi(x) \leq 1$ (Lin, 1990, p. 255). This approximation is slightly less accurate than that of Hamaker in *[3.5.4]* but may be simpler to use.

[3.5.6] $\Phi(x) \geq [1 + \{1 - \exp(-x^2/2)\}^{1/2}]/2$, $x \geq 0$;

$$\Phi(x) \geq \tfrac{1}{2} + x/[2(\pi + 2x^2)]^{1/2}\,, \quad x \geq 0 \quad (\text{Chu, 1955, pp. 263-264})\,.$$

[3.5.7] Derenzo (1977, p. 217) gives a simple approximation to $\Phi(x)$ with integer coefficients. If $0 < x \leq 5.5$ and $0.5 > 1 - \Phi(x) \geq 1.9 \times 10^{-8}$, then

$$1 - \Phi(x) = \tfrac{1}{2}\exp\left[-\frac{(83x + 351)x + 562}{703x^{-1} + 165}\right] + \epsilon(x) ,$$

$$|\epsilon(x)|/\{1 - \Phi(x)\} < 0.42 \times 10^{-4} ;$$

if $x \geq 5.5$ and $1 - \Phi(x) \leq 1.9 \times 10^{-8}$, then

$$1 - \Phi(x) = (2\pi)^{1/2} x^{-1} \exp(-x^2/2 - 0.94/x^2) + \epsilon(x) ,$$

$$|\epsilon(x)|/\{1 - \Phi(x)\} < 0.40 \times 10^{-4} .$$

3.6 APPROXIMATIONS RELATING TO MILLS' RATIO: EXPANSIONS

Mills' ratio is $R(x) = [1 - \Phi(x)]/\phi(x)$. Among the best expressions, in terms of providing good approximations to $R(x)$, are continued fractions. Recent work by Lee (1992, pp. 107-120) has strengthened results dating back to Laplace, and has linked several inequalities for $R(x)$ that were derived between 1940 and 1980.

Denote by $\dfrac{a}{x+}\dfrac{b}{y+}\dfrac{c}{z+}\ldots$ the continued fraction

$$x + \cfrac{a}{y + \cfrac{b}{z + \cfrac{c}{\ldots}}} .$$

The *sth convergent* of the continued fraction $\dfrac{a_1}{x+}\dfrac{a_2}{x+}\ldots\dfrac{a_n}{x+}\ldots$ is

$$\dfrac{a_1}{x+}\dfrac{a_2}{x+}\ldots\dfrac{a_s}{x} .$$

We begin by listing four expansions and continued fractions that have been evaluated extensively.

[3.6.1] Laplace's Continued Fraction

$$R(x) = \dfrac{1}{x+}\dfrac{1}{x+}\dfrac{2}{x+}\dfrac{3}{x+}\ldots \qquad \text{(Laplace, 1805)}$$

Let L_n be the $(n+1)$th convergent, so that

$$L_n = \frac{1}{x+} \frac{1}{x+} \frac{2}{x+} \frac{3}{x+} \cdots \frac{n}{x} \,.$$

The finite continued fraction L_n "is the best approximation to $R(x)$ for $x \geq 3$" (Lee, 1992, p. 107). While L_n converges to $R(x)$ as $n \to \infty$ for each $x > 0$ "many thousand terms may be needed if x is small" (Kerridge and Cook, 1976, p. 401). However, the inequalities

$$L_1 < L_3 < L_5 < \ldots < R(x) < \ldots < L_6 < L_4 < L_2, \quad x > 0 \,,$$

provide sequences of lower and upper bounds of $R(x)$. Thus (Lee, 1992, pp. 113-114) $L_0 = 1/x$ and $L_2 = (x^2 + 2)/(x^3 + 3x)$ are upper bounds and $L_1 = x/(x^2 + 1)$ is a lower bound of $R(x)$; these were derived by Gordon (1941, p. 365) and by Gross and Hosmer (1978, pp. 1354-1355). Partial sums of the Laplace asymptotic expansion of *[3.6.3]* are successively inferior to L_0, L_1, L_2, $L_3, \ldots$; the first is $L_0 = 1/x$. Choices among L_1, L_3, L_4, L_5, and L_6 are preferable to rational bounds of Shenton (1954, p. 188) and of Ruben (1963, pp. 355-364). See also the discussion in Bowman and Shenton (1989, pp. 34-37).

[3.6.2] *Shenton's Continued Fraction*

$$\overline{R}(x) = \frac{\Phi(x) - \frac{1}{2}}{\phi(x)} = \frac{x}{1-} \frac{x^2}{3+} \frac{2x^2}{5-} \frac{3x^2}{7+} \frac{4x^2}{9-} \cdots \frac{nx^2}{2n+1 \pm \ldots}, \quad x > 0 \,,$$

where $\pm$ means $+$ in the fraction $nx^2/(2n+1 \pm \ldots)$ if n is odd, and $-$ if n is even (Shenton, 1954, pp. 177-189). The (n+1)th convergent r_{n+1} stops at $nx^2/(2n+1)$ in this expression. Then

$$r_1 < r_4 < r_5 < r_8 < r_9 < \ldots < r_{4k} < r_{4k+1} < \ldots < \overline{R}(x), \quad x > 0;$$

$$\overline{R}(x) < \ldots < r_{4k+3} < r_{4k+2} < \ldots < r_7 < r_6 < r_3 < r_2, \quad 0 < x < \sqrt{3} \,.$$

This provides sequences of lower and upper bounds of $\overline{R}(x)$. The sequence of upper bounds holds as far as *some* r_{4k+3} (i.e., k exceeding some positive integer $k_0 > 0$) for *any* $x \geq \sqrt{3}$.

The approximants r_n converge rapidly to $\overline{R}(x)$ when x is small

Expansions and Algorithms

(Shenton suggests for $x < \sqrt{3}$). For moderate values of x, convergence seems to be slow at first, but "becomes quite rapid in due course". The rate of convergence "deteriorates slowly" as x increases (Shenton, 1954, p. 186). See also Bowman and Shenton (1989, pp. 34-37).

Shenton also compared his expansion with Laplace's continued fraction of *[3.6.1]*. To achieve an accuracy of 2.5×10^{-7}, the latter would be preferred only when $x \geq 2.5$.

[3.6.3] *Laplace's expansion in $1/x$*

$$R(x) = \frac{1}{x} - \frac{1}{x^3} + \frac{1 \cdot 3}{x^5} - \ldots + (-1)^n \frac{1 \cdot 3 \cdot 5 \ldots (2n-1)}{x^{2n+1}} + R_n(x), \quad x \geq 0$$

(Laplace, 1785). Here, the remainder term satisfies

$$|R_n(x)| < \min\left[\frac{1 \cdot 3 \cdot 5 \ldots (2n-1)}{x^{2n+1}}, \frac{1 \cdot 3 \cdot 5 \ldots (2n+1)}{x^{2n+3}}\right]$$

(Shenton, 1954, p. 180; Stuart and Ord, 1987, Sec. 5.38). The series does not converge, but is "reasonably effective" if x is large (Stuart and Ord, op. cit.). Laplace's continued fraction of *[3.6.1]* can be derived from it (Shenton, op. cit.). The sum of n terms is an upper bound to $R(x)$ when n is even, and a lower bound when n is odd (Feller, 1968, pp. 175, 193-194).

[3.6.4] *Laplace's 1812 expansion*

$$\overline{R}(x) = x + \frac{x^3}{1 \cdot 3} + \frac{x^5}{1 \cdot 3 \cdot 5} + \ldots + \frac{x^{2n-1}}{1 \cdot 3 \cdot 5 \ldots (2n-1)} + \overline{R}_n(x), \quad x > 0$$

(Laplace, 1812, p. 103; Pólya, 1949, p. 66). The sum of n terms is always a lower bound of $\overline{R}(x)$; indeed

$$0 < \overline{R}_n(x), \quad x > 0$$

$$0 < \overline{R}_n(x) < \frac{(2n+3)x^{2n+1}2^{-n-1}}{(2n+3-x^2)(n+\frac{1}{2})_{n+1}}, \quad 0 < x < \sqrt{2n+3},$$

where $(s)_k = s(s-1) \ldots (s-k+1)$ for $k = 1, 2, \ldots$ (Gupta and Waknis, 1965, pp. 144-145).

Bowman and Shenton (1989, p. 35) state that this series for $\overline{R}(x)$ "is convergent and only useful" for small values of x. Shenton's continued fraction of [3.6.2] can be derived from it (Bowman and Shenton, op. cit.).

[3.6.5] The results in this section are by Lee (1992, pp. 107-120).

[3.6.5.1] Let

$$L_n^r = \frac{1}{x+} \frac{1}{x+} \frac{2}{x+} \cdots \frac{n}{x+} (n+1)^{1/2},$$

a refinement of L_n in [3.6.1]. Then

(i) $\quad 1 - \Phi(x) = \phi(x) L_n^r + \epsilon_n$,

where $\epsilon_n \to 0$ as $n \to \infty$; this holds for all $x \geq 0$, "including the origin".

(ii) $\quad L_2^r < L_4^r < \ldots < R(x) < \ldots L_3^r < L_1^r$;

these bounds on R(x) are admissible in the sense defined in [3.6.5.3].

(iii) $L_{2n+1}^r < L_{2n}^r \quad \text{if } x < \sqrt{2n+1}$,

$L_{2n+1}^r < L_{2n+2}^r \quad \text{if } x < \sqrt{2n+2}$.

Thus $\{L_n^r\}$ would be preferred to $\{L_n\}$. Note that

$$L_1^r = \frac{(x + \sqrt{2})}{(x^2 + \sqrt{2}x + 1)} \; ; \quad L_2^r = \frac{x^2 + \sqrt{3}x + 2}{x^3 + \sqrt{3}x^2 + 3x + \sqrt{3}} \; ;$$

L_2^r is superior to the lower bound $\frac{1}{2}\{(x^2 + 4)^{1/2} - x\}$ on Mills' ratio of Birnbaum (1942, pp. 245-246). Further,

$$L_3^r = \frac{x^3 + 2x^2 + 5x + 4}{x^4 + 2x^3 + 6x^2 + 6x + 3} \; ;$$

L_3^r is superior to the upper bound $4/\{3x + (x^2 + 8)^{1/2}\}$ on R(x) of Sampford (1953, pp. 130-132).

Expansions and Algorithms 59

The continued fractions L_n^r and f_n (discussed in *[3.6.5.2]* "provide the sharpest tail bounds" of $R(x)$ in the literature for $x \geq 0.7$ (Lee, 1992, p. 113).

[3.6.5.2] In a further refinement of L_n, let

$$f_n(x; a_n, b_n) = \frac{1}{x+}\frac{1}{x+}\frac{2}{x+}\frac{3}{x+}\cdots\frac{n-1}{x+}\frac{b_n}{x+}a_n.$$

Lee (1992, pp. 112-113) identifies preferred choices of a_n and b_n in terms of accuracy of approximation to $R(x)$ via the (n+1)th convergent of the sequence $\{f_n\}$. This is $f_n(x; a_n^*, b_n^*)$, where

$$b_n^* = \begin{cases} (n^2 + n + 1)^{1/2} + (n-1) & , \quad 0 \leq x < 1, \\ \frac{4}{3}\{n + (n_2 + 3n)^{1/2}\sin(\frac{\theta}{3} - \frac{\pi}{6}), & x \geq 1, \end{cases}$$

and

$$\theta = \arctan\left[\frac{\{27(8n^2 + 13n + 16)/n\}^{1/2}}{9 - 4n}\right];$$

$$a_n^* = 2\{(b_n^*+1)(b_n^* - n)/b_n^*\}^{1/2}.$$

As an approximation to Mills' ratio, one would then prefer $f_n^* = f_n(x; a_n^*, b_n^*)$ to L_n or to L_n^r. The sequence $\{f_n^*\}$ is admissible in the sense described next. See also *[3.8.2]*, i).

[3.6.5.3] Let $P(x)/Q(x)$ and $p(x)/q(x)$ be ratios of polynomials in x in which the degrees of $P(x)$ and of $p(x)$ are both n or less, $P(x)/Q(x)$ is said to be an *admissible lower bound* of $R(x)$ if no other ratio $p(x)/q(x)$ exists such that

$$P(x)/Q(x) \leq p(x)/q(x) \leq R(x) ;$$

admissible upper bounds are defined similarly.

Lee (1992, pp. 109-111) proves the following theorem:
Let n be an odd (even) integer. Then a continued fraction $f_n(x; a_n, b_n)$ is an admissible lower (upper) bound of $R(x)$ (subject to a monotone decreasing absolute error in the family $p(x)/q(x)$ as defined above) if

$$a_n = 2\{(b_n + 1)(b_n - n)/b_n\}^{1/2},$$

$$n \le b_n \le (n^2 + n + 1)^{1/2} + (n-1).$$

It is an admissible upper (lower) bound of $R(x)$ under the same conditions if $a_n = \sqrt{n+1}$ and $b_n = n$, i.e., if $f_n(x; a_n, b_n) = L_n^r$, defined in [3.6.4]. Then

$$L_{2s-1} \le f_{2s+1} \le R(x) \le f_{2s} \le L_{2s-2}$$

when the theorem is satisfied. The sharpest tail bounds on $R(x)$ in the literature when $x \ge 0.7$ are provided by $\{f_n\}$ as in the theorem and by $\{L_n^r\}$; $f_3(x; a_3^*, b_3^*)$ is better than the lower bound of Boyd (1959, pp. 44-46), i.e., $\pi/\{(\pi-1)x + (x^2+2\pi)^{1/2}\}$, if $x \ge 0.1$; $f_2(x; a_2^*, b_2^*)$ improves on Boyd's upper bound $\pi/[2x + \{2\pi + (\pi-2)^2 x^2\}^{1/2}]$ if $x \ge 0.7$.

The smallest absolute error $|f_n(x; a, b) - R(x)|$ at a fixed value of x is given approximately by $b_n(x)$, where

$$b_n(x) = 2n - x\sqrt{n} + (x^2 - 1)/2 + \epsilon_n,$$

$\epsilon_n = 0(n^{-\frac{1}{2}})$ (Lee, 1992, pp. 114-115).

[3.6.6] Shenton (1954, p. 188) quotes the expansion

$$xR(x) = 1 - \frac{1}{x^2+2} + \frac{1}{(x^2+2)(x^2+4)} - \frac{5}{(x^2+2)(x^2+4)(x^2+6)}$$

$$+ \frac{9}{(x^2+2)\ldots(x^2+8)} - \frac{129}{(x^2+2)\ldots(x^2+10)} + \ldots, \quad x > 0.$$

If the general term is $(-1)^n a_n/[(x^2+2)(x^2+4)\ldots(x^2+2n)]$, then (Badhe, 1976, pp. 174-175) $a_6 = 57$, $a_7 = 9141$, $a_8 = 36{,}879$, $a_9 = 1{,}430{,}049$, $a_{10} = 19{,}020{,}019$, and $a_{11} = 1{,}689{,}513{,}233$; see also Abramowitz and Stegun (1964, Sec. 26.2.13). Badhe gives an approximation based on the sum to the term in which $n = 9$, as follows:

Expansions and Algorithms

$$xR(x) = 1 - y\{1 + y(7 + y[55 + y\{445 + 3745\ Q_1(x)y\}])\},$$

$$y = (x^2 + 10)^{-1}, \quad x > 0,$$

$$Q_1(x) = 1 + 8.50\{x^2 - (0.4284,6397,53)x^{-2} + 1.2409,6410,9\}^{-1}.$$

Based on relative error as a criterion, this is more accurate when $x > 4$ than Patry and Keller (1964); see *[3.6.7]*. However, it is not suitable when $x < 2$.

[3.6.7] Patry and Keller (1964, pp. 89-97) give the following continued fraction expansion:

$$R(x) = \left(\sqrt{\tfrac{\pi}{2}}\right)\left[\frac{1}{a_0 w+}\ \frac{1}{a_1 w+}\ \frac{1}{a_2 w+}\ \cdots\right], \quad x \geq 0,\ w = \frac{x}{\sqrt{2}},$$

$$a_0 = \sqrt{\pi},\quad a_1 = 2/\sqrt{\pi},$$

$$a_{2n} = \sqrt{\pi}(2n-1)!!/(2n)!!,$$

$$a_{2n+1} = 2(2n)!!/[(2n+1)!!\sqrt{\pi}] \qquad n = 1,2,\ldots,$$

where $n!! = n(n-2)!!$, and $n!! = 1$ if $n = 0$ or $n = 1$. Convergence is poor near zeroes of the expansion, and so the following approximation was derived from the above (Patry and Keller, 1964, p. 93):

$$R(x) \simeq t/(tx + \sqrt{2}),\quad x \geq 0,\quad t = \sqrt{\pi} + (2-q)x,\quad q = a/b,$$

$$a = 0.8584,0765,7 + x[0.3078,1819,3 + x\{0.0638,3238,91$$

$$- (0.0001,8240,5075)x\}],$$

$$b = 1 + x[0.6509,7426,5 + x\{0.2294,8581,9 + (0.034,3018,23)x\}].$$

As an approximation to $1 - \Phi(x)$, this algorithm has |error| less than 12.5×10^{-9} for the range $0 \leq x \leq 6.38$, and relative error less than 30×10^{-9}, 100×10^{-9}, and 500×10^{-9} if $0 \leq x \leq 2$, $2 < x \leq 4.6$, and $4.6 < x \leq 6.38$, respectively (Patry and Keller, 1964, p. 93). Badhe (1976, p. 174) states that this approximation is better than those of Hart (1966, pp. 600-602), Hastings (1955, pp. 167-169), and Schucany and Gray (1968, pp. 201-

202); see *[3.7.2]*, *[3.7.1]*, and *[3.7.3]*, respectively.

3.7 FURTHER EXPRESSIONS RELATING TO MILLS' RATIO

[3.7.1] The following approximations are based on Hastings (1955, pp. 167-169); see also Abramowitz and Stegun (1964, Sec. 26.2.16).

(a) $\Phi(x) = 1 - \phi(x)[(0.4361,836)t - (0.1201,676)t^2$
$+ (0.9372,980)t^3] + \epsilon(x), \ x \geq 0$,
$t = \{1 + (0.3326,7)x\}^{-1}, \ |\epsilon(x)| < 1.2 \times 10^{-5}$.

(b) $\Phi(x) = 1 - \phi(x)[(0.1806,1682)t + (0.7652,0181)t^2$
$- (0.7616,8891)t^3 + (1.0691,8442)t^4] + \epsilon(x), \ x \geq 0$,
$t = \{1 + (0.2700,90)x\}^{-1}, \ |\epsilon(x)| < 10^{-6}$.

(c) $\Phi(x) = 1 - \phi(x)[(0.3193,8153,0)t - (0.3565,6378,2)t^2$
$+ (1.7814,7793,7)t^3 - (1.8212,5597,8)t^4$
$+ (1.3302,7442,9)t^5] + \epsilon(x), \ x \geq 0$,
$t = \{1 + (0.2316,419)x\}^{-1}, \ |\epsilon(x)| < 7.5 \times 10^{-8}$.

See also the discussion in *[3.6.7]*.

[3.7.2] Hart (1966, p. 600-602) gives the approximation

$$xR(x) \simeq 1 - y[x\sqrt{\pi/2} + \{\pi x^2/2) + y \exp(-x^2/2)\}^{1/2}]^{-1}, \ |x| < \infty,$$

$$y = (1 + 2\pi a^2 x^2)^{1/2}/(1 + ax^2), \quad a = [1 + (1 + 6\pi - 2\pi^2)^{1/2}]/(2\pi).$$

The maximum |error| is 0.00013, near $x = \pm 1$; the maximum |relative error| is 0.00055, near $x = \pm 1.7$; see also the discussion in *[3.6.7]* and *[3.7.3]*.

[3.7.3] Gray and Schucany (1968, p. 718) showed that

Expansions and Algorithms

$$R(x) \simeq \frac{x}{x^2+2}\left[\frac{x^6+6x^4-14x^2-28}{x^6+5x^4-20x^2-4}\right], \quad x > 2,$$

which is uniformly more accurate than that in *[3.7.2]* when $3 \le x \le 10$, in that the |relative error| is smaller. The maximum |relative error| for all $x > 2$ is 0.0020 when $x = 2$, and is 0.00001 when $x = 4$ for all $x \ge 4$. See also the discussion in *[3.6.7]* and *[3.8.2]*.

[3.7.4] Other approximations to $R(x)$ via partial sums appear in the literature as follows:

(a) An expansion involving derivatives (Ruben, 1962, pp. 178-179) having absolute error less than 0.0004 in the range $3 \le x \le 10$ if summation is taken only to the second term.

(b) A rational function expansion by Ruben (1963, pp. 360, 362-363); the sum to three terms compares favorably for $2 \le x \le 10$ with the approximation in *[3.7.3]*, but unfavorably with the convergents L_n of *[3.6.1]*.

(c) An expansion involving square roots (Ruben, 1964, pp. 340-343).

(d) Expansions involving Tchebyshev polynomials (Ray and Pitman, 1963, pp. 893-896) and a Laguerre-Gauss expansion (ibid., pp. 899-901).

(e) Two expansions involving Tchebyshev polynomials (Rabinowitz, 1969, pp. 647-651).

[3.7.5] Shore (1982, pp. 111-112) obtained a simple approximation $\hat{R}(x)$ to $R(x)$ in terms of the cdf $\Phi(x)$:

$$\hat{R}(x) = \begin{cases} 1/\{1.4184[\Phi(x)]^{0.8632}\} & , \quad x \le 0, \\ 1/\{1.4184)\Phi(x)[1-\Phi(x)]^{-0.1368}\} & , \quad x \ge 0. \end{cases}$$

The accuracy of $\hat{R}(x)$ is comparable with that of $\hat{\phi}_1(x)$ in *[3.2.3]*, from which it is derived.

3.8 FROM MILLS' RATIO TO THE DISTRIBUTION FUNCTION

Approximations to Mills' ratio $R(x)$ or to $\overline{R}(x)$ can be reexamined as approximations to $\Phi(x)$ (or to tail probabilities $1 - \Phi(x)$), via the relations

$$\Phi(x) = 1 - \phi(x)R(x) = 0.5 + \phi(x)\overline{R}(x) \ .$$

In this section we reexamine some of the results in [3.6] and [3.7] from this viewpoint.

[3.8.1.1] First note that, if $|e(x)|$ is the absolute error of an approximation to $R(x)$ or $\overline{R}(x)$, then the smaller quantity $|\phi(x)e(x)|$ is the corresponding absolute error of the derived expression when it is used to approximate $\Phi(x)$ or $1 - \Phi(x)$. The relative error is unchanged, whether $R(x)$ or $\Phi(x)$ is of interest. It follows that a preference for one approximation to $R(x)$ or $\overline{R}(x)$ rather than another remains valid when the approximation is adapted for $\Phi(x)$ or $1 - \Phi(x)$, respectively, whenever the preference is made via comparative values of absolute or relative error.

[3.8.1.2] In [3.6.7] we have already discussed the algorithm for $R(x)$ of Patry and Keller (1964, pp. 89-67) in the sense of this section.

[3.8.2] Lee (1992, pp. 115-119) has compared several approximations to $\Phi(x)$. For convenience we will list some of them here again.

(i) $\Phi(x) = 1 - \phi(x)f_n(x; a_n, b_n) + \ldots$,

$$f_n(x; a_n, b_n) = \frac{1}{x+} \frac{1}{x+} \frac{2}{x+} \frac{3}{x+} \cdots \frac{(n-1)}{x+} \frac{b_n}{x+} a_n,$$

$$a_n = 2\{(b_n + 1)(b_n - n)/b_n\}^{1/2},$$

$$b_n = 2n - xn^{1/2} + \tfrac{1}{2}(x^2 - 1)$$

(Lee, op.cit.). This is a particular case of the class of approximations to Mills' ratio discussed in [3.6.5.2].

In this section f_n denotes the version defined above. In some of Lee's discussion b_n is abbreviated at $n^{1/2}$; we denote this by b_n^*, so that

$$b_n^* = \begin{cases} 2n - xn^{1/2} + \tfrac{1}{2}(x^2 - 1) & , \ x \leq n^{1/2}, \\ (3n - 1)/2 & , \ x > n^{1/2}. \end{cases}$$

Expansions and Algorithms

Then we write f_n^* for $f_n(x; a_n, b_n^*)$; the sequence $\{1 - \phi(x)f_n\}$ forms upper bounds of $\Phi(x)$ when n is odd and lower bounds when n is even.

(ii) $\Phi(x) = \frac{1}{2} + \phi(x)\left(\frac{x}{1-} \frac{x^2}{3+} \frac{2x^2}{5-} \frac{3x^2}{7+} \frac{4x^2}{9-} \frac{5x^2}{11+} \cdots \frac{nx^2}{2n+1 \pm} \cdots \right)$

(Shenton, 1954, pp. 177-189), discussed in [3.6.2].

(iii) $\Phi(x) = \frac{1}{2} + \phi(x)\{x + \frac{x^3}{1 \cdot 3} + \frac{x^5}{1 \cdot 3 \cdot 5} + \cdots + \frac{x^{2n+1}}{1 \cdot 3 \cdots (2n+1)}\} + \cdots$

(Laplace, 1812), discussed in [3.6.4].

(iv) $\Phi(x) = \frac{1}{2} + \frac{1}{\sqrt{2\pi}}\{x - \frac{x^3}{6} + \frac{x^5}{40} - \cdots + (-1)^n \frac{x^{2n+1}}{2^n(2n+1)n!}\} + \cdots$

(Laplace, 1785), discussed in [3.3.1].

(v) $\Phi(x) = \frac{1}{2} + x\phi(\frac{1}{2}x) \sum_{t=0}^{n} \frac{\theta_{2t}(\frac{1}{2}x)}{2t+1} + \cdots$,

$\theta_0(x) = 1$, $\theta_1(x) = x^2$, $\theta_n(x) = x^2\{\theta_{n-1}(x) - \theta_{n-2}(x)\}/n$, $n \geq 2$

(Kerridge and Cook, 1976, p. 402), discussed in [3.3.2] and [3.3.3].

(vi) $\Phi(x) = 1 - \phi(x)L_n + \cdots$

$= 1 - \phi(x)\left(\frac{1}{x+} \frac{1}{x+} \frac{2}{x+} \frac{3}{x+} \cdots \frac{n}{x}\right) + \cdots$

(Laplace, 1805), discussed in [3.6.1]; see also [3.8.3].

Then $1 - \phi(x)f_n$ is generally superior to $1 - \Phi(x)L_n$ as an approximation; Lee tabulates the absolute errors of each of these for $x = 0.5(0.5)4.0$ and $n = 1(1)8$; see his Table 2. L_n is preferred to f_n only when $x \geq 3.5$ and for some $n \leq 4$. In i) $1 - \phi(x)f_7$ is preferable for most x to the approximation of Gray and Schucany (1968, p. 718), discussed in [3.7.3]. In his Tables 3 and 4, Lee (1992, pp. 116-119) defines f_n as in i) here (Lee, 1994). He compares i), ii), iii), iv), v) and vi) by tabulating the degree n or number of terms required for accuracy 10^{-d}, with d = 4(1)7 and x = 0.5(0.5)4.0, also d = 10, 15 and x = 1(1)6. By this criterion, the following conclusions can be drawn:

(A) At x = 0.5 ii), iii), iv) and v) are all superior to i) and to vi); vi), based on L_n, as expected performs very poorly when x is small.

(B) Kerridge and Cook's approximation v) requires fewest terms when x = 0.5 (4 ≤ d ≤ 7), 1.0(4 ≤ d ≤ 15), 1.5(4 ≤ d ≤ 7), 2.0(d = 10, 15), 3.0(d = 15). Computationally, however, it may be less attractive than others; see the discussion in [3.3.3] and Moran (1980, pp. 675-676).

(C) When 1.5 ≤ x ≤ 4.0 and 4 ≤ d ≤ 7, i) using f_n requires the fewest terms; however, vi) performs as well when x = 3.5, 4.0. When 3 ≤ x ≤ 6 and d = 10, 15, i) using f_n^* requires the fewest terms, with f_n performing almost as well; vi) performs as well as f_n^* for x = 5,6. See Lee (1992, p. 119) for a further modification of b_n in i) that improves the performance of $f_n(x; a_n, b_n)$ even further when x = 3(1)6.

[3.8.3] Good (1987, pp. 83-84) evaluated the number of terms in the Laplace approximation $\phi(x)$ L_n (see vi) of [3.8.2]), to the tail probability $1 - \Phi(x)$ that are necessary for a proportional accuracy of a) less than 1/1,000 and b) less than 1/10,000. These are:

$$\text{Take } L_n = \frac{1}{x+}\frac{1}{x+}\frac{2}{x+}\frac{3}{x+} \cdots \frac{n\text{-}1}{x+}\frac{n}{x} ,$$

where in case a), $n = n(x) = \begin{cases} [30x - 1.8] , & 1 \leq x < 4.5 , \\ 2 , & x > 4.5 , \end{cases}$

in case b), n = 40.

3.9 QUANTILES

Suppose that Z is a rv with a N(0, 1) distribution, and that $\Phi(z_p)$ = Pr(Z ≤ z_p) = 1 − p. In what follows, z_p is expressed in terms of a given value of p.

[3.9.1] Two approximations are given in Hastings (1955, pp. 191-192); see also Abramowitz and Stegun (1964, Sec. 26.2.22-23). Let t =

$\sqrt{-2\log p}$ and $0 < p \le 0.5$. Then

(a) $z_p = t - \dfrac{(2.30753) + (.27061)t}{1 + (.99299)t + (.04481)t^2} + \epsilon(p)$, $|\epsilon(p)| < 3 \times 10^{-3}$;

(b) $z_p = t - \dfrac{(2.515517) + (0.802853)t + (.010328)t^2}{1 + (1.432788)t + (0.189269)t^2 + (0.001308)t^3} + \epsilon(p)$,

$|\epsilon(p)| < 4.5 \times 10^{-4}$.

[3.9.2] Hastings' algorithm of *[3.9.1(b)]* is of the form

$$z_p = t + S_2(t)/T_3(t),$$

where $S_2(\cdot)$ and $T_3(\cdot)$ are polynomials of degree two and three, respectively. Using constant coefficients to several more decimal places, Odeh and Evans (1974, pp. 96-97) present an algorithm of the form

$$z_p = t + S_4(t)/T_4(t) + \epsilon(p)$$

with polynomials of degree four in t, valid for $10^{-20} < p < 0.5$, and where $|\epsilon(p)| \le 1.5 \times 10^{-8}$. This is an improvement in accuracy over the approximations of *[3.9.1]*; see also *[3.10.3]*.

[3.9.3] Bailey (1981, pp. 275-276) suggests the following as alternatives to Hastings' approximation of *[3.9.1(b)]*. Here let

$$t_1 = \{-\tfrac{1}{2}\pi \log(4p(1-p))\}^{1/2},$$
$$t_2 = \{-2 \log p - \log(2\pi \log(2p))\}^{1/2};$$

(a) $z_p = t_1(1 + 0.0^2 78365 t_1^2 - 0.0^3 28810 t_1^4 + 0.0^5 43728 t_1^6 + |\epsilon_1(p)|,$
$0 < z_p \le 4.98,$

(b) $z_p = t_2 + 0.1633 t_2^{-2} + 0.5962 t_2^{-3} + |\epsilon_2(p)|$, $z_p \ge 4.75$.

Then $|\epsilon_1(p)| < 2.9 \times 10^{-4}$ and $|\epsilon_2(p)| < 1.4 \times 10^{-4}$. In Bailey's notation the roles of p and $1-p$ are switched with ours.

[3.9.4] Wetherill (1965, pp. 202-203) compared three approximations to z_p, noting that the quantiles in the central and the tail regions of $\Phi(\cdot)$ require different approaches to achieve accuracy. Here let $w = 1 - 2p$.

(a) $\quad z_p = (1.253314)w + (0.328117)w^3 + (0.180392)w^5$
$$+ (0.122403)w^7 + (0.113945)w^9 + \epsilon(p) , \quad 0 \le p \le 0.5.$$

Then $|\epsilon(p)| < 1.6 \times 10^{-4}$ if $0.2 < p < 0.3$.

(b) A satisfactory approximation when $|w| \le 0.91$ (corresponding to $0.045 \le p \le 0.5$) is given by

$$z_p \simeq (0.328117)w^3 + (0.180392)w^5 + (0.3630)w^7 - (0.8559)w^9$$
$$+ (1.0480)w^{11}, \quad 0 < p \le 0.5.$$

Then $|\text{relative error}| < 0.002$ if $0.045 < p < 0.50$.

(c) If $0.0001 \le p \le 0.05$, a satisfactory approximation is given by

$$z_p \simeq 0.401703 - (0.625600)v - (0.039811)v^2 - (0.0014146)v^3,$$
$$v = \log[1 - |1 - 2p|] .$$

[3.9.5] A simple approximation to z_p, with a maximum error of 0.02 when $0 \le z_p \le 4.0$ (Hamaker, 1978, pp. 76-77) is given by

$$z_p \simeq t - (0.50 + 0.30t)^{-1} , \quad 0 < p \le 0.5, \quad t = (-2 \log p)^{1/2} .$$

[3.9.6] Hamaker (1978, p. 77) also gives an approximation which holds when $0 < p < 1$, unlike those above, for which adjustments need to be made when $0.50 \le p < 1$:

$$z_p = \text{sgn}(p - 0.50)[(1.238)u\{1 + (0.0262)u\}], \quad 0 < p < 1 ,$$

Expansions and Algorithms

$$u = \{-\log(4p - 4p^2)\}^{1/2}.$$

This gives results almost as accurate as that of *[3.9.1(a)]*.

[3.9.7] Derenzo (1977, pp. 217-218) gives a simple approximation to quantiles, with integer coefficients. Let $\Phi(z_p) = 1 - p$ and $y = -\log(2p)$. Then if $0.50 > p > 10^{-7}$, or $0 < z_p < 5.2$,

$$z_p = \left[\frac{\{(4y + 100)y + 205\}y^2}{\{(2y + 56)y + 192\}y + 131}\right]^{1/2} + \epsilon(p), \quad |\epsilon(p)| < 1.3 \times 10^{-4}.$$

If $10^{-7} > p > 0.50 \times 10^{-112}$, or $5.2 < z_p < 22.6$,

$$z_p = \left[\frac{\{(2y + 280)y + 572\}y}{(y + 144)y + 603}\right]^{1/2} + \epsilon(p), \quad |\epsilon(p)| < 4 \times 10^{-4}.$$

[3.9.8] Schmeiser (1979, pp. 175-176) compares two approximations z^* and $\hat{z}$ to z_p with a third approximation z' of Page (1977, pp. 75-76), where $\Phi(z_p) = 1 - p$:

(a) $z^* = \{(1-p)^{0.135} - p^{0.135}\}/0.1975, \quad 0 < p < 0.5$;

(b) $\hat{z} = 0.20 + \{(1-p)^{0.14} - p^{0.14}\}/0.1596, \quad 0.5 < 1 - p < 1$;

(c) $z' = u - \{(0.134145)u\}^{-1}$,

$u = [\{y + y^2 + 3.31316)^{1/2}\}/0.08943]^{1/3}$,

$y = \log\{(1-p)/p\}/1.59577, \quad 0.5 < 1 - p < 1.$

If $1 - p < 0.5$, z_{1-p} can be approximated by (b) or (c) and the sign of $\hat{z}$ or of z' changed. If $|z_p| \leq 2$, i.e., $p < 0.0228$, then z' is most accurate ($|z' - z_p| \leq 0.0011$) and $|z^* - z_p| \leq 0.0093$; if $|z_p| > 2$, then $\hat{z}$ is most accurate, although the error increases in the tail of the distributiion (if $2 < |z_p| \leq 4$, $|\hat{z} - z_p| < 0.0145$) (Schmeiser, op. cit.).

[3.9.9] The approximation to $1 - \Phi(x)$ of Lin (1990, p. 255; see *[3.5.5]*) can be inverted. If $p = 1 - \Phi(x)$, then the quantile x is approximated via

$$x \simeq 9y/[(4.2)\pi + y], \qquad y = \log(p^{-1} - 1).$$

Here $0 \le p \le 0.50$. The comments given in [3.5.5] on the accuracy of this approximation apply here also.

[3.9.10] Shore (1982, pp. 108-111) developed some simple approximations $z_{(k)}$ to z_p that are useful in some stochastic optimization problems, in the sense that $z_{(k)}$ (for k = 1, 2, 3 below) can be differentiated with respect to ρ, where $\rho = p/(1-p)$; ρ is an increasing function of p. Shore's approximations are based on a "beta-like" distribution. In his notation p becomes our $1 - p$ and vice versa:

(i) $\quad z_{(1)} = -5.5310(\rho^{0.1193} - 1), \quad p < 0.50.$

If $0 \le z_p \le 2.3$, $|z_p - z_{(1)}| \le 0.0083$.

(ii) $\quad z_{(2)} = -0.4115(\rho + \log\rho - 1), \quad p \le 0.50.$

If $0 \le z_p \le 2.6$, $|z_p - z_{(2)}| \le 0.060$. Approximation $z_{(2)}$ is less accurate than $z_{(1)}$ but more useful in some applications, because $dz_{(2)}/dp$ is a linear function of $1/\rho$.

(iii) $\quad z_{(3)} = \begin{cases} -0.495 \log\rho + 0.1337, & p \le 0.50,\ 0.5 \le z_p \le 2.2, \\ -0.4506 \log \rho + 0.2252, & p \le 0.50,\ 2.2 < z_p. \end{cases}$

The accuracy of $z_{(3)}$ is comparable with that of $z_{(2)}$; if we wish to express ρ (and hence p) in terms of z_p, however, $z_{(3)}$ might prove more useful.

(iv) $\quad z_{(4)} = \{-2\log[(3.5554)p^{0.8632}(1-p)]\}^{1/2}, \quad p \le 0.50.$

This form is not a function of ρ; however, only $z_{(1)}$ and $z_{(4)}$ are comparable for accuracy with (a) of [3.9.6].

3.10 APPROXIMATING THE NORMAL BY OTHER DISTRIBUTIONS

A number of writers have sought to fit other distributions to the normal, and then to approximate the normal cdf and quantiles from them.

Expansions and Algorithms

The more successful of these efforts are discussed in *[3.10.3]* and *[3.10.4]*.

[3.10.1] Chew (1968, pp. 22-24) has fitted the uniform, triangular, cosine, logistic, and Laplace distributions, all symmetric about zero, to the standard normal, choosing parameter values to match the second moment, where possible. Near zero, none of these fits very well. Chew claims that beyond about one standard deviation the pdfs of these distributions are "fairly close" together with $\phi(x)$, but the relative errors may be quite large.

[3.10.2] Hoyt (1968, p. 25) approximates the N(0, 1) pdf by that of a sum S of three iid rvs, each having a uniform distribution on $(-1, 1)$. The rv S has pdf g(s), given by

$$g(s) = \begin{cases} (3-s^2)/8, & |s| \leq 1, \\ (3-|s|)^2/16, & 1 \leq |s| \leq 3, \\ 0, & |s| \geq 3. \end{cases}$$

The maximum error in the pdf is 0.024 when $s = 0$, and in the cdf is 0.010 when $s = 0.60$.

[3.10.3] *Approximation by Burr Distributions.* Consider the family of distributions with cdf G given by

$$G(x) = 1 - (1+x^c)^{-k}, \quad x \geq 0, \ c > 0, \ k > 0.$$

Burr (1967, p. 648) approximated $\Phi(x)$ by $G_1(x)$ and $G_2(x)$, where

$$G_1(x) = \begin{cases} 1 - [1 + \{(0.6446,93) + (0.1619,84)x\}^{4.874}]^{-6.158}, & x > -3.9799,80 \\ 0, & x < -3.9799,80 \end{cases}$$

$$G_2(x) = \{G_1(x) + 1 - G_1(-x)\}/2.$$

Then $G_2(x)$ provides the better approximation. The largest difference between $G_2(x)$ and $\Phi(x)$ is 4.6×10^{-4}, when $x = \pm 0.6$.

If we solve the equation $G_2(y_p) = 1-p$ for y_p when p is given, we obtain an approximation to the quantile z_p, where $\Phi(z_p) = 1-p$. Burr (op. cit.) gives the approximation

$$z_p \simeq y_p \simeq [(p^{-1/6.158} - 1)^{1/4.874} - \{(1-p)^{-1/6.158} - 1\}^{1/4.874}]/0.323968.$$

When $0.008 < p < 1/2$, $|\text{error}| < 3.5 \times 10^{-3}$, and when $0.045 < p\ 0.50$, the approximation compares favorably with that of Hastings (1955 p. 191) given in *[3.9.1(a)]*.

[3.10.4] *Approximation by Weibull Distributions.* The Weibull family has cdf W(x) given by

$$W(x) = \Pr(X \le x) = 1 - \exp[-\{(x-a)/\theta\}^m], \ x > a,\ \theta > 0,\ m > 0,$$

where a, θ, and m are location, scale, and shape parameters, respectively. If $c \simeq 3.60$, the third central moment is zero, and if $c \simeq 2.20$, the kurtosis is 3.0, as for a normal pdf (Johnson and Kotz, 1970a, p. 253).

(a) Plait (1962, pp. 23, 26) found that the Weibull distribution for which m = 3.25, a = 0, and θ = 1 fits an $N(0.8963, 0.303^2)$ distribution closely.

(b) By equating several measures of skewness to zero, Dubey (1967, pp. 69-79) arrived at four approximations to $\Phi(x)$, of which we give two that appear to fit the greatest range of values of x in the interval [-3, 3]. The approximations $W_1(x)$ and $W_2(x)$ following were derived by putting the measures (median − mode) and the third central moment μ_3, respectively, equal to zero.

	m	a	θ	x: preferred range
$W_1(x)$	3.25889	− 2.96357	3.30589	[-1.5,-1.1],[0.2,0.9],[2.2,3.0]
$W_2(x)$	3.60232	− 3.24311	3.59893	[-3.0,-1.6],[-1.0,0.1],[1.0,2.1]

The preferred range minimizes |error|, although Dubey (1967, p. 78) finds a few values of x for which his other approximations do slightly better. The preceding combination has a maximum |error| of 0.0078 when x = − 0.9.

(c) Makimo (1984, pp. 1-8) lists these and other references. He notes that when m = 3.43938, the Weibull distribution shares an important property with the normal, that is, the mean and the median are equal. Further, this value of m is almost identical with that for which the mean hazard rates (m.h.r's) of the two distributions are equal, given that they have equal variances. For a pdf g($\cdot$) and cdf G($\cdot$), the m.h.r. is defined as

$$\int_{-\infty}^{\infty} \frac{g(x)}{1 - G(x)} \cdot g(x) dx .$$

[3.10.5] Approximation by Sargan Distributions

The kth-order Sargan distribution has pdf g(x), where

$$g(x) = \lambda_k \exp(-\alpha|x|) p(|x|), \ |x| < \infty, \ \alpha \geq 0 ,$$

where p($\cdot$) is a polynomial of degree k,

$$p(|x|) = 1 + \sum_{j=1}^{k} \gamma_j \alpha^j |x|^j , \ \gamma_j \geq 0 , \ \lambda_k = \tfrac{1}{2}\alpha \sum_{j=1}^{k} (1 + \gamma_j \, j!)^{-1} .$$

First- and second-order distributions have been considered as approximations to the standard normal, with second-order distributions performing better; see Goldfeld and Quandt (1981, pp. 141-155), who found reasonably close agreement using the second-order form, except in the tails. Kafaei and Schmidt (1985, pp. 509-526) have considered some of the ramifications of using this approximation in statistical inference.

REFERENCES

Abramowitz, M., and Stegun, I. A. (1964). *Handbook of Mathematical Functions*, Washington, D.C.: National Bureau of Standards *[3.1.4; 3.2.2; 3.4.2; 3.6.6; 3.7.1;3.9.1]*

Adams, A. G. (1969). Algorithm 39: Areas under the normal curve, *Computer Journal 12*, 197-198. *[3.1.2]*

Allasia, G. (1981). Approssimazione della funzione di distribuzione normale mediante funzion spline, *Statistica 41*, 325-332. *[3.3.4]*

Altman, I. B. (1950). The effect of one-sided truncation on the average of a normal distribution, *Industrial Quality Control* 7(3), 30-31. *[3.1.1]*

Atkinson, A. C. and Pearce, M. C. (1976). The computer generation of beta, gamma and normal random variables (with discussion), *Journal of the Royal Statistical Society A* 139, 431-461. *[3.1.4]*

Badhe, S.K. (1976). New approximation of the normal distribution function, *Communications in Statistics B5*, 173-176. *[3.4.1; 3.6.6, 7]*

Bailey, B.J.R. (1981). Alternatives to Hastings' approximation to the inverse of the normal cumulative distribution function, *Applied Statistics* 30, 275-276. *[3.9.3]*

Beasley, J. D. and Springer, S. G. (1977). Algorithm AS111: The percentage points of the normal distribution, *Applied Statistics* 26, 118-121. *[3.1.3]*

Birnbaum, Z. W. (1942). An equality for Mill's ratio, *Annals of Mathematical Statistics* 13, 245-246. *[3.6.5.1]*

Bowman, K. O. and Shenton, L. R. (1989). *Continued Fractions in Statistical Applications*, New York: Dekker. *[3.6.1, 2, 4]*

Box, G. E. P. and Muller, M. E. (1958). A note on the generation of random normal deviates, *Annals of Mathematical Statistics* 29, 610-611. *[3.1.4]*

Boyd, A. V. (1959). Inequalities for Mills' ratio, *Reports of Statistical Application Research* (Union of Japanese Scientists and Engineers), 6, 44-46. *[3.6.5.3]*

Bratley, P., Fox, B. L. and Schrage, L. (1987). *A Guide to Simulation* (2nd edn.), New York: Springer-Verlag. *[3.1.4]*

Brent, R. P. (1974). A Gaussian pseudo random generator, *Communications of the Association of Computing Machinery* 17, 704-706. *[3.1.4]*

Burr, I. W. (1967). A useful approximation to the normal distribution function, with application to simulation, *Technometrics* 9, 647-651. *[3.10.3]*

Carta, D. G. (1975). Low-order approximations for the normal probability integral and the error function, *Mathematics of Computation* 2976, 856-862. *[3.4.3]*

Chew, V. (1968). Some useful alternatives to the normal distribution, *The American Statistician* 22(3), 22-24. *[3.10.1]*

Chu, J. T. (1955). On bounds for the normal integral, *Biometrika* 42, 263-265. *[3.5.4, 6]*

Craig, R. J. (1984). Normal family distribution functions: FORTRAN and BASIC programs, *Journal of Quality Technology* 16, 232-236. *[3.3.3]*

Dagpunar, J. (1988). *Principles of Random Variate Generation*, New York: Clarendon Press. *[3.1.4]*

Derenzo, S. E. (1977). Approximations for hand calculators using small integer coefficients, *Mathematics of Computation* 31, 214-222. *[3.5.7; 3.9.7]*

Divgi, D. R. (1979). Calculation of univariate and bivariate normal probability functions, *Annals of Statistics* 7, 903-910. *[3.4.4]*

Dubey, S. D. (1967). Normal and Weibull distributions, *Naval Research Logistic Quarterly* 14, 69-79. *[3.10.4]*

Feller, W. (1968). *An Introduction to Probability Theory and Its Applications*, Vol. 1 (3rd ed.), New York: Wiley. *[3.6.3]*

Goldfeld, S. M. and Quandt, R. E. (1981). Econometric modeling with non-normal disturbances, *Journal of Econometrics* 17, 141-155. *[3.10.5]*

Good, I. J. (1987). The calculation of the normal error function with normally acceptable proportional error, *Journal of Statistical Computation and Simulation* 28, 83-84. *[3.3.1; 3.8.3]*

Gordon, R. D. (1941). Values of Mills' ratio of area to bounding ordinate and of the normal probability integral for large values of the argument, *Annals of Mathematical Statistics* 12, 364-366. *[3.6.1]*

Gray, H. L. and Schucany, W. R. (1968). On the evaluation of distribution functions, *Journal of the American Statistical Association* 63, 715-720. *[3.7.3; 3.8.2]*

Gross, A. J. and Hosmer, D. W. (1978). Approximating tail areas of probability distributions, *Annals of Statistics* 6, 1352-1359. *[3.6.1]*

Gupta, S. S. and Waknis, M. N. (1965). A system of inequalities for the incomplete gamma function and the normal integral, *Annals of Mathematical Statistics* 36, 139-149. *[3.6.4]*

Hamaker, H. C. (1978). Approximating the cumulative normal distribution and its inverse, *Applied Statistics* 27, 76-77. *[3.5.4, 5; 3.9.5, 6]*

Hart, R. G. (1966). A close approximation related to the error function, *Mathematics of Computation* 20, 600-602. *[3.6.7; 3.7.2]*

Hastings, C. (1955). *Approximations for Digital Computers*, Princeton, N.J.: Princeton University Press. *[3.2.2; 3.6.7; 3.7.1; 3.9.1; 3.10.3]*

Hill, I. D. (1969). Remark ASR2: A remark on Algorithm AS2 "The normal integral," *Applied Statistics 18*, 299-300. *[3.1.2]*

Hill, I. D. (1973). Algorithm AS66: The normal integral, *Applied Statistics 22*, 414-427. *[3.1.2]*

Hill, I. D. and Joyce, S. A. (1967). Algorithm 304: Normal curve integral, *Communications of the Association for Computing Machinery 10(6)*, 374-375. *[3.1.2]*

Hoyt, J. P. (1968). A simple approximation to the standard normal probability density function, *The American Statistician 22(2)*, 25-26. *[3.10.2]*

Ibbetson, D. (1963). Algorithm 209: Gauss, *Communications of the Association for Computing Machinery 6*, 616. *[3.1.2]*

IMSL (1987). *Stat/Library: User's Manual, Vol. 3*, Version 1.0, Houston, Texas: IMSL Inc. *[3.1.2]*

Johnson, N. L. and Kotz, S. (1970). *Distributions in Statistics-- Continuous Univariate Distributions*, Vol. 1, New York: Wiley. *[3.10.4]*

Kafaei, M.-A. and Schmidt, P. (1985). On the adequacy of the "Sargan distribution" as an approximation to the normal, *Communications in Statistics A 14*, 509-526. *[3.10.5]*

Kennedy, W. J., Jr. and Gentle, J. E. (1980). *Statistical Computing*, New York: Dekker. *[3.1.2, 3, 4]*

Kerridge, D. F. and Cook, G. W. (1976). Yet another series for the normal integral, *Biometrika 63*, 401-403. *[3.3.2, 3; 3.6.1; 3.8.2]*

Knuth, D. E. (1981). *The Art of Computer Programming, Vol. 2: Seminumerical Algorithms* (2nd edn.), Reading, Mass.: Addison-Wesley. *[3.1.4]*

Laplace, P. S. (1785). Mémoire sur les approximations des formules qui sont fonctions de très-grands nombres, *Histoire de l'Academie Royale des Sciences de Paris*, 1-88; reprinted in *Oeuvres Complètes, Vol. 10*, 209-291. *[3.3.1, 2; 3.4.1; 3.6.3; 3.8.2]*

Laplace, P. S. (1805). *Traité de Mécanique Celeste, Vol. 4*, Paris. *[3.3.2; 3.6.1; 3.8.2, 3]*

Laplace, P. S. (1812). *Théorie Analytique de Probabilités, Vol. 2*, Paris: Courcier. *[3.6.4; 3.8.2]*

Lee, C.-I.C. (1992). On Laplace continued fraction for the normal integral, *Annals of the Institute of Statistical Mathematics* 44, 107-120. *[3.3.1; 3.6.1; 3.6.5.1, 2, 3; 3.8.2]*

Lee, C.-I.C. (1994). Personal communications. *[3.8.2]*

Lew, R. A. (1981). An approximation to the cumulative normal distribution with simple coefficients, *Applied Statistics* 30, 299-301. *[3.5.2]*

Lin, J.-T. (1990). A simpler logistic approximation to the normal tail probability and its inverse, *Applied Statistics* 39, 255-257. *[3.5.5; 3.9.9]*

Makino, T. (1984). Mean hazard rate and its application to the normal approximation of the Weibull distribution, *Naval Research Logistics Quarterly* 31, 1-8. *[3.10.4]*

Marsaglia, G. and Bray, T. A. (1964). A convenient method for generating normal variables, *SIAM Review* 6(3), 260-264. *[3.1.4]*

Marsaglia, G., MacLaren, M. D. and Bray, T. A. (1964). A fast procedure for generating normal random variables, *Communications of the Association for Computing Machinery* 7(1), 4-10. *[3.1.3, 4]*

Moran, P. A. P. (1980). Calculation of the normal distribution function, *Biometrika* 67, 675-676. *[3.3.3; 3.8.2]*

Odeh, R. E. and Evans, J. O. (1974). Algorithm AS70: The percentage points of the normal distribution, *Applied Statistics* 23, 96-97. *[3.1.3; 3.9.2]*

Page, E. (1977). Approximations to the cumulative normal function and its inverse for use on a pocket calculator, *Applied Statistics* 26, 75-76. *[3.5.2]*

Patry, J. and Keller, J. (1964). Zur berechnung des fehlerintegrals, *Numerische Mathematik* 6, 89-97. *[3.6.6, 7; 3.8.1.2]*

Plait, A. (1962). The Weibull distribution--with tables, *Industrial Quality Control* 19(5), 17-26. *[3.10.4]*

Pólya, G. (1949). Remarks on computing the probability integral in one and two dimensions, *Proceedings of the First Berkeley Symposium on Mathematical Statistics and Probability*, 63-78, Berkeley: University of California Press. *[3.5.4]*

Rabinowitz, P. (1969). New Chebyshev polynomial approximations to Mills' ratio, *Journal of the American Statistical Association* 64, 647-654. *[3.7.4]*

Ray, W. D. and Pitman, A. E. N. T. (1963). Chebyshev polynomial and other new approximations to Mills' ratio, *Annals of Mathematical Statistics 34*, 892-902. *[3.7.4]*

Ripley, B. D. (1987). *Stochastic Simulation*, Chichester, U.K.: Wiley. *[3.1.4]*

Ruben, H. (1962). A new asymptotic expansion for the normal probability integral and Mills' ratio, *Journal of the Royal Statistical Society B24*, 177-179. *[3.7.4]*

Ruben, H. (1963). A convergent asymptotic expansion for Mills' ratio and the normal probability integral in terms of rational functions, *Mathematische Annalen 151*, 355-364. *[3.6.1; 3.7.4]*

Ruben, H. (1964). Irrational fraction approximations to Mills' ratio, *Biometrika 51*, 339-345. *[3.7.4]*

Sampford, M. R. (1953). Some inequalities on Mills' ratio and related functions, *Annals of Mathematical Statistics 24*, 130-132. *[3.6.5.1]*

Schmeiser, B. (1979). Approximations to the inverse cumulative normal function for use on hand calculators, *Applied Statistics 28*, 175-176. *[3.9.8]*

Schucany, W. R. and Gray, H. L. (1968). A new approximation related to the error function, *Mathematics of Computation 22*, 201-202. *[3.6.7]*

Shah, A. K. (1985). A simpler approximation for areas under the standard normal curve, *American Statistician 39*, 80, 327. *[3.5.1]*

Shenton, L. R. (1954). Inequalities for the normal integral including a new continued fraction, *Biometrika 41*, 177-189. *[3.3.2; 3.6.1, 2, 3, 4; 3.8.2]*

Shore, H. (1982). Simple approximations for the inverse cumulative function, the density function and the loss integral of the normal distribution, *Applied Statistics 31*, 108-114. *[3.2.3; 3.7.5; 3.9.10]*

Strecok, A. J. (1968). On calculation of the inverse of the Error Function, *Mathematics of Computation, 22*, 144-158. *[3.3.3]*

Stuart, A. and Ord, J. K. (1987). *Kendall's Advanced Theory of Statistics, Vol. 1*, New York: Oxford University Press. *[3.3.1,3; 3.6.3]*

Tocher, K. D. (1963). *The Art of Simulation*, London: English Universities Press. *[3.5.3]*

Varnum, E. C. (1950). Normal area nomogram, *Industrial Quality Control 6(4)*, 32-34. *[3.1.2]*

Wetherill, G. B. (1965). An approximation to the inverse normal function suitable for the generation of random normal deviates on electronic computers, *Applied Statistics 14*, 201-205. *[3.9.4]*

Chapter 4
CHARACTERIZATIONS

If a normally distributed variable or if a random sample from a normal distribution has some property P, it may be of interest to know whether such a property characterizes the normal law. Serious research in this area has been done since 1935, but most results date from 1950 or later. Users may find some of the special cases of more practical interest than the most general theorems. Some of the more mathematical and theoretical characterizations have been omitted, but can be found in the books and expository papers on the subject. These include Stuart and Ord (1987, Secs. 15.4, 15.22-29); Kotz (1974, pp. 39-65); Lukacs (1956, pp. 195-214); Lukacs and Laha (1964, Chaps. 5 and 10); Mathai and Pederzoli (1977); and Kagan et al. (1973). The review of the last-named by Diaconis et al. (1977, pp. 583-592) gives an excellent summary and exposition of the field.

While characterizations of the exponential distribution largely involve order statistics (see *[4.8]*, however), those of the normal law frequently feature distributional or independence properties of linear and quadratic forms. Linear forms alone are considered in *[4.1]*, and along with quadratic forms in *[4.2]*; these are based on sets of independent and sometimes iid random variables. Characterizations based on conditional distributions and regression properties are also mainly based on linear and quadratic forms; these are given in *[4.3]*; users may note that the conditioning statistic in all of these characterizations is invariably linear. Special topics are covered in *[4.5]* (characteristic functions), *[4.6]* (polar coordinate transformations), *[4.7]* (characterizations based on sufficiency, information, and concepts from decision theory and statistical inference), and *[4.8]* (order statistics). The question of how close to a characterizing property a distribution can be and be nearly normal is touched upon in *[4.9]* (but see also Diaconis et al., 1977, pp. 590-591).

Several characterizations of the bivariate normal distribution appear in [10.4]; others belong as special cases of characterizations of the multivariate normal distribution, which lies outside the scope of this book. References for these multivariate characterizations include Kagan et al. (1973), Mathai and Pederzoli (1977), Johnson and Kotz (1972, pp. 59-62), Khatri and Rao (1972, pp. 162-173), Khatri and Rao (1976, pp. 81-94), and Khatri (1979, pp. 589-598).

4.1 CHARACTERIZATION BY LINEAR STATISTICS

[4.1.1] (a) Let X_1 and X_2 be independent nondegenerate rvs, not necessarily identically distributed; let $Y = X_1 + X_2$. Then Y is normally distributed if and only if X_1 and X_2 are normally distributed (Mathai and Pederzoli, 1977, p. 6; Cramér, 1936, pp. 405-414).

(b) Let X and Y be rvs. If coefficients a and b exist such that $a^2 + b^2 > 0$, and if X and $aX + bY$ are independent in every coordinate system, then X and Y are independent normal rvs with the same variance (Berk, 1986, pp. 696-701).

Here, "in every coordinate system" means "under every orthogonal rotation of the axes". Berk extends this result: let $X_1, X_2, ..., X_n$ be rvs and $(a_1, ..., a_n)$, $(b_1, ..., b_n)$ nonzero vectors of coefficients. Then if $\sum_i a_i X_i$ $\sum_i b_i X_i$ are independent in every coordinate system in Euclidean n-space, either the X_i's are degenerate or $\sum_i a_i b_i = 0$ and $X_1, ..., X_n$ are mutually independent $N(\cdot, \sigma^2)$ rvs with common variance. See also Hartman and Wintner (1940, pp. 759-779). This result for n = 3 was obtained by James Clerk Maxwell, as discussed in Chapter 1.

(c) The sum $X_1 + ... + X_n$ of n independent rvs (n = 2, 3, ...) is normally distributed if and only if each component variable $X_1, X_2, ..., X_n$ is itself normally distributed (Cramér, 1946, pp. 212-213).

(d) Let $X_1, X_2, ..., X_n$ be a random sample from a certain population, and let $L = a_1 X_1 + a_2 X_2 + ... + a_n X_n$, a linear statistic, where $a_1, ..., a_n$ are nonzero constants. Then the population is normal if and only if L is normally distributed (Lukacs, 1956, p. 198).

(e) Let $X_1, X_2, ..., X_n$ be a random sample from a certain population and let constants $a_1, ..., a_n$ be given, such that

$$A_s = \sum_{j=1}^{n} a_j^s \neq 0,$$

for all positive integer values of s. Then the statistic $a_1X_1 + ... + a_nX_n$ is normally distributed with mean μ and variance σ^2 if and only if each of X_1, ..., X_n is normally distributed with mean μ/A_1 and variance σ^2/A_2 (Lukacs, 1956, pp. 197-198).

[4.1.2] Let $X_1, X_2, ..., X_n$ be a random sample from a certain population and suppose that all absolute moments $E(|X_i|^k)$ exist (i = 1,...,n; k =1,2,...). Consider the linear statistics

$$L_1 = a_1X_1 + ... + a_nX_n, \quad L_2 = b_1X_1 + ... + b_nX_n.$$

Suppose that $\{|a_1|, ..., |a_n|\}$ is not a permutation of $\{|b_1|, ..., |b_n|\}$, and that $\sum_{i=1}^{n} a_i = \sum_{i=1}^{n} b_i$, $\sum_{i=1}^{n} a_i^2 = \sum_{i=1}^{n} b_i^2$. Then the parent population is normally distributed if and only if L_1 and L_2 are identically distributed (Lukacs, 1956, p. 210).

Marcinkiewicz (1939, pp. 612-618) gives a result which generalizes this to infinite sums.

[4.1.3] Let $X_1, ..., X_n$ be a random sample from a certain population. Then every linear statistic $a_1X_1 + ... + a_nX_n$ is distributed like the rv $(a_1^2 + ... + a_n^2)^{1/2}X_1$ if and only if the parent population is normal. This is a special case of a theorem of Lukacs (1956, p. 200). Shimizu (1962, pp. 173-178) has a similar result, requiring the assumption of finite variances.

[4.1.4] (a) Let X_1 and X_2 be independent rvs with finite variances. Then the sum $X_1 + X_2$ and the difference $X_1 - X_2$ are independent if and only if X_1 and X_2 are normally distributed (Bernstein, 1941, pp. 21-22).

(b) Let $X_1, ..., X_n$ be a sample from a certain population and consider the linear statistics

$$L_1 = a_1X_1 + ... + a_nX_n, \quad L_2 = b_1X_1 + ... + b_nX_n$$

such that $\sum_{j=1}^{n} a_j b_j = 0$, while $\sum_{j=1}^{n} (a_j b_j)^2 \neq 0$. Then the parent population is normal if and only if L_1 and L_2 are independently distributed (Lukacs, 1956, p. 205). This result does not require any assumption of the existence of moments.

(c) *The Darmois-Skitovich Theorem.* Let $X_1, ..., X_n$ be mutually independent rvs (not necessarily identically distributed). Let L_1 and L_2 be linear statistics $\sum_{i=1}^{n} a_i X_i$ and $\sum_{i=1}^{n} b_i X_i$, where $a_1, ..., a_n, b_1, ..., b_n$ are constant coefficients.

If L_1 and L_2 are independent, then the rvs X_i for which the product $a_i b_i \neq 0$ are all normally distributed (Kagan et al., 1973, p. 89; Mathai and Pederzoli, 1977, p. 25: Darmois, 1951, pp. 1999-2000; Skitovitch, 1953, pp. 217-219). The form of the converse is given by: If $\sum_{j=1}^{n} \sigma_j^2 a_j b_j = 0$ [where $\text{var}(X_j) = \sigma_j^2$] and those X_i are normally distributed for which $a_j b_j \neq 0$, then L_1 and L_2 are independent.

[4.1.5] An orthogonal transformation of n iid normally distributed rvs preserves mutual independence and normality. Lancaster (1954, p. 251) showed that this property characterizes the normal distribution.

Let $X_1, ..., X_n$ be mutually independent rvs each with zero mean and unit variance, and let n nontrivial linear transformations

$$Y_j = \sum_{i=1}^{n} a_{ij} X_j$$

be made, which exclude those of the type $X_i + a = cY_j$, and such that $Y_1, ..., Y_n$ are mutually independent. Then $X_1, ..., X_n$ are normally distributed and the transformation is orthogonal. Stuart and Ord (1987, Sec. 15.27) state that this characterization of the normal distribution "illuminates the position of central importance which it holds in statistical theory."

[4.1.6] In the following result we consider an infinite set of variables (Kagan et al., 1973, p. 94; Mathai and Pederzoli, 1977, p. 30).

Let $X_1, X_2,...$ be a sequence of independent rvs and $\{a_j\}, \{b_j\}$ be two sequences of real constants such that

(i) The sequences $\{a_j/b_j : a_j b_j \neq 0\}$ and $\{b_j/a_j : a_j b_j \neq 0\}$ are both bounded.

(ii) $\sum_{j=1}^{\infty} a_j X_j$ and $\sum_{j=1}^{\infty} b_j X_j$ converge with probability one to rvs U and V, respectively.

(iii) U and V are independent.

Then for every j such that $a_j b_j \neq 0$, X_j is normally distributed.

4.2 LINEAR AND QUADRATIC CHARACTERIZATIONS

[4.2.1] (a) Let $X_1, X_2, \ldots, X_n$ be a random sample from a certain population; let $\overline{X} = \sum X_i/n$, the sample mean, and let $S^2 = \sum (X_i - \overline{X})^2/(n-1)$, the sample variance. For $n \geq 2$, a necessary and sufficient condition for the independence of $\overline{X}$ and S^2 is that the parent population be normal (Kagan et al., 1973, p. 103; Stuart and Ord, 1987, Sec. 12.15; Lukacs, 1956, p. 200). This result was first obtained by Geary (1936, pp. 178-184) and by Lukacs (1942, pp. 91-93) under more restrictive conditions involving the existence of moments.

(b) Let $X_1, X_2, \ldots, X_n$ be iid random variables with common finite mean and variance. Let the first-order mean square successive differences D_k^2 be defined by

$$D_k^2 = \tfrac{1}{2}(n-k)^{-1} \sum_{i=1}^{n-k} (X_{i+k} - X_i)^2, \quad k = 1, 2, \ldots, n-1,$$

and let $\overline{X} = \sum X_i/n$. Then for each k a necessary and sufficient condition for $\overline{X}$ and D_k^2 to be independent is that the parent population is normal (Geisser, 1956, p. 858).

(c) The following theorem by Rao (1958, p. 915) generalizes results in (a) and (b) above. Let $X_1, \ldots, X_n$ be iid random variables with common finite mean and variance, and let

$$\delta^2 = \left(\sum_{t=1}^{m} \sum_{j=1}^{n} h_{tj}^2 \right)^{-1} \sum_{t=1}^{m} (h_{t1}X_1 + \ldots + h_{tn}X_n)^2, \quad m \geq 1,$$

where $\sum_{j=1}^{n} h_{tj} = 0$ when $t = 1, 2, \ldots, m$. Then a necessary and sufficient condition for the parent population to be normal is that $\overline{X}$ and δ^2 are independent. See also Laha (1953, pp. 228-229).

In (a), $m = n$ and $h_{jj} = 1 - n^{-1}$; $h_{tj} = -n^{-1}$ if $t \neq j$.

In (b), $m = n - k$ and $h_{jj} = -1$, $h_{tj} = 1$ if $j = t + k$ and $h_{tj} = 0$ otherwise.

(d) Let $X_1, \ldots, X_n$ be iid random variables ($n \geq 4$), and let $Y = (X_1 - X_2)/S$. If Y is stochastically independent of the pair $(\bar{X}, S)$, where $\bar{X} = \sum X_i/n$ and $S^2 = \sum(X_i - \bar{X})^2/(n-1)$, then the parent population is normal (Kelkar and Matthes, 1970, p. 1088).

(e) Kaplansky (1943, p. 197) weakened the conditions for the characterization in (a). Let $X_1, \ldots, X_n$ be a random sample from a continuous population with sample mean $\bar{X}$ and sample variance S^2. Then the joint pdf of $X_1, \ldots, X_n$ is a function $h(\bar{x}, s)$ of the values of $\bar{X}$ and S and also the joint pdf of $\bar{X}$ and S is given by $h(\bar{x}, s)s^{n-2}$ if and only if the parent population is normal ($n \geq 3$).

[4.2.2] (a) Let $X_1, \ldots, X_n$ be a random sample from a certain population. Let $L = \sum_{i=1}^n a_i X_i$ and $Q = \sum_{i=1}^n X_i^2 - L^2$, where $\sum_{i=1}^n a_i^2 = 1$. Then a necessary and sufficient condition for the independence of L and Q is that the parent population is normal (Kagan et al., 1973, pp. 105-106; Mathai and Pederzoli, 1977, p. 35).

(b) Let $X_1, \ldots, X_n$ be a random sample from a population where the second moment exists, and let $\bar{X} = \sum X_i/n$, $Q = \sum_{i=1}^n \sum_{j=1}^n a_{ij} X_i X_j$, where $\sum_{i=1}^n a_{ii} \neq 0$ and $\sum_{i=1}^n \sum_{j=1}^n a_{ij} = 0$. Then $\bar{X}$ and Q are independent if and only if the parent population is normal (Mathai and Pederzoli, 1977, p. 38; Kagan et al., 1973, pp. 106-107). See also Lukacs and Laha (1964, p. 81).

(c) Let $X_1, \ldots, X_n$ be iid random variables, and let

$$Q = \sum_{j=1}^n \sum_{k=1}^n a_{tj} X_j X_k + \sum_{j=1}^n b_j X_j,$$

such that $\sum_{j=1}^n a_{jj} \neq 0$ and $\sum_{j=1}^n \sum_{k=1}^n a_{jk} = 0$. Then $X_1 + \ldots + X_n$ and Q are independent if and only if

(i) the parent population is normal,
(ii) $\sum_{j=1}^n a_{jk} = 0$ when $k = 1, \ldots, n$, and
(iii) $\sum_{j=1}^n b_j = 0$ (Lukacs and Laha, 1964, p. 100).

[4.2.3] Based on a standard normal sample, the distribution of $\sum_{i=1}^n (X_i + a_i)^2$ depends on $a_1, \ldots, a_n$ through the noncentral chi-square noncentrality parameter $a_1^2 + \ldots + a_n^2$; see *[5.3.6]*. It turns out that this property is a characterization of the normal law.

(a) If $X_1, \ldots, X_n$ are iid random variables ($n \geq 2$), and the distribution of the statistic $\sum_{i=1}^n (X_i + a_i)^2$ depends on $a_1, \ldots, a_n$ only through $\sum_{i=1}^n a_i^2$, where $a_1, \ldots, a_n$ are real, then the parent population is normal (Kagan et al., 1973, p. 453).

(b) If the random vector $(X_1, \ldots, X_m)$ is independent of the random vector $(X_{m+1}, \ldots, X_n)$, $1 \leq m < n$, and if for arbitrary constants $a_1, \ldots, a_n$, the distribution of $\sum_{j=1}^n (X_j + a_j)^2$ depends on $a_1, \ldots, a_n$ only through $\sum_{j=1}^n a_j^2$, then $X_1, \ldots, X_n$ are mutually independent and normally distributed with zero means and common variance (Kotz, 1974, p. 48).

[4.2.4] The following related results are due to Geisser (1973, p. 492-494).

(a) Let X and Y be independent rvs, and let X have a N(0, 1) distribution. Then the rv $(aX + bY)^2/(a^2 + b^2)$ has a chi-square distribution with one degree of freedom (χ_1^2) for some nonzero a and b if and only if Y^2 is χ_1^2.

(b) Let X and Y be independent with X^2 and Y^2 each having a χ_1^2 distribution. Then the rv $(aX + bY)^2/(a^2 + b^2)$ has a χ_1^2 distribution for some nonzero a and b if and only if at least one of X and Y has a N(0, 1) distribution.

(c) Let X and Y be iid random variables. Then X and Y are N(0, 1) rvs if and only if, for some nonzero a and b, the rvs $(aX + bY)^2/(a^2 + b^2)$ and $(aX - bY)^2/(a^2 + b^2)$ are χ_1^2 rvs.

[4.2.5] The distribution of the sample variance provides a characterization of normality.

(a) Let $X_1, \ldots, X_n$ ($n \geq 2$) be a random sample from a nondegenerate symmetric distribution with finite variance σ^2 and sample mean $\bar{X}$. Then $\sum_{i=1}^n (X_i - \bar{X})^2/\sigma^2$ is distributed as chi-square with $(n-1)$ degrees of freedom (χ_{n-1}^2) if and only if the parent distribution is normal (Ruben, 1974, p. 379).

(b) Let $X_1, X_2, \ldots$ be an iid sequence of nondegenerate rvs with finite variance σ^2, and let m and n be distinct integers not less than 2. Denote $\sum_{i=1}^m X_i/m$ by $\bar{X}_m$, and $\sum_{i=1}^n X_i/n$ by $\bar{X}_n$. Then $\sum_{i=1}^m (X_i - \bar{X}_m)^2/\sigma^2$ and

$\sum_{i=1}^{n}(X_i - \bar{X}_n)^2/\sigma^2$ are distributed as χ^2_{m-1} and χ^2_{n-1}, respectively, if and only if the parent distribution is normal (Ruben, 1975, p. 72; Bondesson, 1977, p. 303-304).

(c) In another characterization when $n \geq 2$, the X_i's as set up in (a) are distributed $N(0, 1)$ if and only if $\sum_{i=1}^{n} X_i^2$ and $n\bar{X}_n^2$ are distributed as χ^2_n and χ^2_1, respectively (Ahsanullah, 1987, pp. 885-888).

(d) Consider a vector of observations following the usual assumptions of the general linear model (Rao, 1973, Chap. 4), having iid and symmetrically distributed error components. Let σ^2 be the error variance and ν the degrees of freedom of the residual sum of squares (RSS). Then the distribution of (RSS/σ^2) is χ^2_ν if and only if the error components are normally distributed. Ruben (1976, p. 186) gives a formal statement and proof.

4.3 CHARACTERIZATIONS BY CONDITIONAL DISTRIBUTIONS AND REGRESSION PROPERTIES

[4.3.1] (a) Let X and Y be rvs, each with a nonzero pdf at zero. If the conditional distribution of X, given $X + Y = x + y$ is the normal distribution with mean $(x + y)/2$ for all x and y, then X and Y are iid normal rvs (Patil and Seshadri, 1964, p. 289).

(b) Let X and Y be iid rvs with density f. Suppose that the conditional pdfs of $X + Y$, given $X - Y = t$ exist and are equal for all t in a Borel set E having positive Lebesgue measure. Then f is a normal density (Ramasubramaniam, 1985, 410-414).

[4.3.2] Let X_1 and X_2 be iid random variables with mean zero. If $E(X_1 - \alpha X_2 | X_1 + \beta X_2) = 0$ and $E(X_1 + \beta X_2 | X_1 - \alpha X_2) = 0$ where $\alpha \neq 0$, $\beta \neq 0$, then X_1 and X_2 are normal (possibly degenerate) when $\beta\alpha = 1$, and degenerate otherwise (Kagan et al., 1973, p. 158, 161).

[4.3.3] Let $X_1, ..., X_k$ be independent rvs and a_j, b_j ($j = 1, ..., k$) be nonzero constants such that, when $i \neq j$, $a_i b_i^{-1} + a_j b_j^{-1} \neq 0$. If the conditional distribution of $\sum_{j=1}^{n} a_j X_j$, given $\sum_{j=1}^{k} b_j X_j$, is symmetric, then $X_1, ..., X_k$ are normal rvs (possibly degenerate). If the characteristic function of X_j is $\exp(itA_j - B_j t^2)$, with A_j real and B_j nonnegative, then

$\sum_{j=1}^{k} A_j a_j = 0$ and $\sum_{j=1}^{k} B_j a_j b_j = 0$ (Kagan et al., 1973, p. 419).

[4.3.4] (a) *The Kagan-Linnik-Rao Theorem* (Kagan et al., 1973, p. 155). Let $X_1,...,X_n$ be iid rvs ($n \geq 3$) with mean zero, and let $\bar{X} = \sum X_i/n$. Then if $E(\bar{X}|X_1 - \bar{X}, ..., X_n - \bar{X}) = 0$, the parent population is normal.

An alternative condition, given by Kagan et al. (1965, p. 405), is that $E(\bar{X}|X_2 - X_1, X_3 - X_1, ..., X_n - X_1) = 0$. If $n = 2$, this property holds for any symmetric parent distribution.

(b) Let $X_1, ..., X_n$ ($n \geq 3$) be mutually independent rvs, each with mean zero, and let $L_1, ..., L_n$ be linearly independent linear functions of $X_1, ..., X_n$, with all the coefficients in L_1 nonzero. Then, if $E(L_1|L_2, ..., L_n) = 0$, the rvs $X_1, ..., X_n$ are all normally distributed (Kagan et al., 1973, p. 156). This is a generalization of the result in (a).

(c) Let $X_1, ..., X_n$ be iid nondegenerate rvs, and $L_i = \sum_{j=1}^{n} a_{ij} X_j$ ($i = 1, ..., n-1$) be linearly independent linear functions of $X_1, ..., X_n$, while $L_n \sum_j a_{nj} X_j$ s a linear form such that the vector $(a_{n1}, ..., a_{nn})$ is not a multiple of any vector with components $0, 1, -1$. Then, from the conditions $E(L_i|L_n) = 0$ ($i = 1, ..., n-1$), it follows that the parent population is normal if and only if $\sum_{j=1}^{n} a_{ij} a_{nj} = 0$ ($i = 1, ..., n-1$) (Kagan et al., 1973, p. 161).

(d) Cacoullos (1967b, p. 1897) gives a result analogous to that in (c); $X_1, ..., X_n$ are iid random variables with mean zero and positive finite variance. The forms $L_1, ..., L_{n-1}$ are linearly independent statistics defined as in (c). Define $L_n = b_1 X_1 + ... + b_n X_n$ (with no restrictions on the coefficients $b_1, ..., b_n$); then if $E(L_i|L_n) = 0$ ($i = 1, ..., n-1$), the parent distribution is normal.

(e) $X_1, ..., X_n$ are iid rvs with mean zero and unit variance. If

$$E(\bar{X}^2|X_2 - X_1, X_3 - X_1, ..., X_n - X_1) = 0 ,$$

then the X_i's are normally distributed (Wesolowski, 1987, pp. 11-12). This result is related to that in (a).

[4.3.5] (a) The following result is related to the Darmois-Skitovich theorem of *[4.1.4]* (c); it weakens the condition of independence of the linear statistics; on the other hand, it requires the assumption of finite variances.

Let $X_1, ..., X_n$ be a random sample from a population with mean zero and variance σ^2. Define linear statistics

$$Y_1 = a_1X_1 + ... + a_nX_n, \qquad Y_2 = b_1X_1 + ... + b_nX_n ,$$

such that $a_n \neq 0$, $|b_n| > \max(|b_1|,...,|b_{n-1}|)$, and $E(Y_1|Y_2) = 0$. If $\sigma^2 < \infty$, $\sum_{i=1}^{n} a_i b_i = 0$ (or $\sigma^2 > 0$), and $a_i b_i/(a_n b_n) < 0$ (i = 1, ..., n − 1), then the parent population is normal (Rao, 1967, p. 5).

(b) a special case of (a) is as follows (Rao, 1967, pp. 1-2, 6). Let $X_1, ..., X_n$ be a random sample from a population with mean zero and finite variance (n ≥ 3), and let $\overline{X}$ be the sample mean. Then if, for *any fixed* value of i (1 ≤ i ≤ n) $E(\overline{X}|X_i - \overline{X}) = 0$, the parent population is normal.

(c) Pathak and Pillai (1968, p. 142) give a characterization which removes the assumption of a finite variance in (a). Let $X_0, X_1, ..., X_n$ be n + 1 iid random variables with common mean zero. Let

$$Y_1 = X_0 - a_1X_1 - ... - a_nX_n, \quad Y_2 = X_0 + b_1X_1 + ... + b_nX_n ,$$

where $a_i b_i > 0$ and $|b_i| < 1$ (i = 1, ..., n), and such that $E(Y_1|Y_2) = 0$. Then the parent distribution is normal if and only if $\sum_{i=1}^{n} a_i b_i = 1$.

(d) Rao (1967, p. 8) extended the characterization in (a) as follows. Let $X_1, ..., X_n$ (n ≥ 3) be mutually independent (but not necessarily identically distributed) rvs, and suppose that there exist n linear statistics

$$Y_i = a_{i1}X_1 + ... + a_{in}X_n, \quad i = 1, ..., n ,$$

such that the determinant $|(a_{ij})| \neq 0$ and $a_{11},...,a_{1n}$ are all nonzero. If $E(X_i) = 0$ (i = 1, ..., n) and $E(Y_1|Y_2, ...,Y_n) = 0$, then $X_1, ..., X_n$ are each normally distributed.

[4.3.6] The following result may be compared with that in *[4.1.6]*. Let $X_1, X_2,...$ be a sequence of iid nondegenerate rvs with mean zero and having moments of all orders. Suppose that $\{a_i\}$ and $\{b_i\}$ are sequences of

Characterizations

real constants such that $\sum_{i=1}^{\infty}|a_i|$ converges, $\sum_{i=1}^{\infty} b_i X_i$ converges with probability one, and

$$E\left(\sum_{i=1}^{\infty} a_i X_i \mid \sum_{i=1}^{\infty} b_i X_i\right) = 0.$$

Then the parent population is normal (Kagan et al., 1973, p. 190).

[4.3.7] *Linear Regression and Homoscedasticity.* Let $X_1, ..., X_n$ be mutually independent rvs with finite variances σ_i^2 ($1 = 1, ..., n$). Let $L_1 = a_1 X_1 + \cdots + a_n X_n$ and $L_2 = b_1 X_1 + \cdots + b_n X_n$ where $a_i b_i \neq 0$ ($i = 1, ..., n$). Then

$$E(L_1|L_2) = \alpha + \beta L_2 \quad \text{and} \quad \text{Var}(L_1|L_2) = \sigma_0^2 \quad \text{(constant)}$$

for some constants α and β, if and only if
  (i) X_i is normal whenever $b_i \neq \beta a_i$, and
  (ii) $\beta = (\sum' a_i b_i \sigma_i^2)/(\sum' a_i^2 \sigma_i^2)$, $\sigma_0^2 = \sum'\{(b_i - \beta a_i)^2 \sigma_i^2\}$

where $\sum'$ denotes summation over all i such that $b_i \neq \beta a_i$ (Kagan et al., 1973, p. 191; Lukacs and Laha, 1964, p. 123).

[4.3.8] Let $X_1, ..., X_n$ be iid random variables with finite variance σ^2. If the conditional expectation of any unbiased quadratic estimator of $c\sigma^2$ ($c \neq 0$), given the fixed sum $X_1 + \cdots + X_n$, does not involve the latter, the distribution of the parent population is normal (Laha, 1953, p. 228). See also *[4.1.2]*(a).

[4.3.9] We shall require the notion of constancy of regression in the remainder of this section. A rv Y with finite mean is defined to have *constant regression* on a rv X if $E(Y|X) = E(Y)$ with probability one with respect to the probability distribution of X (Lukacs, 1956, p. 204). Another way of stating this is that the conditional expectation of Y, given that $X = x$, is free of x almost surely.

[4.3.10] Let $X_1, ..., X_n$ be a random sample from a certain population with finite variance. Let L and Q be linear and quadratic statistics, defined by

$$L = X_1 + \ldots + X_n, \quad Q = \sum_{i=1}^{n} \sum_{j=1}^{n} a_{ij} X_i X_j + \sum_{i=1}^{n} b_i X_i,$$

where $\sum_{i=1}^{n} a_{ii} \neq 0$, $\sum_{i=1}^{n} \sum_{j=1}^{n} a_{ij} = 0$, and $\sum_{i=1}^{n} b_i = 0$. Then the parent population is normal if and only if Q has constant regression on L (Kagan et al., 1973, p. 215; Mathai and Pederzoli, 1977, p. 46; Lukacs and Laha, 1964, p. 106). See also Lukacs (1956, p. 205).

[4.3.11] (a) Let $X_1, \ldots, X_n$ be iid rvs with moments up to order m. Define the linear statistics

$$L_1 = \sum_{i=1}^{n} a_i X_i, \quad L_2 = \sum_{i=1}^{n} b_i X_i,$$

where $\sum_{i=1}^{n} a_i b_i = 0$ and $a_i a_j > 0$ (i, j = 1, ..., n). Then L_2^2 has constant regression on L_1 if and only if the parent distribution is normal (Cacoullos, 1967a, pp. 399-400; see also Cacoullos, 1967b, p. 1895).

(b) Let $X_1, \ldots, X_n$ be as in (a), with a symmetric distribution. Define L_1 and L_2 as in (a), where $a_i \neq 0$ (i = 1, ..., n) and $\sum_{i=1}^{n} a_i b_i = 0$. Then L_2^2 has constant regression on L_1 if and only if the parent distribution is normal (Cacoullos, 1967a, p. 401).

[4.3.12] Let $X_1, \ldots, X_n$ be iid random variables such that $E(|X_1|^p) < \infty$ $(p \geq 2)$. Let $P = P(X_1, \ldots, X_n)$ be a homogeneous polynomial statistic of degree p which is an unbiased estimator of the pth cumulant κ_p. Then P has constant regression on the sum $X_1 + \ldots + X_n$ if and only if the underlying population is normal (Kagan et al., 1973, p. 216).

4.4 INDEPENDENCE OF SOME STATISTICS

[4.4.1] A natural question to ask is whether the independence of the sample mean $\bar{X}$ and some function of $X_1 - \bar{X}, \ldots, X_n - \bar{X}$ in a random sample from a population (see *[5.2.3]*) characterizes the normal law. For a discussion, see Johnson and Kotz (1970, pp. 51-52); particular cases are presented in *[4.2.1]*, *[4.4.2]* and *[4.4.6]*.

[4.4.2] Let $X_1, \ldots, X_n$ be iid random variables and let $M_p = \sum_{i=1}^{n}(X_i - \bar{X})^p / n$, the pth sample moment (p = 2, 3, ...) Suppose that $(p-1)!$ is not divisible by $(n-1)$; then M_p and $\bar{X}$ are independent if and

Characterizations

only if the parent population is normal (Lukacs and Laha, 1964, p. 101).

[4.4.3] A polynomial $P(x_1, ..., x_n)$ of degree p is said to be *nonsingular* if it contains the pth power of at least one variable and if for all positive integers k, $\Pi(k) \neq 0$, where $\Pi(k)$ is the polynomial formed if we replace each positive power x_s^j by $k(k-1) ... (k-j+1)$ in $P(x_1, ..., x_n)$ (Lukacs and Laha, 1964, p. 96).

Let $X_1, ..., X_n$ be iid random variables, and $P = P(X_1, ..., X_n)$ be a nonsingular polynomial statistic. If P and $X_1 + ... + X_n$ are independent, then the parent distribution is normal (Lukacs and Laha, 1964, p. 98).

[4.4.4] Let X and Y be iid random variables. Let the quotient X/Y follow the Cauchy law distributed symmetrically about zero and be independent of $U = X^2 + Y^2$. Then the rvs X and Y follow the normal law (Seshadri, 1969, p. 258). See also *[5.1.1]*.

[4.4.5] Let $X_0, X_1, ..., X_n$ $(n \geq 2)$ be n+1 independent rvs such that $\Pr(X_i = 0) = 0$ $(i = 0, 1, ..., n)$ and with distributions symmetrical about zero. Let

$$Y_1 = X_1/|X_0|, \quad Y_2 = X_2\sqrt{2}/(X_0^2 + X_1^2)^{1/2}, ...,$$

$$Y_n = X_n\sqrt{n}/(X_0^2 + X_1^2 + ... + X_{n-1}^2)^{1/2}.$$

A necessary and sufficient condition for $X_0, ..., X_n$ to be iid $N(0, \sigma^2)$ rvs is that $Y_1, ..., Y_n$ are mutually independent rvs having Student's t distribution with 1, 2, ..., n degrees of freedom, respectively (Kotlarski, 1966, p. 603; Mathai and Pederzoli, 1977, pp. 69, 71).

[4.4.6] For a definition of k-statistics in the following characterization, see Stuart and Ord (1987, Sec. 12.6). Let $X_1, ..., X_n$ be a random sample from a population with cdf G(x), and let p be an integer greater than one, such that the pth moment of G exists. The distribution G is normal if and only if the k-statistic of order p is independent of the sample mean $\overline{X}$ (Lukacs, 1956, p. 201).

4.5 CHARACTERISTIC FUNCTIONS AND MOMENTS

[4.5.1.1] A distribution is characterized if $E[\{\min(X_1, X_2, ..., X_n)\}^k]$ is specified for all $k \geq 1$, where $X_1, X_2, ..., X_n$ is a random sample from the distribution of interest (Chan, 1967, pp. 950-951). See also Arnold and Meeden (1975, pp. 754-758).

[4.5.1.2] A rv X has a $N(\theta, 1)$ distribution if and only if, for all real-valued differentiable functions $f(\cdot)$ such that $E|f'(X)| < \infty$, the identity

$$E\{(X - \theta)f(X)\} = E\{f'(X)\}$$

holds (Prakasa Rao, 1979, pp. 903-909).

[4.5.1.3] Let X be a continuous rv with pdf $f(\cdot)$, mean μ, and variance σ^2. Let g be real-valued, absolutely continuous and differentiable. Let

$$U(X) = \sup_g \left\{\frac{\text{Var}[g(X)]}{\sigma^2 \, E[g'(X)]^2}\right\}, \quad V(X) = \inf_g \left\{\frac{\text{Var}[g(X)]}{\sigma^2 \, E^2[g'(X)]}\right\}.$$

Then if $U(X) \equiv 1$, X has a normal distribution (Borovkov and Utev, 1983, pp. 219-228). If $V(X) \equiv 1$, then X has a normal distribution (Cacoullos and Papathanasiou, 1989, pp. 351-356). See also Konwar (1991, pp. 287-295). These properties resulted from research that was stimulated by the publication of the inequality presented in [2.3.5].

[4.5.2] Let m_1, m_2, and n be integers such that $0 \leq m_i \leq n$, $n > 1$ ($i = 1, 2$). Suppose $\psi_1(t)$ and $\psi_2(t)$ are two characteristic functions such that

$$\psi_1(t)\psi_2(t) = \exp(ict)\{\psi_1(t/\sqrt{n})\}^{m_1}\{\psi_1(-t/\sqrt{n})\}^{n-m_1}$$

$$\cdot \{\psi_2(t/\sqrt{n})\}^{m_2}\{\psi_2(-t/\sqrt{n}/)\}^{n-m_2}$$

for all values of t. Then $\psi_1(t)$ and $\psi_2(t)$ correspond to normal distributions (Blem, 1956, pp. 59, 61).

Characterizations

[4.5.3] The following result is useful in the study of particle size distributions. Let Y be a rv with pdf p(x) and moment-generating function M(t), and let X be the rv with pdf g(x), where

$$g(x) = \exp(tx)p(x)/M(t), \quad -\infty < x < \infty.$$

Then $\text{Var}(X) = \text{Var}(Y)$ for all values of t if and only if Y has a normal pdf (Ziegler, 1965, p. 1203). Under these conditions, if Y is a $N(\mu, \sigma^2)$ rv, then X has a $N(\mu + \sigma^2 t, \sigma^2)$ distribution.

[4.5.4] Let ψ be a nontrivial characteristic function such that, in some neighborhood of the origin where it does not vanish,

$$\psi(t) = \prod_{j=1}^{\infty} [\{\psi(\pm \beta_j t)\}^{v_j}], \quad 0 < \beta_j < 1, \quad v_j > 0 \ (j = 1, 2, ...).$$

Then (a) $\sum_{j=1}^{\infty} v_j \beta_j^2 \leq 1$ and (b) ψ is the characteristic function of a normal law, if and only if $\sum_{j=1}^{\infty} v_j \beta_j^2 = 1$ (Kagan et al., 1973, p. 178).

[4.5.5] The rv W^2 has a χ_1^2 distribution (see *[5.3]*) if and only if the characteristic function of W, $\psi(t)$, satisfies the equation

$$\psi(t) + \psi(-t) = 2\exp(-t^2/2)$$

(Geisser, 1973, p. 492). The distribution of W is then $N(0, 1)$ (see *[2.3.1]*).

[4.5.6] Let Z_1, Z_2, and Z_3 be three mutually independent rvs, symmetrically distributed about zero with cdf continuous at zero. Then the joint characteristic function of the ratios Z_1/Z_3 and Z_2/Z_3 is $\exp[-\{t_1^2 + t_2^2\}^{1/2}]$ if and only if Z_1, Z_2, and Z_3 are iid normal rvs with mean zero (Kotz, 1974, p. 47; Pakshirajan and Mohan, 1969, p. 532).

[4.5.7] Let X, Y and Z be iid rvs, having a finite absolute moment of specified order p, $p > 0$, p odd. Suppose for some positive constant A that the equation

$$E(|aX + bY + cZ|^p) = A(a^2 + b^2 + c^2)^{p/2}$$

is satisfied for any real numbers a, b and c. Then X, Y and Z each are normally distributed with mean zero (Braverman, 1985, pp. 465-474).

4.6 CHARACTERIZATIONS FROM PROPERTIES OF TRANSFORMATIONS

[4.6.1] Let $x_1, x_2, ..., x_n$ be the Cartesian coordinates, not all zero, of a point in Euclidean space of n dimensions. We can transform to *polar coordinates* $(r, \theta_1, \theta_2, ..., \theta_{n-1})$, if $n \geq 2$, as follows: Define the transformation T by

$$x_1 = r \cos \theta_1, \quad x_2 = r \sin \theta_1 \cos \theta_2,$$
$$x_3 = r \sin \theta_1 \sin \theta_2 \cos \theta_3, \ldots,$$
$$x_{n-1} = r \sin \theta_1 \sin \theta_2 \ldots \sin \theta_{n-2} \cos \theta_{n-1},$$
$$x_n = r \sin \theta_1 \ldots \sin \theta_{n-2} \sin \theta_{n-1}, \quad -\infty < x_i < \infty \ (i = 1, ..., n),$$
$$r \geq 0, 0 \leq \theta_i < 2\pi \quad (i = 1, ..., n-1).$$

If $X_1, ..., X_n$ are rvs with joint pdf $p_1(x_1, ..., x_n)$, transformation T leads to *polar rvs* $R, \Theta_1, ..., \Theta_{n-1}$ with joint pdf $p_2(r, \theta_1, ..., \theta_{n-1})$ given by

$$p_2(r, \theta_1, ..., \theta_{n-1}) = p_1(x_1, ..., x_n) r^{n-1} |\sin^{n-2} \theta_1 \sin^{n-3} \theta_2 \ldots \sin \theta_{n-2}|.$$

We can speak of R as the *radial variable*, and of $\Theta_1, ..., \Theta_{n-1}$ as the *angular variables;* note that $R^2 = X_1^2 + ... + X_n^2$.

(a) Let $X_1, ..., X_n$ be a random sample from a certain population, which is absolutely continuous. Then the joint pdf of the sample, $p_1(x_1, ..., x_n)$, is constant on the "sphere" $x_1^2 + x_2^2 + ... + x_n^2 = r^2$ if and only if the parent population is standard normal (Bartlett, 1934, pp. 327-340; Stuart and Ord, 1987, Sec. 15.25).

(b) Let $X_1, ..., X_n$ $(n \geq 2)$ be rvs and suppose that under T the ratio

$$p_2(r, \theta_1, ..., \theta_{n-1})/(r^{n-1} \sin^{n-2} \theta_1 ... \sin \theta_{n-2})$$

is well defined and nonzero everywhere, continuous in r and equal to $p_1(x_1,$

Characterizations 97

..., x_n), which is in turn continuous in each x_i ($i = 1,...,n$). Then $X_1, ..., X_n$ are mutually independent and R is independent of $\Theta_1, \Theta_2, ..., \Theta_{n-1}$ if and only if $X_1, ..., X_n$ are iid $N(0, \sigma^2)$ random variables (Tamhanker, 1967, p. 1924; Kotz, 1974, p. 50; Johnson and Kotz, 1970, p. 52). This result essentially gives conditions under which the independence of the radial and (joint) angular variables characterizes normality. Tamhanker gives the transformed pdf $p_2(r, \theta_1, ..., \theta_n)$ without absolute value notation.

For a closely related characterization see Alam (1971, p. 523).

[4.6.2] Let X_1 and X_2 be iid random variables with (continuous) pdf $g(x)$. Let Y_1 and Y_2 be rvs defined by the transformation

$$Y_1 + iY_2 = (X_1 + iX_2)^k (X_1^2 + X_2^2)^{-(k-1)/2}, \quad i = \sqrt{-1},$$

where k is an integer not less than 2. Then X_1 and X_2 are iid $N(0, \sigma^2)$ rvs if and only if Y_1 and Y_2 are independent with the same distribution as X_1 and X_2 (Beer and Lukacs, 1973, pp. 100, 103; Kotz, 1974, pp. 49-50).

[4.6.3] Let X_1 and X_2 be iid random variables with (continuous) pdf $g(x)$. Let Y_1 and Y_2 be rvs defined by the transformation

$$Y_1 = X_1 \cos\{a(X_1^2 + X_2^2)\} + X_2 \sin\{a(X_1^2 + X_2^2)\},$$

$$Y_2 = -X_1 \cos\{a(X_1^2 + X_2^2)\} + X_2 \cos\{a(X_1^2 + X_2^2)\},$$

where a is a nonzero constant. Then X_1 and X_2 are iid $N(0, \sigma^2)$ rvs if and only if Y_1 and Y_2 are independent with the same distribution as X_1 and X_2 (Beer and Lukacs, 1973, pp. 102, 106-107).

[4.6.4] (a) The following result was designed for testing normality. Let $n = 2k + 3$, where $k \geq 2$; let $X_1, ..., X_n$ be iid random variables with (unknown) mean μ and (unknown) variance σ^2, $|\mu| < \infty$, $\sigma^2 > 0$. Let

$$Z_1 = (X_1 - X_2)/\sqrt{2}, \quad Z_2 = (X_1 + X_2 - 2X_3)\sqrt{3.2}, \quad ...,$$

$$Z_{n-1} = \{X_1 + ... + X_{n-1} - (n-1)X_n\}/\sqrt{n(n-1)},$$

$$Z_n = X_1 + \ldots + X_n \ ;$$
$$Y_1 = Z_1^2 + Z_2^2, \quad Y_2 = Z_3^2 + Z_4^2, \ \ldots,$$
$$Y_{k+1} = Z_{n-2}^2 + Z_{n-1}^2, \quad S_{k+1} = Y_1 + Y_2 + \ldots + Y_k \ ,$$

where $k + 1 = \frac{1}{2}(n-1)$. Finally, let

$$U_{(r)} = (Y_1 + \ldots + Y_r)/S_{k+1} \ , \quad r = 1, \ldots, k.$$

Then $U_{(1)}, U_{(2)}, \ldots, U_{(k)}$ behave like the k order statistics of k iid random variables having a uniform distribution over $(0, 1)$, if and only if the parent distribution of $X_1, \ldots, X_n$ is $N(\mu, \sigma^2)$ (Csörgö and Seshadri, 1971, pp. 333-339; Kotz, 1974, p. 49).

(b) The sources in (a) also give an analogous characterization which is suitable when μ is known; here $n = 2k$, $Z_i = X_i - \mu$ ($i = 1, \ldots, n$), the Y_j are similarly defined, with $k + 1$ replaced by k; $S_k = Y_1 + \ldots + Y_k$ and $U_{(r)} = (Y_1 + \ldots + Y_r)/S_k$, where now $r = 1, \ldots, k-1$. The conclusion is directly analogous, replacing k by $k - 1$.

[4.6.5] Suppose that a nonsingular linear transformation

$$\begin{pmatrix} W \\ Z \end{pmatrix} = \begin{pmatrix} a & b \\ c & d \end{pmatrix} \begin{pmatrix} X \\ Y \end{pmatrix}$$

exists of a pair of independent rvs (X, Y) to another pair of independent rvs (W, Z). Then

i) each variable has a finite variance,

ii) all four variables are normally distributed or all are degenerate (Lancaster, 1987, pp. 101-106). This characterization does not require the assumption of finite variances.

[4.6.6] In the following two characterizations X and Y are iid random variables, and $U = \min(X, Y)$. For both results see Ahsanullah and Hamedani (1988, pp. 95-99).

(a) X and Y are symmetric about zero, with pdf $g(\cdot)$. If U^2 is distributed as chi-square with one degree of freedom (χ_1^2), then X and Y are standard normal.

(b) X and Y have an absolutely continuous distribution. If U^2 is distributed as χ_1^2 and X/Y as Cauchy with pdf $[\pi(1 + x^2)]^{-1}$, then X and Y are standard normal.

4.7 SUFFICIENCY, ESTIMATION, AND TESTING

The following material concerns the role of the normal distribution in statistical inference rather than distributional properties, but is given here for convenience. Terms such as information, sufficiency, completeness, and admissibility are defined and their properties are developed in sources such as Bickel and Doksum (1977), Ferguson (1967), and Lehmann (1986). See also Mood et al. (1974) for a discussion of concepts used in the theory of estimation and testing of hypotheses.

[4.7.1] It turns out that, among all continuous distributions depending on a location parameter, and having certain regularity conditions, the location parameter is estimated worst of all in the case of normal samples. This important result is due to Kagan et al. (1973, pp. 405-406). Formally, let $p(x - \theta)$ denote the pdf of a family of continuous distributions with location parameter θ; assume $p(\cdot)$ to be continuously differentiable, having fixed finite variance σ^2, and such that $|x|p(x) \to 0$ as $|x| \to \infty$. The *Fisher information measure* $I_p(\theta)$ is given here by

$$I_p(\theta) = E[\{\partial \log p(x - \theta)/\partial \theta\}^2|\theta] = I_p(0), = I_p,$$

say. Then in the class of all such densities p, $\min_p I_p$ is attained when X has a $N(\theta, \sigma^2)$ distribution.

[4.7.2] If $p(x)$ is a pdf, then a measure of its closeness to a uniform distribution (Mathai and Pederzoli, 1977, p. 15) is given by its *entropy*, defined as

$$J = -\int_a^b \log p(x) \cdot p(x) \, dx,$$

where $\Pr(a < X < b) = 1$. Let X be a rv with pdf $p(x)$, and with given mean and variance. Among all such continuous distributions, the $N(\mu, \sigma^2)$

pdf maximizes the entropy (Mathai and Pederzoli, 1977, p. 16; Kagan et al., 1973, pp. 408-410).

[4.7.3] Let p(x; θ) be a pdf indexed by a parameter θ. If $\theta_1 \neq \theta_2$, but p($\cdot$; θ_1) = p($\cdot$; θ_2), where θ, θ_1, θ_2 lie in the parameter space, we say that θ is *unidentifiable*; otherwise, θ is *identifiable* (Bickel and Doksum, 1977, p. 60).

Let (X_1, X_2) be (jointly) normally distributed, and let Y be a rv independent of (X_1, X_2). Consider a linear structure given by

$$U = Y + X_1, \quad V = \beta Y + X_2.$$

Then the parameter β is identifiable if and only if Y is not normally distributed (Lukacs and Laha, 1964, p. 126).

[4.7.4] (a) The following result modernizes a result of Gauss (1809; 1906). Let $\{G(x - \theta); |\theta| < \infty\}$ be the cdf of a location parameter family of absolutely continuous distributions on the real line and let the pdf g(x) be lower semicontinuous at zero. If for all random samples of size 2 and 3, a maximum likelihood estimator (MLE) of θ is $\overline{X}$, then G($\cdot$) is a normal distribution with mean zero (Teicher, 1961, p. 1215).

(b) Let $\{G(x/\sigma); \sigma > 0\}$ be a scale parameter family of absolutely continuous distributions with the pdf g(x) satisfying
  (i) g($\cdot$) is continuous on $(-\infty, \infty)$,
  (ii) $\lim_{y \to 0}[g(\lambda y)/g(y)] = 1$, whenever $\lambda > 0$.

If for all sample sizes a MLE of σ is $\left(\sum_{i=1}^{n} X_i^2/n\right)^{1/2}$, then G($\cdot$) has a N(0, 1) distribution (Teicher, 1961, p. 1221).

[4.7.5] (a) Let $X_1, X_2, ..., X_n$ (n $\geq$ 2) be mutually independent nondegenerate rvs; θ is a location parameter, so that X_i has cdf $G_i(x - \theta)$, (i = 1, 2, ..., n), $-\infty < \theta < \infty$. A necessary and sufficient condition for $b_1 X_1 + ... + b_n X_n$ (b_1, ..., b_n all nonzero) to be a sufficient statistic for θ is that each X_i is a normal variable with variance a/b_i for some constant a (Kelkar and Matthes, 1970, p. 1086).

(b) Let $X_1, ..., X_n$ (n $\geq$ 2) be defined as in (a). Then the sample mean $\overline{X}$ is sufficient for θ if and only if $X_1, ..., X_n$ are iid normal rvs

(Dynkin, 1961, pp. 17-40; Mathai and Pederzoli, 1977, p. 75). This is the case in (a) for which $b_1 = \ldots = b_n = 1/n$.

[4.7.6] (a) Let $X_1, \ldots, X_n$ ($n \geq 2$) be mutually independent rvs with a common scale parameter σ, so that X_i has cdf $G_i(x/\sigma)$ ($i = 1, \ldots, n$), $\sigma > 0$. Let G_i be absolutely continuous in a neighborhood of zero, and let $\partial g_i/\partial x$ be nonzero and continuous when $x = 0$. Then if $\sum_{i=1}^n X_i^2$ is sufficient for σ, $X_1, \ldots, X_n$ have normal distributions with mean zero (Kelkar and Matthes, 1970, p. 1087).

(b) Let $X_1, X_2, \ldots, X_n$ ($n \geq 4$) be iid random variables, each X_i having a location parameter θ ($-\infty < \theta < \infty$) and a scale parameter σ ($\sigma > 0$). Let $\overline{X} = \sum_{i=1}^n X_i/n$ and $S^2 = \sum_{i=1}^n (X_i - \overline{X})^2/(n-1)$. If $(\overline{X}, S^2)$ is a sufficient statistic for (θ, σ), then each X_i has a normal distribution (Kelkar and Matthes, 1970, p. 1087).

[4.7.7] If $X_1, X_2, \ldots, X_n$ are independent random variables such that their joint density function involves an unknown location parameter θ, then a necessary and sufficient condition for $\sum_{i=1}^n b_i X_i$ to be a boundedly complete sufficient statistic is that $b_i > 0$ and that X_i is a normal variable with mean θ and variance proportional to $1/b_i$ ($i = 1, \ldots, n$) (from Basu, 1955, p. 380). In particular, $\overline{X}$ is boundedly complete sufficient if and only if $X_1, \ldots, X_n$ are iid normal rvs.

[4.7.8] Let $X_1, \ldots, X_n$ be continuous iid random variables with pdf $g(x)$ and variance σ^2. Let $\overline{X}$ be the sample mean, and $h(X_1, \ldots, X_n)$ be any distribution-free unbiased estimator of σ^2.

(i) If $\overline{X}$ and $h(X_1, \ldots, X_n)$ are independent, $g(\cdot)$ is the normal pdf.

(ii) If g is the normal pdf, $\overline{X}$ and $h(X_1, \ldots, X_n)$ are uncorrelated (Shimizu, 1961, p. 53).

[4.7.9] The following characterizations hold under a *quadratic loss* function $(\hat{\theta} - \theta)^2$, where $\hat{\theta}$ estimates a parameter θ.

(a) Let θ be a location parameter and $X_1, \ldots, X_n$ ($n \geq 3$) be mutually independent rvs such that X_j has cdf $G_j(x - \theta)$, $E(X_j) = 0$ and $Var(X_j) = \sigma_j^2 < \infty$; $j = 1, \ldots, n$. The optimal linear estimator of θ under quadratic loss, given by

$$\hat{L} = \left(\sum_{i=1}^n X_i/\sigma_i^2\right)\left(\sum_{i=1}^n 1/\sigma_i^2\right)^{-1},$$

is admissible in the class of all unbiased estimators of θ if and only if all the cdfs $G_j(\cdot)$ are normal (j = 1, ..., n) (Kagan et al., 1973, p. 227).

(b) In particular in (a), when $X_1, ..., X_n$ are iid random variables (n $\geq$ 3), $\hat{L} = \overline{X}$, the sample mean. Hence the admissibility under quadratic loss of $\overline{X}$ in the class of unbiased estimators of θ characterizes the parent distribution as normal (Kagan et al., 1973, p. 228).

(c) Let $X_1, ..., X_n$ and $Y_1, ..., Y_n$ be mutually independent rvs (n $\geq$ 3) with zero means and finite variances, where X_j has cdf $G_j(x - \theta - \Delta)$, Y_j has cdf $H_j(x - \theta)$, (j=1, ..., n) and $|\theta| < \infty$, $|\Delta| < \infty$. In order that an estimator of Δ of the form $\sum a_j(X_j - Y_j)$ be admissible under quadratic loss in the class of unbiased estimators of Δ, it is necessary and sufficient that $X_1, ..., X_n, Y_1, ..., Y_n$ all be normally distributed and that $\text{Var}(X_j)/\text{Var}(Y_j)$ be constant; j = 1, ..., n (Kagan et al., 1973, p. 228).

(d) Kagan et al. (1973, Chap. 7) give further characterizations of normality through admissibility of certain estimators under quadratic loss; these include polynomials, least squares estimators, and linear estimators of shift parameters in autoregressive processes.

[4.7.10] *Absolute error loss* for an estimator $\hat{\theta}$ of θ is $|\hat{\theta} - \theta|$. Let $X_1,...,X_n$ (n $\geq$ 6) be a random sample from a unimodal distribution with cdf $f(x - \theta)$ and a continuously differentiable pdf. Then the sample mean $\overline{X}$ is admissible as an estimator of θ under absolute error loss if and only if the parent distribution is normal (Kagan et al., 1973, p. 247; see also p. 251).

[4.7.11] Let $X_1, ..., X_n$ be a random sample (n $\geq$ 3) from a population with continuous cdf $F(x - \theta)$ and finite variance. We wish to test the null hypothesis H_0: $\theta = 0$ against the composite alternative H_1: $\theta > 0$. For every α (0 < α < 1), the test having critical region determined by the relation

$$\overline{X} > c_\alpha, \quad c_\alpha = \max\{c: \Pr(\overline{X} > c|\theta = 0) = \alpha\}$$

is uniformly most powerful of size α for testing H_0 against H_1 if and only if the parent population is normal (Kagan et al., 1973, p. 451).

[4.7.12] Consider the general linear model

$$Y = X\beta + e$$

for n observations $Y_1, ..., Y_n$ denoted by column vector **Y** and n iid random errors $e_1, ..., e_n$, denoted by column vector **e** with $Ee_i = 0$ and $0 < Var(e_i) < \infty$, $i = 1, ..., n$; **X** is the design matrix and β a vector of p unknown parameters. Stepniak (1991, pp. 115-117) simplified the conditions of Theorem 7.4.1 of Kagan et al. (1973, p. 227) as follows:

Let $\hat{\psi}$ be the least squares estimator of an estimable linear parametric function $c'\beta$ in this set up, with $n > 2$, such that the rank of X is not sensitive to the deletion of any two rows. Then the admissibility of $\hat{\psi}$ among all unbiased estimators of $c'\beta$ with quadratic loss characterizes the normal law.

4.8 MISCELLANEOUS CHARACTERIZATIONS

We begin this section with some characterizations derived through order statistics.

[4.8.1] Let $X_1, ..., X_n$ be a random sample from a continuous population, and let $Y_i = \sum_{j=1}^{n} a_{ij} X_j$ ($i = 1, ..., n$) be an orthogonal transformation. Further, let $W_1 = \max(X_1, ..., X_n) - \min(X_1, ..., X_n)$ and $W_2 = \max(Y_1, ..., Y_n) - \min(Y_1, ..., Y_n)$ be the corresponding sample ranges. Then W_1 and W_2 are identically distributed if and only if the parent population of $X_1, ..., X_n$ is normal (Kotz, 1974, p. 50).

[4.8.2] In what follows below, $X_1, ..., X_n$ are iid nontrivial rvs, having cdf G and finite variance. The order statistics are given by $X_{(1)} \leq X_{(2)} \leq ... \leq X_{(n)}$, $n \geq 2$, and Φ is the N(0, 1) cdf. These characterizations are due to Govindarajulu (1966, pp. 1011-1015).

(a) $E[X_{(n)}^2 - X_{(n-1)} X_{(n)}] = 1$ if and only if there is an extended real number A ($-\infty \leq A < \infty$) such that

$$G(x) = \begin{cases} 0, & x \leq A \\ \{\Phi(x) - \Phi(A)\}/\{1 - \Phi(A)\}, & A < x < \infty, \end{cases}$$

a normal distribution truncated from below.

(b) $E[X_{(1)}^2 - X_{(1)}X_{(2)}] = 1$ if and only if there is an extended real number $B(-\infty < B \leq \infty)$ such that

$$G(x) = \begin{cases} \Phi(x)/\Phi(B), & -\infty < x < B, \\ 1, & x \geq B, \end{cases}$$

a normal distribution truncated from above.

(c) Assume that $G(0) = 1$. Then $\sum_{i=1}^{n} E[X_{(i)}X_{(n)}] = 1$ if and only if

$$G(x) = \begin{cases} 2\Phi(x), & -\infty < x \leq 0, \\ 1, & x > 0, \end{cases}$$

a normal distribution, folded on the left.

(d) Suppose that the parent distribution has mean at zero. Then for $i = 1, 2, ..., n$, $\sum_{j=1}^{n} E[X_{(i)}X_{(j)}] = 1$ if and only if $G(x) = \Phi(x)$, $|x| < \infty$.

[4.8.3] (a) Let X_1, X_2, X_3 be mutually independent rvs, symmetrical about the origin, and such that $\Pr(X_k = 0) = 0$, $k = 1, 2, 3$. Let $Y_1 = X_1/X_3$, $Y_2 = X_2/X_3$. The necessary and sufficient condition for X_1, X_2, and X_3 to be normally distributed with a common standard deviation σ is that the joint distribution of (Y_1, Y_2) is the bivariate Cauchy distribution with joint pdf $g(y_1, y_2)$, where

$$g(y_1, y_2) = [2\pi(1 + y_1^2 + y_2^2)^{3/2}]^{-1}, \quad -\infty < y_1, y_2 < \infty$$

(Kotlarski, 1967, p. 75).

(d) Define X_1, X_2, and X_3 as in (a); and let

$$V_1 = X_1(X_1^2 + X_2^2)^{-1/2}, \quad V_2 = (X_1^2 + X_2^2)^{1/2}(X_1^2 + X_2^2 + X_3^2)^{-1/2}.$$

Then X_1, X_2, and X_3 are iid normally distributed rvs if and only if V_1 and V_2 are independent with respective pdfs $h_1(u)$ and $h_2(u)$ given by

Characterizations

$$h_1(u) = \pi^{-1}(1-u^2)^{-1/2}, \quad |u| < 1 \; ; \; h_2(u) = u(1-u^2)^{-1/2}, \quad 0 < u < 1$$

(Kotlarski, 1967, pp. 75-76).

[4.8.4] Let f(x) be a differentiable pdf. Then f(·) is normal if and only if

$$f'(x) = -f(x)(\gamma x + \delta)$$

for some positive γ and some δ (Hombas, 1985, pp. 387-388). Geometrically, the subtangent at each point (x, f(x)) is $1/(\gamma x + \delta)$.

4.9 NEAR-CHARACTERIZATIONS

Suppose that a random sample $X_1, ..., X_n$ has some property P which characterizes a distribution G (or family of distributions). The question then arises: if the sample *nearly* has the property P, is the parent distribution nearly G, and vice versa? If the answer is yes, then (with formal definitions) the characterization is termed *stable*. Work in this field is limited, but we can give a few results.

[4.9.1] Two rvs X and Y are ϵ-independent if, for all x and y,

$$|\Pr(X < x, Y < y) - \Pr(X < x)\Pr(Y < y)| < \epsilon .$$

The rv X is ϵ-normal with parameters μ and σ if for its cdf G

$$|G(x) - \Phi\{(x-\mu)/\sigma\}| < \epsilon, \quad |x| < \infty$$

(Nhu, 1968, p. 299). See also Meshalkin (1968, p. 1747).

[4.9.2] (a) Let X_1 and X_2 be iid random variables with mean zero, variance one, and $E|X_i|^3 < M < \infty$ (i = 1, 2). If $X_1 + X_2$ and $X_1 - X_2$ are ϵ-independent, then the cdf G of X_1 and X_2 satisfies the inequality

$$\sup_x |G(x) - \Phi(x)| < C_1 \, \epsilon^{1/3} ,$$

where C_1 depends on M only (Meshalkin, 1968, p. 1748; Kagan et al., 1973, p. 298; see also Nhu, 1968, p. 300). This result relates to that of Bernstein

(1941), given in [4.1.4](a).

(b) Let X_1 and X_2 be defined as in (a), with cdf G, and let H be the cdf of $(X_1 + X_2)/\sqrt{2}$. If $\sup_x |G(x) - H(x)| \leq \epsilon$, then

$$\sup_x |G(x) - \Phi(x)| < C_2 \epsilon^{1/3},$$

where C_2 depends on M only (Meshalkin, 1968, p. 1748). This result relates to that of Cramér (1936), given in [4.1.1](a).

[4.9.3] Let $X_1, ..., X_n$ be iid random variables with mean μ, variance σ^2, and $E[|X_i|^{2(1+\delta)}] = \beta_\delta < \infty$; $0 < \delta \leq 1$. If $\bar{X}$ and S^2 are ϵ-independent, then the parent population is $A(\epsilon)$-normal with parameters μ and σ ([4.9.1]), where $A(\epsilon) = C_3/\sqrt{\log(1/\epsilon)}$ and C_3 depends only on μ, σ, δ, and β_δ (Nhu, 1968, p. 300). This result relates to the fundamental characterization in [4.2.1](a).

REFERENCES

Ahsanullah, M. (1987). A note on the characterization of the normal distribution, *Biometrika Journal 29*, 885-888. [4.2.5]

Ahsanullah, M. and Hamedani, G. G. (1988). Some characterizations of normal distribution, *Calcutta Statistical Association Bulletin 37*, 95-99. [4.6.6]

Alam, K. (1971). A characterization of normality. *Annals of the Institute of Statistical Mathematics 23*, 523-525. [4.6.1]

Arnold, B. C. and Meeden, G. (1975). Characterization of distributions by sets of moments of order statistics, *Annals of Statistics 3*, 754-758. [4.5.1.1]

Bartlett, M. S. (1934). The vector representation of a sample. *Proceedings of the Cambridge Philosophical Society 30*, 327-340. [4.6.1]

Basu, D. (1955). On statistics independent of a complete sufficient statistic, *Sankhyā 15*, 377-380. [4.7.7]

Beer, S. and Lukacs, E. (1973). Characterizations of the normal distribution by suitable transformations, *Journal of Applied Probability 10*, 100-108. [4.6.2, 3]

Berk, R. H. (1986). Sphericity and the normal law, *Annals of Probability* **14**, 696-701. *[4.1.1]*

Bernstein, S. (1941). Sur une propriété caractéristique de la loi de Gauss, *Transactions of the Leningrad Polytechnic Institute* **3**, 21-22; reprinted (1964) in *Collected Works, Vol. 4*, 314-315. *[4.1.4; 4.9.2]*

Bickel, P. J. and Doksum, K. A. (1977). *Mathematical Statistics: Basic Ideas and Selected Topics*, San Francisco: Holden-Day. *[4.7; 4.7.3]*

Blum, J. R. (1956). On a characterization of the normal distribution, *Skandinavisk Aktuarietidskrift* **39**, 59-62. *[4.5.2]*

Bondesson, L. (1977). The sample variance, properly normalized, is χ^2-distributed for the normal law only, *Sankyā* **A39**, 303-304. *[4.2.5]*

Borovkov, A. A. and Utev, S. A. (1983). On an inequality and a related characterization of the normal distribution, *Theory of Probability and its Applications* **28**, 219-228. *[4.5.1.3]*

Braverman, M. Sh. (1985). The characteristic properties of normal and stable distributions, *Theory of Probability and its Applications* **30**, 465-474. *[4.5.7]*

Cacoullos, T. (1967a). Some characterizations of normality, *Sankhyā* **A29**, 399-404. *[4.3.11]*

Cacoullos, T. (1967b). Characterizations of normality by constant regression of linear statistics on another linear statistic, *Annals of Mathematical Statistics* **38**, 1894-1898. *[4.3.4, 11]*

Cacoullos, T. and Papathanasiou, V. (1989). Characterization of distributions by variance bounds, *Statistics and Probability Letters* **7**, 351-356. *[4.5.1.3]*

Chan, L. K. (1967). On a characterization of distribution by expected values of extreme order statistics, *American Mathematical Monthly* **74**, 950-951. *[4.5.1.1]*

Cramér, H. (1936). Über eine Eigenschaft der normalen Verteilungsfunktion, *Mathematische Zeitschrift* **41**, 405-414. *[4.1.1; 4.9.2]*

Cramér, H. (1946). *Mathematical Methods of Statistics*, Princeton, N.J.: Princeton University Press. *[4.1.1]*

Csörgö, M. and Seshadri, V. (1971). Characterizing the Gaussian and exponential laws via mappings onto the unit interval, *Zeitschrift für Wahrscheinlichkeitstheorie und Verwandte Gebiete* **18**, 333-339. *[4.6.4]*

Darmois, G. (1951). Sur une propriété caractéristique de la loi de probabilité de Laplace, *Comptes Rendus de l'Académie des Sciences*

(Paris) *232*, 1999-2000. *[4.1.4; 4.3.5]*

Diaconis, P., Olkin, I. and Ghurye, S. G. (1977). Review of *Characterization Problems in Mathematical Statistics* (by A. M. Kagan, Yu. V. Linnik, and C. R. Rao), *Annals of Statistics* 5, 583-592. *[Introduction]*

Dynkin, E. B. (1961). Necessary and sufficient statistics for a family of probability distributions, *Selected Translations in Mathematics, Statistics, and Probability* 1, 17-40. *[4.7.5]*

Ferguson, T. S. (1967). *Mathematical Statistics--A Decision Theoretic Approach*, New York: Academic. *[4.7]*

Gauss, C. F. (1809). *Theoria Motus Corporum Coelestium;* reprinted (1906) in *Werke*, Liber 2, Sectio III, pp. 240-244. *[4.7.4]*

Geary, R. C. (1936). Distribution of "Student's" ratio for nonnormal samples, *Journal of the Royal Statistical Society B3*, 178-184. *[4.2.1]*

Geisser, S. (1956). A note on the normal distribution, *Annals of Mathematical Statistics* 27, 858-859. *[4.2.1]*

Geisser, S. (1973). Normal characterizations via the square of random variables, *Sankyā A35, 494-494*. *[4.2.4; 4.5.5]*

Govindarajulu, Z. (1966). Characterization of normal and generalized truncated normal distributions using order statistics, *Annals of Mathematical Statistics* 37, 1011-1015. *[4.8.2]*

Hartman, P. and Wintner (1940). On the spherical approach to the normal law, *American Journal of Mathematics* 62, 759-779. *[4.1.1]*

Hombas, V. C. (1985). Characterizing the normal density as a solution of a differential equation, *Statistica Neerlandica* 39, 387-388. *[4.8.4]*

Johnson, N. L. and Kotz, S. (1970). *Distributions in Statistics: Continuous Univariate Distributions, Vol. 1*, New York: Wiley. *[4.4.1; 4.6.1]*

Johnson, N. L. and Kotz, S. (1972). *Distributions in Statistics: Continuous Multivariate Distributions*, New York: Wiley. *[Introduction]*

Kagan, A. M., Linnik, Yu. V. and Rao, C. R. (1965). On a characterization of the normal law based on a property of the sample average. *Sankyā A27*, 405-406. *[4.3.4]*

Kagan, A. M., Linnik, Yu. V. and Rao, C. R. (1973). *Characterization Problems in Mathematical Statistics* (trans. B. Ramachandran), New York: Wiley. *[4.1.4, 6; 4.2.1, 2, 3; 4.3.2, 3, 4, 6, 7, 10, 12; 4.5.4; 4.7.1, 2, 9, 10, 11; 4.9.2]*

Kaplansky, I. (1943). A characterization of the normal distribution, *Annals of Mathematical Statistics* **14**, 197-198. *[4.2.1]*

Kelkar, D. and Matthes, T. K. (1970). A sufficient statistics characterization of the normal distribution, *Annals of Mathematical Staitistics* **41**, 1086-1090. *[4.2.1; 4.7.5, 6]*

Khatri, C. G. (1979). Characterization of multivariate normality, II: Through linear regressions, *Journal of Multivariate Analysis* **9**, 589-598. *[Introduction]*

Khatri, C. G. and Rao, C. R. (1972). Functional equations and characterization of probability laws through linear functions of random variables, *Journal of Multivariate Analysis* **2**, 162-173. *[Introduction]*

Khatri, C. G. and Rao, C. R. (1976). Characterizations of multivariate normality, I: Through independence of some statistics, *Journal of Multivariate Analysis* **6**, 81-94. *[Introduction]*

Konwar, R. M. (1991). On characterizations of distributions by mean absolute deviation and variance bounds, *Annals of the Institute of Statistical Mathematics* **43**, 287-295. *[4.5.1.3]*

Kotlarski, I. (1966). On characterizing the normal distribution by Student's law, *Biometrika* **53**, 603-606. *[4.4.5]*

Kotlarski, I. (1967). On characterizing the gamma and the normal distribution, *Pacific Journal of Mathematics* **20**, 69-76. *[4.8.3]*

Kotz, S. (1974). Characterization of statistical distributions: A supplement to recent surveys, *International Statistical Review* **42**, 39-65. *[4.2.3; 4.5.6; 4.6.1, 2, 4; 4.8.1]*

Laha, R. G. (1953). On an extension of Geary's theorem, *Biometrika* **40**, 228-229. *[4.2.1; 4.3.8]*

Lancaster, H. O. (1954). Traces and cumulants of quadratic forms in normal variables, *Journal of the Royal Statistical Society* **B16**, 247-254. *[4.1.5]*

Lancaster, H. O. (1987). Finiteness of the variances in characterizations of the normal distribution, *Australian Journal of Statistics,* **29**, 101-106. *[4.6.5]*

Lehmann, E. L. (1986). *Testing Statistical Hypotheses* (2nd edn.), New York: Wiley. *[4.7]*

Lukacs, E. (1942). A characterization of the normal distribution, *Annals of Mathematical Statistics* **13**, 91-93. *[4.2.1]*

Lukacs, E. (1956). Characterization of populations by properties of suitable statistics, *Proceedings of the Third Berkeley Symposium on Mathematical Statistics and Probability 2*, 195-214. *[4.1.1, 2, 3, 4; 4.2.1; 4.3.9, 10; 4.4.6]*

Lukacs, E. and Laha, R. G. (1964). *Applications of Characteristic Functions*, New York: Hafner. *[4.2.2; 4.3.7, 10; 4.4.2, 3; 4.7.3]*

Marcinkiewicz, J. (1939). Sur une propriété de la loi de Gauss, *Mathematische Zeitschrift 44*, 612-618. *[4.1.2]*

Mathai, A. M. and Pederzoli, G. (1977). *Characterizations of the Normal Probability Law*, New York: Halsted/Wiley. *[4.1.1, 4; 4.2.2; 4.3.10; 4.4.5; 4.7.2, 5]*

Meshalkin, L. D. (1968). On the robustness of some characterizations of the normal distribution, *Annals of Mathematical Statistics 39*, 1747-1750. *[4.9.1, 2]*

Mood, A. M., Graybill, F. A. and Boes, D. C. (1974). *Introduction to the Theory of Statistics*, New York: McGraw-Hill. *[4.7]*

Nhu, H. H. (1968). On the stability of certain characterizations of a normal population, *Theory of Probability and Its Applications 13*, 299-304. *[4.9.1, 2, 3]*

Pakshirajan, R. P. and Mohan, N. R. (1969). A characterization of the normal law, *Annals of the Institute of Statistical Mathematics 21*, 529-532. *[4.5.6]*

Pathak, P. K. and Phillai, R. N. (1968). On a characterization of the normal law, *Sankyā A30*, 141-144. *[4.3.5]*

Patil, G. P. and Seshadri, V. (1964). Characterization theorems for some univariate probability distributions, *Journal of the Royal Statistical Society B26*, 286-292. *[4.3.1]*

Prakasa Rao, B.L.S. (1979). Characterizations of distributions through some identities, *Journal of Applied Probability 16*, 903-909. *[4.5.1.2]*

Ramasubramaniam, S. (1985). A characterisation of the normal distribution, *Sankhyā A*, 410-414. *[4.3.1]*

Rao, C. R. (1967). On some characterizations of the normal law, *Sankyā A29*, 1-14. *[4.3.5]*

Rao, C. R. (1973). *Linear Statistical Inference and Its Applications* (2nd edn.), New York: Wiley. *[4.2.5]*

Rao, J. N. K. (1958). A characterization of the normal distribution, *Annals of Mathematical Statistics 29*, 914-919; addendum (1959), op.

cit. 30, 610. *[4.2.1]*

Ruben, H. (1974). A new characterization of the normal distribution through the sample variance, *Sankyā A36*, 379-388. *[4.2.5]*

Ruben, H. (1975). A further characterization of normality through the sample variance, *Sankyā A37*, 72-81. *[4.2.5]*

Ruben, H. (1976). A characterization of normality through the general linear model, *Sankyā A38*, 186-189. *[4.2.5]*

Seshadri, V. (1969). A characterization of the normal and Weibull distributions, *Canadian Mathematical Bulletin 12*, 257-260. *[4.4.4]*

Shimizu, R. (1961). A characterization of the normal distribution, *Annals of the Institute of Statistical Mathematics 13*, 53-56. *[4.7.8]*

Shimizu, R. (1962). Characterization of the normal distribution, II, *Annals of the Institute of Statistical Mathematics 14*, 173-178. *[4.1.3]*

Skitovitch, V. P. (1953). On a property of the normal distribution, *Doklady Akademii Nauk SSSR 89*, 217-219 (in Russian). *[4.1.4; 4.3.5]*

Stepniak, C. (1991). On characterization of the normal law in the Gauss Markov model, *Sankyā A*, 115-117. *[4.7.12]*

Stuart, A. and Ord, J. K. (1987). *Kendall's Advanced Theory of Statistics, Vol. 1* (5th edn.), New York: Oxford University Press. *[4.1.5; 4.2.1; 4.6.1]*

Tamhanker, M. V. (1967). A characterization of normality, *Annals of Mathematical Statistics 38*, 1924-1927. *[4.6.1]*

Teicher, H. (1961). Maximum likelihood characterization of distributions, *Annals of Mathematical Statistics 32*, 1214-1222. *[4.7.4]*

Wesolowski, J. (1987). A regressional characterization of the normal law, *Statistics and Probability Letters 6*, 11-12. *[4.3.4]*

Ziegler, R. K. (1965). A uniqueness theorem concerning moment distributions, *Journal of the American Statistical Association 60*, 1203-1206. *[4.5.3]*

Chapter 5
SAMPLING DISTRIBUTIONS

We have already listed in Section [2.3] a few distributions arising from some transformations of normal random variables; others appear in Chapter 9. In this chapter we summarize distributions and their properties when they are based on random samples from normal populations, that is, from mutually independent normal random variables $X_1, ..., X_n$, which will usually but not in every case be identically distributed. One class of sampling distributions which will be considered in Chapter 8 rather than here is that involving the order statistics. Also covered separately in Chapter 7 are approximations based on the normal law to the distributions introduced in Sections [5.3] to [5.5]. The abbreviation *iid* in what follows means "independently and identically distributed."

5.1 SAMPLES NOT LARGER THAN FOUR

[5.1.1] (a) Let X_1 and X_2 be iid $N(0, \sigma^2)$ rvs. Then the rv Y given by $Y = X_1/X_2$ has a *Cauchy distribution* with pdf

$$g(y) = [\pi(1 + y^2)]^{-1}, \quad -\infty < y < \infty.$$

This is also a Student's t distribution with one degree of freedom; see [5.4.1]. The converse is false (Steck, 1958, p. 604; Fox, 1965, p. 631); but see [4.6.6].

(b) If $Y = (a + Z_1)/(b + Z_2)$, where Z_1 and Z_2 are independent $N(0, 1)$ rvs, then the pdf of Y is $g(y; a, b)$, where

$$g(y; a, b) = \frac{\exp\{-(a^2 + b^2)/2\}}{\pi(1 + t^2)}\left[1 + \frac{q}{\phi(q)}\left\{\Phi(q) - \tfrac{1}{2}\right\}\right],$$

113

$$q = (b + qy)/(1 + y^2)^{1/2} .$$

Marsaglia (1965, pp. 194-199) gives representations of the cdf of this distribution in terms of bivariate normal probabilities. If $b > 3$, then

$$\Pr(Y \leq t) \simeq \Phi\{(bt - a)/(1 + t^2)^{1/2}\} .$$

(c) For the distribution of X_1/X_2, where $X_i \sim N(\mu_i, \sigma_i^2)$ independently, $i = 1, 2$, see Springer (1979, Sec. 4.6).

[5.1.2] Let X_1 and X_2 be independent normal rvs with common variance. Then the rvs $aX_1 + bX_2$ and $cX_1 + dX_2$ are independent if and only if $ac + bd = 0$. In particular, $X_1 + X_2$ and $X_1 - X_2$ are independent.

[5.1.3] Let X_1 and X_2 be iid $N(0, 1)$ rvs. Then the random variable Y, where $Y = X_1 X_2$, has pdf $g(y)$ given by

$$g(y) = k_0(|y|)/\pi, \quad -\infty < y < \infty ;$$

here $k_0(y)$ is the modified Bessel function of the third kind (Epstein, 1948, pp. 375-377; Stuart and Ord, 1987, Sec. 11.11; Craig, 1936, pp. 1-15; Aroian et al., 1978, pp. 165-172). See also *[5.8.7]* and *[10.5.3]*, where references to tables of fractiles and moments are discussed.

[5.1.4] Let X_1 and X_2 be independent $N(0, \sigma_1^2)$ and $N(0, \sigma_2^2)$ rvs, respectively, and let

$$Y = X_1 X_2 / (X_1^2 + X_2^2)^{1/2} .$$

Then Y has a normal distribution with variance $[(\sigma_1^2)^{-1} + (\sigma_2^2)^{-1}]^{-1}$. If, in addition, $\sigma_1^2 = \sigma_2^2$, and $W = (X_1^2 - X_2^2)/(X_1^2 + X_2^2)$, then W has a normal distribution; further, Y and W are independent (Shepp, 1964, p. 459).

[5.1.5] *Laplace Distribution.* Let Z_1, Z_2, Z_3, Z_4 be iid standard normal random variables. Then the rv $Y = Z_1 Z_2 + Z_3 Z_4$ has a Laplace distribution with pdf (Mantel, 1973, p. 31)

$$g(y) = \tfrac{1}{2} \exp(-|y|), \quad -\infty < y < \infty .$$

Sampling Distributions

[5.1.6] The *Birnbaum-Saunders distribution* arises as a fatigue life model. Let X have a $N(0, \alpha^2/4)$ distribution, and let

$$T = \beta[1 + 2X^2 + 2X(1 + X^2)^{1/2}].$$

Then T has pdf $g(t; \alpha, \beta)$ and cdf $G(t; \alpha, \beta)$, where

$$g(t; \alpha, \beta) = \frac{1}{2\sqrt{2\pi}\,\alpha^2\,\beta t^2} \frac{t^2 - \beta^2}{\sqrt{t/\beta} - \sqrt{\beta/t}} \exp\left[-\frac{1}{2\alpha^2}\left(\frac{t}{\beta} + \frac{\beta}{t} - 2\right)\right],$$

$$t > 0,\ \alpha,\ \beta > 0,$$

$$G(t; \alpha, \beta) = \Phi\{\alpha^{-1}(\sqrt{t/\beta} - \sqrt{\beta/t})\},\quad \alpha, \beta > 0.$$

The mean and variance of T are $\beta(1 + \tfrac{1}{2}\alpha^2)$ and $\alpha^2\beta^2(1 + 5\alpha^2/4)$, respectively (Birnbaum and Saunders, 1969, pp. 323, 324; Mann et al., 1974, pp. 150-155). Further, the rv $1/T$ has cdf $G(\cdot\,; \alpha, \beta^{-1})$ in our notation, and the rv aT has cdf $G(\cdot\,; \alpha, a\beta)$ if $a > 0$.

5.2 THE SAMPLE MEAN: INDEPENDENCE

[5.2.1] Let X_i ($i = 1, 2, ..., n$) be n independent $N(\mu_i, \sigma_i^2)$ rvs. Then the random variable $Y = \sum_{i=1}^{n} a_i X_i$, where $a_1, ..., a_n$ are real numbers, has a $N(\sum_{i=1}^{n} a_i\mu_i, \sum_{i=1}^{n} a_i^2\sigma_i^2)$ distribution. In particular, let $\overline{X} = \sum X_i/n$, where $a_1 = ... = a_n = 1$ and $\sigma_1^2 = ... = \sigma_n^2 = \sigma^2$. Then $\overline{X}$ is the *sample mean* of $X_1, ..., X_n$ and has a $N(\mu, \sigma^2/n)$ distribution.

[5.2.2] (a) Let X_i ($i = 1, 2, ..., n$) be iid $N(\mu, \sigma^2)$ rvs. Then the rvs $a_1X_1 + a_2X_2 + ... + a_nX_n$ and $b_1X_1 + b_2X_2 + ... + b_nX_n$ are independent if and only if $a_1b_1 + a_2b_2 + ... + a_nb_n = 0$.

(b) Let $\underline{a}_1, \underline{a}_2, ..., \underline{a}_r$ be any $n \times 1$ mutually orthogonal vectors ($2 \leq r \leq n$). Then the rvs $\underline{a}_1'\underline{X}, \underline{a}_2'\underline{X}, ..., \underline{a}_r'\underline{X}$ are mutually independent, where $\underline{X}' = (X_1, X_2, ..., X_n)$. If, in addition, $\underline{a}_1, ..., \underline{a}_r$ are orthonormal, then $(\underline{a}_1'\underline{X}, ..., \underline{a}_r'\underline{X})$ are iid $N(\mu, \sigma^2)$ rvs also (Rao, 1973, pp. 183-184).

[5.2.3] Let X_i ($i = 1, 2, ..., n$) be iid $N(\mu, \sigma^2)$ rvs. Let $\overline{X}$ be the *sample mean* $\sum X_i/n$, $S^2 = \sum_{i=1}^{n}(X_i - \overline{X})^2/(n-1)$ be the *sample variance*,

$Y = \sum_{i=1}^{n}|X_i - \overline{X}|/n$ be the *sample absolute deviation*, and $W = \max(X_i) - \min(X_i)$ be the *sample range*. Then

(i) $\overline{X}$ and S^2 are independent.

(ii) $\overline{X}$ and Y are independent.

(iii) $\overline{X}$ and W are independent.

(iv) $\overline{X}$ and any function of the set of statistics $(X_1 - \overline{X}, ..., X_n - \overline{X})$ are independent.

(v) Any function $g(x_1, x_2, ..., x_n)$ of a sample of n independent observations from a normally distributed population and the sample mean $\overline{X}$ are independent, provided only that $g(x_1, x_2, ..., x_n) = g(x_1 + a, x_2 + a, ..., x_n + a)$ (Daly, 1946, p. 71; Herrey, 1965, pp. 261-262). Another result for obtaining certain uncorrelated sample statistics from any symmetric distribution is given by Hogg (1960, pp. 265-266).

[5.2.4] Let $X_1, X_2, ...$ be an iid sequence of $N(\mu, \sigma^2)$ random variables and define $Y_n = [nX_{n+1} - (X_1 + \cdots + X_n)]/\sqrt{n(n+1)}$; $n = 1, 2, ...$. Then $Y_1, Y_2, ...$ is an iid sequence of $N(0, \sigma^2)$ random variables. This property follows from those in *[5.2.2]*(b) and *[5.2.3]*(v).

5.3 SAMPLING DISTRIBUTIONS RELATED TO CHI-SQUARE

Of particular interest in what follows is the statistic S^2, or $\sum(X_i - \overline{X})^2/(n-1)$, the sample variance in a random sample of size n from a $N(\mu, \sigma^2)$ distribution.

[5.3.1] Let $Z_1, Z_2, ..., Z_n$ be iid $N(0, 1)$ random variables. Then the rv Y, where $Y = \sum_{i=1}^{n} Z_i^2$, has a *chi-squared distribution* with n degrees of freedom, denoted by χ_n^2. The quantiles of Y are $\chi_{n;\beta}^2$, where $\Pr(Y \leq \chi_{n;\beta}^2) = 1 - \beta$, and Y has pdf $g(y; n)$, given by

$$g(y; n) = [2^{n/2}\Gamma(n/2)]^{-1} y^{(n-2)/2} e^{-y/2}; \quad y > 0, n = 1, 2, ...$$

The mean, variance, skewness, and kurtosis of a χ_ν^2 rv are, respectively, ν, 2ν, $\sqrt{8/\nu}$, and $3 + 12/\nu$. See Johnson and Kotz (1970a, Chap. 17) and Lancaster (1969) for a fuller discussion.

Sampling Distributions 117

[5.3.2] *Some Basic Properties of Chi-Square from Normal Samples.*
Let $X_1, ..., X_n$ be iid $N(\mu, \sigma^2)$ random variables.
(a) The rv $\sum_{i=1}^{n}(X_i - \mu)^2/\sigma^2$ has a χ_n^2 distribution.
(b) The rv $\sum_{i=1}^{n}(X_i - \bar{X})^2/\sigma^2 = (n-1)S^2/\sigma^2$ has a χ_{n-1}^2 distribution.
(c) See *[5.3.7]* for applicable properties of quadratic forms, i.e., when $\mu = \underline{0}$.

[5.3.3] Let $G(y; \nu)$ be the cdf and $g(y; \nu)$ the pdf of a χ_ν^2 rv. Then

(i) $1 - G(y; 2m+1) = 2[1 - \Phi(\sqrt{y})] + 2\sum_{r=1}^{m} g(y; 2r+1)$

$= 2[1 - \Phi(\sqrt{y})] + 2\phi(\sqrt{y}) \sum_{r=1}^{m} \frac{y^{(2r-1)/2}}{1 \cdot 3 \cdot 5 \cdots (2r-1)}$;

(ii) $1 - G(y; 2m) = 2\sum_{r=1}^{m} g(y; 2r)$

$= \sqrt{2\pi}\phi(\sqrt{y}) \left[1 + \sum_{r=1}^{m-1} \frac{y^r}{2 \cdot 4 \cdots (2r)}\right]$;

(iii) $\Phi(\sqrt{y}) = \sum_{r=1}^{\infty} g(y; 2r+1) + \frac{1}{2} = \sum_{r=2}^{\infty} g(y; r)$

(Puri, 1973, p. 63; Lancaster, 1969, p. 21; Abramowitz and Stegun, 1964, Secs. 26.4.4, 5).

[5.3.4] The *sample variance* S^2 has a $\{\sigma^2/(n-1)\}\chi_{n-1}^2$ distribution, so that $\Pr(S^2 \leq y) = \Pr(\chi_{n-1}^2 \leq (n-1)y/\sigma^2)$. Note that the second sample moment m_2 is sometimes termed the sample variance, where $m_2 = \sum(X_i - \bar{X})^2/n = (n-1)S^2/n$. If $\nu = n-1$, the pdf of S^2 is $g(y; \nu)$, where

$$g(y; \nu) = (\tfrac{1}{2}\nu/\sigma^2)^{\nu/2} y^{\nu/2-1} \exp\{-\nu y/(2\sigma^2)\}/\Gamma(\nu/2), \quad y > 0.$$

The first four moments and the shape factors of the sampling distribution of S^2 are:

$$E(S^2) = \sigma^2, \qquad \text{Var}(S^2) = 2\sigma^4/(n-1),$$
$$\mu_3(S^2) = 8\sigma^6/(n-1)^2, \qquad \mu_4(S^2) = 12(n+3)\sigma^8/(n-1)^3 :$$

Skewness: $\quad \mu_3(S^2)/\{\text{Var}(S^2)\}^{3/2} = 2\sqrt{2/(n-1)}$,

Kurtosis: $\quad \mu_4(S^2)/\{\text{Var}(S^2)\}^2 = 3 + 12/(n-1)$.

Generally,
$$\mu'_r(S^2) = \{\sigma^2/(n-1)\}^r E\{(\chi^2_{n-1})^r\},$$

$$\mu_r(S^2) = \{\sigma^2/(n-1)\}^r E\{(\chi^2_{n-1} - n + 1)^r\},$$

where χ^2_{n-1} denotes the chi-square distribution with $n-1$ degrees of freedom.

[5.3.5] The sampling distribution of the *sample standard deviation* S, where $S = \{\sum_{i=1}^{n}(X_i - \overline{X})^2/(n-1)\}^{1/2}$, has pdf $g(s; n)$ given by

$$g(s; n) = \frac{s^{\nu-1}\nu^{\nu/2} \exp\{-\nu s^2/(2\sigma^2)\}}{2^{(\nu-2)/2}\sigma^\nu \Gamma(\nu/2)}, \quad s > 0, \; \nu = n-1 \geq 1.$$

This is the pdf of $(\sigma/\sqrt{n-1}) \times$ (chi with $n-1$ degrees of freedom). Probability statements about S are most easily made from the distribution of S^2. Thus $\Pr(a \leq S \leq b) = \Pr(a^2 \leq S^2 \leq b^2)$; $b > a > 0$. From Johnson and Welch (1939, pp. 216-218) we find the moments of S about the origin to be given by

$$E(S) = \sigma\sqrt{2/(n-1)} \; \Gamma(n/2)/\Gamma\{(n-1)/2\}, \quad E(S^2) = \sigma^2,$$
$$E(S^3) = n\sigma^2 E(S)/(n-1), \quad E(S^4) = \sigma^4(n+1)/(n-1),$$
$$E(S^r) = \{2\sigma^2/(n-1)\}^{r/2}\Gamma\{(n+r-1)/2\}/\Gamma\{(n-1)/2\}, \quad r = 1, 2, \ldots.$$

Further, Jensen's inequality gives

$$E(S) = E(\{S^2\}^{1/2}) < \{E(S^2)\}^{1/2} = \sigma.$$

There are no concise expressions for the central moments, but David (1949, p. 390) gives expansions accurate to $O(n^{-2})$: if $\nu = n-1$,

Sampling Distributions

$$E(S) \simeq \sigma(1 - \nu^{-1}/4 + \nu^{-2}/32), \quad Var(S) \simeq \sigma^2(\tfrac{1}{2}\nu^{-1} - \tfrac{1}{8}\nu^{-2}),$$

$$\mu_3(S) \simeq \tfrac{1}{4}\sigma^3 \nu^{-2}, \quad \mu_4(S) \simeq \tfrac{3}{4}\sigma^4 \nu^{-2}.$$

Pearson and Hartley (1966, Table 35) tabulate values of $E(S/\sigma)$, $\sqrt{Var(S/\sigma)}$, $\mu_3(S)/\{Var(S)\}^{3/2}$, and $\mu_4(S)/\{Var(S)\}^2$ to 6, 5, 4, and 4 decimal places, respectively, for $\nu = 1(1)20(5)50(10)100$.

[5.3.6] Let $X_1, X_2, ..., X_n$ be mutually independent rvs such that X_i has a $N(\mu_i, 1)$ distribution, and let $\lambda = \sum_{i=1}^{n} \mu_i^2$. Then the rv Y, where $Y = \sum_{i=1}^{n} X_i^2$ has a *noncentral chi-square distribution* with n degrees of freedom and noncentrality parameter λ, denoted by $\chi_n^2(\lambda)$. The pdf $g(y; n, \lambda)$ of Y is given by

$$g(y; n, \lambda) = \frac{e^{-(\lambda+y)/2}}{2^{n/2}} \sum_{j=0}^{\infty} \frac{y^{(1/2)n+j-1} \lambda^j}{\Gamma(\tfrac{1}{2}n + j) 2^{2j} j!}, \quad y > 0, \lambda \geq 0$$

(Johnson and Kotz, 1970b, p. 132).

A word of caution needs to be given about the noncentrality parameter. The majority of sources going back to Patnaik (1949, pp. 202-232), use the definition above. But Searle (1971, p. 49) and Graybill (1961, p. 74) define it as $\tfrac{1}{2}\sum \mu_i^2$, while Guenther (1964, pp. 957-960) define $\sqrt{\sum \mu_i^2}$ as the noncentrality parameter. These discrepancies may lead to some confusion, for example, in determining moments.

The mean, variance, skewness, and kurtosis of a $\chi_\nu^2(\lambda)$ rv (as defined above) are $\nu + \lambda$, $2(\nu + 2\lambda)$, $\sqrt{8}(\nu + 3\lambda)/(\nu + 2\lambda)^{3/2}$, and $3 + 12(\nu + 4\lambda)/(\nu + 2\lambda)^2$, respectively; see Johnson and Kotz (1970b, p. 134) and their Chap. 28 for further discussion.

[5.3.7] Let $G(y; \nu, \lambda)$ be the cdf corresponding to the pdf $g(y; \nu, \lambda)$ of [5.3.6]. Then if $a = \sqrt{\lambda} + \sqrt{y}$ and $b = \sqrt{\lambda} - \sqrt{y}$ (Han, 1975, pp. 213-214),

$$G(y; 1, \lambda) = \Phi(a) - \Phi(b),$$

$$G(y; 3, \lambda) = \Phi(a) - \Phi(b) + \{\phi(a) - \phi(b)\}/\sqrt{\lambda},$$

$$G(y; 5, \lambda) = \Phi(a) - \Phi(b) + (2 - \lambda^{-1})\{\phi(a) - \phi(b)\}/\sqrt{\lambda}$$
$$- \lambda^{-1}\{a\phi(a) - b\phi(b)\} ,$$

$$G(y; 7, \lambda) = \Phi(a) - \Phi(b) + (3 - 4\lambda^{-1} + 3\lambda^{-2})\{\phi(a) - \phi(b)\}/\sqrt{\lambda}$$
$$- 3(\lambda^{-1} - \lambda^{-2})\{a\phi(a) + b\phi(b)\} + \lambda^{-3/2}\{a^2\phi(a) + b^2\phi(b)\} .$$

[5.3.8] If X has a noncentral χ^2 distribution with two degrees of freedom and noncentrality parameter λ, then

$$\Pr(X \leq r) = (2\pi)^{-1} \int\int_\Gamma \exp\{-(x^2 + y^2)/2\} \, dy \, dx ,$$

where Γ is the circle $(x - \sqrt{\lambda})^2 + y^2 \leq r^2$. This is the probability that (X, Y) lies in an offset circle of radius r at a distance $\sqrt{\lambda}$ from the origin, when X and Y are independent N(0, 1) rvs. See *[10.5]* and Owen (1962, pp. 172-180), where these probabilities are tabulated.

[5.3.9] *Some Basic Properties of $\chi^2_\nu(\cdot)$ Relating to Normal Samples.*

Let $X_1, ..., X_n$ be defined as in *[5.3.6]* above, writing $\underline{X}' = (X_1, ..., X_n)$ and $\underline{\mu}' = (\mu_1, \mu_2, ..., \mu_n)$. Let $\underline{A}, \underline{A}_1, \underline{A}_2, ..., \underline{A}_k$ be n×n symmetric matrices. The following theorems give properties of linear combinations of the sample variables and of quadratic forms; see Rao (1973, pp. 185-188) and Searle (1971, pp. 55-64). For some approximations see *[7.14.1]*.

(a) The quadratic form $\underline{X}'\underline{A}\underline{X}$ has a noncentral chi-square distribution if and only if $\underline{A}$ is idempotent (that is $\underline{A}^2 = \underline{A}$), in which case $\underline{X}'\underline{A}\underline{X}$ has degrees of freedom equal to the rank of $\underline{A}$, and noncentrality parameter $\underline{\mu}'\underline{A}\underline{\mu}$.

(b) Let $\underline{X}'\underline{A}_1\underline{X}$ and $\underline{X}'\underline{A}_2\underline{X}$ have noncentral chi-square distributions. Then they are independent if and only if $\underline{A}_1\underline{A}_2 = \underline{0}$.

(c) Let $\underline{B}$ be a q×n matrix. Then $\underline{X}'\underline{A}\underline{X}$ and $\underline{B}\underline{X}$ are independent if and only if $\underline{B}\underline{A} = \underline{0}$. This result holds whether or not $\underline{X}'\underline{A}\underline{X}$ has a noncentral chi-square distribution.

(d) *Cochran's Theorem.* Let the rank of $\underline{A}_j$ be r_j (j = 1, 2, ..., k) and $Q_j = \underline{X}'\underline{A}_j\underline{X}$, such that

Sampling Distributions

$$\sum_{i=1}^{n} X_i^2 = \underline{X}'\underline{X} = Q_1 + \ldots + Q_k \ .$$

Then Q_j has a $\chi^2_{r_j}(\lambda_j)$ distribution ($j = 1, \ldots, k$) and $Q_1, \ldots, Q_k$ are mutually independent if and only if either

$$r_1 + \ldots + r_k = n$$

or

$\underline{A}_j$ is idempotent, $j = 1, \ldots, k$ and $\underline{A}_j\underline{A}_s = 0$ for all $s \neq j$.

Then $\lambda_j = \underline{\mu}'\underline{A}_j\underline{\mu}$ and $\sum_{i=1}^{n}\mu_i^2 = \sum_{i=1}^{k}\lambda_j$.

(e) If $\underline{X}'\underline{X} = Q_1 + Q_2$, where Q_1 has χ_a^2 distribution, then Q_2 has a χ^2_{n-a} distribution, $n > a > 0$.

(f) If $Q = Q_1 + Q_2$, where Q has a χ_a^2 distribution, Q_1 has a χ_b^2 distribution ($b < a$), and Q_2 is nonnegative, then Q_2 has a χ^2_{a-b} distribution.

(g) $E(\underline{X}'\underline{A}\underline{X}) = \underline{\mu}'\underline{A}\underline{\mu} + \text{trace }(\underline{A})$. This holds whether or not $\underline{X}$ is a vector of (independent) normal rvs.

5.4 SAMPLING DISTRIBUTIONS RELATED TO t

[5.4.1] Let Z be a standard normal rv and U a χ_k^2 random variable, independent of Z. Then the rv Y defined by $Y = Z\sqrt{k/U}$ has a *Student t distribution* with k degrees of freedom, denoted by t_k. The quantiles are $t_{k;\beta}$, where $\Pr(Y \leq t_{k;\beta}) = 1 - \beta$, and the pdf g(y; k) of Y is given by

$$g(y; k) = (k\pi)^{-1/2} \frac{\Gamma\{(k+1)/2\}}{\Gamma(k/2)} \left(1 + \frac{y^2}{k}\right)^{-(k+1)/2}, \ |y| < \infty, \ k = 1, 2, \ldots \ .$$

This is a Pearson Type VII distribution symmetrical about zero; the odd moments vanish, when they exist. The rth moment exists if and only if k > r. The variance and kurtosis of Y are $k/(k-2)$ and $3 + 6/(k-4)$, respectively, and they exist only if k > 2 and k > 4, respectively (Johnson and Kotz, 1970b, p. 96); see also *[5.1.1]*.

[5.4.2] With the notation of *[5.4.1]*,

$$\lim_{k \to \infty} g(y; k) = \phi(y), \quad \lim_{k \to \infty} \Pr(Y \leq y) = \Phi(y).$$

[5.4.3] (a) Let $\bar{X}$ be the mean $\sum X_i/n$ and S^2 be the sample variance $\sum(X_i - \bar{X})^2/(n-1)$ of a sample of size n under normality ($n \geq 2$). If $T = \sqrt{n}(\bar{X} - \mu)/S$, then T has a t_{n-1} distribution.

(b) If $U = \sqrt{n}(\bar{X} - \mu)/\{\sum_{i=1}^{n}(X_i - \mu)^2/n\}$, then U does not have a t distribution because $\bar{X}$ and $\sum(X_i - \mu)^2$ are not independent. In fact, $\Pr(|U| \leq \sqrt{n}) = 1$, since $U = \sqrt{n}T/(T^2 + n - 1)^{1/2}$, where T is defined in (a) above (D. B. Owen, personal communication).

(c) If X_0 and X_1 are independent χ_k^2 rvs, then the random variable Y has a t_k distribution, where (Cacoullos, 1965, p. 528)

$$Y = \tfrac{1}{2}\sqrt{k}\,(X_1 - X_0)/\sqrt{X_0 X_1}.$$

[5.4.4] Let $X_1, X_2, \ldots, X_m$ and $Y_1, Y_2, \ldots, Y_n$ be independent iid samples of $N(\mu, \sigma^2)$ and $N(\xi, \sigma^2)$ rvs, respectively. If

$$T = \frac{\sqrt{mn/(m+n)}\,\{\bar{X} - \bar{Y} - (\mu - \xi)\}}{\left[\{\sum(X_i - \bar{X})^2 + \sum(Y_j - \bar{Y})^2\}/(m + n - 2)\right]^{1/2}},$$

where $\bar{X} = \sum X_i/m$ and $\bar{Y} = \sum Y_j/n$, then T has a t_{m+n-2} distribution.

[5.4.5] Koehler (1983, pp. 103-105) derived a simple approximation to the quantiles $t_{\nu,\beta}$ of Student's t, given by

$$(t_{\nu,\beta})^{-1} \simeq -0.0953 - 0.631(\nu+1)^{-1} + (0.81)\left[-\ell n\{4\beta(1-\beta)\}\right]^{1/2}$$

$$+ 0.076\left[(2\sqrt{2\pi} - 1)\beta\sqrt{\nu}\right]^{1/\nu};$$

in Koehler's form, replace β by $\alpha/2$ to obtain the two-sided version. When $0.000005 \leq \beta \leq 0.1$, the relative error does not exceed 1.4% and 0.6% if $\nu \geq 8$ and $\nu \geq 50$, respectively. It compares favorably with other approximations, although the three-term expansion of Fisher and Cornish (1960, pp. 205-225) is more accurate when ν is large; this is given by

Sampling Distributions

$$t_{\nu,\beta} \simeq z_\beta + \tfrac{1}{4}z_\beta(z_\beta^2 + 1)\nu^{-1} + \tfrac{1}{96}z_\beta(5z_\beta^4 + 16z_\beta^2 + 3)\nu^{-2},$$

z_β being the corresponding standard normal percentile.

[5.4.6] Chu (1956, pp. 783-784) derived the following inequalities for the cdf $G(y; \nu)$ of a t_ν distribution; if $a \geq 0$, $b \geq 0$, and $\nu \geq 3$,

$$\frac{\nu}{\nu+1}\left\{\Phi\left(b\sqrt{\tfrac{\nu+1}{\nu}}\right) - \Phi\left(-a\sqrt{\tfrac{\nu+1}{\nu}}\right)\right\} \leq G(b;\nu) - G(-a;\nu)$$

$$\leq \sqrt{\tfrac{7\nu-3}{7\nu-14}}\left\{\Phi\left(b\sqrt{\tfrac{\nu-2}{\nu}}\right) - \Phi\left(-a\sqrt{\tfrac{\nu-2}{\nu}}\right)\right\}.$$

Putting one of the arguments equal to zero gives for the cdf,

$$\frac{\nu}{\nu+1}\Phi\left(a\sqrt{\tfrac{\nu+1}{\nu}}\right) + \frac{1}{2(\nu+1)} \leq G(a;\nu)$$

$$\leq \sqrt{\tfrac{7\nu-3}{7\nu-14}}\Phi\left(a\sqrt{\tfrac{\nu-2}{\nu}}\right) + \tfrac{1}{2}\left(1 - \sqrt{\tfrac{7\nu-3}{7\nu-14}}\right),$$

where $a \geq 0$ and $\nu \geq 3$. If $\overline{\Phi}(x) = 1 - \Phi(x)$ and $\overline{G}(a, \nu) = 1 - G(a, \nu)$, these inequalities are reversed when Φ and G are replaced by $\overline{\Phi}$ and $\overline{G}$, respectively. Sharper inequalities than the above are given by

$$c_\nu\sqrt{\tfrac{2}{\nu+1}}\left\{\Phi\left(b\sqrt{\tfrac{\nu+1}{\nu}}\right) - \Phi\left(-a\sqrt{\tfrac{\nu+1}{\nu}}\right)\right\} \leq G(b;\nu) - G(-a;\nu)$$

$$\leq c_\nu\sqrt{\tfrac{2}{\nu-2}}\left\{\Phi\left(b\sqrt{\tfrac{\nu-2}{\nu}}\right) - \Phi\left(-a\sqrt{\tfrac{\nu-2}{\nu}}\right)\right\},$$

$$c_\nu = \frac{\Gamma\{(\nu+1)/2\}}{\Gamma(\nu/2)}, \quad a \geq 0, b \geq 0, \nu \geq 3.$$

[5.4.7] Let X be a $N(\mu, \sigma^2)$ rv and U/σ^2 a χ_k^2 rv, independent of X. Then the rv Y defined by $Y = X\sqrt{k/U}$ has a *noncentral t distribution* with k degrees of freedom and noncentrality parameter δ, where $\delta = \mu/\sigma$; we denote this by $t_k(\delta)$. The pdf is $g(y; k, \delta)$, where

$$g(y; k, \delta) = \frac{k^{k/2} \exp(-\delta^2/2)}{\sqrt{\pi} \,(k+y^2)^{(k+1)/2}} \sum_{i=0}^{\infty} \left[\frac{\Gamma\{(k+i+1)/2\}}{i!\Gamma(k/2)} \left(\frac{2^{i/2}}{k+y^2}\right) (\delta y)^i \right],$$

$$-\infty < y < \infty, \ -\infty < \delta < \infty, \ k = 1, 2, \ldots$$

(Johnson and Kotz, 1970b, p. 205). Merrington and Pearson (1958, p. 485) give the first four moments of Y about the origin:

$$E(Y) = \frac{(\nu/2)^{1/2} \delta \, \Gamma\{(\nu-1)/2\}}{\Gamma(\nu/2)}, \quad \mu_2'(Y) = \frac{(1+\delta^2)\nu}{\nu-2},$$

$$\mu_3'(Y) = \frac{(\nu/2)^{3/2} \delta (3+\delta^2) \Gamma\{(\nu-3)/2\}}{\Gamma(\nu/2)}, \quad \mu_4'(Y) = \frac{(3+6\delta^2+\delta^4)\nu^2}{(\nu-2)(\nu-4)}.$$

[5.4.8] Let $X_1, \ldots, X_n$ be iid $N(\mu, \sigma^2)$ rvs, and let S^2 be the sample variance. If $Y = \sqrt{n}\,\bar{X}/S$, then Y has a $t_{n-1}(\sqrt{n}\mu/\sigma)$ distribution.

[5.4.9] Let $G(y; k, \delta)$ be the cdf corresponding to the pdf $g(y; k, \delta)$ of *[5.4.7]*. Then (Owen, 1968, pp. 465-466)

$$G(0; k, \delta) = \Phi(-\delta), \quad G(1; 1, \delta) = 1 - \left[\Phi(\delta/\sqrt{2})\right]^2,$$

$$G(y; k, \delta) = \frac{\sqrt{2\pi}}{\Gamma(k/2) 2^{(k-2)/2}} \int_0^\infty \Phi\left(\frac{yx}{\sqrt{k}} - \delta\right) x^{k-1} \phi(x)\, dx.$$

See also Hawkins (1975, p. 43).

[5.4.10] Let $V = \sigma/\mu$, the *coefficient of variation* in the population. In a random sample of size n, the sample coefficient of variation (SCV) is defined in two ways;

$$v = S/\bar{X}, \quad v_n = S_n/\bar{X} = (\sqrt{(n-1)/n})\, v,$$

where $S^2 = \sum_i (X_i - \bar{X})^2/(n-1)$ and $S_n^2 = \sum_i (X_i - \bar{X})^2/n$. Then $\sqrt{n}/v$ has a noncentral $t_{n-1}(\sqrt{n}/V)$ distribution; see Johnson and Welch (1940, pp. 362-363) and Owen (1968, pp. 445-478). A simple approximation due to McKay (1932, pp. 695-698) is most succinctly expressed via

$$B\,\frac{v_n^2}{1+v_n^2} = \left(\frac{n-1}{n}\right) B\,\frac{v^2}{1+v^2 - (v^2/n)} \sim \chi_{n-1}^2$$

approximately, where $B = n(1 + 1/V^2)$. This approximation turns out to be "not nearly as good as indicated" in early confirmatory studies by Pearson and by Fieller; see Warren (1982, pp. 659-666), who recommends using the exact relationship of the SCV to noncentral t, or the normal approximation to noncentral t discussed in *[7.11.2]*(b), which performs well in the upper tail for the cases examined. There has been some confusion in the literature over the choice of v or v_n for SCV, and consequently over the application of McKay's approximation; see for example Umphrey (1983, pp. 629-635). Dropping the term v^2/n (usually negligible in practice) from the denominator of the second form, one obtains approximately,

$$\left(\frac{n-1}{n}\right) B \frac{v^2}{1+v^2} \sim \chi^2_{n-1} ,$$

a version used by David (1949, p. 387); see *[5.4.13]*.

[5.4.11] Iglewicz et al. (1968, p. 581) give the following approximation when $V \leq 0.5$ to the percentiles of the sample CV v, where $\Pr(v \leq v_p) = 1 - p$ and $\chi^2_{n-1;p}$ is defined similarly in *[5.3.1]*:

$$v_p \simeq V\left(\frac{\chi^2_{n-1;p}}{n-1}\right)^{1/2} [1 + \frac{V^2}{2n}\{\chi^2_{n-1;p} - (n-2)\} + \frac{V^4}{8n^2}\{3(\chi^2_{n-1;p})^2$$

$$- 8(n-2)\chi^2_{n-1;p} + (n-2)(5n-12)\}] .$$

This approximation is accurate to four decimal places if $n \geq 5$ and if *either* $V \leq 0.3$ *or* the first two terms only are used and $V \leq 0.2$. If $0.3 \leq v \leq 0.5$, the approximation is still "quite good".

[5.4.12] Iglewicz and Myers (1970, pp. 166-169) compared six approximations to percentiles of the SCV v. Two of the best of these follow.

(a) Pearson distributions were fitted to the approximate first four moments of v, and approximate percentage points were obtained from values tabulated by Johnson et al. (1963, pp. 459-498). For moderate or large values of n this approach gives the best results, but is cumbersome to use.

(b) The approximation of McKay given in *[5.4.10]* leads to

$$v_p \simeq \sqrt{n/(n-1)}\sqrt{\chi^2_{n-1;p}/(B-\chi^2_{n-1;p})} \; ,$$

if the probability of v being negative is negligible. If (a) seems too cumbersome, this approach is uniformly more accurate than others considered by the authors. Exact and approximate values of v_p are tabulated for $p = 0.99, 0.95, 0.05$ and 0.01, for $V = 0.1(0.1)0.4$, and for $n = 10, 20, 30, 50$.

[5.4.13] David (1949, p. 387) gives approximations to the first four moments of the distribution of v under normality: to terms of order n^{-2},

$$E\left(\frac{v}{V}\right) \simeq 1 + \frac{V^2}{n} - \frac{1}{4(n-1)} + \frac{3V^4}{n^2} - \frac{V^2}{4n(n-1)} + \frac{1}{32(n-1)^2} \; ,$$

$$\mathrm{Var}\left(\frac{v}{V}\right) \simeq \frac{V^2}{n} + \frac{1}{2(n-1)} + \frac{8V^4}{n^2} + \frac{V^2}{n(n-1)} - \frac{1}{8(n-1)^2} \; ,$$

$$\mu_3\left(\frac{v}{V}\right) \simeq \frac{6V^4}{n^2} + \frac{3V^2}{n(n-1)} + \frac{1}{4(n-1)^2} \; ,$$

$$\mu_4\left(\frac{v}{V}\right) \simeq \frac{3V^4}{n^2} + \frac{3V^2}{n(n-1)} + \frac{3}{4(n-1)^2} \; .$$

These approximations assume that V is not large. If μ is close to zero, however, V could be large, and a correspondingly large value of n would be required.

[5.4.14] Let v be the sample CV and r the number of negative sample values out of n. Then $v \geq \{n/(n-1)\}\{r/(n-r)\}^{1/2}$. This inequality does not depend upon normality (Summers, 1965, p. 67).

5.5 DISTRIBUTIONS RELATED TO F

[5.5.1] Let U_1 and U_2 be independent χ^2_m and χ^2_n rvs, respectively. Then the rv Y defined by $Y = U_1 m^{-1}/(U_2 n^{-1})$ has an *F distribution* with (m, n) degrees of freedom, denoted by $F_{m,n}$. The quantiles are $F_{m,n;\beta}$, where $\Pr(Y \leq F_{m,n;\beta}) = 1 - \beta$, and the pdf $g(y; m, n)$ is given by

$$g(y; m, n) = m^{m/2} n^{n/2} \{B(\tfrac{1}{2}m, \tfrac{1}{2}n)\}^{-1} y^{(1/2)m-1}(my + n)^{-(1/2)(m+n)},$$

$$y > 0, \quad m, n = 1, 2, \ldots$$

$$\Pr(Y \leq y) = I_\gamma(m/2, n/2),$$

where $B(\cdot, \cdot)$ is the beta function, $\gamma = my/(n + my)$ and $I_\gamma(\cdot, \cdot)$ is the incomplete beta function ratio, tabulated by Pearson (1934). The mean and variance of Y are $n/(n-2)$ for $n > 2$, and $2n^2(m + n - 2)/\{m(n-2)^2(n-4)\}$ for $n > 4$, respectively.

[5.5.2] Let $X_1, X_2, \ldots, X_m$ and $Y_1, Y_2, \ldots, Y_n$ be independent iid random samples of $N(\mu, \sigma^2)$ and $N(\xi, \sigma^2)$ rvs, respectively, with sample mean $\bar{X}$ and $\bar{Y}$, so that $\bar{X} = \sum X_i/m$ and $\bar{Y} = \sum Y_j/n$. Then the rv W defined by

$$W = \frac{\sum_{i=1}^m (X_i - \bar{X})^2 (m-1)^{-1}}{\sum_{j=1}^n (Y_j - \bar{Y})^2 (n-1)^{-1}}$$

has a $F_{m-1, n-1}$ distribution. Further, the rv $\sum(X_i - \bar{X})^2 / \{\sum(X_i - \bar{X})^2 + \sum(Y_j - \bar{Y})^2\}$, which is equal to $(m-1)W/\{n-1+(m-1)W\}$, has a beta distribution with pdf $x^{(m-3)/2}(1-x)^{(n-3)/2}/B(\tfrac{1}{2}(m-1), \tfrac{1}{2}(n-1))$, $0 \leq x \leq 1$; $m, n \geq 2$ (Johnson and Kotz, 1970b, p. 78).

[5.5.3] If T is a rv with a t_ν distribution, then the rv T^2 has a $F_{1,\nu}$ distribution. Cacoullos (1965, pp. 529, 1249), showed that

$$F_{n, n; \alpha} = 1 + 2n^{-1} t_{n;\alpha}^2 + 2t_{n;\alpha}(n^{-1} + n^{-2} t_{n;\alpha}^2)^{1/2},$$

$$t_{n;\alpha} = \tfrac{1}{2}\sqrt{n}\,(F_{n,n;\alpha}^{1/2} - F_{n,n;\alpha}^{-1/2});$$

in the correction to his paper (ibid., p. 1249), the author acknowledges that J. O. Irwin (1953, p. 228) "was apparently aware of the relation between t and F in question".

[5.5.4] Let U have a noncentral $\chi_m^2(\lambda)$ distribution, and V a χ_n^2 distribution, where U and V are independent. Then the rv Y defined by $Y = Um^{-1}/(Vn^{-1})$ has a *noncentral F distribution* with (m, n) degrees of freedom and noncentrality parameter λ. See *[5.3.6]* for a caution regarding various differences in defining λ in the literature; here it is based on that in *[5.3.6]*. Denote noncentral F by $F_{m,n}(\lambda)$, and its quantiles by $F_{m,n;\beta}(\lambda)$, where $\Pr(Y \leq F_{m,\ n;\ \beta}(\lambda)) = 1 - \beta$. The pdf $g(y; m, n, \lambda)$ is given by

$$g(y; m, n, \lambda) = \sum_{j=0}^{\infty} \left[\frac{e^{-(1/2)\lambda}[(1/2)\lambda]^j}{j! B[(1/2)m + j, (1/2)n]} \left(\frac{m}{n}\right)^{(1/2)m+j} y^{(1/2)m+j-1} \right.$$

$$\left. \times \left(1 + \frac{m}{n}y\right)^{-(1/2)(m+n)-j} \right], y > 0, \lambda > 0, m,n = 1, 2, \ldots,$$

where $B(\cdot, \cdot)$ is the beta function. The mean and variance of Y are, respectively, $n(m + \lambda)/\{m(n-2)\}$ for $n > 2$ and $2(n/m)^2\{(m + \lambda)^2 + (m + 2\lambda)/(n-2)\}/\{(n-2)^2(n-4)\}$ for $n > 4$ (Johnson and Kotz, 1970b, pp. 190-191).

[5.5.5] Let U, V, and Y be defined as in *[5.5.4]*, except that V has a noncentral $\chi_n^2(\eta)$ distribution. Then Y has a *doubly noncentral F distribution* with (m, n) degrees of freedom and noncentrality parameters (λ, η). See Johnson and Kotz (1970b, Chap. 30) for further discussion.

[5.5.6] If the rv X has a noncentral $F_{m,n}(\lambda)$ distribution, and if $U = mX/(n + mX)$, then the rv U has a *noncentral beta distribution* denoted by $B[(\frac{1}{2}m, \frac{1}{2}n; \lambda]$, with pdf $g(u; m, n, \lambda)$ given by

$$g(u; m, n, \lambda) = \frac{e^{-\lambda/2} \sum_{j=0}^{\infty} \left(\frac{1}{2}\lambda\right)^j (j!)^{-1} u^{(1/2)m+j-1}(1-u)^{(1/2)n-1}}{B[(1/2)m + j, (1/2)n]},$$

$$0 \leq u \leq 1, \lambda > 0; m, n = 1, 2, \ldots.$$

See Graybill (1961, p. 79), where his noncentrality parameter λ is half the corresponding λ here; the caution given in *[5.3.6]* applies.

The cdf of the above distribution is $G(u; m, n, \lambda)$, where

$$G(u; m, n, \lambda) = e^{-\lambda/2} \sum_{j=0}^{\infty} \left(\frac{1}{2}\lambda\right)^j (j!)^{-1} I_u\left(\frac{1}{2}m + j, \frac{1}{2}n\right), 0 \leq u \leq 1;$$

$I_u(\cdot, \cdot)$ is the incomplete beta function ratio, tabulated by Pearson (1934). An alternative form for the pdf is given by Seber (1963, p. 542).

Sampling Distributions 129

[5.5.7] Let X_i be a $N(\mu_i, 1)$ rv $(i = 1, ..., m)$ and Y_j a $N(0, 1)$ rv $(j = 1, ..., n)$, where $X_1, ..., X_m, Y_1, ..., Y_n$ are mutually independent. Let $S = \sum X_i^2$, $T = \sum Y_j^2$, and $\lambda = \sum \mu_i^2$. Then the rv $S/(S + T)$ has a noncentral $B(\frac{1}{2}m, \frac{1}{2}n; \lambda)$ distribution, given above in *[5.5.6]*. See also Hodges (1955, pp. 648-653), who defines noncentral beta in terms of $T/(S + T)$, that is, $1 - S(S + T)^{-1}$, rather than by $S/(S + T)$.

5.6 THE SAMPLE MEAN DEVIATION

[5.6.1] The *mean deviation* d of a normal sample of size n is defined by

$$d = \sum |X_i - \bar{X}|/n .$$

Godwin (1945, pp. 254-256) obtained the distribution of d, but it is not simple or easily tractable. Pearson and Hartley (1966, Table 21) give upper and lower 0.1, 0.5, 1.0, 2.5, 5.0, and 10.0 percentiles of the distribution of d in a normal sample when $\sigma = 1$, for $n = 2(1)10$, to three decimal places. Pearson and Hartley (1972, Table 8) give values of the cdf of d when $\sigma = 1$ for $d = 0.0(0.01)3.0$ and $n = 2(1)10$, to five decimal places. A method for approximating to the distribution when $n > 10$ is given in Pearson and Hartley (1966, p. 89). Cadwell (1954, pp. 12-17) shows that $c(d/\sigma)^{1.8}$ has an approximate χ_q^2 distribution for values of c and q tabulated when $n = 4(1)10(5)50$. The distribution of d has a $N(\sigma/\sqrt{2/\pi}, (1 - 2\pi^{-1})\sigma^2/n)$ distribution, approximately, when n is very large.

[5.6.2] *Moments of d* (Herrey, 1965, p. 259; Kamat, 1954, pp. 541-542). The first three moments of d/σ under normality are as follows:

$$E(d) = \sigma \sqrt{2(n-1)/(n\pi)} ,$$

$$\text{Var}(d) = \frac{2\sigma^2(n-1)[(\pi/2) + (n^2 - 2n)^{1/2} - n + \arcsin\{(n-1)^{-1}\}]}{n^2 \pi} ,$$

$$\mu_3(d) = \sigma^3\left(\frac{n-1}{n}\right)^{3/2}\left[\sqrt{\frac{2}{\pi}}\frac{(4-n)}{n^2(n-1)} + \left(\frac{2}{\pi}\right)^{3/2}\left\{\frac{(n-2)}{n}\left(\frac{n-3}{n-1}\right)^{1/2} + 2\right.\right.$$
$$\left.\left. + \frac{3(n-2)^2}{n^2(n-1)}\arcsin\left(\frac{1}{n-2}\right) - \frac{3(n-1)}{n}\left[1 - \frac{1}{(n-1)^2}\right]^{1/2} - \frac{3}{n}\arcsin\left(\frac{1}{n-1}\right)\right\}\right].$$

Kamat also gives an expression for the fourth moment.

Geary (1936, p. 301) gave expansions for several moments; with a refinement for μ_3 and μ_4 by Pearson (1945, p. 252), and with $\nu = n-1$ these are:

$$\sigma^{-2}\text{Var}(d) = \{(n-1)/n\}\{(0.0450,70)\nu^{-1} - (0.1246,48)\nu^{-2}$$
$$+ (0.0848,59)\nu^{-3} + (0.0063,23)\nu^{-4}\} + O(n^{-5}),$$

$$\sigma^{-3}\mu_3(d) = \{(n-1)/n\}^{3/2}\{(0.2180,14)\nu^{-2} - (0.0741,70)\nu^{-3}$$
$$+ (0.0573,13)\nu^{-4} - (0.0404,57)\nu^{-5}\} + O(n^{-6}),$$

$$\sigma^{-4}\mu_4(d) = 3\{\text{Var}(d)/\sigma^2\}^2 + \{(n-1)/n\}^2\{(0.1147,71)\nu^{-3}$$
$$- (0.0685,09)\nu^{-4} + (0.0333,71)\nu^{-5}\} + O(n^{-6}).$$

Johnson (1958, p. 481) gives the approximation

$$\sigma^{-2}\text{Var}(d) \simeq n^{-1}(1 - 2/\pi)\{1 - (0.12)n^{-1}\}.$$

See Pearson and Hartley (1966, pp. 41-42) for references to fuller information for small samples; in their Table 20, values of E(d), Var(d), s.d.(d), and the shape factors are given for n = 2(1)20,30,60 to decimal places varying from three to six.

[5.6.3] The ratio of the sample mean deviation to sample standard deviation, d/S, is used for detecting changes in kurtosis, i.e., as a test of normality. The statistic that has been studied is *Geary's a*, where

$$a = \sum|X_i - \bar{X}|/[n\sum\{(X_i - \bar{X})^2\}]^{1/2} = d/\sqrt{m_2} = \{\sqrt{n/(n-1)}\}d/S.$$

The corresponding value in the normal population is $\nu_1/\sqrt{\nu_2} = \nu_1/\sqrt{\mu_2} = \sqrt{2/\pi}$, or 0.7978,8456. Geary (1936, p. 303) gives the following

Sampling Distributions

approximation to the pdf g(a) of a: if $E(a) = \eta$ and $\mu_3(a)/\{Var(a)\}^{3/2} = \lambda$, the shape factor for skewness of the distribution of a, then

$$g(a) \simeq \frac{1}{\sigma\sqrt{2\pi}}\left[1 - \frac{\lambda}{6}\left\{\frac{3(a-\eta)}{\sigma} - \frac{(a-\eta)^3}{\sigma^3}\right\}\right]\exp\left\{-\frac{(a-\eta)^2}{2\sigma^2}\right\}, \; a > 0.$$

Expressions for the moments appearing in this approximation are given in [5.6.6] below; the approximation is the basis of Pearson and Hartley's (1966) Table 34A, which gives upper and lower 1, 5, and 10 percent points of g(a) to four decimal places for $n-1 = 10(5)50(10)100(100)1000$, and corresponding values of E(a) and s.d.(a) to five decimal places. Geary (1936, pp. 304-305) gives charts from which the values of these percent points can be read off for sample sizes between 11 and 1001.

[5.6.4] Geary's a and the sample variance S^2 are independent (Geary, 1936, p. 296).

[5.6.5] The moments of Geary's a satisfy the relation

$$E(a^r) = \{n/(n-1)\}^{r/2}E(d^r)/E(S^r),$$

where d is the sample mean deviation and S^2 the sample variance (Geary, 1936, p. 296).

[5.6.6] The exact first four moments of Geary's a are given by

$$E(a) = ((n-1)/\pi)^{1/2}\Gamma\{(n-1)/2\}/\Gamma(n/2),$$

$$E(a^2) = n^{-1} + \{2/(n\pi)\}[(n^2-2n)^{1/2} + \arcsin\{(n-1)^{-1}\}],$$

$$E(a^3) = \{n/(n-1)\}^{1/2}E(d^3/\sigma^3)/E(S/\sigma),$$

$$E(a^4) = \{n^2/(n^2-1)\}E(d^4/\sigma^4).$$

Geary (1936, p. 301) gives the following expansions in powers of ν^{-1}, where $\nu = n-1$ and $\sqrt{2/\pi} \simeq 0.7978,846$:

$$E(a) = (\sqrt{2/\pi}) + (0.1994,71)\nu^{-1} + (0.0249,34)\nu^{-2}$$

$$- (0.0311{,}68)\nu^{-3} - (0.0081{,}82)\nu^{-4} + O(n^{-5}),$$

$$\text{Var}(a) = (0.0450{,}70)\nu^{-1} - (0.1246{,}48)\nu^{-2} + (0.1098{,}49)\nu^{-3} \\ + (0.0063{,}23)\nu^{-4} + O(n^{-5}),$$

$$\mu_3(a) = -(0.0168{,}57)\nu^{-2} + (0.0848{,}59)\nu^{-3} - (0.2418{,}25)\nu^{-4} + O(n^{-5}),$$

$$\mu_4(a) = 3\{\text{Var}(a)\}^2 + (0.0110{,}51)\nu^{-3} - (0.1454{,}43)\nu^{-4} + O(n^{-5});$$

for the shape factors,

$$\mu_3(a)/\{\text{Var}(a)\}^{3/2} = -1.7618\{1 - (2.3681)\nu^{-1} - (8.8646)\nu^{-2}\}\sqrt{\nu} + O(n^{-7/2}),$$

$$\mu_4(a)/\{\text{Var}(a)\}^2 = 3 + 5.441\{1 - (7.628)\nu^{-1}\}/\nu + O(\nu^{-3}).$$

5.7 THE MOMENT RATIOS $\sqrt{b_1}$ AND b_2

The sample rth moment m_r of a random sample $X_1, ..., X_n$ from any population is $\sum_{i=1}^{n}(X_i - \bar{X})^r/n$, calculated about the sample mean $\bar{X}$. Of interest are the standardized third and fourth moments $\sqrt{b_1}$ and b_2, respectively; these have figured in tests for departure from normality. A useful source that expands upon the summary following and that gives further references is Bowman and Shenton (1986, pp. 279-329).

Many of the properties of $\sqrt{b_1}$ and b_2 were obtained after R.A. Fisher had developed the structure that he called k-statistics. These are symmetric functions of the observations in a sample. The rth k-statistic κ_r has as its expected value the rth cumulant k_r of the underlying population; see Stuart and Ord (1987, Chapters 12 and 13).

[5.7.1] The third sample moment m_3, where $m_3 = \sum(X_i - \bar{X})^3/n$, has been studied in standardized form rather than on its own. This is $\sqrt{b_1}$, a measure of skewness of the sample, i.e.,

Sampling Distributions 133

$$\sqrt{b_1} = \frac{m_3}{m_2^{3/2}} = \frac{\sqrt{n}\,\Sigma(X_i-\bar{X})^3}{\left[\Sigma(X_i-\bar{X})^2\right]^{3/2}}.$$

[5.7.1.1] The sampling distribution of $\sqrt{b_1}$ under normality has been approximated by a Pearson Type VII (Student t-type) and by a symmetric Johnson S_u distribution (Pearson, 1963, pp. 95-111), judged acceptable for $n \geq 30$. D'Agostino and Tietjen (1973, pp. 169-173) compared these approximations for smaller sample sizes and found them to be satisfactory for $n \geq 8$.

[5.7.1.2] D'Agostino (1970, pp. 679-681) showed that Johnson's S_u for $n \geq 8$ leads to the following simple normal approximation, which agrees between the first and 99th percent points with computer simulation results to two or more decimal places: Z is approximately N(0, 1), where

$$Z = \delta \log\left[U + \sqrt{1 + U^2}\right],$$

$$\delta = \frac{1}{\sqrt{\log W}}, \quad U = \left[\frac{(W^2-1)(n+1)(n+3)}{12(n-2)}\right]^{1/2} \sqrt{b_1},$$

$$W^2 = \left[2(\beta-1)\right]^{1/2} - 1, \quad \beta = \frac{3(n^2+27n-70)(n+1)(n+3)}{(n-2)(n+5)(n+7)(n+9)}.$$

See also Section 5 of D'Agostino and Pearson (1973, pp. 618-621).

[5.7.1.3] Pearson and Hartley (1966, Table 34B) tabulate the first and fifth percentiles and values of the standard deviation of $\sqrt{b_1}$ under normality for $n = 25(5)50(10)100(25)200(50)1000(200)2000(500)5000$. Note, however, that while the third edition of *Biometrika Tables for Statisticians,* Vol. 1 (Pearson and Hartley, 1966) contains corrections by Pearson (1965), the earlier editions do not have these.

[5.7.2] *Moments of $\sqrt{b_1}$ under Normality* (Fisher, 1930, pp. 16-28; Geary, 1947, p. 68). The distribution being symmetric about the origin, the odd moments all vanish.

$$\text{Var}(\sqrt{b_1}) = 6(n-2)/\{(n+1)(n+3)\},$$

$$\mu_4(\sqrt{b_1}) = \frac{108(n-2)(n^2 + 27n - 70)}{(n+1)(n+3)(n+5)(n+7)(n+9)}.$$

Pearson (1930, p. 242) has given expansions in powers of $1/n$ of the standard deviation and kurtosis:

$$\text{s.d.}(\sqrt{b_1}) = \sqrt{(6/n)} \,(1 - 3n^{-1} + 6n^{-2} - 15n^{-3}) + O(n^{-9/2}),$$

$$\mu_4(\sqrt{b_1})/\{\text{Var}(\sqrt{b_1})\}^2 = 3 + 36n^{-1} - 864n^{-2} + 1{,}2096n^{-3} + O(n^{-4}).$$

The statistic $[(n+1)(n+3)/\{6(n-2)\}]^{1/2}\sqrt{b_1}$ has unit variance and fourth moment $3 + 36n^{-1} + 0(n^{-2})$, indicating a reasonably rapid approach to normality (Stuart and Ord, 1987, Sec. 12.18).

[5.7.3] A measure of the *kurtosis* of a sample is b_2, where

$$b_2 = n\sum\{(X_i - \bar{X})^4\}/[\sum\{(X_i - \bar{X})^2\}]^2 = m_4/m_2^2,$$

and $m_4 = \sum\{(X_i - \bar{X})^4\}/n$, the sample fourth moment. This statistic has been used along with $\sqrt{b_1}$ in tests of normality.

[5.7.3.1] The sampling distribution of b_2 under normality ranges from 0 to ∞, and is highly skewed in general. It can be approximated by a skew Pearson Type IV or skew Johnson S_u distribution by equating the first four moments (Pearson, 1963, pp. 106-109). For $25 \leq n \leq 50$, the S_u curve gives an adequate fit except in the tails, and for $n \geq 200$ the better fit is provided by the Pearson Type IV curves (D'Agostino and Pearson, 1973, pp. 614-615). D'Agostino and Tietjen (1971, pp. 669-672) note that curve-fitting techniques for b_2 can be problematic, but that Cornish-Fisher, Gram-Charlier and Edgeworth expansions using up to the first seven moments are ineffective even when $n = 100$.

[5.7.3.2] Anscombe and Glynn (1983, pp. 227-230) fitted a Pearson Type V distribution — essentially a linear function of the reciprocal of a chi-square variable — to b_2 under normality. This is done by equating the first three moments; see *[5.7.4]*. Let

$$A = 6 + \frac{8}{\sqrt{\beta_1}}\left[\frac{2}{\sqrt{\beta_1}} + \sqrt{1 + \frac{4}{\beta_1}}\,\right],$$

Sampling Distributions

$$\sqrt{\beta_1} = \frac{\mu_3(b_2)}{\{\text{Var}(b_2)\}^{3/2}} = \frac{6(n^2 - 5n + 2)}{(n+7)(n+9)} \left\{ \frac{6(n+3)(n+5)}{n(n-2)(n-3)} \right\}^{\frac{1}{2}}, \quad n \geq 4.$$

Here A denotes degrees of freedom of the chi-square variable. For $n \geq 4$, A is smallest when $n = 24$, and thereafter is "just over 18". Finally, convert the chi-square variable to approximate normality via the Wilson-Hilferty transformation *[7.8.4]*. If $x = [b_2 - \mu_1'(b_2)]/\sqrt{\text{Var}(b_2)}$, then the variable

$$\left[\left(1 - \frac{2}{9A}\right) - \left\{ \frac{1 - (2/A)}{1 + x\sqrt{2/(A-4)}} \right\}^{\frac{1}{3}} \right] \bigg/ \sqrt{\frac{2}{9A}}$$

is distributed $N(0,1)$, approximately. Anscombe and Glynn found that for all percent points from 5% up through 99.75% and for $30 \leq n \leq 200$ the errors in approximation never exceeded 0.04, and usually did not exceed 0.03, when compared with exact percentage points of D'Agostino and Pearson (1973, pp. 613-622).

[5.7.3.3] D'Agostino and Pearson (1973, pp. 613-622) provide percent points of b_2 for $20 \leq n \leq 200$, extending from 0.01% to 99.9%. These are presented via charts and are based on computer simulations. D'Agostino and Tietjen (1971, pp. 669-672) have tabulated the kth percentiles of b_2 to two decimal places, for $1 \leq k \leq 99$ and $n = 7(1)10,12,15(5)50$.

Pearson and Hartley (1966, Table 34C) tabulate upper and lower 5 and 1 percentiles of the distribution of b_2 under normality for $n = 50(25)150(50)700(100)1000(200)2000(500)5000$. However, while the third edition of *Biometrika Tables for Statisticians*, Vol. 1 (Pearson and Hartley, 1966) contains the percentage points for $25 < n < 200$ given by Pearson (1965, p. 284), the earlier editions do not have these.

[5.7.4] *Moments of b_2 under normality* (Fisher, 1930, pp. 16-28; Hsu and Lawley, 1939, p. 246). Recall that for a $N(\mu, \sigma^2)$ rv, the kurtosis is equal to 3.

$$E(b_2) = \frac{3(n-1)}{n+1}, \quad \text{Var}(b_2) = \frac{24n(n-2)(n-3)}{(n+1)^2(n+3)(n+5)},$$

$$\mu_3(b_2) = \frac{1728n(n-2)(n-3)(n^2-5n+2)}{(n+1)^3(n+3)(n+5)(n+7)(n+9)},$$

$$\mu_4(b_2) = \frac{1728n(n-2)(n-3)(n^5+207n^4-1707n^3+4105n^2-1902n+720)}{(n+1)^4(n+3)(n+5)(n+7)(n+9)(n+11)(n+13)}.$$

Pearson (1930, p. 243) gave the following expansions:

$$\text{s.d.}(b_2) = \sqrt{\tfrac{24}{n}}\left(1 - \tfrac{15}{2n} + \tfrac{271}{8n^2} - \tfrac{2319}{16n^3}\right) + O(n^{-9/2}),$$

$$\text{Skewness} = \frac{\mu_3(b_2)}{\{\mu_2(b_2)\}^{3/2}} = \frac{216}{n}\left(1 - \tfrac{29}{n} + \tfrac{519}{n^2} - \tfrac{7637}{n^3}\right) + O(n^{-5}),$$

$$\text{Kurtosis} = \frac{\mu_4(b_2)}{\{\mu_2(b_2)\}^2} = 3 + \frac{540}{n} - \frac{20{,}196}{n^2} + \frac{470{,}412}{n^3} + O(n^{-4}).$$

[5.7.5] The moment ratios $\sqrt{b_1}$ and b_2 are uncorrelated but not independent under normality (D'Agostino and Pearson, 1973, pp. 613-622; correction, 1974, p. 647). However, each of $\sqrt{b_1}$ and b_2 is independent of s^2 (Fisher, 1930, pp. 17-28). Bowman and Shenton (1986, pp. 280-287, 295-306) discuss the joint behavior of $(\sqrt{b_1}, b_2)$ under normal sampling; see also Shenton and Bowman (1977, Section 3).

5.8 MISCELLANEOUS RESULTS

[5.8.1] *Sampling Distributions under Linear Regression.* Suppose that $Y_1, Y_2, ..., Y_n$ are iid and that Y_i has an $N(\alpha + \beta(x_i - \bar{x}), \sigma^2)$ distribution; $i = 1, 2, ..., n$, where $\bar{x} = \sum x_i/n$. Here $x_1, ..., x_n$ may be predetermined constants such as temperature, pressure, thickness, etc., or may be realizations of one variable in a bivariate normal sample (see Chapter 10). Let Σ denote summation from $i = 1$ to $i = n$, and

Sampling Distributions

$$a = \bar{Y} = \sum Y_i/n, \quad b = \sum\{(x_i - \bar{x})Y_i\}/\sum\{(x_i - \bar{x})^2\},$$

$$S^2 = \sum_{i=1}^{n}\{Y_i - a - b(x_i - \bar{x})^2/(n-2).$$

Then $\sum\{Y_i - a - b(x_i - \bar{x})\}^2/\sigma^2$ has a χ^2_{n-2} and $(b-\beta)^2\sum(x_i - \bar{x})^2/\sigma^2$ has independently a χ^2_1 distribution; thus $(b-\beta)\sqrt{\{\sum(x_i - \bar{x})^2\}}/S$ has a t_{n-2} distribution. If Y is a future observation, with mean $\alpha + \beta(x - \bar{x})$, then (Brownlee, 1965, pp. 335-342)

$$\{Y - \alpha - \beta(x - \bar{x})\}/(S[n^{-1} + (x - \bar{x})^2/\sum\{(x_i - x)^2\}]^{1/2})$$

has a t_{n-2} distribution, and so also does

$$\{Y - a - b(x - \bar{x})\}/(S[1 + n^{-1} + (x - \bar{x})^2/\sum\{(x_i - \bar{x})^2\}]^{1/2}).$$

[5.8.2] If X has a $N(\theta, 1)$ distribution and Y has independently a χ^2_n distribution, let $Q = X/(X^2 + Y^2)^{1/2}$. Then $\sqrt{n}\, Q/(1-Q^2)^{1/2}$ has a noncentral $t_n(\theta)$ distribution, and Q^2 a noncentral $B(\frac{1}{2}, \frac{1}{2}n; \theta^2)$ distribution (see *[5.5.6]*). Hogben et al. (1964, pp. 298-314, 316) derive moments and an approximation to the distribution of Q. The first four central moments are tabulated to six decimal places for $\theta = 0.1(0.1)1.0(0.2)2(0.5)6(1)10$ and $n = 1(1)24(5)40,50,60,80,100$.

[5.8.3] If X has a $N(\theta, 1)$ distribution and Y has independently a noncentral χ^2_n distribution, then the variables X/Y and $X/(X^2 + Y^2)^{1/2}$ are *generalized noncentral t* and *generalized noncentral beta* variables, respectively. Park (1964, pp. 1584, 1588) has obtained approximations to the first few moments, and asymptotic expressions for the pdfs in the tails of these distributions.

[5.8.4] Let $X_1, ..., X_n$ be iid $N(0, 1)$ random variables, $S_r = X_1 + \cdots + X_r$ ($r = 1, 2, ..., n$), and U_n the maximum of the partial sums $S_1, ..., S_n$. The first two moments about the origin of U_n are given by

$$\mu'_1(U_n) = (2\pi)^{-1/2} \sum_{s=1}^{n-1} s^{-1/2}$$

$$\mu_2'(U_n) = (n+1)/2 + (2\pi)^{-1} \sum_{r=1}^{n-2} \sum_{s=1}^{r} \{s(r-s+1)\}^{-1/2} .$$

See, for example, Anis (1956, pp. 80-83), where the first four moments about the mean are tabulated for $n = 2(1)15$, or Solari and Anis (1957, pp. 706-716), where the moments of the maximum of adjusted partial sums $\{S_r - r\bar{X}\}$ are discussed. These statistics may be of interest in storage problems.

[5.8.5] Let $X_1, ..., X_n$ be iid $N(0, \sigma^2)$ rvs, and let

$$D^2 = \nu^{-1} \sum_{j=1}^{\nu} (X_{j+1} - X_j)^2, \quad \nu = n-1 ,$$

the *mean square successive difference*. Shah (1970, pp. 193-198) represents the cdf of $D^2/(2\sigma^2)$ in terms of Laguerre polynomials. Harper (1967, p. 421) derived moments of q^2, where $q^2 = D^2/2$:

$$E(q^2) = 1, \quad Var(q^2) = \nu^{-2}(3\nu - 1), \quad \nu \geq 1 ,$$
$$\mu_3(q^2) = 4\nu^{-3}(5\nu - 3), \quad \nu \geq 1,$$
$$\mu_4(q^2) = 3\nu^{-4}(9\nu^2 + 64\nu - 57), \quad \nu \geq 2.$$

Harper (1967) gives the first eight moments and cumulants, obtains a table of percentage points of the q^2 distribution, and considers some normal and chi-square approximations.

[5.8.6] The following results are useful in determining prediction limits for a future observation X when we have an iid sample $X_1, ..., X_n$ from a $N(\mu, \sigma^2)$ population available, with μ and σ generally unknown; see, e.g., Whitmore (1986, pp. 141-143).

Let $\bar{X}$ and S^2 be the mean and sample variance of $X_1, ..., X_n$. Then
i) $E(X - \bar{X}) = 0$ and $Var(X - \bar{X}) = (1 + 1/n)\sigma^2$.
ii) $X - \bar{X}$ is normally distributed, $N(0, (1 + 1/n)\sigma^2)$.
iii) $X - \bar{X}$ and S^2 are independent.

iv) $\dfrac{X - \bar{X}}{S(1 + 1/n)^{1/2}} \sim \;'t'_{n-1}$.

[5.8.7] Springer (1979, Secs. 4.5, 4.6) derives the distribution of the product of n independent $N(\mu_i, \sigma_i^2)$ variables, and the distribution of the quotient of independent normal rvs having different means and variances, via techniques that employ the Mellin transform.

REFERENCES

Abramowitz, M. and Stegun, I. A. (1964). *Handbook of Mathematical Functions*, Washington, D.C: National Bureau of Standards. *[5.3.3]*

Anis, A. A. (1956). On the moments of the maximum of partial sums of a finite number of independent normal variates, *Biometrika 43*, 79-84. *[5.8.4]*

Anscombe, F. J. and Glynn, W. (1983). Distribution of the kurtosis statistic b_2 for normal samples, *Biometrika 70*, 227-234. *[5.7.3.2]*

Aroian, L. A., Taneja, V. S. and Cornwell, L. W. (1978). Mathematical forms of the distribution of the product of two normal variables, *Communications in Statistics A7(2)*, 165-172. *[5.1.3]*

Birnbaum, Z. W. and Saunders, S. C. (1969). A new family of life distributions, *Journal of Applied Probability 6*, 319-327. *[5.1.6]*

Bowman, K. O. and Shenton, L. R. (1986). Moment ($\sqrt{b_1}$, b_2) techniques, *Goodness-of-Fit Techniques* (R. B. D'Agostino and M.A. Stephens, eds.), pp. 279-329, New York: Dekker. *[5.7; 5.7.5]*

Brownlee, K. A. (1965). *Statistical Theory and Methodology in Science and Engineering* (2nd edn.), New York: Wiley. *[5.8.1]*

Cacoullos, T. (1965). A relation between t and F distributions, *Journal of the American Statistical Association 60*, 528-531; correction, *ibid.*, 1249. *[5.4.5; 5.5.3]*

Cadwell, J. H. (1954). The statistical treatment of mean deviation, *Biometrika 41*, 12-18. *[5.6.1]*

Chu, J. T. (1956). Errors in normal approximations to t, τ, and similar types of distribution, *Annals of Mathematical Statistics 27*, 780-789. *[5.4.6]*

Craig, C. C. (1936). On the frequency function of xy, *Annals of Mathematical Statistics 7*, 1-15. *[5.1.3]*

D'Agostino, R. B. (1970). Transformation to normality of the null distribution of g_1, *Biometrika 57*, 679-681. *[5.7.1.2]*

D'Agostino, R. B. and Pearson, E. S. (1973). Tests for departure from normality. Empirical results for the distribution of b_2 and $\sqrt{b_1}$, *Biometrika 60*, 613-622. *[5.7.1.2; 5.7.3.1, 2, 3; 5.7.5]*

D'Agostino, R. B. and Tietjen, G. L. (1971). Simulation probability points of b_2 for small samples, *Biometrika 58*, 669-672. *[5.7.3.1, 3]*

D'Agostino, R. B. and Tietjen, G. L. (1973). Approaches to the null distribution of $\sqrt{b_1}$, *Biometrika 60*, 169-173. *[5.7.1.1]*

Daly, J. F. (1946). On the use of the sample range in an analogue of Student's t-test, *Annals of Mathematical Statistics 17*, 71-74. *[5.2.3]*

David, F. N. (1949). Note on the application of Fisher's k-statistics, *Biometrika 36*, 383-393. *[5.3.5; 5.4.10, 13]*

Epstein, B. (1948). Some applications of the Mellin transform in statistics, *Annals of Mathematical Statistics 19*, 370-379. *[5.1.3]*

Fisher, R. A. (1930). The moments of the distribution for normal samples of measures of departures from normality, *Proceedings of the Royal Society of London A130*, 16-28. *[5.7.2, 4, 5]*

Fisher, R. A. and Cornish, E. A. (1960). The percentile points of distributions having known cumulants, *Technometrics 2*, 209-225. *[5.4.5]*

Fox, C. (1965). A family of distributions with the same ratio property as normal distribution, *Canadian Mathematical Bulletin 8*, 631-636. *[5.1.1]*

Geary, R. C. (1936). Moments of the ratio of the mean deviation to the standard deviation for normal samples, *Biometrika 28*, 295-305. *[5.6.2, 3, 4, 5, 6]*

Geary, R. C. (1947). The frequency distribution of $\sqrt{b_1}$ for samples of all sizes drawn at random from a normal population, *Biometrika 34*, 68-97. *[5.7.2]*

Godwin, H. J. (1945). On the distribution of the estimate of mean deviation obtained from samples from a normal population, *Biometrika 33*, 254-256. *[5.6.1]*

Graybill, F. A. (1961). *An Introduction to Linear Statistical Models*, Vol. 1, New York: McGraw-Hill. *[5.3.6; 5.5.6]*

Guenther, W. C. (1964). Another derivation of the non-central chi-square distribution, *Journal of the American Statistical Association 59*, 957-960. *[5.3.6]*

Han, C. P. (1975). Some relationships between noncentral chi-squared and normal distributions, *Biometrika* 62, 213-214. *[5.3.7]*

Harper, W. M. (1967). The distribution of the mean half-square successive difference, *Biometrika* 54, 419-433. *[5.8.5]*

Hawkins, D. M. (1975). From the noncentral t to the normal integral, *American Statistician* 29, 42-43. *[5.4.9]*

Herrey, E. M. J. (1965). Confidence intervals based on the mean absolute deviation of a normal sample, *Journal of the American Statistical Association* 60, 257-269. *[5.2.3; 5.6.2]*

Hodges, J. L. (1955). On the non-central beta distribution, *Annals of Mathematical Statistics* 26, 648-653. *[5.5.7]*

Hogben, D., Pinkham, R. S., and Wilk, M. B. (1964). (1) The moments of a variate related to the non-central "t"; (2) An approximation to the distribution of Q, *Annals of Mathematical Statistics* 35, 298-318. *[5.8.2]*

Hogg, R. V. (1960). Certain uncorrelated statistics, *Journal of the American Statistical Association* 62, 265-267. *[5.2.3]*

Hsu, C. T. and Lawley, D. N. (1939). The derivation of the fifth and sixth moments of the distribution of b_2 in samples from a normal population, *Biometrika* 31, 238-248. *[5.7.4]*

Iglewicz, B. and Myers, R. H. (1970). Comparisons of approximations to the percentage points of the sample coefficient of variation, *Technometrics* 12, 166-169. *[5.4.12]*

Iglewicz, B., Myers, R. H. and Howe, R. B. (1968). On the percentage points of the sample coefficient of variation, *Biometrika* 55, 580-581. *[5.4.11]*

Irwin, J. O. (1953). Discussion of Hotelling, H., New light on the correlation coefficient and its transforms, *Journal of the Royal Statistical Society B* 15, 228. *[5.5.3]*

Johnson, N. L. (1949). Systems of frequency curves generated by methods of translations, *Biometrika* 36, 149-176. *[5.7.1, 3]*

Johnson, N. L. (1958). The mean deviation, with special reference to samples from a Pearson Type III population, *Biometrika* 45, 478-483. *[5.6.2]*

Johnson, N. L. and Kotz, S. (1970a). *Distributions in Statistics: Continuous Univariate Distributions*, Vol. 1, New York: Wiley. *[5.3.1]*

Johnson, N. L. and Kotz, S. (1970b). *Distributions in Statistics: Continuous Univariate Distributions*, Vol. 2, New York: Wiley. *[5.3.6; 5.4.1, 7; 5.5.2, 4, 5]*

Johnson, N. L., Nixon, E., Amos, D. E. and Pearson, E. S. (1963). Tables of percentage points of Pearson curves, for given $\sqrt{\beta_1}$ and β_2, expressed in standard measure, *Biometrika* 50, 459-498. *[5.4.12]*

Johnson, N. L. and Welch, B. L. (1939). On the calculation of the cumulants of the χ-distribution, *Biometrika* 31, 216-218. *[5.3.5]*

Johnson, N. L. and Welch, B. L. (1940). Applications of the non-central t distribution, *Biometrika* 31, 362-389. *[5.4.10]*

Kamat, A. R. (1954). Moments of the mean deviation, *Biometrika* 41, 541-542. *[5.6.2]*

Koehler, K. J. (1983). A simple approximation for the percentiles of the t distribution, *Technometrics* 25, 103-105. *[5.4.5]*

Lancaster, H. O. (1969). *The Chi-Squared Distribution*, New York: Wiley. *[5.3.1, 3]*

McKay, A. (1932). Distribution of the coefficient of variation and the extended "t" distribution, *Journal of the Royal Statisticial Society* 95, 695-698. *[5.4.10, 12]*

Mann, N. R., Schafer, R. E. and Singpurwalla, N. D. (1974). *Methods for Statistical Analysis of Reliability and Life Data*, New York: Wiley. *[5.1.6]*

Mantel, N. (1973). A characteristic function exercise, *The American Statistician* 27(1), 31. *[5.1.5]*

Marsaglia, G. (1965). Ratios of normal variables and ratios of sums of uniform variables, *Journal of the American Statistical Association* 60, 193-204. *[5.1.1]*

Merrington, M. and Pearson, E. S. (1958). An approximation to the distribution of noncentral "t", *Biometrika* 45, 484-491. *[5.4.7]*

Owen, D. B. (1962). *Handbook of Statistical Tables*, Reading, Mass.: Addison-Wesley. *[5.3.8]*

Owen, D. B. (1968). A survey of properties and applications of the noncentral t-distribution, *Technometrics* 10, 445-478. *[5.4.9, 10]*

Park, J. H. (1964). Variations of the non-central t and beta distributions, *Annals of Mathematical Statistics* 35, 1583-1593. *[5.8.3]*

Patnaik, P. B. (1949). The non-central χ^2- and F-distributions and their applications, *Biometrika 36*, 202-232. *[5.3.6]*

Pearson, E. S. (1930). A further development of tests for normality, *Biometrika 22*, 239-249. *[5.7.2, 4]*

Pearson, E. S. (1945). The probability integral of the mean deviation, *Biometrika 33*, 252-253. *[5.6.2]*

Pearson, E. S. (1963). Some problems arising in approximating to probability distributions, using moments, *Biometrika 50*, 95-112. *[5.7.1.1; 5.7.3.1]*

Pearson, E. S. (1965). Tables of percentage points of $\sqrt{b_1}$ and b_2 in normal samples: A rounding off, *Biometrika 52*, 282-285. *[5.7.1.3]*

Pearson, E. S. and Hartley, H. O. (1966). *Biometrika Tables for Statisticians*, Vol. 1 (3rd edn.), London: Cambridge University Press. *[5.3.5; 5.6.1, 2, 3; 5.7.1.3; 5.3.3]*

Pearson, E. S. and Hartley, H. O. (1972). *Biometrika Tables for Statisticians*, Vol. 2, London: Cambridge University Press. *[5.6.1]*

Pearson, K., ed. (1934). *Tables of the Incomplete Beta Function*, London: Cambridge University Press. *[5.5.1, 6]*

Puri, P. S. (1973). On a property of exponential and geometric distributions and its relevance to multivariate failure rate, *Sankhyā A35*, 61-78. *[5.3.3]*

Rao, C. R. (1973). *Linear Statistical Inference and Its Applications* (2nd edn.), New York: Wiley. *[5.2.2; 5.3.9]*

Searle, S. R. (1971). *Linear Models*, New York: Wiley. *[5.3.6, 9]*

Seber, G. A. F. (1963). The noncentral chi-squared and beta distributions, *Biometrika 50*, 542-545. *[5.5.6]*

Shah, B. K. (1970). On the distribution of half the mean square successive difference, *Biometrika 57*, 193-198. *[5.8.5]*

Shenton, L. R. and Bowman, K. O. (1977). A bivariate model for the distribution of $\sqrt{b_1}$ and b_2, *Journal of the American Statistical Association 72*, 206-211. *[5.7.5]*

Shepp, L. (1964). Problem 62-9: Normal functions of normal random variables, *SIAM Review 6*, 459. *[5.1.4]*

Solari, M. E. and Anis, A. A. (1957). The mean and variance of the maximum of the adjusted partial sums of a finite number of independent normal variates, *Annals of Mathematical Statistics 28*, 706-716. *[5.8.4]*

Springer, M. D. (1979). *The Algebra of Random Variables*, New York: Wiley. *[5.1.1; 5.8.7]*

Steck, G. P. (1958). A uniqueness property not enjoyed by the normal distribution, *Annals of Mathematical Statistics 29*, 604-606. *[5.1.1]*

Stuart, A. and Ord, J. K. (1987). *Kendall's Advanced Theory of Statistics, Vol. 1* (5th edn.), New York: Oxford University Press. *[5.1.3; 5.7; 5.7.2]*

Summers, R. D. (1965). An inequality for the sample coefficient of variation and an application to variables sampling, *Technometrics 7*, 67 *[5.4.14]*

Umphrey, G. J. (1983). On the adequacy of the chi-squared approximation for the coefficient of variation, *Communications in Statistics B 12*, 629-635. *[5.4.10]*

Warren, W. G. (1982). On the adequacy of the chi-squared approximations for the coefficient of variation, *Communications in Statistics B 11*, 659-666.

Whitmore, G. A. (1986). Prediction limits for a univariate normal observation, *American Statistician 40*, 141-143. *[5.8.6]*

Chapter 6
LIMIT THEOREMS AND EXPANSIONS

The normal distribution holds a chief place among statistical distributions on account of the Central Limit Theorem, discussed in Sections *[6.1]* and *[6.2]*. The most widely used version of this theorem states that the standardized sample mean $\bar{X}_n$ of a random sample $X_1,...,X_n$ of size n from an infinite population with common mean μ and finite variance tends to the standard normal distribution as $n \to \infty$. That is, if $\bar{G}_n$ is the cdf of the standardized sample mean, then $\bar{G}_n(x) \to \Phi(x)$ for every value of x. In a more general set up that does not assume the existence of any moments, are there constants a_n and b_n such that $\Pr\left[(S_n - a_n)/b_n \leq x\right] \to \Phi(x)$ as $n \to \infty$, where $S_n = \sum_{i=1}^{n} X_i$? Here $(S_n - a_n)/b_n$ is a *normed sum*. In this chapter we shall be mainly interested in the behavior of such normed sums, but some results are based on *triangular arrays* $\{X_{in};\ i= 1,\ 2,\ ...,\ r_n;\ n = 1, 2, ...\}$, and others on increments in certain stochastic processes. In Section 6.3 we consider the *asymptotic normality* of more general sequences $\{T_n\}$ of rvs, such as statistics that are functions of a random sample $X_1, ..., X_n$.

For some sequences of distributions, convergence to normality may be rapid, while for others it can be very slow. This is an important consideration in trying to use a normal approximation based on the Central Limit Theorem, which in itself gives no information about the rapidity of convergence. The Berry-Esseen theorem, given in Section *[6.4]*, gives some information in the form of upper bounds to $\sup_x |\bar{G}_n(x) - \Phi(x)|$ that are proportional to $1/\sqrt{n}$, but there are also more recent results that give lower bounds to this quantity as well. We have listed results for quantiles separately in Section *[6.5]*. A number of series expansions of $\bar{G}_n(x)$ and of other asymptotically normal statistics about $\Phi(x)$ also provide approximations to $\bar{G}_n(x)$ for large or even moderate values of n (Section *[6.6]*).

In the next chapter, we consider normal approximations to the more commonly used distributions in statistics; many of these approximations are based on the series expansions in *[6.5]* or on asymptotic normality.

6.1 CLASSICAL CENTRAL LIMIT THEOREMS

Unless otherwise stated in this section X_1, X_2, ... is a sequence of mutually independent rvs with partial sums $S_n = \sum_{i=1}^{n} X_i$, n = 1, 2, ... Following the early history of the central limit theorem up to 1900, outlined in Chapter 1, the important breakthroughs occurred between 1920 and 1937, and included research by S. Bernstein, W. Feller, A. Y. Khinchine, A. N. Kolmogorov, P. Lévy and J. W. Lindeberg. LeCam (1986, pp. 78-96, with discussion) gives a fascinating account of the emergence of these results. Here in historical sequence we shall state many of them, in addition to early results, citing both original sources and some modern references. For other discussions, see Gnedenko (1968, pp. 94-121, 302-317), Gnedenko and Kolmogorov (1968, pp. 125-132), Cramér (1970, pp. 53-69) and Heyde (1983, pp. 651-655).

[6.1.1] *The de Moivre-Laplace Limit Theorem.* Let Y_n have a binomial distribution such that $\Pr(Y_n = y) = \binom{n}{x} p^y (1-p)^{n-y}$; y = 0, 1, 2, ..., n; 0 < p < 1. Let $Z_n = (Y_n - np)/\sqrt{np(1-p)}$, so that Z_n is standardized, with mean zero and variance one; let Z_n have cdf $G_n(\cdot)$. Then, if a < b,

$$\lim_{n \to \infty} \Pr(a \leq Z_n \leq b) = \Phi(b) - \Phi(a),$$

$$\lim_{n \to \infty} \Pr(a \leq Z_n < b) = \Phi(b) - \Phi(a),$$

$$\lim_{n \to \infty} \left[\frac{\Pr(a \leq Z_n \leq b)}{\Phi\{b + \tfrac{1}{2}(np(1-p))^{-1/2}\} - \Phi\{a - \tfrac{1}{2}(np(1-p))^{-1/2}\}} \right] = 1.$$

This last form is more accurate if the denominator is used as an approximation to $\Pr(a \leq Z_n \leq b)$ (de Moivre, 1756; Feller, 1968a, pp. 182-186; Gnedenko, 1968, p. 104; Woodroofe, 1975, pp. 97-106).

The proofs of this result usually rely on Stirling's approximation to n!

or on characteristic functions. A straightforward proof that relies only on the relative slopes of suitably normed line segments has been given by Jensen and Rootzén (1986, pp. 231-232).

[6.1.2] *The de Moivre-Laplace Local Limit Theorem.* With the conditions and notation of *[6.1.1]* (Gnedenko, 1968, p. 98; Woodroofe, 1975, pp. 98, 105).

$$\lim_{n\to\infty}\left[\frac{\sqrt{np(1-p)}\,\Pr(Y_n=m)}{\phi\{(m-np)/\sqrt{np(1-p)}\}}\right]=1;\quad m=0,1,2,\ldots,n.$$

[6.1.3] *Tchebyshev's Central Limit Theorem.* Let $X_1, X_2, \ldots$ have zero means and finite moments of all orders, and let $\sigma_i^2 = \text{Var}(X_i)$. Further, suppose that

(i) $\lim_{n\to\infty}\left[(\sigma_1^2 + \sigma_2^2 + \cdots + \sigma_n^2)/n\right]$ exists and is finite,

(ii) $|E(X_i^r)| < A_r < \infty,\quad r = 2, 3, \ldots$.

Then

$$\lim_{n\to\infty}\Pr\{a \leq S_n/(\sigma_1^2 + \cdots + \sigma_n^2)^{1/2} \leq b\} = \Phi(b) - \Phi(a).$$

Condition (i) is necessary and was added by Kolmogorov (Tchebyshev, 1890, pp. 305-315; Maistrov, 1974, pp. 202-203).

[6.1.4] *The Markov-Tchebyshev Central Limit Theorem* (Modern Version). Suppose that $E(X_i) = \xi_i$ and $\text{Var}(X_i) = \sigma_i^2$; $i = 1, 2, \ldots$; if

$$\lim_{n\to\infty}\left[\sum_{k=1}^{n} E\bigl(|X_k - \xi_k|^r\bigr)/\bigl(\sigma_1^2 + \cdots + \sigma_n^2\bigr)\right] = 0,$$

then (Maistrov, 1974, pp. 202-203)

$$\lim_{n\to\infty}\Pr\left[\{S_n - (\xi_1 + \cdots + \xi_n)\}/\bigl(\sigma_1^2 + \cdots + \sigma_n^2\bigr)^{1/2} < x\right] = \Phi(x).$$

[6.1.5] *Lyapunov's Theorem* (Lyapunov, 1901, pp. 1-24; Gnedenko, 1968, p. 310). Let $X_1, X_2, \ldots$ have means $E(X_i) = \mu_i$ and variances $\text{Var}(X_i) = \sigma_i^2 < \infty$; $i = 1, 2, \ldots$. Let $S_n = X_1 + \cdots + X_n$, $m_n = \mu_1 + \cdots + \mu_n$, and $V_n^2 = \sigma_1^2 + \cdots + \sigma_n^2$. If there exists a positive number δ such that

$$\lim_{n\to\infty} \left[V_n^{-(2+\delta)} \sum_{k=1}^{n} E\left(|X_k - \mu_k|^{2+\delta} \right) \right] = 0 \,,$$

then as $n \to \infty$ $\overline{G}_n(x)$ converges to $\Phi(x)$ uniformly in x, where $\overline{G}_n(\cdot)$ is the cdf of $(S_n - m_n)/V_n$.

[6.1.6] *The Central Limit Theorem* – iid case (Lévy, 1925, p. 233; Lindeberg, 1922, pp. 211-225; Cramér, 1970, pp. 53-55; Woodroofe, 1975, p. 251; Feller, 1971, p. 259). Let $X_1, X_2, \ldots$ be iid with common mean μ and common finite variance σ^2, and let $\overline{G}_n(x)$ be the cdf of the rv $(X_1 + \cdots + X_n - n\mu)/(\sigma\sqrt{n})$. Then as $n \to \infty$ $\overline{G}_n(x)$ converges to $\Phi(x)$ uniformly in x.

[6.1.7] *Lindeberg's Theorem.* Let $X_1, X_2, \ldots$ be a sequence of mutually independent rvs with S_n, V_n, m_n, and $\overline{G}_n(\cdot)$ defined as in *[6.1.5]*, and let G_k be the cdf of X_k; $k = 1, 2, \ldots$.

(i) Given $\epsilon > 0$, if

$$\lim_{n\to\infty} \left[V_n^{-2} \sum_{k=1}^{n} \int_{|x-\mu_k| \geq \epsilon B_n} (x - \mu_k)^2 dG_k(x) \right] = 0 \,, \qquad (*)$$

then (Lindeberg, 1922, pp. 211-225) as $n \to \infty$ $\overline{G}_n(x)$ converges to $\Phi(x)$ for all x .

The condition (*) is known as the *Lindeberg condition*, and it is not only sufficient, but in a certain sense necessary for the central limit property to hold; see *[6.1.9]*.

(ii) LeCam (1986, p. 79) stated Lindeberg's theorem in a form that includes a bound obtainable by the considerations of Sec. *[6.3]*: Let I(A) equal 1 or 0 according as some event A occurs or not, i.e., I(A) is the indicator function of A; suppose that the X_i's have mean zero and variances σ_i^2 ($i = 1, 2, \ldots$). Then whenever $\sum_{i=1}^{n} E\left[|X_i/V_n|^2 I(|X_i| > \epsilon V_n) \right] < \epsilon$,

$$\sup_x |\overline{G}_n(x) - \Phi(x)| \leq 5\epsilon \,.$$

Remarks. The Lindeberg condition (∗) implies that $\overline{G}_n(x) \to \Phi(x)$ as above in several cases (Ash, 1972, pp. 336-338):

- $\Pr(|X_k| \leq A < \infty) = 1$ and $B_n \to \infty$; $k = 1, 2, \ldots$. This is the "uniformly bounded independent rvs" case.
- The "iid" case of *[6.1.6]*.
- The "independent Bernoulli rvs" case, leading to the de Moivre-Laplace theorem of *[6.1.1]*.
- The Lyapunov condition of *[6.1.5]*, leading to Lyapunov's theorem.

[6.1.8] LeCam (1986, pp. 83-84) attributes to Bernstein (1926, pp. 1-59) the first central limit theorem that does not assume the existence of any moments. The next, by Lévy (1931, pp. 119-155), requires the concept of *equivalence*: Let $X_1, X_2, \ldots$ and $X'_1, X'_2, \ldots$ be infinite sequences of independent rvs. The two sequences are equivalent if

$$\sum_{j=1}^{\infty} \Pr(X_j \neq X'_j) < \infty.$$

LeCam states Lévy's theorem as follows: If there exists a sequence equivalent to $X_1, X_2, \ldots$ to which Lindeberg's theorem *[6.1.7]* applies, then one can express $S_n = A_n + B_n Z_n$, where A_n and B_n are nonrandom, and where Z_n converges to $N(0, 1)$.

[6.1.9] In 1935 Lévy and Feller, working independently of one another, determined *necessary and sufficient conditions* for the central limit theorem to hold for a sequence $X_1, X_2, \ldots$ of independent rvs. See LeCam (1986, pp. 85-89) for an interesting discussion of priorities; the references are Feller (1935, pp. 512-559) and Lévy (1935a, pp. 357-402).

(i) LeCam (op. cit., p. 87) states Feller's theorem, where again there is no assumption of moments:

Suppose that each of $X_1, X_2, \ldots$ has median equal to zero. For given $\delta > 0$, let $p_n(\delta)$ be the smallest number such that

$$\sum_{i=1}^{n} \Pr\left[|X_i| > p_n(\delta)\right] \leq \delta.$$

Then there exist numbers a_n and b_n such that, if $Z_n = (S_n - a_n)/b_n$, then Z_n converges in distribution to $N(0, 1)$ if and only if for every choice of $\delta > 0$,

$$\lim_{n\to\infty}\left\{\sum_{i=1}^{n} E\left[X_i^2\, I(|X_i| < p_n(\delta))\right] \Big/ p_n^2(\delta)\right\} = \infty.$$

(ii) Consider again the Lindeberg theorem of *[6.1.7]*. Suppose that the conditions

$$V_n \to \infty, \quad \sigma_n/V_n \to 0 \qquad (**)$$

hold as $n \to \infty$. Then (Feller, op. cit.) the condition $(*)$ is necessary for $\overline{G}_n(x)$ to converge to $\Phi(x)$; see also Feller (1971, p. 520) and Ash (1972, pp. 336-342).

Cramér (1970, pp. 57-60) points out that the conditions $(**)$ are jointly equivalent to the condition that $\sigma_\nu/V_n \to 0$ as $n \to \infty$ uniformly for $\nu = 1, 2, \ldots, n$, and that the *Lindeberg-Feller condition* $(*)$ implies the conditions $(**)$. Note what conditions $(**)$ assert; that the total standard deviation (s.d.) of the sum S_n tends to infinity, while each component s.d. σ_n contributes only a small fraction of the total s.d. Where the parent distribution has no moments, analogous conditions on other measures of dispersion such as the interquartile range can be shown to hold; see, for example Theorem 2 of Le Cam (1986, p. 80). Ash (1972, pp. 341-342) states that if, for all $\epsilon > 0$,

$$\Pr\{|X_k - \mu_k|/V_n \geq \epsilon\} \to 0$$

as $n \to \infty$, uniformly for $k = 1, 2, \ldots, n$, then condition $(*)$ is necessary and sufficient for the central limit property to hold.

[6.1.10] While central limit theorems are stated in terms of standardized partial sums as as $(S_n - m_n)/V_n$ in the notation of *[6.1.7]*, applications in practice often refer to S_n being approximately normal for large n, with mean m_n and variance B_n^2. This is true in the following sense. Let Y_n be a rv with a $N(m_n, V_n^2)$ distribution. Then (Ash, 1972, pp. 356-357)

$$|\Pr(S_n \leq y) - \Pr(Y_n \leq n)| = \left|\bar{G}_n\left(\frac{y-m_n}{B_n}\right) - \Phi\left(\frac{y-m_n}{B_n}\right)\right| \to 0$$

as $n\to\infty$ for all y, because $\bar{G}_n \to \Phi$ uniformly over the real line.

6.2 FURTHER CENTRAL LIMIT THEOREMS

In this section we give central limit theorems for sums of rvs that may not be independent, as well as limit theorems for probability densities that converge to the standard normal pdf $\phi(\cdot)$.

[6.2.1] Lévy (1935b, pp. 627-629; 1937) proved that under certain conditions the central limit theorem holds for martingales; see also Le Cam (1986, pp. 84, 85, 89, 90). The latter are usually treated in a set- and measure-theoretic framework, which we shall avoid here. The sequence of rvs $X_1, X_2, \ldots$ is a *martingale* sequence if $E(X_i|X_1, X_2, \ldots, X_{i-1}) = 0$ (i = 1, 2, ...). Let $S_0 \equiv 0$ and $S_i = X_1 + \cdots + X_i$, i = 1, 2, ... Then alternatively $E(S_i|S_{i-1}) = S_{i-1}$ and $E(X_i|S_{i-1}) = 0$. A recent version of the theorem was given by Kir'yanova and Rotar' (1991, pp. 289-302):

In the preceding set up, let $S_n^* = S_n/\sqrt{n}$. Assume that, almost surely and for some k > 2,

$$E(X_i|S_{i-1}) = 0, \quad E(X_i^2|S_{i-1}) = 1, \quad \sup_i(E|X_i|^k) < \infty; \ i = 1, 2, \ldots.$$

Then $\sup_z |\Pr(S_n^* < z) - \Phi(z)| \to 0$ as $n\to\infty$. See *[6.4.7]* for a discussion of the rate of convergence.

[6.2.2] *A Conditioned Central Limit Theorem* (Rényi, 1958, pp. 215-228). Let $X_1, X_2, \ldots$ be an iid sequence of rvs with mean zero and unit variance, and let B be an event such that $\Pr(B) > 0$. Let $S_n^* = \sum_{i=1}^n X_i/\sqrt{n}$ (n = 1, 2, ...); then $\lim_{n\to\infty} \Pr(S_n^* < x|B) = \Phi(x)$ for all x. See also *[6.4.8]* and Landers and Rogge (1977, p. 595).

[6.2.3] *Limit Theorem for Dependent Variables*. A sequence $X_1, \ldots, X_2, \ldots$ of rvs is called *m-dependent* if and only if $\{X_b, X_{b+1}, \ldots, X_{b+s}\}$ and $\{X_{a-r}, X_{a-r+1}, \ldots, X_a\}$ are independent sets of variables when $b-a > m$

(Serfling, 1968, p. 1162).

(i) X_1, X_2, ... is a sequence of m-dependent, uniformly bounded rvs and $S_n = X_1 + \cdots + X_n$, with standard deviation V_n. Then if $V_n/n^{1/3} \to \infty$ as $n \to \infty$, $\overline{G}_n(x) \to \Phi(x)$ for all x as $n \to \infty$, where $\overline{G}_n$ is the cdf of $\{S_n - E(S_n)\}/V_n$ (Chung, 1974, p. 214).

(ii) Let $T_a = \sum_{i=a+1}^{a+n} X_i/\sqrt{n}$, where X_1, X_2, ... is an m-dependent sequence of rvs such that $E(X_i) = 0$, $E(|X_i|^{2+\delta}) \le M < \infty$ for some $\delta > 0$, and $E(T_a^2) \to A^2 > 0$ uniformly in a as $n \to \infty$. Then the limiting distribution of $(nA^2)^{-1/2} \sum_{i=1}^n X_i$ is $N(0, 1)$ (Serfling, 1968, pp. 1158-1159; Hoeffding and Robbins, 1948, pp. 773-780).

Serfling (1968, op. cit.), Brown (1971, pp. 59-66) and Serfling (1980, Sec. 1.11) give several central limit theorems under various dependency conditions, including some for stationary sequences, bounded sequences, and martingales.

[6.2.4] *Random Number of Terms.* Let X_1, X_2, ... be a sequence of iid random variables with mean zero and variance one, let $S_n = \sum_{i=1}^n X_i$, and let $N(1)$, $N(2)$, ... be a sequence of positive integer-valued rvs independent of X_1, X_2, ... and such that $N(n)/n \to c$ in probability, for some positive constant c. Then $\Pr(S_{N(n)}/\sqrt{N(n)} \le x) \to \Phi(x)$ for all x as $n \to \infty$ (Anscombe, 1952, p. 601; Feller, 1971, p. 265).

[6.2.5] Let X_1, X_2, ... be an iid sequence of rvs with mean zero and variance one, and let $Y_n = \max(S_1, S_2, ..., S_n)$, where $S_n = X_1 + \cdots + X_n$. Then $Y_n/\sqrt{n}$ converges in distribution to the half-normal distribution, that is, for all $y \ge 0$, as $n \to \infty$ (Chung, 1974, p. 222),

$$\Pr(Y_n/\sqrt{n} \le y) \to 2\Phi(y) - 1.$$

[6.2.6] A general central limit theorem for *triangular arrays*. The rvs $X_{n,k(n)}$ for $k = 1, 2, ..., h(n)$ are said to be *infinitesimal* if

$$\sup_{1 \le k \le h(n)} \Pr\{|X_{n,k}| \ge \epsilon\} \to 0$$

as $n \to \infty$ for every positive ϵ, where $h(n)$ is a positive-integer-valued function of the positive integer n. Suppose further that $X_{n,1}$, $X_{n,2}$, ...,

$X_{n,h(n)}$ are mutually independent for each value of n, and that

$$Y_n = X_{n,1} + \cdots + X_{n,h(n)}.$$

Then, if $X_{n,k}$ has cdf $G_{n,k}$, and if the sequence of rvs $\{Y_n\}$ (n = 1, 2, ...) converges to a limit as n→ ∞,

$$\sum_{k=1}^{h(n)} \int_{|x| \geq \epsilon} dG_{n,k}(x) \to 0$$

as n→∞ for every $\epsilon > 0$ if and only if the limiting distribution of Y_n is normal (Gnedenko and Kolmogorov, 1968, pp. 95, 126).

[6.2.7] *Limit Theorems for Densities.*

(a) (Gnedenko, 1968, p. 317). Let $X_1, X_2, \ldots$ be an iid sequence of absolutely continuous rvs with common mean μ and finite variance σ^2. For all $n \geq n_0$, let the standardized sum $\{X_1 + \cdots + X_n) - n\mu\}/(\sqrt{n}\sigma)$ have pdf $u_n(x)$. Then $u_n(x) \to \phi(x)$ as $n \to \infty$ uniformly in x ($|x| < \infty$) if and only if there exists a number m such that $u_m(x)$ is bounded.

This result also holds if the boundedness condition for $u_m(x)$ is replaced by the condition that the common pdf g(x) of $X_1, X_2, \ldots$ is bounded (Rényi, 1970, p. 449).

(b) (Gnedenko and Kolmogorov, 1968, p. 224). Let $X_1, X_2, \ldots$ be an iid sequence of absolutely continuous rvs with common mean zero, finite variance σ^2, and pdf g(x). Let the pdf of the sum $X_1 + \cdots + X_m$ be $g_m(\cdot)$. If for some $m \geq 1$, and for any r such that $1 < r \leq 2$,

$$\int_{-\infty}^{\infty} \{g_m(x)\}^r \, dx < \infty,$$

then $\sigma\sqrt{n}\, g_n(\sigma\sqrt{n}x) \to \phi(x)$ uniformly in x; $|x| < \infty$. See also Feller (1971, p. 516).

6.3 ASYMPTOTIC NORMALITY

In this section we consider the convergence to normality of more general sequences $\{T_n\}$ of rvs. Formally, $\{T_n\}$ is *asymptotically normal* if there exist sequences of constants $\{a_n\}$ and $\{b_n\}$, $b_n > 0$, such that for the sequence $\{Z_n\}$ with

$$Z_n = (T_n - a_n)/b_n \; ,$$

$$\lim_{n\to\infty} G_n(x) = \Phi(x)$$

for all x, where $G_n(\cdot)$ is the cdf of Z_n. For general reading on this topic, see Serfling (1980, Sec. 1.5.5) and Hoeffding (1982, pp. 139-147). We write

$$T_n \sim AN(a_n, b_n^2) \; .$$

[6.3.1] If $T_n \sim AN(a_n, b_n^2)$, a_n and b_n are not unique; we also have that $T_n \sim AN(\mu_n^*, \sigma_n^{*2})$ if and only if $\sigma_n^*/b_n \to 1$ and $(\mu_n^* - a_n)/b_n \to 0$ as $n \to \infty$ (Serfling, 1980, Sec. 1.5.5).

[6.3.2] We first state two results on the asymptotic normality of the sample moments of a random sample $X_1, ..., X_n$ having a common distribution function $G(\cdot)$. Let

$$m_k' = \sum_{i=1}^n X_i^k/n, \quad m_k = \sum_{i=1}^n (X_i - \bar{X})^k/n, \quad \mu_k' = E(m_k'), \quad \mu_k = E(m_k) \; .$$

(a) If $\mu_{2k}' < \infty$, the random vector $\sqrt{n}\,(m_1' - \mu_1', ..., m_k' - \mu_k')$ is asymptotically multivariate normal with mean vector $\underline{0}$ and $k \times k$ covariance matrix in which

$$\text{Cov}(m_r', m_s') = \mu_{r+s}' - \mu_r'\mu_s' \; ; \quad 1 \leq r, s \leq k$$

(Cramér, 1946, p. 364; Serfling, 1980, p. 68).

In particular, $\sqrt{n}\,(m_r' - \mu_r') \sim AN(0, \mu_{2r}' - \mu_r'^2)$. This result is an application of the Lindeberg-Lévy Central Limit Theorem of *[6.1.6]* to the sample $X_1^r, ..., X_n^r$.

(b) If $\mu_{2k}' < \infty$, the random vector $\sqrt{n}\,(m_2 - \mu_2, ..., m_k - \mu_k)$ of the first k central sample moments is asymptotically multivariate normal with

Limit Theorems and Expansions

mean vector $\underline{0}$ and $(k-1) \times (k-1)$ covariance matrix in which

$$\text{Cov}(m_r, m_s) = \mu_{r+s} - \mu_r \mu_s - r\mu_{r-1}\mu_{s+1} - s\mu_{r+1}\mu_{s-1}$$
$$+ rs\mu_2\mu_{r-1}\mu_{s-1}, \quad 2 \leq r, s \leq k$$

(Serfling, 1980, p. 72). In particular,

$$\sqrt{n}\,(m_r - \mu_r) \sim \text{AN}(0, \mu_{2r} - \mu_r^2 - 2r\mu_{r-1}\mu_{r+1} + r^2\mu_2\mu_{r-1}^2).$$

See also Cramér (1946, p. 365).

(c) Writing $\mu_2 = \sigma^2$, the joint asymptotic distribution of the sample mean $\overline{X}$ and sample variance s^2 can be expressed via

$$\sqrt{n}\,(\overline{X} - \mu, s^2 - \sigma^2) \sim \text{AN}\!\left((0,0), \begin{bmatrix} \sigma^2 & \mu_3 \\ \mu_3 & \mu_4 - \sigma^4 \end{bmatrix}\right)$$

(Serfling, 1980, pp. 72-73).

[6.3.3] A very useful property of asymptotic normality is that for an AN sequence $\{T_n\}$ of statistics, the sequence of $\{h(T_n)\}$ is also AN for certain well-behaved transformations $h(\cdot)$.

[6.3.3.1] Let $\{T_n\}$ be $\text{AN}(\mu, \sigma_n^2)$, where $\sigma_n \to 0$ as $n \to \infty$. Let $h(\cdot)$ be a real-valued function such that $h'(t)$ exists at $t = \mu$ and $h'(\mu) \neq 0$. Then $\{h(T_n)\} \sim \text{AN}(h(\mu), \{h'(\mu)\}^2 \sigma_n^2)$ (Serfling, 1980, p. 118; Hoeffding, 1982, p. 142; Rao, 1973, pp. 385-386).

[6.3.3.2] Rao (1973, pp. 385-386) gives several variations on the main result in *[6.3.3.1]*, of which we quote the following:

If $T_n \sim \text{AN}(\theta, \sigma^2(\theta)/n)$, if $h(\cdot)$ is a real-valued function with a continuous derivative $h'(\cdot)$, and if $\sigma(\theta)$ is also continuous, then

$$Z = \frac{\sqrt{n}\,[h(T_n) - h(\theta)]}{h'(T_n)\sigma(T_n)}$$

converges in distribution to $N(0, 1)$.

[6.3.4] To illustrate the usefulness of these results, let $Z_1, \ldots, Z_n$ be independent $N(\mu, 1)$, so that $Z_1^2, \ldots, Z_n^2$ each are independent noncentral chi-square with 1 degree of freedom, and $U_n = Z_1^2 + \cdots + Z_n^2 \sim \chi^2(n, \lambda)$, $\lambda = n\mu^2$. By the Lindeberg-Lévy Central Limit property of *[6.1.6]* $\chi^2(n, \lambda)$ is asymptotically normal and by the result in *[6.2]*,

$$Y_n = \left(\frac{U_n + b}{E(U_n)}\right)^h$$

is also asymptotically normal. One can then choose constants h and b to improve the accuracy of the approximation to normality, for example by stabilizing the variance and/or rendering the third moment $\mu_3(T_n)$ as close to zero as possible; see *[7.15.1, 2]*. Here, if b is set equal to zero, the best choice of h leads to the Wilson-Hilferty approximations to chi-square and noncentral chi-square; see *[7.8.4; 7.9.3]*. The additional flexibility in choosing b leads to an improved approximation to normality for noncentral chi-square; see *[7.9.5]*.

[6.3.5] The sequence $\{X_1, X_2, ...\}$ of rvs is called *stationary* if for all r the joint distribution of X_i, X_{i+1}, ..., X_{i+r} does not depend on i; m-dependent sequences are defined in *[6.1.12]*.

Let $\{X_1, X_2, ...\}$ be a stationary m-dependent sequence of rvs such that $E(X_1) = \mu$ and $E\{|X_1|^3\}$ exists. Then (Fraser, 1957, p. 219) $(1/\sqrt{n})\sum_{j=1}^{n} X_j$ is $AN(\sqrt{n}\mu, \sigma^2)$, where

$$\sigma^2 = \text{Var}(X_1) + 2\left[\text{Cov}(X_1, X_2) + \cdots + \text{Cov}(X_1, X_{m+1})\right].$$

[6.3.6] Let $X_1, X_2, ...$ be an iid sequence of rvs with mean μ, finite variance σ^2, and finite fourth central moment μ_4. Let $\overline{X}_n = \left(\sum_{i=1}^{n} X_i\right)/n$ and $s_n^2 = \sum_{i=1}^{n}(X_i - \overline{X}_n)^2/(n-1)$, the sample variance, so that $s_n/\overline{X}_n$ is the *sample coefficient of variation*. Then

(i) $s_n^2 \sim AN(\sigma^2, (\mu_4 - \sigma^4)/n)$,

(ii) $s_n \sim AN(\sigma, (\mu_4 - \sigma^4)/(4\sigma^2 n))$,

(iii) $\dfrac{s_n}{\overline{X}_n} \sim AN\left(\dfrac{\sigma}{\mu}, \dfrac{1}{n}\left[\dfrac{\sigma^2 \mu_2}{\mu^4} - \dfrac{\mu_3}{\mu^3} + \dfrac{\mu_4 - \mu_2^2}{4\mu^2 \sigma^2}\right]\right)$, $\mu \neq 0$;

Limit Theorems and Expansions

$\frac{1}{\sqrt{n}} \frac{S_n}{X_n} \to \frac{1}{N(0,1)}$ in distribution if $\mu = 0$ (Serfling, 1980, pp. 119, 137).

[6.3.7] *Asymptotic normality of U-statistics.*

(i) (Hoeffding, 1948, p. 305; Lehmann, 1975, pp. 366-368). Let $X_1, ..., X_n$ be iid rvs, and let $h(x, y)$ be a symmetric function, i.e., $h(x, y) = h(y, x)$. Further, let

$$E\, h(X_i, X_j) = \theta, \quad h_1(x) = E\, h(X_i, x) = E\, h(x, X_i),$$

$$\sigma_1^2 = Var\{h_1(X_i))\} > 0; \quad i, j = 1,...,n\, ; \quad i \neq j,$$

$$U = \sum\sum_{i<j} \{h(X_i, X_j)\}\, ;$$

then $E(U) = \theta$ and $\sqrt{n}\,(U - \theta)$ is asymptotically $N(0, 4\sigma_1^2)$ as $n \to \infty$.

(ii) More general U-statistics are generated from symmetric functions, involving subsets of r variables ($3 \leq r < n$). The asymptotic normality of such U-statistics was established under certain conditions by Hoeffding (1948, p. 305); see Serfling (1980, pp. 172-199). The following form was given by Korolyuk and Borovskikh (1985, pp. 439-450): let $K(\cdot, \cdot, ..., \cdot)$ be a symmetric function of m variables, i.e., K is invariant when its arguments are permuted. If $X_1, ..., X_n$ are iid rvs, let

$$U = \binom{n}{m}^{-1} \sum_{1 \leq i_1 < i_2 <...< i_m \leq n} \cdots \sum K(X_{i_1}, X_{i_2}, ..., X_{i_m}),$$

a U-statistic. If $g(x) = E\{K(X_1, ..., X_m)|\, X_1 = x\}$, assume that

$$E\{K(X_1, ..., X_m)\} = 0, \quad \sigma_1^2 = E\{g(X_1)\}^2 > 0.$$

Then $U \sim AN(0, m^2\sigma_1^2/n)$.

(iii) A two-sample version (Lehmann, 1975, pp. 362-365). Let $X_1, ..., X_m$ and $Y_1, ..., Y_n$ be iid samples, each sample from possibly different distributions, and independent of one another. Let $f(x, y)$ be a function of two variables, let $m \leq n$, and suppose that $m/n \to \lambda$ as m and $n \to \infty$, where λ may be zero. Further let

$$E[f(X_i, Y_j)] = \theta, \qquad f^*(X_i, Y_j) = f(X_i, Y_j) - \theta,$$

$$f_{10}(x) = E[f^*(x, Y_j)], \qquad f_{01}(y) = E[f^*(X_i, y)],$$

$$\sigma_{10}^2 = \text{Var } f_{10}(X_i), \qquad \sigma_{01}^2 = \text{Var } f_{01}(Y_j),$$

$$U = (mn)^{-1} \sum_{i=1}^{m} \sum_{j=1}^{n} f(X_i, Y_j);$$

then $E(U) = \theta$ and $\sqrt{m}(U - \theta)$ is asymptotically $N(0, \sigma_{10}^2 + \sigma_{01}^2)$ as m and n → ∞. See also Serfling (1980, p. 193). For an extension to functions involving subsets of r and s variables ($3 \leq r < m$, $3 \leq s < n$), and leading to more general U-statistics, see Fraser (1957, pp. 229-230).

(iv) The theorems stated here have applications to rank statistics that are useful in nonparametric one- and two-sample hypothesis-testing problems. For example, the Wilcoxon one- and two-sample rank-sum statistics and the Spearman rank correlations coefficient can be shown to be asymptotically normal; see Lehmann (1975, pp. 365-366, 368-371); Fraser (1957, pp. 231, 234-235). A number of test statistics of this kind have asymptotic normality, as described in Hollander and Wolfe (1973).

[6.3.8] The following result establishes asymptotic normality in a *hypergeometric distribution*. Let N = population size, M = number of "successes" in the population, and n = number drawn at random from the population without replacement. Let X = number of "successes" in the sample, so that X has a probability mass function

$$\Pr(X = k) = g(k; N, M, n) = \binom{M}{k}\binom{N-M}{n-k} / \binom{M}{n}.$$

Suppose further that $1 \leq M \leq N/2$, and $1 \leq n \leq N/2$. Then, if $p = M/N$ and $\lambda = n/N$,

$$\lim_{n p \lambda \to +\infty} \sum{}' g(k; N, M, n) = \Phi(x),$$

where $\sum'$ denotes summation over the set $k \leq np + x\sqrt{\{np(1-p)(1-\lambda)\}}$ (Rényi, 1970, p. 466).

Limit Theorems and Expansions

[6.3.9] A sequence $\{G_n(\cdot)\}$ of cumulative distribution functions converges to the standard normal cdf $\Phi(\cdot)$ if and only if the sequence $\{\psi_n(t)\}$ of the characteristic functions of $\{G_n(\cdot)\}$ converges to $\exp(-t^2/2)$, the characteristic function of the standard normal distribution. The convergence of $\psi_n(t)$ to $\exp(-t^2/2)$ is uniform over every finite interval (Feller, 1971, p. 508).

6.4 RAPIDITY OF CONVERGENCE TO NORMALITY

None of the results in this chapter so far gives us any information on how large the value of n needs to be for the cdf $\overline{G}_n(\cdot)$ of the standardized sum of n random variables to approach within a specified amount δ, say, of the standard normal cdf Φ. The first to derive an upper bound to $|\overline{G}_n(x) - \Phi(x)|$ was Tchebyshev; see Adams (1974, p. 75). We give the bound obtained by Lyapunov for historical interest, followed by the important Berry-Esseen theorem and its ramifications; a crucial assumption is the existence of finite absolute third moments, although William Feller proved in one of his last published papers (see *[6.4.6]*) that a form of the main result holds without the third moment assumption. Recent research has established the existence of lower bounds to the rate of convergence of a sequence of cumulative distribution functions to $\Phi(\cdot)$. We discuss these results briefly in *[6.4.9]*.

Throughout, our interest is in upper and lower bounds in the framework of asymptotic normality of general forms of statistics.

[6.4.1] Let $X_1, X_2, \ldots$ be an iid sequence of rvs with common mean μ, finite variance σ^2, and finite absolute third moment $\nu_3 = E(|X_i - \mu|^3)$. Let $\overline{G}_n(\cdot)$ be the cdf of $(X_1 + \cdots + X_n - n\mu)/(\sqrt{n}\sigma)$. Then there is a positive constant γ such that, for all real x (Lyapunov, 1901, pp. 1-24; Gnedenko and Kolmogorov, 1968, p. 201),

$$|\overline{G}_n(x) - \Phi(x)| < \gamma(\nu_3/\sigma^3) \log n / \sqrt{n}, \quad n = 1, 2, \ldots.$$

[6.4.2] *The Berry-Esseen Theorem.* The following improves upon Lyapunov's result of *[6.4.1]* by removing the factor log n from the upper bound for $|\bar{G}_n(x) - \Phi(x)|$.

(a) Let X_1, X_2, ... be an iid sequence of rvs under the conditions of *[6.4.1]*. Then, for some positive constant C, and for all x (Feller, 1971, p. 542; Gnedenko and Kolmogorov, 1968, p. 201),

$$\sqrt{n}\, |\bar{G}_n(x) - \Phi(x)| \leq C\nu_3/\sigma^3, \quad n = 1, 2, \ldots.$$

(b) Let X_1, X_2, ... be a sequence of mutually independent rvs, such that $E(X_i) = \mu_i$, $\text{Var}(X_i) = \sigma_i^2$, and $E(|X_i - \mu_i|^3) = \beta_i < \infty$; $i = 1, 2, \ldots$. Further, let $\rho_n = \sum_{i=1}^n \beta_i / (\sum_{i=1}^n \sigma_i^2)^{3/2}$, and let $\bar{G}_n(\cdot)$ be the cdf of the standardized sum

$$\{(X_1 - \mu_1) + \cdots + (X_n - \mu_n)\}/(\sigma_1^2 + \cdots + \sigma_n^2)^{1/2}.$$

Then, for all real x and for n = 1, 2, ..., there is a positive constant C such that

$$|\bar{G}_n(x) - \Phi(x)| \leq C\rho_n$$

(Berry, 1941, pp. 122-136; Esseen, 1942, pp. 1-19; Cramér, 1970, p. 78).

(c) Under the conditions of (a) or (b) (Esseen 1956, p. 160-170; Beek, 1972, pp. 188, 196),

$$(\sqrt{10} + 3)/(6\sqrt{2\pi}) = 0.4097{,}32 \leq C < 0.7975.$$

The infimum value of C is attained for a particular Bernoulli distribution, and hence it cannot be improved upon (Bhattacharya and Rao, 1976, p. 240). Values of C may be derived for particular families or kinds of distributions, but the upper bound above is the most recent result in a series of progressively sharper bounds obtained by a number of workers using the assumptions in (a) or (b) only, the earliest of which was Essen's (1956, pp. 160-170) bound of 7.5 for C.

Beek's upper bound for C of 0.7975 has not been improved upon for iid

Limit Theorems and Expansions

sequences of rvs; prior to his paper of 1972, however, it appeared (Zolotarev, 1966, pp. 95-105) that sharper upper bounds for C might be found in this case.

[6.4.3] Zahl (1966, pp. 1225-1245) obtained a modified Berry-Esseen result. Using the notation of *[6.4.2](b)*, let

$$\beta_i' = \begin{cases} \beta_i & , \quad \beta_i \geq 3\sigma_i^3/\sqrt{2} \\ \sigma_i^3(0.7804 - 0.1457\beta_i/\sigma_i^3)^{-1} , & \text{otherwise, } i = 1, 2, \ldots, \end{cases}$$

$$\rho_n' = \sum_{i=1}^{n} \beta_i' / \left(\sum_{i=1}^{n} \sigma_i^2\right)^{3/2}.$$

Then

$$\sup_x |\bar{G}_n(x) - \Phi(x)| \leq (0.650)\rho_n'.$$

This result is sharper than that of Beek (1972, pp. 185, 196) if and only if $\sum_{i=1}^{n}\beta_i'/\sum_{i=1}^{n}\beta_i < 0.7975/0.650$ (Beek, 1972).

[6.4.4] Let $X_1, X_2, \ldots$ be an iid sequence of rvs with the conditions and notation of *[6.4.1]*. If the rvs of the sequence are symmetrically distributed about μ and their common cdf is continuous at μ, then (Gnedenko and Kolmogorov, 1968, p. 218)

$$\lim_{n\to\infty} \sup_{|x|<\infty} \left[\sqrt{n}\,|\bar{G}_n(x) - \Phi(x)|\right] \leq 1/\sqrt{2\pi} = 0.3989{,}423.$$

Equality holds when $\Pr(X_i = -a) = 1/2 = \Pr(X_i = a)$ for some real number a.

[6.4.5] Let $X_1, X_2, \ldots$ be an iid sequence of rvs. Under the conditions of *[6.4.1]* and using the same notation,

$$\lim_{n\to\infty} \inf_{a,b} \sup_{|x|<\infty} \left[\sqrt{n}\,\left|\bar{G}_n(x) - \Phi\left(\frac{x-a}{b}\right)\right|\right] \leq (2\pi)^{-1/2}\nu_3/\sigma^3$$

with equality if $\Pr(X_i = -h) = 1/2 = \Pr(X_i = h)$ for some real h (Rogozin, 1960, pp. 114-117).

[6.4.6] The following theorem gives forms of the Berry-Esseen theorem from truncation of the variables, and is due to Feller (1968b, pp. 261-263). Notice that it is independent of the method of truncation and, more important, does not require the assumptions in *[6.4.1]* to *[6.4.5]* of third moments.

Let $X_1, X_2, \ldots$ be a sequence of mutually independent rvs, and let

$$X'_k = \begin{cases} X_k, & \text{if } -\tau_k < X_k < \tau'_k, \\ 0, & \text{otherwise}. \end{cases} \quad -\infty \leq -\tau_k < 0 < \tau'_k \leq \infty,$$

Suppose that $E(X_k) = 0$ and $E(X_k^2) = \sigma_k^2 < \infty$; $k = 1, 2, \ldots$. Let

$$\beta'_k = E(X_k - X'_k)^2, \qquad \gamma_k = E(|X'_k|^3), \qquad V_n^2 = \sigma_1^2 + \cdots + \sigma_n^2.$$

$$b'_n = \beta'_1 + \cdots + \beta'_n, \qquad c_n = \gamma_1 + \cdots + \gamma_n.$$

Then if $\overline{G}_n(\cdot)$ is the cdf of the normalized sum of $X_1, X_2, \ldots, X_n$ as in *[6.4.2]*(b),

$$\sup_x |\overline{G}_n(x) - \Phi(x)| \leq 6\{(c_n/V_n^3 + (b'_n/V_n^2)\}, \quad n = 1, 2, \ldots.$$

[6.4.7] In *[6.2.1]* we stated a martingale central limit theorem. In order to state some results on rates of convergence we will require the following framework: $X_1, X_2, \ldots$ is a sequence of rvs, and for $j = 1, 2, \ldots$, $\mathcal{F}_j = \sigma(X_1, \ldots, X_j)$ is the σ-field generated by $X_1, \ldots, X_j$. As in *[6.2.1]* $S_0 = 0$, $S_n = X_1 + \cdots + X_n$, $S_n^* = S_n/\sqrt{n}$ ($n = 1, 2, \ldots$). Assume

(1) $E(X_j|\mathcal{F}_{j-1}) = 0$ almost surely, $j = 1, 2, \ldots$,

(2) $E(X_j^2|\mathcal{F}_{j-1}) = 1$ almost surely, $j = 1, 2, \ldots$,

(3) $\gamma_\ell = \sup_j E|X_j|^\ell < \infty$ for some $\ell > 2$.

The martingale property follows from (1). Let

$$\sup_z |\Pr(S_n^* < z) - \Phi(z)| = \delta_n.$$

Then, if (1), (2) and (3) all hold, $\delta_n = O(n^{-\frac{1}{2}\frac{\ell-1}{\ell+1}})$; see Hausler (1988, pp. 275-279). Bolthausen (1982, pp. 672-688) gives conditions under which $\delta_n = O(n^{-1/2}\log n)$, and also under which $\delta_n = O(n^{-1/2})$. See also Kir'yanova and Rotar' (1991, pp. 289-302).

[6.4.8] This result pertains to the conditioned central limit theorem of *[6.2.2]*; see Landers and Rogge (1977, p. 598). Let X_1, X_2, ... be an iid sequence of rvs with mean zero and variance one, where $E(|X_i|^q) < \infty$ for some $q \geq 3$; and let B_k be an event depending on X_1, ..., X_k only, $1 \leq k < n$; so B_k is a member of the σ-algebra generated by X_1, ..., X_k. Then for each r such that $2 \leq r \leq q$, there exists a constant c_r such that whenever $\Pr(B_k) > 0$,

$$\sup_x |\Pr(\overline{X}_n < x|B_k) - \Phi(x)| \leq c_r(k/n)^{1/2}/\{\Pr(B_k)\}^{1/r}.$$

[6.4.9] While the Berry-Esseen inequality of *[6.4.2]* gives an upper bound to $\Delta_n = \sup_x |\overline{G}_n(x) - \Phi(x)|$, research has been done to determine the rate of convergence of $\overline{G}_n$ to Φ more precisely. This has been done in several directions (Hall, 1982):

(a) Studies of the behavior of $\overline{G}_n(x) - \Phi(x) - L_n(x)$, where $L_n(x)$ is the leading term in an asymptotic expansion of Edgeworth-type leading term in an asymptotic expansion of Edgeworth-type (see *[6.6]* and Hall (1982, Chap. 3)).

(b) Changing the norming constants so that $(S_n - c_n)/d_n \to N(0, 1)$, where $c_n/E(S_n) \to 0$ and $d_n/\sqrt{\text{Var}(S_n)} \to 1$ as $n \to \infty$, when the X_i's in each row of a triangular array are iid. The sequences $\{c_n\}$ and $\{d_n\}$ can be chosen to make the rate of convergence fastest, in some sense (Hall, 1982, Chaps. 2, 4).

(c) Establishing some sort of *lower bound* to Δ_n, which along with an upper bound is needed in order to describe the rate of convergence of $\overline{G}_n(\cdot)$ to $\Phi(\cdot)$ adequately.

We confine ourselves here to the lower bound to D_n. Early results on the problem of providing a sharp estimate of the rate of convergence go back to the mid-1960s. Hall (1982, p. 9) provides several references.

One of the most straightforward presentations of the lower bound property in the literature is that of Hall and Barbour (1984, pp. 107-110). Let $X_1, ..., X_n$ be independent rvs with zero means and variances $\sigma_1^2, ..., \sigma_n^2$. Suppose that X_i's form the nth row of a triangular array, and that Lindeberg's condition (∗) of *[6.1.7]* holds for the array, so that $S_n = \sum_{i=1}^{n} X_i$ converges in distribution to N(0, 1). Let

$$\Delta_n = \sup_x |\Pr(S_n \leq x) - \Phi(x)|,$$

$$\delta_n = \sum_{i=1}^{n} E\left[X_i^2 I(|X_i| > 1)\right] + \sum_{i=1}^{n} E\left[X_i^4 I(|X_i| \leq 1)\right] + \left|\sum_{i=1}^{n} E\left[X_i^3 I(|X_i| \leq 1)\right]\right|$$

with I(A) as the indicator function of an event A. Then there exists a universal constant C (i.e., free of n or of distributional properties of the X_i's), such that

$$C(\Delta_n + \sum_{i=1}^{n} \sigma_i^4) \geq \sigma_n.$$

Then one has

$$\Delta_n \geq (\delta_n/C) - \sum_{i=1}^{n} \sigma_i^4, \quad C > 0.$$

If the X_i's are iid, then $\sum_{i=1}^{n} \sigma_i^4 = 1/n$ and $(\delta_n + n^{-1/2})/(\Delta_n + n^{-1/2})$ is bounded away from zero and from infinity as $n \to \infty$. The authors write: "The Berry-Esseen inequality gives an upper bound to the rate of convergence in the central limit theorem, and our inequality gives a lower bound." (Note, however, that if the X_i's are normally distributed, $\Pr(S_n \leq x) = \Phi(x)$ and the lower bound is zero.) Hall (1982, pp. 20-21) restates the Lindeberg-Lévy-Feller theorem of *[6.1.7, 9]* in terms of δ_n, rather than via condition (∗).

[6.4.10] Callaert and Janssen (1978, pp. 417-418) derive a Berry-Esseen bound for *U-statistics* of order two (see *[6.2.14]*), as follows: let h(x, y) be a symmetric function, so that h(x, y) = h(y, x) and let $X_1, ..., X_n$ be a random sample from a common distribution. Suppose further that $E\, h(X_i, X_j) = 0$, $i \neq j$, and that $\sigma_1^2 > 0$, where σ_1^2 is defined as in *[6.2.14]*. Let

Limit Theorems and Expansions

$$U = \binom{n}{2}^{-1} \sum_{1 \leq i < j \leq n} \cdots \sum h(X_i, X_j).$$

If $\nu_3 = E|h(X_1, X_2)|^3 < \infty$, then for some positive constant C,

$$\sqrt{n}|\Pr(\sigma_n^{-1} U \leq x) - \Phi(x)| \leq C\nu_3 \sigma_1^{-3}, \quad n = 1, 2, \ldots,$$

where $\text{Var}(U) = \sigma_n^2$. The authors indicate that the result remains valid for symmetric function $h(\cdot)$ of higher order than two, and also for generalized U-statistics based on two or more samples (op. cit., p. 420). See also Ghosh (1985, pp. 255-270).

Lower bounds for $\sqrt{n}|\Pr(\sigma_n^{-1} U \leq x) - \Phi(x)|$ in the spirit of *[6.4.9]* have been derived; see, for example, Maesono (1991, pp. 37-50).

[6.4.11] Korolyuk and Borovskikh (1985, pp. 439-450) obtained a Berry-Esseen bound for U-statistics involving higher-order symmetric functions. Under the conditions of *[6.3.7]* (ii) and with the same notation,

$$\sup_x \left| \Pr\left(\frac{\sqrt{n}}{m\sigma_1} U < x\right) - \Phi(x) \right| \leq C_{K,m}/\sqrt{n}$$

for some positive constant $C_{K,m}$ depending only on K and on m.

6.5 LIMIT THEOREMS FOR SAMPLE FRACTILES

[6.5.1] Let $X_1, X_2, \ldots$ be a sequence of iid random variables having common cdf $F(\cdot)$, and let x_p ($0 < p < 1$) denote a pth fractile of F, so that $F(x_p) = p$. (Note that this notation differs from that in *[2.1]*). For discrete X, x_p is any value such that $\Pr(X \leq x_p) \geq p$ and $\Pr(X \geq x_p) \geq 1 - p$.

In the iid sample $X_1, \ldots, X_n$ the *pth fractile statistic* $\hat{x}_p$ is a quantity such that the number of sample values not exceeding $\hat{x}_p$ is at least $[np]$ and the number at least equal to $\hat{x}_p$ is at least $[n(1-p)]$, where $[x]$ denotes the greatest integer not exceeding x.

Let F have pdf $f(x)$, where $f(x)$ is continuous, and suppose that x_p is unique with $f(x_p) > 0$. Then (Rao, 1973, pp. 422-423)

$$\sqrt{n}\,(\hat{x}_p - x_p) \sim AN(0, p(1-p)/\{f(x_p)\}^2)\ .$$

An elementary proof of this result appears in Angus (1992, pp. 304-306), with the mild difference that [np] is replaced by a sequence of integers k(n): $k(n)/n \to p$ as $n\to\infty$, and for some constant C, $|k(n) - np| \le C$ for all n. If $k(n) = [np]$, then $C = 1$.

[6.5.2] If $0 < p(1) < p(2) < \cdots < p(k) < 1$, with corresponding unique fractiles $x_{p(1)} < \cdots < x_{p(k)}$ such that $f(x_{p(1)}) > 0$, ..., $f(x_{p(k)}) > 0$, then the asymptotic joint distribution of $\sqrt{n}(\hat{x}_{p(1)} - x_{p(1)})$, ..., $\sqrt{n}(\hat{x}_{p(k)} - x_{p(k)})\}$ is multivariate normal with mean vector (0, ..., 0); the asymptotic covariance between $\sqrt{n}\,\hat{x}_{p(i)}$ and $\sqrt{n}\,\hat{x}_{p(j)}$ is

$$p(i)\{1-p(j)\} \Big/ \Big[f(x_{p(i)})f(x_{p(j)})\Big],\ 1 \le i < j \le k$$

(Mosteller, 1946, pp. 383-384). See also Wilks (1962, pp. 271-274).

[6.5.3] Let $X_1, X_2, \ldots$ be iid rvs with common finite mean μ, variance σ^2 and third central moment μ_3. Let

$$S_n = \sum_{i=1}^n X_i,\ \ \bar{X}_n = S_n/n,\ \ Y_n = (S_n - n\mu)/(\sqrt{n}\sigma) = \sqrt{n}(\bar{X}_n - \mu)/\sigma\ ,$$

so that $Y_n \sim AN(0, 1)$. Shore (1986, pp. 242-246) approximates the fractiles y_p of Y_n via

$$\hat{y}_p = \begin{cases} (1 - 0.41781)z_p - \tfrac{1}{3}\kappa_3, & z_p < 0\ , \\ (1 + 0.41781)z_p - \tfrac{1}{3}\kappa_3, & z_p \ge 0\ , \end{cases}$$

where z_p is the corresponding N(0, 1) fractile and $\kappa_3 = \mu_3/(\sqrt{n}\sigma^3)$; see also *[7.16]*. We can then approximate the error of the approximation via

$$\delta_{n,p} = \hat{y}_p - z_p = \begin{cases} -[(0.41781)z_p + \tfrac{1}{3}\kappa_3]\ , & z_p < 0\ , \\ (0.41781)z_p - \tfrac{1}{3}\kappa_3\ , & z_p \ge 0\ . \end{cases}$$

The corresponding error of approximation of fractiles of S_n is $\sqrt{n}\sigma\,\delta_{n,p}$. If the X_i's are discrete, add $-1/(2\sqrt{n}\sigma)$ to the expression for δ_n. If we want to bound the error so that $|\delta_{n,p}| \le d$, say, we require

$$n \geq \left(\frac{\mu_3}{\epsilon\sigma^3}\right)^2, \quad \epsilon = \frac{d}{|\text{sgn}(z_p)(0.41781)z_p - \frac{1}{3}|}.$$

[6.5.4] In the same paper Shore (1986, pp. 242-246) estimates the error in a more general linear set up. Let $X_1, X_2, \ldots$ be a sequence of independent random variables, such that X_i has finite mean μ_i, variance σ_i^2 and third central moment μ_{3i}. Let

$$S_n^* = \sum_{i=1}^n a_i X_i, \quad Y_n = (S_n^* - E(S_n^*))/\sigma(S_n^*),$$

and suppose that $Y_n \sim AN(0, 1)$. Here $a_1, a_2, \ldots$ is a sequence of finite constants,

$$E(S_n^*) = \sum_{i=1}^n a_i \mu_i, \quad \sigma^2(S_n^*) = \sum_{i=1}^n a_i^2 \sigma_i^2.$$

If κ_3 is the third cumulant of Y_n, so that

$$\kappa_3 = \frac{\sum_{i=1}^n a_i^3 \mu_{3i}}{(\sum_{i=1}^n a_i^2 \sigma_i^2)^{3/2}},$$

then the expressions for $\hat{y}_p$ and for $\delta_{n,p}$ given in *[6.5.3]* may be applied. If the X_i's are discrete add $-1/(2\sigma(S_n))$ to the expression for $\delta_{n,p}$. If we require that $|\delta_{n,p}| \leq d$, choose an initial n such that

$$n \geq \left(\frac{\max_i |a_i^3 \mu_{3i}|}{\epsilon \min_i |a_i^3 \sigma_i^3|}\right)^2,$$

where ϵ is defined in *[6.5.3]*, and adjust to obtain the smallest value of n for which $|\delta_{n,p}| \leq d$.

6.6 EXPANSIONS

[6.6.1] Consider a cumulative distribution function $G(x)$ with mean zero and variance one. We give expansions for $G(x)$ in terms of $\Phi(x)$ and its derivatives, and for the important case where $\overline{G}_n(x)$ is the cdf of a normalized sum of iid rvs, in powers of $1/\sqrt{n}$ and in terms of $\Phi(x)$ and of

$\phi(x)$. The genesis for such expansions is a paper by Tchebyshev (1890; 1962), whose ideas lead to the Edgeworth series (Sec. *[6.6.2]*). We also give expansions for quantiles of G in terms of those of Φ, and vice versa, the Cornish-Fisher series is discussed in *[6.6.12]* to *[6.6.14]*.

Notice that in *[6.6.2]* to *[6.6.5]*, *[6.6.8]*, and *[6.6.9]* below, the *formal* expansions are given, but the question of their validity (i.e., do they converge, and if so, how rapidly?) is a different matter. Some conditions for convergence appear in *[6.6.6]*, and the discussion in *[6.6.7]*, *[6.6.10]*, and *[6.6.11]* is also relevant. A more meaningful approach to these expansions is to view them as *approximations* when only a limited number of terms are used, and then to see whether an approximation in any given instance is uniformly good over some interval for x, and (in the case of $\overline{G}_n$) for what values of n.

For more detailed discussion, see Cramér (1946, pp. 221-230), Cramér (1970, pp. 81, 86-88), Gnedenko and Kolmogorov (1968, pp. 190-196, 220-222), Draper and Tierney (1973, pp. 495-524), Johnson and Kotz (1970, pp. 16-19, 33-35), Stuart and Ord (1987, Secs. 6.17-20, 23) and Wilks (1962, pp. 262-266).

[6.6.2] Let X_1, X_2, ... be an iid sequence of rvs with absolutely continuous cdf $G(x)$ and let $\overline{G}_n(x)$ be the cdf of the normalized sum $(X_1 + \cdots + X_n - n\mu)/(\sqrt{n}\sigma)$, where μ and σ^2 are the common mean and variance of G; n = 1, 2, ... Suppose further that G has cumulants $\kappa_1 = \mu$, κ_2, κ_3, ..., and let $\lambda_r = \kappa_r/\sigma^r$, r = 1, 2, ...

The formal *Edgeworth expansion* for $\overline{G}_n(x)$ is given by

$$\overline{G}_n(x) = \Phi(x) - \phi(x) \left[\frac{\lambda_3 H_2(x)}{3!\sqrt{n}} + \left\{ \frac{\lambda_4 H_3(x)}{4!} + \frac{10\lambda_3^2 H_5(x)}{6!} \right\} \frac{1}{n} \right.$$

$$\left. + \left\{ \frac{\lambda_5 H_4(x)}{5!} + \frac{35\lambda_3\lambda_4 H_6(x)}{7!} + \frac{280\lambda_3^3 H_8(x)}{9!} \right\} \frac{1}{n^{3/2}} + \cdots + O(n^{-5/2}) \right]$$

(Edgeworth, 1905, pp. 36-65, 113-141), where $H_r(x)$ is the rth Hermite polynomial, as defined in *[2.1.9]*: see Cramér (1970, pp. 86-87) and Draper and Tierney (1973, pp. 499, 502-507), where coefficients of terms in $1/n^{r/2}$ are given up to r = 10. We can write the above in terms of the central

Limit Theorems and Expansions

moments of G and with explicit expressions for the polynomials:

$$\bar{G}_n(x) = \Phi(x) - \phi(x)\left[\frac{\mu_3(x^2-1)}{\sigma^3 3!\sqrt{n}} + \left\{\left(\frac{\mu_4}{\sigma^4} - 3\right)\frac{x^3 - 3x}{4!}\right.\right.$$
$$\left.\left. + \frac{10\mu_3^2(x^5 - 10x^3 + 15x)}{\sigma^6 6!}\right\}\frac{1}{n} + \cdots\right].$$

[6.6.3] If, in the expressions in *[6.4.2]*, we put $n = 1$, we obtain the Edgeworth expansion for the cdf of $(X_1 - \mu)/\sigma$.

[6.6.4] Let $g_n(x)$ be the pdf of $\bar{G}_n(x)$ as defined in *[6.6.2]*. The formal Edgeworth expansion of $g_n(x)$ is obtained by differentiating both sides of the expansion for $\bar{G}_n(x)$:

$$g_n(x) = \phi(x)\left[1 + \frac{\lambda_3 H_3(x)}{3!\sqrt{n}} + \left\{\frac{\lambda_4 H_4(x)}{4!} + \frac{10\lambda_3^2 H_6(x)}{6!}\right\}\frac{1}{n}\right.$$
$$\left. + \left\{\frac{\lambda_5 H_5(x)}{5!} + \frac{35\lambda_3\lambda_4 H_7(x)}{7!} + \frac{280\lambda_3^3 H_9(x)}{9!}\right\} n^{3/2} + O(n^{-5/2})\right],$$

$$= \phi(x)\left[1 + \frac{\mu_3(x^3 - 3x)}{\sigma^3 3!\sqrt{n}} + \left\{\left(\frac{\mu_4}{\sigma^4} - 3\right)\frac{x^4 - 6x^2 + 3}{4!}\right.\right.$$
$$\left.\left. + \frac{10\mu_3^2(x^6 - 15x^4 + 45x^2 - 15)}{\sigma^6 6!}\right\}\frac{1}{n} + \cdots\right]$$

explicitly. See Cramér (1946, p. 229), Draper and Tierney (1973, pp. 499, 502-503) for coefficients of $1/n^{r/2}$ up to $r = 10$, and Feller (1971, p. 535).

[6.6.5] If we put $n = 1$ in the expressions in *[6.6.4]*, we obtain the Edgeworth expansion for the pdf of $(X_1 - \mu)/\sigma$. In applications, only terms up to that in $1/n$ (with $n = 1$ here) are generally used (Johnson and Kotz, 1970, p. 19). See *[6.6.11]* for further discussion.

[6.6.6] In the Edgeworth expansions of $\bar{G}_n(x)$ in *[6.6.2]* and of $g_n(x)$ in *[6.6.4]*, the highest moment of $X_1, X_2, \ldots$ appearing in the coefficient of $n^{-r/2}$ is μ_{r+2}. If higher moments than μ_{r+2} do not exist, then in a sense to be stated presently, we can write the expansion up to terms in $n^{-r/2}$, with a remainder equal to $O(n^{-(r+2)/2})$. The following is due to Cramér (1970, pp. 81-82); see also Gnedenko and Kolmogorov (1968, p. 220):

Let X_1, X_2, ... be an iid sequence of rvs with finite absolute moments ν_s of the sth order, where $s > 3$, and let the sequence have common characteristic function $\psi(t) = E[\exp(itX_1)]$. If $\lim \sup |\psi(t)| < 1$ as $|t| \to \infty$, then there exist polynomials $Q_1(x)$, $Q_2(x)$, ... such that

$$\overline{G}_n(x) - \Phi(x) = \phi(x) \left\{ \frac{Q_1(x)}{n^{1/2}} + \frac{Q_2(x)}{n} + \cdots + \frac{Q_{s-3}(x)}{n^{(s-3)/2}} \right\} + R_{s,n},$$

where $|R_{s,n}| < M/n^{(s-2)/2}$ uniformly in x, and where M depends on s and the cdf G, but is functionally independent of n and of x.

If the expansion is taken to the term in $Q_{s-2}(x)/n^{(s-2)/2}$, the remainder is $O(1/n^{(s-2)/2})$. The polynomials are derived as in *[6.6.2]*; see Draper and Tierney (1973, pp. 499-508). The condition on the characteristic function $\psi(t)$ holds for all absolutely continuous distributions (Cramér, 1970, p. 81), but not for lattice distributions.

[6.6.7] If for the iid sequence X_1, X_2, ..., the cdf G is absolutely continuous and $\mu_3 = 0$, $\overline{G}_n(x)$ is approximated by $\Phi(x)$ except for terms of order $1/n$: if, in addition, the kurtosis $\mu_4/\sigma^4 = 3$, $\overline{G}_n(x)$ is approximated by $\Phi(x)$ except for terms of order $1/n^{3/2}$ (Wilks, 1962, pp. 265-266).

[6.6.8] Let $G(x)$ be a cumulative distribution function having pdf $g(x)$. The formal *Gram-Charlier expansion of Type A* for $g(x)$ is given by

$$g(x) = \sum_{j=0}^{\infty} c_j H_j(x) \phi(x) = \sum_{j=0}^{\infty} b_j D(j)\{\phi(x)\},$$

where $H_j(x)$ is the Hermite polynomial of order j (see *[2.1.9]*), $D(j)\{\phi(x)\}$ is the jth derivative of $\phi(x)$ with respect to x, and $\{c_j\}$, $\{b_j\}$ are sequences of constants (Stuart and Ord, 1987, Secs. 6.17, 20; Johnson and Kotz, 1970, pp. 16-17). If the mean is zero, then

$$g(x) = \phi(x)\{1 + \tfrac{1}{2}(\mu_2 - 1)H_2(x) + \mu_3 H_3(x)/6 \\ + (\mu_4 - 6\mu_2 + 3)H_4(x)/24 + \cdots \}.$$

If additionally the variance of $g(x)$ is one, then

Limit Theorems and Expansions 171

$$g(x) = \phi(x)\{1 + \mu_3 H_3(x)/6 + (\mu_4 - 3)H_4(x)/24 + \cdots\}.$$

Integration gives the Gram-Charlier expansion of $G(x)$; again, $\mu_2 = 1$:

$$G(x) = \Phi(x) - \phi(x)\{\mu_3 H_2(x)/6 + (\mu_4 - 3)H_3(x)/24 + \cdots\}.$$

(See Cramér, 1946, pp. 222-223; Charlier, 1905, 1-35.)

[6.6.9] Cramér (1970, 87-88) gives the Gram-Charlier Type A expansion of the cdf $\overline{G}_n(x)$ of the normalized sum of n iid random variables having a common pdf (see *[6.6.2]*). Using notation from *[6.6.2]*,

$$\overline{G}_n(x) = \Phi(x) - \phi(x) \left[\frac{\lambda_3}{3!\sqrt{n}} H_2(x) + \frac{\lambda_4}{4!n} H_3(x) + \frac{\lambda_5}{5!n^{3/2}} H_4(x) \right.$$
$$\left. + \frac{1}{6!}\left(\frac{\lambda_6}{n^2} + \frac{10\lambda_3^2}{n}\right) H_5(x) + \frac{1}{7!}\left(\frac{\lambda_7}{n^{5/2}} + \frac{35\lambda_3\lambda_4}{n^{3/2}}\right) H_6(x) + \cdots \right];$$

$$g_n(x) = \phi(x) \left[1 + \frac{\lambda_3}{3!\sqrt{n}} H_3(x) + \frac{\lambda_4}{4!n} H_4(x) + \frac{\lambda_5}{5!n^{3/2}} H_5(x) \right.$$
$$\left. + \frac{1}{6!}\left(\frac{\lambda_6}{n^2} + \frac{10\lambda_3^2}{n}\right) H_6(x) + \frac{1}{7!}\left(\frac{\lambda_7}{n^{5/2}} + \frac{35\lambda_3\lambda_4}{n^{3/2}}\right) H_7(x) + \cdots \right].$$

[6.6.10] We have given the expansions in *[6.6.9]* to enough terms to illustrate that they differ from those in *[6.6.2]* and *[6.6.4]*. When they converge, the Edgeworth and Gram-Charlier series are essentially the same, differing only in order. The Edgeworth expansions order the terms in ascending powers of $(1/\sqrt{n})$; the Gram-Charlier expansions order terms in derivatives of $\Phi(x)$ in ascending order, or equivalently, in Tchebyshev-Hermite polynomials in increasing order. It therefore seems natural to employ Edgeworth series for $\overline{G}_n(x)$ or $g_n(x)$ as a device for approximations to the distribution of $(X_1 + \cdots + X_n - n\mu)/(\sigma/\sqrt{n})$ if n is large or even moderately large, and the highest moment appearing in the coefficient of $n^{k/2}$ is μ_{k+2} (Johnson and Kotz, 1970, p. 17).

[6.6.11] If $n = 1$, the above expressions yield expansions for $G(x)$ and its pdf $g(x)$ in terms of $\Phi(x)$ and $\phi(x)$. For some intervals in x, either the

Edgeworth or Gram-Charlier approximations or both may give negative density functions, particularly in the tails of the distribution. In the same way, they may not give unimodal curves, even when g(x) is unimodal. Barton and Dennis (1952, pp. 425-427) discuss these shortcomings when only terms containing moments up to μ_4 are used; they give a diagram in the (β_1, β_2) plane ($\beta_1 = \mu_3^2/\sigma^6$, $\beta_2 = \mu_4/\sigma^4$) showing regions in which each type of expansion of g(x) is unimodal, and also regions in which the approximation to g(x) is nonnegative for all values of x. The Gram-Charlier approximation does better than the Edgeworth by both criteria in their study; the approximations tend to perform best by those criteria when the skewness $\sqrt{\beta_1}$ and kurtosis β_2 are near to zero and do not exceed 5, respectively. See Stuart and Ord (1987, Sec. 6.23) and Johnson and Kotz (1970, pp. 18-20), where the diagram of Barton and Dennis (op. cit.) is reproduced.

Note also that the sum to k terms, say, of the Gram-Charlier series for g(x) in [6.6.8] may fluctuate irregularly from one value of k to the next. It has been customary in both series for g(x) to include terms involving μ_2, μ_3, and μ_4 only, and sometimes with the Gram-Charlier series for g(x), to include terms as far as that in $H_6(x)$.

[6.6.12] Cornish and Fisher (1937, pp. 307-320) developed expansions for quantiles of continuous distributions in terms of corresponding quantiles of a standard normal distribution, and vice versa; these were further extended by Fisher and Cornish (1960, pp. 209-226). We shall give these here in the context of normalized sums $(X_1 + \cdots + X_n - n\mu)/(\sqrt{n}\sigma)$ with cdf $\overline{G}_n(x)$, where $X_1, X_2, \ldots$ is an iid sequence of rvs with common pdf g(x), mean μ, variance σ^2, and cumulants $\kappa_3, \kappa_4, \ldots$. As in [6.6.2], $\overline{G}_n$ has cumulants λ_{rn}, where $\lambda_{rn} = \kappa_r/\sigma^r n^{1-r/2}) = \lambda_r/n^{1-r/2}$; r = 3, 4, ...

Let $\Phi(z_p) = \overline{G}_n(x_p)$. Then

$$x_p = z_p + \frac{\kappa_3}{6\sigma^3 n^{1/2}}(z_p^2 - 1) + \left[\frac{\kappa_4}{24\sigma^4}(z_p^2 - 3z_p) + \frac{\kappa_3^2}{36\sigma^6}(-2z_p^3 + 5z_p)\right]\frac{1}{n}$$
$$+ \left[\frac{\kappa_5}{120\sigma^5}(z_p^4 - 6z_p^2 + 3) + \frac{\kappa_3\kappa_4}{24\sigma^7}(-z_p^4 + 5z_p^2 - 2)\right.$$
$$\left.+ \frac{\kappa_3^3}{324\sigma^9}(12z_p^4 - 53z_p^2 + 17)\right]\frac{1}{n^{3/2}} + \left[\frac{\kappa_6}{720\sigma^6}(z_p^5 - 10z_p^3 + 15z_p)\right.$$

Limit Theorems and Expansions

$$+ \frac{\kappa_3\kappa_5}{180\sigma^8}\left(-2z_p^5 + 17z_p^3 - 21z_p\right) + \frac{\kappa_4^2}{384\sigma^8}\left(-3z_p^5 + 24z_p^3 - 29z_p\right)$$

$$+ \frac{\kappa_3^2\kappa_4}{288\sigma^{10}}\left(14z_p^5 - 103z_p^3 + 107z_p\right) + \frac{\kappa_4^3}{7776\sigma^{12}}\left(-252z_p^5 + 1688z_p^3\right.$$

$$\left.- 1511z_p\right)\Bigg]\frac{1}{n^2} + \ldots \ .$$

The inverse expansion of z_p in terms of x_p is given by

$$z_p = x_p + \frac{\kappa_3}{6\sigma^3 n^{1/2}}\left(-x_p^2 + 1\right) + \left[\frac{\kappa_4}{24\sigma^4}\left(-x_p^3 + 3x_p\right) + \frac{\kappa_3^2}{36\sigma^6}\left(4x_p^3 - 7x_p\right)\right]\frac{1}{n}$$

$$+ \left[\frac{\kappa_5}{120\sigma^5}\left(-x_p^4 + 6x_p^2 - 52\right) + \frac{\kappa_3\kappa_4}{144\sigma^7}\left(11x_p^4 - 42x_p^2 + 15\right)\right.$$

$$+ \frac{\kappa_3^3}{648\sigma^9}\left(-69x_p^4 + 187x_p^2 - 52\right)\Bigg]\frac{1}{n^{3/2}} + \left[\frac{\kappa_6}{720\sigma^6}\left(x_p^5 + 10x_p^3 - 15x_p\right)\right.$$

$$+ \frac{\kappa_3\kappa_5}{360\sigma^8}\left(7x_p^5 - 48x_p^3 + 51x_p\right) + \frac{\kappa_4^2}{384\sigma^8}\left(5x_p^5 - 32x_p^3 + 35x_p\right)$$

$$+ \frac{\kappa_3^2\kappa_4}{864\sigma^{10}}\left(-111x_p^5 + 547x_p^3 - 456x_p\right)$$

$$+ \frac{\kappa_4^3}{7776\sigma^{12}}\left(948x_p^5 - 3628x_p^3 + 2473x_p\right)\Bigg]\frac{1}{n^2} + \ldots$$

(Draper and Tierney, 1973, pp. 503-516; Johnson and Kotz, 1970, p. 34).

[6.6.13] Cornish-Fisher expansions for the quantile x_p of a *standardized* random variable $(X - \mu)/\sigma$, where X is a rv with pdf $g(x)$, mean μ, variance σ^2, and cumulants $\kappa_3, \kappa_4, \ldots$, are obtained from *[6.6.12]* by putting n equal to one, and similarly for the inverse expansion of z_p. The former then becomes a transformation to normality.

[6.6.14] The remarks in *[6.6.7]* apply with $\bar{G}_n(x)$ and $\Phi(x)$ replaced by x_p and z_p, respectively, and vice versa. It has been more common, however, for Cornish-Fisher expansions to be used as transformations of a unit normal rv (with n = 1 in *[6.6.12]*), but the form in which they appear in *[6.6.12]* should make them easier to use for the normalized sums with which this chapter has been mainly concerned.

The negative frequencies and multimodalities of some Edgeworth series do not apply here; terms up to those in $1/n^2$ (involving κ_6) are more frequently included in approximations than they are in [6.6.2], say (Johnson and Kotz, 1970, p. 35). Draper and Tierney (1973, pp. 503-518) give coefficients of terms involving cumulants up to κ_{10}, accounting in both forms of Cornish-Fisher expansion for terms up to those in $1/n^4$.

Fisher and Cornish (1960, pp. 211-213) give values to twelve decimal places of Hermite polynomials $H_r(x_p)$; r = 1(1)7; p = 0.0005 ($\times 10, 10^2, 10^3$), 0.001($\times$ 10,10^2), 0.0025($\times 10, 10^2$); and for the same values of p, numerical values of the coefficients of the adjusted polynomials appearing in the first expansion in [6.6.12], to five decimal places.

REFERENCES

Adams, W. J. (1974). *The Life and Times of the Central Limit Theorem*, New York: Caedmon. [6.4]

Angus, J. E. (1992). An alternative derivation of asymptotic normality for sample quantiles, *SIAM Review 34*, 304-306. [6.5.1]

Anscombe, F. J. (1952). Large-sample theory of sequential estimation, *Proceedings of the Cambridge Philosophical Society 48*, 600-607. [6.2.4]

Ash, R. B. (1972). *Real Analysis and Probability*, New York: Academic. [6.1.7, 9, 10]

Barton, D. E. and Dennis, K. E. (1952). The conditions under which Gram-Charlier and Edgeworth curves are positive definite and unimodal, *Biometrika, 39*, 452-427. [6.6.11]

Beek, P. van (1972). An application of the Fourier method to the problem of sharpening the Berry-Esseen inequality. *Zeitschrift für Wahrscheinlichkeitstheorie und Verwandte Gebiete 23*, 187-197. [6.4.2, 3]

Berry, A. C. (1941). The accuracy of the Gaussian approximation to the sum of independent variates, *Transactions of the American Mathematical Society 49*, 122-136. [6.4.2]

Bernstein, S. (1926). Sur-l'extension du théorème limite du calcul des probabilitiés aux sommes de quantités dépendantes, *Math. Ann. 97*, 1-59. [6.1.8]

Bhattacharya, R. N. and Rao, R. R. (1976). *Normal Approximation and Asymptotic Expansions*, New York: Wiley. [6.4.2]

Bolthausen, E. (1982). Exact convergence rates in some martingale central limit theorems, *Annals of Probability 10*, 672-688. *[6.4.7]*

Brown, B. M. (1971). Martingale central limit theorems, *Annals of Mathematical Statistics 42*, 59-66. *[6.2.3]*

Callaert, H. and Janssen, P. (1978). The Berry-Esseen theorem for U-statistics, *Annals of Statistics 6*, 417-421. *[6.4.10]*

Charlier, C. V. L. (1905). Über die Darstellung willkürlicher Funktionen, *Arkiv för Matematik, Astronomi, och Fysik 2(20)*, 1-35. *[6.6.8]*

Chung, K. L. (1974). *A Course in Probability Theory*, New York: Academic. *[6.1.3, 5, 6]*

Cornish, E. A. and Fisher, R. A. (1937). Moments and cumulants in the specification of distributions, *Review of the International Statistical Institute 5*, 307-320. *[6.6.12]*

Cramér, H. (1946). *Mathematical Methods of Statistics*, Princeton, N.J.: Princeton University Press. *[6.3.2; 6.6.1, 4, 8]*

Cramér, H. (1970). *Random Variables and Probability Distributions*, 3rd ed. (1st ed., 1937), London: Cambridge University Press. *[6.1; 6.1.6, 9,; 6.4.2; 6.6.1, 2, 6, 9]*

De Moivre, A. (1738, 1756). *The Doctrine of Chances*; reprint (1967), New York: Chelsea. *[6.1.1]*

Draper, N. R. and Tierney, D. E. (1973). Exact formulas for additional terms in some important series expansions, *Communications in Statistics 1*, 495-524. *[6.6.1, 2, 4, 6, 12, 14]*

Edgeworth, F. Y. (1905). The law of error, *Transactions of the Cambridge Philosophical Society 20*, 36-65, 113-141. *[6.6.2]*

Esseen, C. G. (1942). On the Liapounoff limit of error in the theory of probability, *Arkiv för Matematik, Astronomi, och Fysik 28A*, 1-19. *[6.4.2]*

Esseen, C. G. (1956). A moment inequality with an application to the central limit theorem, *Skandinavisk Aktuarietidskrift 39*, 160-170. *[6.4.2]*

Feller, W. (1935). Über den zentralen Grenzwertsatz der Wahrscheinlichkeitsrechnung, *Mathematische Zeitschrift 40*, 521-559. *[6.1.9]*

Feller, W. (1968a). *An Introduction to Probability Theory and Its Applications*, Vol. 1 (3rd ed.), New York: Wiley. *[6.1.1]*

Feller, W. (1968b). On the Berry-Esseen theorem. *Zeitschrift für Wahrscheinlichkeitstheorie und verwandte Gebiete 10*, 261-268. *[6.4.6]*

Feller, W. (1971). *An Introduction to Probability Theory and Its Applications*, Vol. 2 (2nd ed.), New York: Wiley. *[6.1.6, 9; 6.2.4, 7; 6.3.9; 6.4.2; 6.6.4]*

Fisher, R. A. and Cornish, E. A. (1960). The percentile points of distributions having known cumulants, *Technometrics 2*, 209-225. *[6.6.12, 14]*

Fraser, D. A. S. (1957). *Nonparametric Methods in Statistics*, New York: Wiley. *[6.3.5, 7]*

Ghosh, M. (1985). Berry-Esseen bounds for functionals of U-statistics, *Sankhyā A 47*, 255-270. *[6.4.10]*

Gnedenko, B. V. (1962, 1968). *The Theory of Probability* (4th ed., trans. B. D. Seckler), New York: Chelsea. *[6.1; 6.1.1, 2; 6.2.7]*

Gnedenko, B. V. and Kolmogorov, A. N. (1968). *Limit Distributions for Sums of Independent Random Variables* (rev. ed., trans. K. L. Chung), Reading, Mass.: Addison-Wesley. *[6.1; 6.2.6, 7; 6.4.1, 2, 4; 6.6.1, 6]*

Hall, P. (1982). *Rates of Convergence in the Central Limit Theorem*, Boston: Pitman. *[6.4.9]*

Hall, P. and Barbour, A. D. (1984). Reversing the Berry-Esseen inequality, *Proceedings of the American Mathematical Society 90*, 107-110. *[6.4.9]*

Hausler, E. (1988). On the rate of convergence in the central limit theorem for martingales with discrete and continuous time, *Annals of Probability 16*, 275-279. *[6.4.7]*

Heyde, C. C. (1983). Limit theorem, central, *Encyclopedia of Statistical Sciences, Vol. 3* (Kotz, S., Johnson, N. L. and Read, C. B., eds.), 651-655. New York: Wiley. *[6.1]*

Hoeffding, W. (1948). A class of statistics with asymptotically normal distribution, *Annals of Mathematical Statistics 19*, 293-325. *[6.3.7]*

Hoeffding, W. (1982). Asymptotic normality, *Encyclopedia of Statistical Sciences, Vol. 1* (Kotz, S., Johnson, N. L. and Read, C. B., eds.) 13-147. Wiley, New York. *[6.3; 6.3.3.1]*

Hoeffding, W. and Robbins, H. (1948). The central limit theorem for dependent random variables, *Duke Mathematical Journal 15*, 773-780. *[6.2.3]*

Hollander, M. and Wolfe, D. A. (1973). *Nonparametric Statistical Methods*, New York: Wiley. *[6.3.7]*

Jensen, E. L. and Rootzén, H. (1986). A note on De Moivre's limit theorems: Easy proofs, *Statistics and Probability Letters* 4, 231-232. *[6.1.1]*

Johnson, N. L. and Kotz, S. (1970). *Distributions in Statistics: Continuous Univariate Distributions*, Vol. 1, New York: Wiley. *[6.6.1, 5, 8, 10, 11, 12, 14]*

Kir'yanova, L. V. and Rotar', V. I. (1991). Estimates for the rate of convergence in the central limit theorem for martingales, *Theory of Probability and its Applications* 36, 289-302. *[6.4.7]*

Korolyuk, V. S. and Borovskikh, Yu. V. (1985). Approximation of nondegenerate U-statistics, *Theory of Probability and its Applications* 30, 439-450. *[6.3.7; 6.4.11]*

Landers, D. and Rogge, L. (1977). Inequalities for conditioned normal approximations, *Annals of Probability* 5, 595-600. *[6.2.2; 6.4.7]*

LeCam, L. (1986). The central limit theorem around 1935, *Statistical Science* 1, 78-91. *[6.1; 6.1.7, 8, 9; 6.2.1]*

Lehmann, E. L. (1975). *Nonparametrics: Statistical Methods Based on Ranks*, San Francisco: Holden-Day. *[6.3.7]*

Lévy, P. (1925). *Calcul des Probabilités.* Paris. *[6.1.6]*

Lévy, P. (1931). Sur les séries dont les termes sont des variables éventuelles indépendantes, *Studia Math.* 3, 119-155. *[6.1.8]*

Lévy, P. (1935a). Propriétés asymptotiques des sommes de variables indépendantes ou enchaînées, *J. Math. Pures Appl.*, 347-402. *[6.1.9]*

Lévy, P. (1935b). Propriétés asymptotiques des sommes de variables aléatoires enchaînées, *Comptes Rendus de l'Académie des Sciences de Paris* 199, 627-629. *[6.2.1]*

Lévy, P. (1937). *Théorie de l'Addition des Variables Aléatoires*, Paris: Gauthier-Villars. *[6.2.1]*

Lindeberg, J. W. (1922). Eine neue Herleitung des Exponentialgesetzes in der Wahrscheinlichkeitsrechnung, *Mathematische Zeitschrift* 15, 211-225. *[6.1.6, 7]*

Lyapunov, A. M. (1901). Nouvelle forme du théorème sur la limite de probabilité, *Mémoires de l'Académie Impériale des Sciences de St. Pétersbourg* 12, 1-24. *[6.1.5; 6.4.1]*

Maesono, Y. (1991). On the normal approximation of U-statistics of degree two, *Journal of Statistical Planning and Inference* 27, 37-50. *[6.4.10]*

Maistrov, L. E. (1974). *Probability Theory: A Historical Sketch* (trans. S. Kotz), New York: Academic. *[6.1.3, 4]*

Mosteller, F. (1946). On some useful "inefficient" statistics, *Annals of Mathematical Statistics* 17, 377-408. *[6.5.2]*

Rao, C. R. (1973). *Linear Statistical Inference and Its Applications* (2nd ed.), New York: Wiley. *[6.3.3.1, 2; 6.5.1]*

Rényi, A. (1958). On mixing sequences of sets, *Acta Mathematica* (Hungarian Academy of Sciences) 9, 215-228. *[6.2.2]*

Rényi, A. (1970). *Probability Theory*, Budapest: Akadémiai Kiadó. *[6.2.7; 6.3.8]*

Rogozin, B. A. (1960). A remark on Esseen's paper: "A moment inequality with an application to the central limit theorem," *Theory of Probability and Its Applications* 5, 114-117. *[6.4.5]*

Serfling, R. J. (1968). Contributions to central limit theory for dependent variables, *Annals of Mathematical Statistics* 39, 1158-1175. *[6.2.3]*

Serfling, R. J. (1980). *Approximation Theorems of Mathematical Statistics*, New York: Wiley. *[6.2.3; 6.3; 6.3.1,2; 6.3.3.1; 6.3.6, 7]*

Shore, H. (1986). An approximation for the error of the normal approximation to a linear combination of independently distributed random errors, *Transactions of the Institute of Industrial Engineers* 20, 242-246. *[6.5.3, 4]*

Stuart, A. and Ord, J. K. (1987). *Kendall's Advanced Theory of Statistics*, Vol. 1 (5th edn.) New York: Oxford University Press. *[6.6.1, 8, 11]*

Tchebyshev, P. L. (1890). Sur deux théorèmes relatifs aux probabilités, *Acta Mathematica* 14, 305-315; reprinted (1962) in *Oeuvres*, Vol. 2, New York: Chelsea. *[6.1.3]*

Wilks, S. S. (1962). *Mathematical Statistics*, New York: Wiley. *[6.5.2; 6.6.1, 7]*

Woodroofe, M. (1975). *Probability with Applications*, New York: McGraw-Hill. *[6.1.1, 2, 6]*

Zahl, S. (1966). Bounds for the central limit theorem error, *SIAM Journal on Applied Mathematics* 14, 1225-1245. *[6.4.3]*

Zolotarev, V. M. (1966). An absolute estimate of the remainder term in the central limit theorem, *Theory of Probability and Its Applications* 11, 95-105. *[6.4.2]*

Chapter 7
NORMAL APPROXIMATIONS TO DISTRIBUTIONS

Many of the distributions which cannot be evaluated in closed form can be approximated, and it is not surprising that many of these approximations are based on the normal law. The best are those that are asymptotic, converging in some sense to normality; for example, the binomial, Poisson, and gamma distributions can be represented as the sum of iid random variables, so that the central limit theorem can be used to provide suitable approximations.

Some distribution functions can be expressed directly in terms of the incomplete gamma and beta function ratios. The latter, for example, is the right-tail probability of a binomial rv (which has the central limit property), and can be directly related to the cdfs of the negative binomial, beta, Student t, and F distributions; see Abramowitz and Stegun (1970, Secs. 26.5.23-28) details. It is often the case, then, that those approximations which are good for the binomial are good in some sense for these other distributions.

We have not listed all normal approximations. A poor approximation may be given if it is of historical interest, but the aim has been to present those combining accuracy with simplicity. Criteria for accuracy via absolute and relative error are mentioned whenever a clear statement can be given.

In each section, the rough order of the material is as follows: simple approximations to the cdf, accurate approximations to the cdf, exact normal deviates, and approximations to percentiles. Bounds are included when these are based on normality properties.

Distributions covered are as follows: the binomial *[7.1]*, Poisson *[7.2]*, negative binomial *[7.3]*, hypergeometric *[7.4]*, miscellaneous discrete *[7.5]*,

179

beta *[7.6]*, von Mises *[7.7]*, chi-squared and gamma *[7.8]*, noncentral chi-squared *[7.9]*, Student's t *[7.10]*, noncentral t *[7.11]*, F *[7.12]*, noncentral F *[7.13]*, and miscellaneous continuous *[7.14]*. There is a discussion of normalizing transformations in *[7.15]*, including quantiles. The source materials contain much more information than is given here. See, in particular, Johnson et al. (1992), Johnson and Kotz (1970a, 1970b), Molenaar (1970; 1985, pp. 340-347) and Peizer and Pratt (1968, pp. 1416-1456).

Ling (1978, pp. 274-283) compares the accuracy of various approximations to tail probabilities of the t, chi-square and F families of distributions. Molenaar (1985, pp. 340-347), compares approximations to well-known discrete distributions. Alfers and Dinges (1984, pp. 399-419) present normal approximations for beta- and related distributions (such as binomial, Student t, Poisson and gamma) based on the approach of Peizer and Pratt, and compare various approximations for tail probabilities.

7.1 THE BINOMIAL DISTRIBUTION

[7.1.1] Notation. Let the probability function (pf) of a binomial rv Y be $g(y; n, p)$, where $0 < p < 1$ and

$$\Pr(Y = y) = g(y; n, p) = \binom{n}{y} p^y (1-p)^{n-y}, \; y = 0, 1, ..., n,$$

and let $q = 1 - p$. Denote the cdf of Y by $G(y; n, p) = \Pr(Y \leq y)$, and note that $G(k; n, p) = 1 - G(n - k - 1); n, q)$, a result which may give a choice for each approximation, in trying to reduce error. In practice Y is the number of occurences of an outcome A of interest, in n independent and identical trials; in each trial A occurs with probability p.

[7.1.2.1] Classical Normal Approximation with Continuity Correction. This approximation is derived directly from the de Moivre-Laplace limit theorem (see *[6.1.1]*):

$$G(y; n, p) \simeq \Phi\{(y + \tfrac{1}{2} - np)/\sqrt{npq}\}, \; y = 0, 1, 2, ..., n,$$

Normal Approximations to Distributions 181

$$\Pr(a \leq Y \leq b) \simeq \Phi\{(b + \tfrac{1}{2} - np)/\sqrt{npq}\} - \Phi\{(a - \tfrac{1}{2} - np)/\sqrt{npq}\},$$

$0 \leq a \leq b \leq n$, a and b being integers. If a, b, or y are not integers (but $0 \leq a, b, y \leq n$ still holds), they should be replaced in the normal approximation by [a], [b], [y], where [a], for example, denotes "the largest integer not greater than a".

[7.1.2.2] Gram-Charlier Approximation. A Gram-Charlier series expansion leads to an improvement of the approximation in *[7.1.2.1]*;

$$G(y; n, p) \simeq \Phi(z) - \tfrac{1}{6}(q-p)(z^2-1)\phi(z)/\sqrt{npq},$$

$$z = (k + \tfrac{1}{2} - np)/\sqrt{npq}\,;\quad k = 0, 1, 2, \ldots, n.$$

If $M(n, p)$ is the maximum absolute error, then $M(n, p) \leq 0.056/\sqrt{npq}$; see Raff (1956, pp. 293-303). Ghosh (1980, p. 430) shows that $M(n, p) = O(1/n)$.

[7.1.3] From the local de Moivre-Laplace limit theorem of *[6.1.2]*,

$$g(y; n, p) = \binom{n}{y} p^y (1-p)^{n-y} \simeq \phi\{(y-np)/\sqrt{npq}\}/\sqrt{npq}$$
$$\simeq f(y; np, npq).$$

Then $\sum_{k=0}^{n} |g(y; n, p) - f(y; np, npq)| = \{1 + 4\exp(-3/2)\}/(3\sqrt{2\pi npq}) + O(1/(npq))$ (Johnson et al., 1992, p. 115; Govindarajulu, 1965, p. 152). Alternatively, $g(y; n, p)$ can be approximated as in *[7.1.2.1]*, with $b = y$, $a = y - 1$.

[7.1.4] Recommended. For a combination of accuracy and simplicity Molenaar (1970, pp. 7-8, 86-88, 109-110) recommends approximating $G(k; n, p)$ via versions of $\Phi\left[\sqrt{q(4k+a)} - \sqrt{p(4n-4k+b)}\right]$; see Freeman and Tukey (1950, pp. 607-611) where $a = 4$, $b = 0$.

Suppose that p is close to $\tfrac{1}{2}$; by this is meant

$0.25 \leq p \leq 0.75 \quad (n = 3),$
$0.40 \leq p \leq 0.60 \quad (n = 30),$
$0.46 \leq p \leq 0.54 \quad (n = 300).$

Then if $0.45 \leq |G(k; n, p) - \frac{1}{2}| \leq 0.495$,

$$G(k; n, p) \simeq \Phi\left[\sqrt{4k + 3}\sqrt{q} - \sqrt{4n - 4k - 1}\sqrt{p}\right];$$

and if $0.05 \leq G(k; n, p) \leq 0.93$,

$$G(k; n, p) \simeq \Phi\left[\sqrt{4k + 2.5}\sqrt{q} - \sqrt{4n - 4k - 1.5}\sqrt{q}\right].$$

When p is not close to $\frac{1}{2}$ in the sense above, then if $0.45 \leq |G(k; n, p) - \frac{1}{2}| \leq 0.495$,

$$G(k; n, p) \simeq \Phi\left[\sqrt{4k + 4}\sqrt{q} - \sqrt{4n - 4k}\sqrt{p}\right];$$

and if $0.05 \leq G(k; n, p) \leq 0.93$,

$$G(k; n, 1) \simeq \Phi\left[\sqrt{4k + 3}\sqrt{q} - \sqrt{4n - 4k - 1}\sqrt{p}\right].$$

Molenaar (1970, pp. 111-114) tabulates values of the relative error in these approximations for values of n between 5 and 100, when p = 0.05, 0.20, 0.40, and 0.50. The absolute error is $O(1/\sqrt{n})$ if $p \neq 1/2$, and $O(1/n)$ if $p = 1/2$ (Molenaar, 1970, p. 92).

[7.1.5.1] Improved classical normal approximations can be given by

$$G(k; n, p) \simeq \Phi\left[(k + c - np)\{(n + d)pq + \delta\}^{-1/2}\right];$$

thus, in *[7.1.2.1]*, $c = \frac{1}{2}$, $d = \delta = 0$. If $c = (2 - p)/3$, $d = \frac{1}{3}$ and $\delta = 0$, the maximum error in the approximation is reduced to about that in *[7.1.2.2]* unless p is close to $\frac{1}{2}$ (Gebhardt, 1969, pp. 1641-1642). The problem of choosing c, d, and δ optimally in some sense, however, is complicated; see Molenaar (1970, pp. 76-79).

[7.1.5.2] Ghosh (1980, pp. 427-438) provides two approximations to $G(\cdot)$ that improve on the classical versions of *[7.1.2]*; the first applies when $p \neq \frac{1}{2}$ and the second when $p = \frac{1}{2}$.

(a) $G(k; n, p) \simeq \Phi(\delta\{1 + 2zc + c^2\}^{1/2} - c)$, $p \neq \frac{1}{2}$,

$c = 3(q-p)^{-1}\sqrt{npq}$, $\delta = \text{sgn}(q-p)$,

and z is defined in [7.1.2.2]. This improves considerably on [7.1.2.1] and slightly on [7.1.2.2]. As $p \to \frac{1}{2}$ the three approximations become identical. The Ghosh approximation has maximum absolute error of order $1/n$.

(b) $G(k; n, p) \simeq \Phi(a)$, $p = \frac{1}{2}$,

$a = 2c \cos\{\frac{1}{3}\pi + \cos(\frac{3}{2} bc^{-1})\}$, $b = z(1 - d^{-1})$,

$c = \sqrt{(4n + \frac{1}{3})}$, $d = 12n + 1$,

and z is defined in [7.1.2.2]. Alternatively, if n is large and z is bounded,

$a = b + b^3 d^{-1} + 3b^5 d^{-2} + 12b^7 d^{-3} + 55b^9 d^{-4} + 273b^{11} d^{-5}$

$+ 1428 b^{13} d^{-6} + O(b^{15} d^{-7})$.

The maximum absolute error is of order $n^{-3/2}$. The same property holds for all p for a more complicated approximation (Ghosh, 1980, pp. 434-435) of which the special case when $p = \frac{1}{2}$ is given here. The Ghosh approximation when $p = \frac{1}{2}$ is compared with the Camp-Paulson approximation of [7.1.7] and appears to be consistently better. For $n \geq 100$, the former "can often achieve a six-decimal accuracy." However, the latter is simpler.

[7.1.6] Angular Transformations. These transformations involve the trigonometric inverse sine function, and are useful because they may stabilize the variance (see [7.1.5]), freeing the latter of its dependence on the Bernoulli probability of occurrence p. The function $2 \arcsin\sqrt{p}$ is tabulated in Dixon and Massey (1983, Table A-28).

[7.1.6.1] Let Y have the binomial distribution of [7.1.1] and let

$$Z = 2\sqrt{n+\delta}\left[\arcsin\sqrt{\frac{Y + \frac{1}{2} + \beta}{n + \gamma}} - \arcsin\sqrt{p + \frac{\alpha}{n}}\right].$$

Then Z has an approximate N(0, 1) distribution in the following (and other) cases:

(a) α an arbitrary constant, $\beta + \frac{1}{2} = \alpha$, $\gamma = \delta = 0$ (Curtiss, 1943, pp. 116-117). Z is set equal to zero if $Y < -\alpha$ or $Y > n - \alpha$.

(b) $\alpha = \beta = \gamma = \delta = 0$ (Raff, 1956, pp. 293-303). If M(n, p) is the maximum error as defined in *[7.1.2]*, then $M(n, p) \leq (0.140)/\sqrt{npq}$. The Raff approximation is preferable to the classical approximation in *[7.1.2]* unless p is close to 0.5.

(c) $\alpha = 0$, $\beta = p\gamma$, $\beta = q\gamma$ or $\beta = \frac{1}{2}\gamma$ (Molenaar, 1970, pp. 82-84). Suitable choices of β, γ, and δ lead to an approximation more accurate than that in *[7.1.5.1]*.

(d) $\alpha = 0$, $\beta = \frac{1}{6}$, $\gamma = \delta = \frac{1}{3}$ (Gebhardt 1969, p. 1642). When np is small, the maximum error is one-half to one-tenth of that in the configuration in (b).

(e) $\alpha = 0$, $\beta = \frac{3}{8}$, $\gamma = \frac{3}{4}$, $\delta = 0$ or $\delta = \frac{1}{2}$ (Anscombe, 1948, pp. 246-254).

[7.1.6.2] A variation on configuration (e) in *[7.1.6.1]* that gives a variance of $1 + O(n^{-2})$ is given by

$$Z = \begin{cases} 2 \arcsin(1/\sqrt{4n}) & , Y = 0 \\ 2\sqrt{n + \frac{1}{2}} \arcsin\left[\{(Y + \frac{3}{8})/(Y + \frac{3}{4})\}^{1/2}\right] & , 1 \leq Y \leq n-1 \\ \pi - 2 \arcsin(1/\sqrt{4n}) & , Y = n \; ; \end{cases}$$

see Anscombe (1948, pp. 246-254) and Hoyle (1973, p. 209).

[7.1.6.3] The *average angular transformation* is given by

$$T = \frac{1}{2}\left(\arcsin\sqrt{\frac{Y}{n+1}} + \arcsin\sqrt{\frac{Y+1}{n+1}}\right) ;$$

$$Z = 2\sqrt{n + \frac{1}{2}}\left(T - \arcsin\sqrt{p}\right)$$

then is approximately N(0, 1). Dixon and Massey (1983, p. 373) note that this transformation produces an approximately constant variance if $np(1-p) > 0.8$.

[7.1.7] The *Camp-Paulson* approximation is more accurate than those listed above (excepting that in *[7.1.5.2]*), but also more complex (Camp, 1951, pp. 130-131):

Normal Approximations to Distributions

$$G(k; n, p) \simeq \Phi\{-x/(3\sqrt{z})\}, \quad z = \left[\frac{(n-k)p}{(k+1)q}\right]^{2/3}\left(\frac{1}{n-k}\right) + \frac{1}{k+1},$$

$$x = \left[\frac{(n-k)p}{(k+1)q}\right]^{1/3}\left(9 - \frac{1}{n-k}\right) + \frac{1}{k+1} - 9.$$

See Molenaar (1970, pp. 92-94). The error is $O((npq)^{-1})$ for all p, and is $O((npq)^{-3/2})$ if $p = 1/2$ or $p = 0.042$ (Molenaar, 1970, p. 94). The maximum absolute error M(n, p), defined in the same way as in *[7.1.2]*, is less than 0.0122 for any values of n and p, decreasing as np increases above a value of 0.02. Further, $M(n, p) \leq 0.007/\sqrt{npq}$ (Raff, 1956, pp. 293-303), and $M(n, p) \leq \{z(z^2 - 3)/(108n)\}\phi(z) + O(n^{-3/2})$ (Ghosh, 1980, p. 435); see also *[7.1.5.2]*, and remarks below in *[7.1.8]*.

[7.1.8] The Borges Approximation (Borges, 1970, pp. 189-199). Let k be an integer, $0 \leq k < n$. Then if $h(k, n) = (k + \frac{1}{6})/(n + \frac{1}{3})$,

$$G(k; n, p) \simeq \Phi(y_{k+\frac{1}{2}}), \quad y_k = (pq)^{-1/6}\sqrt{n + \tfrac{1}{3}} \int_p^{h(k,n)} \{s(1-s)\}^{-1/3} ds.$$

The maximum absolute error M(n, p), defined as in *[7.1.2]* is $O(1/(npq))$ for all values of p, and is $O(1/(npq)^{3/2})$ if $p = \frac{1}{2}$ (Molenaar, 1970, p. 95).

Let $J(x) = \frac{3}{2}x^{2/3}[1 + 8x/(60 - 25x)]$. Molenaar suggests the approximation

$$\int_0^x [s(1-s)]^{1/3} ds \simeq \begin{cases} J(x), & 0 < x \leq \tfrac{1}{2} \\ 2.0533902 - J(1-x), & \tfrac{1}{2} < x < 1. \end{cases}$$

The agreement is good to 2 decimal places if $x = 1/2$, improving to 5 decimal places if $x = 0.10$.

Both the Borges and Camp-Paulson approximations are superior to those of *[7.1.5.1]* with $c = (2-p)/3$, $d = 1/3$, and $\delta = 0$, and to those of *[7.1.6]* (arcsine) with $\alpha = 0$, $\beta = 1/6$, $\gamma = 1/3$, and $\delta = 1/3$ (Gebhardt, 1969, p. 1645). The Camp-Paulson is better than the Borges approximation (by not very much) for all values of p when $n \leq 20$, say, and if $0.2 < p < 0.8$ when $n \geq 50$, say (Molenaar, 1970, p. 98; Gebhardt, 1969, pp. 1643, 1645), although these conclusions consider the amount of computation as well as relative error as criteria; based on maximum absolute error alone, the Borges approximation is preferable if n ≤ 20 and np is near to 2 or 3.

[7.1.9] A more accurate approximation than any described so far, but also more cumbersome, is due to Peizer and Pratt (1968, pp. 1417-1420, 1448); this is accurate to $O(1/(npq)^{3/2})$ if $p \neq \frac{1}{2}$, and to $O(1/(npq)^2)$ if $p = \frac{1}{2}$. It is given by $G(k; n, p) \simeq \Phi(z)$, where

$$z = \frac{d\left[1 + qT\{(k+\tfrac{1}{2})/(np)\} + pT\{(n-k-\tfrac{1}{2})/(nq)\}\right]^{1/2}}{\sqrt{pq(n+\tfrac{1}{6})}},$$

$$d = k + \tfrac{2}{3} - (n + \tfrac{1}{3})p + 0.02\left(\frac{q}{k+1} - \frac{p}{n-k} + \frac{q-\tfrac{1}{2}}{n+1}\right),$$

$$T(x) = (1 - x^2 + 2x \log x)/(1-x)^2, \quad x \neq 1; \quad T(1) = 0.$$

See Molenaar (1970, p. 102) for a refinement of d. For the above (Peizer and Pratt, 1968, p. 1418),

$$|G(k; n, p) - \Phi(z)| \leq \begin{cases} 0.001 & \text{if } k \geq 1, n-k \geq 2, \\ 0.01 & \text{if } k \geq 0, n-k \geq 1. \end{cases}$$

See also Alfers and Dinges (1984, pp. 399-419) on the approximation of tail probabilities.

[7.1.10] A *recommended* approximation, more accurate than that given in *[7.1.4]*, has been proposed by Molenaar (1970, pp. 100, 105, 110):

$$G(k; n, p) \simeq \Phi\{2\sqrt{(k+1)q + A} - 2\sqrt{(n-k)p + B}\}, \quad p \neq \tfrac{1}{2},$$

$$A = (4 - 10p + 7p^2)(k + \tfrac{1}{2} - np)^2/(36npq) - \tfrac{1}{18}(8 - 11p + 5p^2),$$

$$B = (1 - 4p + 7p^2)(k + \tfrac{1}{2} - np)^2/(36npq) - \tfrac{1}{18}(2 + p + 5p^2);$$

$$G(k; n, \tfrac{1}{2}) \simeq \Phi(\sqrt{2k+2+\beta} - \sqrt{2n-2k+\beta}),$$

$$\beta = \tfrac{1}{12}n^{-1}(2k+1-n)^2 - \tfrac{5}{6}.$$

Normal Approximations to Distributions

These are accurate to $O(1/(npq)^{3/2})$ if $p \neq \frac{1}{2}$, and to $O(1/(npq)^2)$ if $p = \frac{1}{2}$. See Molenaar (1970, pp. 100-101, 104) for slightly improved but more cumbersome alternatives.

[7.1.11] Let $u = (k + \frac{1}{2} - np)/\sigma$ and $\sigma = \sqrt{npq}$. Then (Molenaar, 1970, pp. 72, 76) the *exact normal deviate* z defined by $G(k; n, p) = \Phi(z)$ satisfies

$$z = u + \tfrac{1}{6}\sigma^{-1}(q-p)(-u^2 + 1) + \tfrac{1}{72}\sigma^{-2}\{(5 - 14pq)u^3 + (-2 + 2pq)u\}$$

$$+ \tfrac{1}{6480}\sigma^{-3}(q-p)\{(-249 + 438pq)u^4 + (79 - 28pq)u^2 + 128 - 26pq\}$$

$$+ O(\sigma^{-4});$$

$$u = z + \tfrac{1}{6}\sigma^{-1}(1 - 2p)(z^2 - 1) + \tfrac{1}{72}\sigma^{-2}\{z^3(2p^2 - 2p - 1)$$

$$+ z(-14p^2 + 14p - 2)\} + \tfrac{1}{1620}\sigma^{-3}(3z^4 + 7z^2 - 16)$$

$$\times (2p^3 - 3p^2 - 3p + 2) + O(\sigma^{-4}).$$

See also Pratt 1968, p. 1467).

[7.1.12] If an approximation is desired to be *accurate near an assigned probability* α or $1 - \alpha$, and if $\Phi(z_\alpha) = 1 - \alpha$, $0 < \alpha \leq 1/2$, then we have that $G(k; n, p) = \Phi(z) \simeq \Phi(u)$, where

$$u = 2\sqrt{(k+1+b)q} - 2\sqrt{(n-k+b)p}, \quad b = \tfrac{1}{12}(z_\alpha^2 - 4).$$

The error is $(q-p)(z_\alpha^2 - z^2)/(12\sqrt{npq}) + O(1/(npq))$. See Molenaar (1970, pp. 88-91) for discussion and possible refinements.

[7.1.13] *Bounds.* Slud (1977, pp. 404-410) gives several inequalities involving the normal cdf ($k = 0, 1, 2, \ldots, n$):

(a) If $0 \leq p \leq \tfrac{1}{4}$, $np \leq k \leq n$, or $np \leq k \leq nq$, then

$$\Pr(Y \geq k) = 1 - G(k-1; n, p) \geq 1 - \Phi\{(k - np)/\sqrt{npq}\}.$$

(b) If $np \le k \le nq$ and $k \ge 2$, then

$$g(k; n, p) > \Phi\{(k-np+1)/\sqrt{npq}\} - \Phi\{(k-np)/\sqrt{npq}\}.$$

(c) If $p \le \frac{1}{4}$ and either (i) $k \ge nq$ or (ii) $k \ge n/3$, $n \ge 27$, then

$$g(k; n, p) - \left[\Phi\left(\frac{k-np+1}{\sqrt{npq}}\right) - \Phi\left(\frac{k-np}{\sqrt{npq}}\right)\right] > \frac{\gamma}{\sqrt{npq}} \phi\left(\frac{k-np}{\sqrt{npq}}\right),$$

where, if (ii) holds, $\gamma = \delta = \{(k-np)^2/(2npq)\}(k^{-1} - (3npq)^{-1})$, and if (i) holds, $\gamma = \max(\delta, 0.16)$.

7.2 THE POISSON DISTRIBUTION

[7.2.1] Notation. Let the pf of a Poisson rv Y be $g(y; \lambda)$, where

$$\Pr(Y = y) = g(y; \lambda) = e^{-\lambda}\lambda^y/y!, \quad y = 0, 1, 2, \ldots.$$

Denote the cdf of Y by $G(y; \lambda)$, $= \Pr(Y \le y)$. In addition to what follows here, see also *[7.15.3]*.

[7.2.2] Classical Normal Approximation with Continuity Correction.

$$G(k; \lambda) \simeq \Phi\left[(k + \tfrac{1}{2} - \lambda)/\sqrt{\lambda}\right], \quad k = 0, 1, 2, \ldots$$

Asymptotic behavior indicates that, for large λ, the correction of $\frac{1}{2}$ should only be used for probabilities between 0.057 and 0.943; if $\lambda = 10$, for probabilities between 0.067 and 0.953; and if $\lambda = 4$, between 0.073 and 0.958. The approximation may overestimate probabilities less than 0.16 and underestimate $1 - \Pr(Y \le y)$ when the latter is less than 0.16. See Molenaar (1970, pp. 34-36) for discussion of these and other points.

With these cautions in mind, for nonnegative integers a and b,

$$\Pr(a \le Y \le b) \simeq \Phi\left[(b + \tfrac{1}{2} - \lambda)/\sqrt{\lambda}\right] - \Phi\left[(a + \tfrac{1}{2} - \lambda)/\sqrt{\lambda}\right],$$

$$g(k; \lambda) \simeq \Phi\left[(k + \tfrac{1}{2} - \lambda)/\sqrt{\lambda}\right] - \Phi\left[(k - \tfrac{1}{2} - \lambda)/\sqrt{\lambda}\right].$$

Normal Approximations to Distributions 189

[7.2.3] *Recommended Simple Approximation*, accurate to $O(\lambda^{-1/2})$ (Molenaar, 1970, pp. 8-9, 38-40, 64).

$$G(k; \lambda) \simeq \Phi(2\sqrt{k + \tfrac{3}{4}} - 2\sqrt{\lambda}), \quad \begin{array}{l} 0.09 < G(k; \lambda) < 0.94; \ \lambda \geq 15, \\ 0.05 < G(k; \lambda) < 0.93; \ \lambda < 15, \end{array}$$

$$G(k; \lambda) \simeq \Phi(2\sqrt{k+1} - 2\sqrt{\lambda}), \quad \text{otherwise.}$$

This approximation is more accurate than that of *[7.2.2]*, and is based on that of Freeman and Tukey (1950, p. 607). Molenaar (1970, p. 38-41) discusses $\Phi(2\sqrt{k+\alpha} - 2\sqrt{\lambda+\beta})$ for various choices of α and β as an approximation to $G(k; \lambda)$.

[7.2.4] There are also variance-stabilizing transformations $\sqrt{Y+\alpha}$ and $\sqrt{Y} + \sqrt{Y+1}$, described in *[7.15.3]*, which can be used as normal approximation.

[7.2.5] (a) An approximation of Makabe and Morimura (1955, pp. 37-38) has bounds which also give accuracy to $O(1/\lambda)$ as $\lambda \to \infty$; see Govindarajulu (1965, p. 156). Let k and m be nonnegative integers; then if $k < m$ and $\lambda \geq 1$,

$$G(m; \lambda) - G(k; \lambda) = \Phi(b) - \Phi(a)$$
$$+ \{\phi(b)(1 - b^2) - \phi(a)(1 - a^2)\}/(6\sqrt{\lambda}) + R_1 ,$$
$$a = (k - \tfrac{1}{2} - \lambda)/\sqrt{\lambda}, \ b = (m + \tfrac{1}{2} - \lambda)/\sqrt{\lambda} ,$$
$$|R_1| < (0.0544)/\lambda + (0.0108)/\lambda^{3/2} + (0.2743)/\lambda^2$$
$$+ (0.0065)/\lambda^{5/2} + (1 + \tfrac{1}{2}\lambda^{-1/2})\exp(-2\sqrt{\lambda}) ;$$

see Molenaar (1970, p. 53) and (c) below.

(b) Cheng (1949, p. 39) gives, similarly,

$$G(m; \lambda) = \Phi(c) + \frac{\phi(c)}{6\sqrt{\lambda}}(1 - c^2) + R_2, \ c = \frac{1}{\sqrt{\lambda}}(k + \tfrac{1}{2} - \lambda) ,$$

$$|R_2| < 0.076/\lambda + 0.043/\lambda^{3/2} + 0.13/\lambda^2 .$$

(c) $g(k; \lambda) = \phi(y)\left[\dfrac{1}{\sqrt{\lambda}} + \dfrac{3y - y^3}{6\lambda}\right] + R_3, \ y = \dfrac{k - \lambda}{\sqrt{\lambda}} ,$

$$|R_3| < (0.0748)/\lambda^{3/2} + (0.00554)/\lambda^2 + (0.3724)/\lambda^{5/2}$$

$$- (0.5595)/\lambda^3 + \{1 + (\tfrac{1}{2})\lambda^{-1/2}\}\lambda^{-1/4} \exp(-2\sqrt{\lambda}) .$$

This result applies if $\lambda \geq 1$; see sources listed in (a).

[7.2.6] The Wilson-Hilferty approximation to chi-square (see *[7.8.4]*) leads here to

$$G(k; \lambda) \simeq \Phi\left[3\sqrt{k+1} - \{3\sqrt{k+1}\}^{-1} - 3\lambda^{1/3}(k+1)^{1/6}\right].$$

The error is $\phi(z)(3z - z^3)/(108\lambda) + O(\lambda^{-3/2})$ if $G(k; \lambda) = \Phi(z)$ (Molenaar, 1970, pp. 48-49).

[7.2.7] *A Recommended Accurate Approximation* (Molenaar, 1970, pp. 62, 64).

$$G(k; \lambda) \simeq \Phi\left[2\sqrt{k + \tfrac{1}{9}(t+4)} - 2\sqrt{\lambda + \tfrac{1}{36}(t-8)}\,\right], \quad t = (k - \lambda + \tfrac{1}{6})^2/\lambda ,$$

with error $\tfrac{1}{6480}\phi(z)(-6z^4 + 26z^2 + 7)\lambda^{-3/2} + O(\lambda^{-2})$, if $G(k; \lambda) = \Phi(z)$.

[7.2.8] Another accurate approximation (except near $k = 0$ when $\lambda \leq 5$) is that of *Peizer and Pratt* (1968, pp. 1417-1420); see also Molenaar (1970, pp. 59-61). Here $G(k; \lambda) \simeq \Phi(u)$, where

$$u = \{k - \lambda + \tfrac{2}{3} + \epsilon/(k+1)\}\left[1 + T\{(k + \tfrac{1}{2})/\lambda\}\right]^{1/2}\sqrt{\lambda} ,$$

$$T(x) = (1 - x^2 + 2x \log x)/(1-x)^2, \quad x \neq 1; \quad T(1) = 0.$$

If tail probabilities are being approximated, ϵ increases optimally from 0.02077 to 0.02385 as α (or $1 - \alpha$) decreases from 0.10 to 0.005. The error is $\phi(z)(-z^2 + 1620\epsilon - 32)/(1620\lambda^{3/2}) + O(\lambda^{-2})$ if $G(k; \lambda) = \Phi(z)$. When $\epsilon = 0.02$, the maximum absolute error is 0.001 if $k \geq 1$ and is 0.01 if $k \geq 0$ (Peizer and Pratt, op. cit. pp. 1417-1424)

[7.2.9] (a) If $\Pr(Y \leq k - 1) = G(k - 1; \lambda) = \Phi(x)$, then the *exact normal deviate* x is given (Riordan 1949, pp. 417, 420; Molenaar, 1970, p. 32) by

Normal Approximations to Distributions 191

$$x = u + \frac{u^2 - 1}{3k^{1/2}} + \frac{7u^3 - u}{36k} + \frac{219u^4 - 14u^2 - 13}{1620k^{3/2}}$$

$$+ \frac{3993u^5 - 152u^3 + 119u}{40320k^2} + O(k^{-5/2}), \quad u = \frac{k - \lambda}{\sqrt{k}} .$$

(b) Molenaar (1970, p. 33) gives two expressions in powers of $\lambda^{-1/2}$, and we reproduce one of them here. If $G(k; \lambda) = \Phi(z)$, then

$$z = v + \frac{-v^2 + 1}{6\lambda^{1/2}} + \frac{5v^3 - 2v}{7\lambda} + \frac{-249v^4 + 79v^2 + 128}{6480\lambda^{3/2}} + O(\lambda^{-2}),$$

$$v = (k + \tfrac{1}{2} - \lambda)/\sqrt{\lambda} .$$

[7.2.10] For an approximation designed to be *most accurate near an assigned probability* α or $1 - \alpha$, where $0 < \alpha < \tfrac{1}{2}$ and $\Phi(z_\alpha) = 1 - \alpha$, we have $G(k; \lambda) = \Phi(z) \simeq \Phi(u)$, where

$$u = 2\{k + \tfrac{1}{18}(z_\alpha^2 + 11)\}^{1/2} - 2\{\lambda - \tfrac{1}{36}(z_\alpha^2 + 2)\}^{1/2} .$$

The error is $(z_\alpha^2 - z^2)\phi(z)/(12\sqrt{\lambda}) + O(\lambda^{-3/2})$. See Molenaar (1970, pp. 41-44) for a discussion of this and other possible approximations.

[7.2.11] $G(k; \lambda) \leq \Phi\{(k + 1 - \lambda)/\sqrt{\lambda}\}$ (Bohman 1963, pp. 47-52).

7.3 THE NEGATIVE BINOMIAL DISTRIBUTION

[7.3.1] The pf of a negative binomial rv Y is $h(y; s, p)$ where

$$\Pr(Y = y) = h(y; s, p) = \binom{s + y - 1}{y} p^s q^y, \quad \begin{array}{l} s > 0, q = 1 - p, \\ 0 < p < 1; y = 0, 1, 2, \ldots . \end{array}$$

Let the cdf of Y be $H(y; s, p)$. The parameter s is usually, but need not be, a positive integer; alternative forms of the distribution are given in Johnson et al. (1992, pp. 199-200); when $s = 1$, we have a *geometric distribution*, in which case probabilities can be computed exactly in simple form. In addition to what follows, see also *[7.15.4]*(b).

[7.3.2] Approximations to H(y; s, p) when s is a positive integer can be deduced from those for binomial probabilities discussed in *[7.1]*. Thus

$$H(y; s, p) = 1 - G(s-1; s+y, p) = G(y; s+y, q),$$

where G is the cdf of the binomial as in *[7.1]* (Patil 1960, p. 501).

Hence from the approximations to G(k; n, p) in *[7.1]* we can construct approximations to H(y; s, p) as follows:
  (i) Replace k by y.
  (ii) Replace n by y + s and n − k by s.
  (iii) Interchange p and q.
  (iv) If $G(k; n, p) \simeq \Phi(z)$, then $H(y; s, p) \simeq \Phi(z)$, where z is expressed in terms of y, s, and p from (i), (ii), and (iii).

[7.3.3] To illustrate the procedure of *[7.3.2]*, we adapt Molenaar's recommended simple yet accurate binomial approximation, given in *[7.1.4]*. Suppose that p is close to $\frac{1}{2}$; that is,

if s + y = 3, $0.25 \le p \le 0.75$,
if s + y = 30, $0.40 \le p \le 0.60$,
if s + y = 300, $0.46 \le p \le 0.54$.

Then if $0.45 \le |H(y; s, p) - \frac{1}{2}| \le 0.495$;

$$H(y; s, p) \simeq \Phi\left[\sqrt{4y+3}\sqrt{p} - \sqrt{4s-1}\sqrt{q}\right];$$

and if $0.07 \le H(y; s, p) \le 0.95$,

$$H(y; s, p) \simeq \Phi\left[\sqrt{4y+2.5}\sqrt{p} - \sqrt{4s-1.5}\sqrt{q}\right],$$

When p is not close to $\frac{1}{2}$ in the sense above, then if $0.45 \le |H(y; s, p) - \frac{1}{2}| \le 0.495$

$$H(y; s, p) \simeq \Phi\left[\sqrt{4y+4}\sqrt{p} - \sqrt{4sq}\right];$$

and if $0.07 \leq H(y; s, p) \leq 0.95$,

$$H(y; s, p) \simeq \Phi\left[\sqrt{4y+3}\sqrt{p} - \sqrt{4s-1}\sqrt{q}\,\right].$$

The absolute error is $O(1/\sqrt{s+y})$ if $p \neq \frac{1}{2}$, and $O(1/(s+y))$ if $p = \frac{1}{2}$.

[7.3.4] As a second illustration of the rule in *[7.3.2]*, we adapt the *recommended* accurate approximation of Molenaar, given in *[7.1.10]*:

$$H(y; s, p) \simeq \Phi\{2\sqrt{(y+1)p + A} - 2\sqrt{sq + B}\,\}, \quad p \neq \tfrac{1}{2},$$

$$A = \frac{(4 - 10q + 7q^2)(py - sq + \tfrac{1}{2})^2}{36(s+y)pq} - \tfrac{1}{18}(8 - 11q + 5q^2),$$

$$B = \frac{(1 - 4q + 7q^2)(py - sq + \tfrac{1}{2})^2}{36(s+y)pq} - \tfrac{1}{18}(2 + q + 5q^2);$$

$$H(y; s, \tfrac{1}{2}) \simeq \Phi(\sqrt{2y + 2 + \beta} - \sqrt{2s + \beta}\,,$$
$$\beta = \{(s - y - 1)^2 - 10(s+y)\}/\{12(s+y)\}\,.$$

These are accurate to $O(\{(s+y)pq\}^{-3/2})$ if $p \neq 1/2$, and to $O(\{(s+y)pq\}^{-2})$ if $p = 1/2$.

[7.3.5] The Camp-Paulson approximation of *[7.1.7]* was adapted by Bartko (1966, p. 349; 1967, p. 498):

$$H(y; s, p) \simeq \Phi\{-x/(3\sqrt{z})\},$$

$$x = (9 - \tfrac{1}{s})\left[\frac{sq}{p(y+1)}\right]^{1/3} - \frac{9y+8}{y+1}, \quad z = \tfrac{1}{s}\left[\frac{sq}{p(y+1)}\right]^{2/3} + \frac{1}{y+1}.$$

This result also follows from *[7.3.2]* and *[7.1.7]*. Bartko's table of maximum absolute errors in selected cases is reproduced in Johnson et al. (1992, p. 211). If $0.05 \leq p \leq 0.95$ and $s = 5, 10, 25$, or 50, this error is no greater than 0.004, 0.005, 0.003, or 0.002, respectively; for some values of p, it may be considerably smaller.

[7.3.6] From *Peizer and Pratt* (1968, pp. 1417-1422) comes the following accurate approximation, also obtained from *[7.3.2]* and *[7.1.9]*:

$$H(y; s, p) \simeq \Phi(u),$$

$$u = d\left[\frac{1 + pT\left(\frac{y+\frac{1}{2}}{yq+sq}\right) + qT\left(\frac{s-\frac{1}{2}}{yp+sp}\right)}{pq(y+s+\frac{1}{6})}\right]^{1/2},$$

$$d = y + \tfrac{2}{3} - (y+s+\tfrac{1}{3})q + 0.02\left(\frac{p}{y+1} - \frac{q}{s} + \frac{p-\frac{1}{2}}{y+s+1}\right),$$

$$T(x) = (1 - x^2 + 2x \log x)/(1-x)^2, \quad x \neq 1; \quad T(1) = 0.$$

$$|H(y; s, p) - \Phi(u)| \leq \begin{cases} 0.001, & y \geq 1, \ s \geq 2, \\ 0.01, & y \geq 0, \ s \geq 1. \end{cases}$$

The approximation is accurate to $O(1/\{(y+s)pq\}^{3/2})$ if $p \neq 1/2$, and to $O(1/\{(y+x)pq\}^2)$ if $p = 1/2$ (see *[7.1.9]*).

[7.3.7] From *[7.3.2]* and *[7.1.10]*, the *exact normal deviate* defined by $H(y; s, p) = \Phi(z)$, where $u = py - qs + \frac{1}{2}$ and $\sigma = \sqrt{(y+s)pq}$, is given by

$$z = u + \tfrac{1}{6}(p-q)(1-u^2)/\sigma + \tfrac{1}{72}\left[(5-14pq)u^3 + (-2+2pq)u\right]/\sigma^2$$

$$+ \tfrac{1}{6480}(p-q)\left[(-249 + 438pq)u^4 + (79-28pq)u^2 + 128 - 26pq\right]/\sigma^3$$

$$+ O(\sigma^{-4}).$$

[7.3.8] From *[7.3.2]* and *[7.1.11]*, we obtain an approximation desired to be accurate near probability α or $1-\alpha$. If $\Phi(z_\alpha) = 1-\alpha$ ($0 < \alpha \leq 1/2$), then $H(y; s, p) = \Phi(z) \simeq \Phi(u)$, where

$$u = 2\sqrt{(y+1+b)p} - 2\sqrt{(s+b)q}, \quad b = \tfrac{1}{12}(z_\alpha^2 - 4).$$

The error is $(q-p)(z_\alpha^2 - z^2)/(12\sqrt{(s+y)pq}) + O(\{(s+y)pq\}^{-1})$.

Normal Approximations to Distributions 195

7.4 THE HYPERGEOMETRIC DISTRIBUTION

[7.4.1] Let the pf of a hypergeometric rv Y be g (y; n, M, N), where

$$g(y; n, M, N) = \binom{M}{y}\binom{N-M}{n-y} \bigg/ \binom{N}{n} ;$$

$\max(0, n - N + M) \leq y \leq \min(n, M)$, y a nonnegative integer. We assume that $n \leq M \leq N/2$; the notation of the distribution can always be arranged so that these inequalities hold (Molenaar, 1970, p. 116). Let G(y; n, M, N) be the cdf of Y.

In practice, a random sample is taken without replacement from a finite population of size N, of which M objects have a certain attribute and the remaining $N - M$ objects do not. The sample is of size n; g(y; n, M, N) is the probability that the number Y of objects in the sample having the attribute is equal to y.

[7.4.2] Classical Normal Approximation with Continuity Correction. If n is large, but M/N is not small, then $G(k; n, M, N) \simeq \Phi(u)$, where

$$u = \frac{k + \tfrac{1}{2} - nM/N}{\{(N-n)(N-1)^{-1}n(M/N)(1-M/N)\}^{1/2}} = \frac{k + \tfrac{1}{2} - E(Y)}{\sqrt{\text{Var}(Y)}}$$

(Hemelrijk, 1967, pp. 226-227; Molenaar, 1970, p. 119).

[7.4.3] A Recommended Simple Approximation (Molenaar, 1970, pp. 6-7, 132-133, 148). $G(k; n, M, N) \simeq \Phi(u)$, where

$$u = \begin{cases} 2\left[\sqrt{(k+1)(N-M-n+k+1)} - \sqrt{(n-k)(M-k)}\right]\big/\sqrt{N-1} \\ \qquad\qquad\qquad\qquad\qquad\qquad \text{for tail probabilities;} \\ 2\left[\sqrt{(k+\tfrac{3}{4})(N-M-n+k+\tfrac{3}{4})} - \sqrt{(n-k-\tfrac{1}{4})(m-k-\tfrac{1}{4})}\right]\big/\sqrt{N}, \\ \qquad\qquad\qquad\qquad 0.05 < G(k; n, M, N) < 0.93. \end{cases}$$

This approximation is generally superior to that in *[7.4.2]*, but not to those in *[7.4.6]* and *[7.4.7]*.

[7.4.4] Let $n/N \to t$, $M/N \to p$, $(N-M)/N \to q$, $h(k-np) \to x$ as $N \to \infty$, where $1/h = \{Npqt(1-t)\}^{1/2}$. Then, when N is large (Feller 1968, p. 194), $g(k; n; M, N) \simeq h\phi(x)$.

[7.4.5] Nicholson (1956, pp. 474-475) derived upper and lower bounds to $\Pr(b \leq Y \leq c)$, the difference between these bounds being $O(\sigma^{-1})$, where $\sigma^2 = np^*(1-p^*)(1-n/N)$, and $p^* = M/N$; see also Molenaar (1970, p. 135).

[7.4.6] *A recommended accurate approximation* is given by Molenaar (1985, p. 345): let $p^* = M/N$, $s^* = n/N$, $\tau^2 = (N-n)nM(N-M)/N^3 = N \operatorname{Var}(Y)/(N-1)$, and $z_k = (k + \tfrac{1}{2} - np^*)/\tau$. Then $G(c; n, M, N) \simeq \Phi(u)$, where

$$u = z_c + (z_c^2 - 1)\left[-\frac{(2s^* - 1)(2p^* - 1)}{6\tau} + z_c \frac{1 - 3s^* + 3s^{*2}}{48\tau^2}\right].$$

This approximation is recommended as being rather accurate, unless $s^* \leq p^* \leq 0.25$; for the latter case, see Molenaar (1970, pp. 126, 136). See also Ling and Pratt (1984, pp. 55-57).

[7.4.7] *Recommended Most Accurate Approximation.* Ling and Pratt (1984, pp. 49-60) published the following approximation, based on unpublished notes handwritten by David Peizer before August 1966; it will be convenient to alter our notation to follow theirs; see Fig. 1.

Fig. 1

y	n − y	n		a	b	n
M − y	(N − M) − (n − y)	N − n	→	c	d	m
M	N − M	N		r	s	N

We thus seek an approximation to $P(Y \leq a \mid n, r, N)$, $= G(a; n, r, N)$, which equivalently is $1 - P(Y \leq b - 1 \mid n, s, N)$. Without loss of generality the table can be arranged so that $a \leq d$ and $a < b \leq c$, i.e., so that $2a + 1 \leq n \leq r \leq N - n$. Three adjustments are made to the table, in which the margins may not quite equal the marginal totals; see Fig. 2. Define

Fig. 2

$A=a+\frac{1}{2}$ $B=b-\frac{1}{2}$	n	$\rightarrow$	$A'=A+\frac{1}{6}$ $B'=B+\frac{1}{6}$	$n'=n+\frac{1}{6}$
$C=c-\frac{1}{2}$ $D=d+\frac{1}{2}$	m		$C'=C+\frac{1}{6}$ $D'=D+\frac{1}{6}$	$m'=m+\frac{1}{6}$
r s	N		$r'=r+\frac{1}{6}$ $s'=s+\frac{1}{6}$	$N'=N-\frac{1}{6}$

$\rightarrow$

$A'' = A' + \frac{.02}{A+.5} + \frac{.01}{n+1} + \frac{.01}{r+1}$ $B'' = B' + \frac{.02}{B+.5} + \frac{.01}{n+1} + \frac{.01}{s+1}$	n'
$C'' = C' + \frac{.02}{C+.5} + \frac{.01}{m+1} + \frac{.01}{r+1}$ $D'' = D' + \frac{.02}{D+.5} + \frac{.01}{m+1} + \frac{.01}{s+1}$	m'
r' s'	N'

$$L = A \log\left(\frac{AN}{nr}\right) + B \log\left(\frac{BN}{ns}\right) + C \log\left(\frac{CN}{mr}\right) + D \log\left(\frac{DN}{ms}\right).$$

(a) Then $G(a; n, r, N) \simeq \Phi(u)$, where

$$u = \frac{A''D'' - B''C''}{|AD - BC|} \left(2L \, \frac{mnrs \, N'}{m'n'r's'N}\right)^{1/2};$$

(b) $G(a; n, r, N) \simeq \Phi(v)$, where v is defined like u in (a), with $A''D'' - B''C''$ replaced by $A'D' - B'C'$.

Approximations (a) and (b) are more accurate than those in *[7.4.2]*, *[7.4.3]* or *[7.4.6]*; (a) is more accurate than (b) and is better than a number of approximations of Peizer type considered in the same study. See also Johnson et al. (1992, p. 261).

[7.4.8] The *exact normal deviate* z defined by $G(c; n, M, N) = \Phi(z)$ satisfies

$$z = z_c + (2s^* - 1)(2p^* - 1)(1 - z_c^2)/(6\tau)$$
$$+ [z_c^3\{5 - 14s^*(1 - s^*) - 14p^*(1 - p^*) + 38s^*(1 - s^*)p^*(1 - p^*)\}$$

$$+ \; z_c\{-2 + 2s^*(1-s^*) + 2p^*(1-p^*)$$
$$+ \; 10s^*(1-s^*)p^*(1-p^*)\}]/(72\tau^2) + O(\tau^{-2}) \; ,$$

where $z_c = (c + \frac{1}{2} - np^*)/\tau$, and where p*, s* and τ are defined as in [7.4.6]; see Molenaar (1970, p. 124), who also inverts this expansion to express z_c in terms of z.

7.5 MISCELLANEOUS DISCRETE DISTRIBUTIONS

[7.5.1] *Neyman's Type A* distribution, obtained by compounding two Poisson distributions, has pdf g(k; λ, η) given by

$$g(k; \lambda, \eta) = \begin{cases} \sum_{j=0}^{\infty} e^{-\lambda}(\lambda^j/j!)e^{-j\eta}\{(j\eta)^k/k!\}, & k = 1, 2, \ldots, \\ \exp\{-\lambda(1-e^{-\eta})\} & , k = 0 \, . \end{cases}$$

If λ is large and η is not too small, the standardized rv $(Y - \lambda\eta)/\sqrt{\lambda\eta(1+\eta)}$ has an approximate N(0, 1) distribution, where Y has the pdf above (Neyman, 1939, p. 46; Martin and Katti, 1962, pp. 355-359).

[7.5.2] *Absorption Distributions.* Suppose that a large number n of particles cross or fail to cross an infinite slab (in two dimensions) of width b, containing a large number M of absorption points, and with an initial expected number θ of absorption points to be encountered by each particle. Let Y be the number of absorptions. Then Y has pf g(k; θ, n, M), where

$$g(k; \theta, n, M) = \frac{q^{(M-k)(n-k)} \Pi_{j=n-k+1}^{n}(1-q^j)\Pi_{j=M-k+1}^{M}(1-q^j)}{\Pi_{j=1}^{k}(1-q^j)} \, .$$

$q = 1 - \theta/M$, k = 0, 1, ..., min(M, n). Then the standardized rv $\{Y - (1-q^M)(1-q^n)\}/\sigma$ tends rapidly to normality, where

$$\sigma^2 = (1-q^M)q^M(1-q^n)q^n/\left[(1-q)\{1-(1-q^M)(1-q^n)\}^2\right]$$

(Borenius, 1953, pp. 151-157).

Normal Approximations to Distributions 199

[7.5.3] Let $X_1, \ldots, X_n$ be a sequence of iid random variables, usually assumed continuous, and let the rv Y be the total number of *runs* of increasing and of decreasing values in the sequence $X_1, \ldots, X_n$. If $n > 12$, the distribution of the standardized rv $\{Y - \frac{1}{3}(2n-1)\}\sqrt{90/(16n-29)}$ is approximately $N(0, 1)$. Wallis and Moore (1941, p. 405) suggest replacing $Y - (2n-1)/3$ by $\text{sgn}\{|Y - \frac{1}{3}(2n-1)| - \frac{1}{2}\}$ for a continuity correction.

[7.5.4] Let $X_1, \ldots, X_N$ be a sequence of rvs with realized values $x_1, \ldots, x_N$. The value of x_j is a *record* if, for some $j \leq k \leq N$, x_j is the largest (or smallest) of $x_1, \ldots, x_k$. Let Y_1 be the total number of records (largest and smallest) in $X_1, \ldots, X_N$, and let

$Y_2 =$ (total number of upper records) $-$ (total number of lower records).

Then as $N \to \infty$, Y_1 and Y_2 are asymptotically independent; the asymptotic distributions of Y_1 and Y_2 are $N(\sum_{j=2}^{N} j^{-1}, 2\sum_{j=2}^{N} j^{-1} - 4\sum_{j=2}^{N} j^{-2})$ and $N(0, 2\sum_{j=2}^{N} j^{-1})$, respectively. Foster and Stuart (1954, p. 6) tabulate exact and approximate cdfs of Y_1 and Y_2 when $N = 15$ and $N = 6$, respectively.

7.6 THE BETA DISTRIBUTION

[7.6.1] Let the probability density function (pdf) of a beta rv Y be $h(y; a, b)$, given by

$$h(y; a, b) = y^{a-1}(1-y)^{b-1}/B(a, b), \quad 0 < y < 1, a > 0, b > 0,$$

where $B(a, b)$ is the beta function. Let $H(y; a, b)$ be the cdf of Y.

[7.6.2] When a and b are positive integers, approximations to $H(y; a, b)$ can be deduced from those for binomial probabilities discussed in *[7.1]*. Using the notation G for the binomial cdf in *[7.1]*,

$$H(y; a, b) = G(b-1; a+b-1, 1-y) = 1 - G(a-1; a+b-1, y).$$

Either of these forms give rise to approximations, the choice depending on which reduces the error most. Using the first form, we can construct approximations to H(y; a, b) from those for G(k; n, p) as follows:
  (i) Replace k by b − 1.
  (ii) Replace n by a + b − 1, and n − k by a.
  (iii) Replace p by 1 − y and q by y.
  (iv) If G(k; n, p) $\simeq$ $\Phi(z)$, then H(y; a, b) $\simeq$ $\Phi(z)$, where z is expressed in terms of y, a, and b from (i), (ii), and (iii).

[7.6.3] From *[7.6.2]* and Molenaar's *recommended* simple yet accurate approximation in *[7.1.4]*, we derive the following: Let y be close to 1/2 in the sense that, if a + b = 4, 31, or 301, then |y − 1/2| ≤ 0.25, 0.1, or 0.04, respectively. Then if 0.45 ≤ |H(y; a, b) − $\frac{1}{2}$| ≤ 0.495,

$$H(y; a, b) \simeq \Phi\left[\sqrt{(4b-1)y} - \sqrt{(4a-1)(1-y)}\right] ;$$

and if 0.05 ≤ H(y; a, b) ≤ 0.93,

$$H(y; a,b) \simeq \Phi\left[\sqrt{(4b-\tfrac{3}{2})y} - \sqrt{(4a-\tfrac{3}{2})(1-y)}\right].$$

When y is not close to $\frac{1}{2}$ as above, and if 0.45 ≤ |H(y; a, b) − $\frac{1}{2}$| ≤ 0.495,

$$H(y; a, b) \simeq \Phi\left[\sqrt{4by} - \sqrt{4a(1-y)}\right] ;$$

and if 0.05 ≤ H(y; a, b) ≤ 0.93,

$$H(y; a, b) \simeq \Phi\left[\sqrt{(4b-1)y} - \sqrt{(4a-1)(1-y)}\right].$$

The absolute error is $O(1/\sqrt{a+b})$ if y ≠ $\frac{1}{2}$ and is $O(1/(a+b))$ if y = $\frac{1}{2}$.

[7.6.4] From *[7.6.2]* and the *Camp-Paulson* approximation of *[7.1.7]*, the following approximation obtains: H(y; a, b) $\simeq$ $\Phi\{-x/(3\sqrt{z})\}$, where

$$x = \left(9 - \tfrac{1}{a}\right)\left[\frac{a(1-y)}{by}\right]^{1/3} + \tfrac{1}{b} - 9, \quad z = \tfrac{1}{a}\left[\frac{a(1-y)}{by}\right]^{2/3} + \tfrac{1}{b} ;$$

the error is $O(\{(a + b)y(1 - y)\}^{-1})$ for all y, but if $y = \frac{1}{2}$ or $y = 0.042$, it is $O(\{(a + b)y(1 - y)\}^{-3/2})$.

[7.6.5] The *Peizer-Pratt* approximation can be obtained from *[7.6.2]* and *[7.1.9]*; or see Peizer and Pratt (1968, pp. 1417-1422); for $0 < y < 1$, we have $H(y; a, b) \simeq \Phi(u)$, where

$$u = d \left[\frac{1 + yT\{\frac{b - \frac{1}{2}}{(a + b - 1)(1 - y)}\} + (1 - y)T\{\frac{a - \frac{1}{2}}{(a + b - 1)y}\}}{y(1 - y)(a + b - \frac{5}{6})} \right]^{1/2},$$

$$d = (a + b - \frac{2}{3})y - (a - \frac{1}{3}) + \frac{1}{50}\left(\frac{y}{b} - \frac{1 - y}{a} + \frac{y - \frac{1}{2}}{a + b}\right),$$

$$T(x) = (1 - x^2 + 2x \log x)/(1 - x)^2, \quad x \neq 1; \quad T(1) = 0.$$

$$|H(y; a, b) - \Phi(u)| \leq \begin{cases} 0.001 & \text{if } a \geq 2, \ b \geq 2, \\ 0.01 & \text{if } a \geq 1, \ b \geq 1. \end{cases}$$

Generally, the approximation is accurate to $O(1/\{(a + b)y(1 - y)\}^{3/2})$ if $y \neq \frac{1}{2}$ and to $O(1/\{(a + b)y(1 - y)\}^2)$ if $y = \frac{1}{2}$ (see *[7.1.9]*). See also Alfers and Dinges (1984, pp. 399-419) for approximations to tail probabilities.

[7.6.6] Molenaar's *recommended more accurate* approximation of *[7.1.10]* adapts with *[7.6.2]* to give the following:

$$H(y; a, b) \simeq \Phi\{2\sqrt{by + A} - 2\sqrt{a(1 - y) + B}\}, \quad 0 < y < 1, y \neq \frac{1}{2},$$

$$A = \frac{1}{\sigma}(1 - 4y + 7y^2)\{a - \frac{1}{2} - (a + b - 1)y\}^2 - \frac{1}{18}(2 + y + 5y^2),$$

$$B = \frac{1}{\sigma}(4 - 10y + 7y^2)\{a - \frac{1}{2} - (a + b - 1)y\}^2 - \frac{1}{18}(8 - 11y + 5y^2),$$

$$\sigma^2 = 36(a + b - 1)y(1 - y);$$

$$H(\tfrac{1}{2}; a, b) \simeq \Phi(\sqrt{2b + \beta} - \sqrt{2a + \beta}),$$

$$\beta = \{(a-b)^2 - 10(a+b-1)\}/\{12(a+b-1)\} .$$

If $\eta = (a+b)y(1-y)$, this is accurate to $O(\eta^{-3/2})$ when $y \neq \frac{1}{2}$, and to $O(\eta^{-2})$ when $y = \frac{1}{2}$ (see [7.1.10]).

[7.6.7] Suppose that $(a+b-1)(1-y) \geq 0.8$ and $a+b > 6$. Then (Abramowitz and Stegun, 1964, Sec. 26.5.21) $H(y; a, b) = \Phi(3x/\sqrt{z}) + \epsilon$,

$$x = (by)^{1/3}\{1 - (9b)^{-1}\} - \{a(1-y)\}^{1/3}\{1 - (9a)^{-1}\} ,$$

$$z = (by)^{2/3}b^{-1} + \{a(1-y)\}^{2/3}a^{-1}, \ |\epsilon| < 5 \times 10^{-3} .$$

7.7 THE VON MISES DISTRIBUTION

[7.7.1] A *von Mises* or *circular normal* rv Y has pdf $g(y; a)$ where $g(y; a) = \{2\pi I_0(a)\}^{-1} \exp(a \cos y)$, $-\pi < y < \pi$, $a > 0$. Let $G(y; a)$ be the cdf of Y. Normal approximations to $G(y; a)$ improve as $a \to \infty$, so we assume that a is large.

[7.7.2] A number of approximations have been compared by Upton (1974, pp. 369-371), who also introduced the last [(iv) below]:

(i) $G(y; a) \simeq \Phi(y\sqrt{a})$;
(ii) $G(y; a) \simeq \Phi(y\sqrt{a-\frac{1}{2}})$, $a > 10$;
(iii) $G(y; a) \simeq \Phi[y\sqrt{a}\{1 - (8a)^{-1}\}]$;
(iv) $G(y; a) \simeq \Phi[y\sqrt{a}[1 - (8a)^{-1}\} - \frac{1}{24} y^3\sqrt{a}\{1 + (4a)^{-1}\}]$.

Based on comparisons for which $0.6 \leq G(y; a) \leq 0.99$ and $1 \leq a \leq 15$, (iii) is better than (i), (ii) has errors in the tails to $O(a^{-1})$, but (iv) is recommended as most accurate, with two-, three- or four-place decimal-place accuracy if $a \geq 3$, $a \geq 6$, or $a \geq 12$, respectively.

[7.7.3] G. W. Hill (1976, p. 674) gives an approximation that improves on those in [7.7.2]. Let $w = 2\sqrt{a} \sin(y/2)$, i.e., $y = 2 \arcsin(\frac{1}{2} w/\sqrt{a})$. Then $G(y; a) \simeq \Phi(u)$, where

$$u = w - \frac{w}{8a} - \frac{2w^3 + 7w}{128a^2} - \frac{8w^5 + 46w^3 + 177w}{3072a^3} - \cdots ,$$

Normal Approximations to Distributions

$$w = u + \frac{u}{8a} + \frac{2u^3 + 9u}{128a^3} + \frac{8u^5 + 70u^3 + 225u}{3072a^3} + \cdots.$$

[7.7.4] *An Approximation Accurate to* $O(a^{-3})$ (G. W. Hill, 1976, pp. 674-675): $G(y; a) \simeq \Phi(u)$, where

$$u = (\text{sgn } y)\left[(2a - \tfrac{1}{2} - \tfrac{3}{64a})(1 - \cos y)(1 - \tfrac{1 - \cos y}{16a})\right]^{1/2}, \quad y \neq 0.$$

[7.7.5] G. W. Hill (1976, p. 675) gives an approximation that is accurate to $O(a^{-5})$, but it requires access to tables of the modified Bessel function $I_0(a)$; see G. W. Hill (op. cit.), however, for details of the algorithm:

$$I_0(a) \simeq (2\pi a)^{-1/2} e^a \left[1 + \frac{1}{8a} + \frac{1 \cdot 9}{2!(8a)^2} + \frac{1 \cdot 9 \cdot 25}{3!(8a)^3} + \cdots\right],$$

when a is large. The following form is suggested: $G(y; a) \simeq \Phi(u)$, where

$$u \simeq z - z^3 \left\{8a - \frac{2z^2 + 16}{3} - \frac{4z^4 + 7z^2 + 334}{96a}\right\}^{-2},$$

$$z \simeq u - u^3 \left\{8a - \frac{2u^2 + 16}{3} - \frac{4u^4 + 25u^2 + 334}{96a}\right\}^{-2};$$

$$z = \sin(y/2)\sqrt{2\pi} \; e^a / I_0(a).$$

7.8 THE CHI-SQUARED AND GAMMA DISTRIBUTIONS

[7.8.1] (a) Let the pdf of a chi-squared rv Y be $g(y; \nu)$, so that

$$g(y; \nu) = \{2^{\nu/2}\Gamma(\nu/2)\}^{-1} y^{(1/2)\nu - 1} e^{-y/2}, \quad y > 0, \quad \nu = 1, 2, \ldots.$$

Let the cdf of Y be $G(y; \nu)$, such that $G(y_p; \nu) = 1 - p = \Phi(z_p)$ (see *[5.3.1]* to *[5.3.3]*). The distribution is that of the sum of squares of ν iid $N(0, 1)$ rvs.

(b) The most general form of the pdf of a gamma random variable X is given by $h(x; a, b, c) = \{b^a \Gamma(a)\}^{-1}(x-c)^{a-1} \exp\{-(x-c)/b\}$; $x > c$, a >

0, b > 0, where in most applications c = 0. Probability statements about the distribution can be taken from the transformation $Y = 2(X-c)/b$, since Y has a χ^2_{2a} distribution. The approximations in this section are presented in the context of chi-square rvs, but this transformation can easily be used to apply the results to a general gamma distribution. Thus $\Pr(X \leq x) = G(2(y-c)/b;\ 2a)$ and $y_p = 2(x_p-c)/b$ with obvious notation for quantiles x_p of X.

For exact computation, Ling (1978, p. 278) indicates that the expansions for $G(y;\ \nu)$ given in [5.3.3] give superior performance, both in accuracy and efficiency.

[7.8.2] The simple normal approximation

$$G(y;\ \nu) \simeq \Phi\{(y-\nu)/\sqrt{2\nu}\}$$

follows from the Central Limit Theorem (Chap. 6). It is not very accurate, unless the degrees of freedom ν are very large.

[7.8.3] *The Fisher Approximation* is given by

$$G(y;\ \nu) \simeq \Phi(\sqrt{2y} - \sqrt{2\nu - 1})$$

(Stuart and Ord, 1987, Secs. 16.5, 6). This is an improvement over that of [7.8.2], but is not as accurate as that of [7.8.4]; it is satisfactory if $\nu > 100$ (Severo and Zelen, 1960, p. 411); see Zar (1978, pp. 280-290) for tables comparing the accuracies of the Fisher and Wilson-Hilferty approximations.

[7.8.4] (a) *The Wilson-Hilferty Approximation* (Wilson and Hilferty, 1931, pp. 684-688):

$$G(y;\ \nu) \simeq \Phi(x) = \Phi\left\{\sqrt{\frac{9\nu}{2}}\left[\left(\frac{y}{\nu}\right)^{1/3} - 1 + \frac{2}{9\nu}\right]\right\}.$$

The first two moments and cumulants of W and its shape factors, where $W = (Y/\nu)^{1/3}$, are

$$\mu(W) = \kappa_1(W) = 1 - 2/(9\nu) + O(\nu^{-3}),$$

$$\sigma^2(W) = \kappa_2(W) = 2/(9\nu) + O(\nu^{-3}) \ ;$$

$$\gamma_1(W) = \frac{\kappa_3(W)}{\{\kappa_2(W)\}^{3/2}} = \frac{4}{27}(\frac{2}{\nu})^{3/2} + O(\nu^{-5/2}) \ ,$$

$$\gamma_2(W) = \frac{\kappa_4(W)}{\{\kappa_2(W)\}^2} = -\frac{4}{9\nu} + O(\nu^{-2}) \ .$$

The approximation depends on the closeness of W to a $N(1-2/(9\nu), 2/(9\nu))$ distribution; see [6.3.3] and the discussion in [6.3.4].

Its superior performance arises because $(Y/\nu)^{1/3}$ tends to symmetry and to normality more rapidly than does $\sqrt{2Y}$ in [7.8.3], and $\sqrt{2Y}$ tends to normality more rapidly than does Y in [7.8.2] (Stuart and Ord, 1987, Secs. 16.5, 7, 8; Johnson and Kotz, 1970a, pp. 176-177). The minimum absolute error when $\nu \geq 3$ is less than 0.007, and is less than 0.001 when $\nu \geq 15$ (Mathur, 1961, pp. 103-105).

(b) In (a), replace x by $x + h_\nu$ or $x + h'_\nu$, where

$$h_\nu = -\frac{2}{27\nu}\left\{\frac{2\sqrt{2}(x^2-1)}{3\sqrt{\nu}} - \frac{x^3-3x}{4}\right\},$$

$$h'_\nu = \frac{1}{486\nu^2}\left\{(9\nu+16)(x^3-3x) - 24(x^2-1)\sqrt{2\nu}\right\}$$

(Severo and Zelen, 1960, pp. 411-416). These approximations give improvements over that in (a). Severo and Zelen suggest approximating h_ν by $(60/\nu)h_{60}$ when $\nu \geq 30$, say; and by $(20/\nu)h_{20}$ when $5 \leq \nu \leq 20$. Values of h_{60} and h_{20} are tabulated to five decimal places for $x = -3.5(0.05)3.5$. See also Abramowitz and Stegun (1964, Sec. 26.4.15) for values of h_{60}.

[**7.8.5**] The approximations of [7.8.2, 3, 4, 7] are all members of the Box-Cox family $Y^* = (Y^\lambda - 1)\lambda$ for $\lambda \neq 0$ and $Y^* = \log Y$ for $\lambda = 0$, for $\lambda = 1, \frac{1}{2}, \frac{1}{3}, 0$, respectively. The *fourth-root transformation* $W = Y^{1/4}$ corresponds to $\lambda = \frac{1}{4}$, and its transforms chi-square "to very near normality for all degrees of freedom", although [7.8.4] is slightly more accurate if $\nu > 2$ (Hawkins and Wixley, 1986, pp. 296-298).

Goria (1992, pp. 55-64) considered linear combinations of $(Y/\nu)^{1/2}$ and

$(Y/\nu)^{1/4}$. By seeking to bring the third and fourth cumulants close to zero when ν is large, Goria derived the normal approximation

$$L = (Y/\nu)^{1/4} + 4(Y/\nu)^{1/2}$$

as being most accurate in this class. L has moments, cumulants and shape factors

$$\mu(L) = \kappa_1(L) = 5 - \frac{1}{\nu} - \frac{3}{128\nu^2} + O(\nu^{-3}),$$

$$\sigma^2(L) = \kappa_2(L) = \frac{9}{2\nu} + \frac{1}{8\nu^2} + O(\nu^{-3});$$

$$\gamma_1(L) = \frac{\kappa_3(L)}{\{\kappa_2(L)\}^{3/2}} = \frac{199}{1728}\left(\frac{2}{\nu}\right)^{3/2} + O(\nu^{-5/2}),$$

$$\gamma_2(L) = \frac{\kappa_4(L)}{\{\kappa_2(L)\}^2} = -\frac{1}{3\nu} + O(\nu^{-2}).$$

Comparison with $\gamma_1(W)$ and $\gamma_2(W)$ in *[7.8.4](a)* indicates that the shape factors of L are closer to zero than those of the Wilson-Hilferty variable W in *[7.8.4]*.

Goria tabulates the smallest df ν giving rise to $|e| \leq 0.1, .007, .002, .001$ and $.0001$, where e is the local maximum absolute error resulting from the normality assumption. If we require e to be bounded above by $.007, .002, .001$ and $.0001$, respectively, then the accuracy of L obtains at $\nu \geq 3, 7, 12$ and 87, respectively, while that of the Wilson-Hilferty variable W of *[7.8.4](a)* only obtains at $\nu \geq 3, 8, 14$ and 110, respectively. By this criterion the Goria approximation is to be preferred.

[7.8.6] **Peizer-Pratt Approximation** (Peizer and Pratt, 1968, pp. 1418, 1421, 1424). $G(y; \nu) \simeq \Phi(u)$, where

$$u = \frac{d}{\sqrt{2y}}\left[1 + T\left(\frac{\nu-1}{y}\right)\right]^{1/2}, \quad d = y - \nu + \frac{2}{3} - \frac{0.08}{\nu},$$

$$T(x) \simeq (1 - x^2 + 2x \log x)/(1-x)^2, \quad x \neq 1; \; T(1) = 0;$$

Normal Approximations to Distributions 207

$$|G(y; \nu) - \Phi(u)| \leq 0.001 \text{ (resp. 0.01) if } \nu \geq 4 \text{ (resp. } \nu \geq 2).$$

Ling (1978, p. 278) shows that the Peizer-Pratt approximation gives "very accurate results, even for fairly small ν", and is better for $5 \leq \nu \leq 240$ than the Wilson-Hilferty approximation of *[7.8.4]* or those of *[7.8.2, 3]* by several orders of magnitude.

[7.8.7] If Y has a χ_ν^2 distribution, then the distribution of log Y is more nearly normal than that of Y (Johnson and Kotz, 1970a, p. 181). The resulting approximation is not used very much in practice, but it could be useful in the approximation to the distribution of the sample generalized variance (see *[7.14.4]*). Using the Kullback-Leibler criterion Hawkins and Wixley (1986, pp. 296-298) showed that this transformation fails to provide accuracy even when ν is large.

[7.8.8] In the next few sections we give expressions for the *quantiles* y_p of χ_ν^2 in terms of z_p, where ideally

$$G(y_p;\nu) = \Phi(z_p) = 1 - p.$$

Some of these follow from results in *[7.8.2]-[7.8.4]* or extensions of these. Note, however, the exact results

$$y_p = z_p^2, \nu = 1; \quad y_p = -2\ln p, \nu = 2.$$

A useful survey is that of Zar (1978, pp. 280-290), who compares these approximations for accuracy; the best are given in *[7.8.11]* (both extensions of the Wilson-Hilferty result), and in *[7.8.13]* (the Cornish-Fisher expansion). Zar's criterion is to tabulate the smallest df ν for which the relative error of the approximation is no greater than 1%, .05%, 0.1% and 0.05%, when min(p, 1−p) = 0.01, .005, .01, .025, .05, .10, .25 and .50. All approximations perform better in the right tail of χ_ν^2 (p close to 1) than in the left tail (p close to 0); see also the comparative study by Sahai and Thompson (1974, pp. 86-88).

[7.8.9] $\quad y_p \simeq \frac{1}{2}(z_p + \sqrt{2\nu - 1})^2$ (Fisher, from *[7.8.3]*).

[7.8.10] $y_p \simeq \nu(z_p\sqrt{\frac{2}{9\nu}} + 1 - \frac{2}{9\nu})^3$ (Wilson-Hilferty, from *[7.8.4]*(a)). This approximation is uniformly more accurate than that in *[7.8.9]*.

[7.8.11] Two *recommended approximations* for accuracy (Goldstein, 1973, pp. 483-485); in these x denotes z_p:

(a) $(\frac{y_p}{\nu})^{1/3} \approx 1 - \frac{2}{9\nu} + \frac{(4x^4 + 16x^2 - 28)}{1215\nu^2} + \frac{(8x^6 + 720x^4 + 3216x^2 + 2904)}{229635\nu^3}$

$+ (\frac{2}{\nu})^{1/2} \{\frac{x}{3} + \frac{-x^3 + 3x}{162\nu} - \frac{3x^5 + 40x^3 + 45x}{5832\nu^2}$

$+ \frac{301x^7 - 1519x^5 - 32769x^3 - 79349x}{7873200\nu^3}\}$

(b) $(\frac{y_p}{\nu})^{1/3} \approx 1.0000886 - \frac{0.2237368}{\nu} - \frac{0 \cdot 01513904}{\nu^2}$

$+ \nu^{-1/2}x\left(0.4713941 + \frac{0 \cdot 02607083}{\nu} - \frac{0 \cdot 008986007}{\nu^2}\right)$

$+ \nu^{-1}x^2\left(0 \cdot 0001348028 + \frac{0 \cdot 01128186}{\nu} + \frac{0 \cdot 02277679}{\nu^2}\right)$

$+ \nu^{-3/2}x^3\left(-0 \cdot 008553069 - \frac{0 \cdot 01153761}{\nu} - \frac{0 \cdot 01323293}{\nu^2}\right)$

$+ \nu^{-2}x^4\left(0 \cdot 00312558 + \frac{0 \cdot 005169654}{\nu} - \frac{0 \cdot 006950356}{\nu^2}\right)$

$+ \nu^{-5/2}x^5\left(-0 \cdot 0008426812 + \frac{0 \cdot 00253001}{\nu} + \frac{0 \cdot 001060438}{\nu^2}\right)$

$+ \nu^{-3}x^6\left(0 \cdot 00009780499 - \frac{0 \cdot 001450117}{\nu} + \frac{0 \cdot 001565326}{\nu^2}\right).$

These approximations are extensions of *[7.8.10]*. They are more accurate than those of *[7.8.9, 10, 12]* if Zar's criterion of *[7.8.8]* is followed.

Approximation (b) is only marginally better than (a); (a) is simpler, however, and some may prefer it for that reason (Zar, 1978, Table 2).

[7.8.12] *Severo-Zelen extensions* of the Wilson-Hilferty approximation. Defining h_ν and h'_ν as in *[7.8.4]*(b) (see Severo and Zelen, 1960, pp. 411-416),

$$y_p \simeq y_p(h) = \nu\left[(z_p - h)\sqrt{\frac{2}{9\nu}} + 1 - \frac{2}{9\nu}\right]^3,$$

where h = h_ν or h = h'_ν. Zar (1978, Table 2) shows that these improve on [7.8.10] considerably, but are less accurate than either of those in [7.8.11]; $y_p(h'_\nu)$ is slightly more accurate than $y_p(h_\nu)$. Writing

$$h_\nu \approx (c/\nu)h_c = H/\nu, \quad h'_\nu \approx (c/\nu)h'_\nu = H'/\nu$$

for some constant c, Severo and Zelen found optimum choices of c for given values of p that lead to an approximation $y_p(H/\nu)$ or $y_p(H'/\nu)$ that improves on $y_p(h_\nu)$ or $y_p(h'_\nu)$; see the discussion in Zar (op. cit., Secs. 3, 4).

[7.8.13] *Cornish-Fisher expansions* (CF). Denoting z_p by z,

$$y_p \simeq \nu + \sqrt{\nu}(z\sqrt{2}) + \tfrac{2}{3}(z^2 - 1) + \frac{1}{9\sqrt{2\nu}}(z^3 - 7z)$$

$$- \frac{1}{405\nu}(6z^4 + 14z^2 - 32) + \frac{1}{4860\sqrt{2}\,\nu^{3/2}}(9z^5 + 256z^3 - 433z)$$

$$+ \frac{1}{22515\nu^2}(12z^6 - 243z^4 - 923z^2 + 1472) + O(\nu^{-5/2}) \,.$$

Peiser (1943, pp. 56-62) and Goldberg and Levine (1946, pp. 216-225) derived approximations to the quantiles y_p which are essentially the first four and six terms, respectively, of the above CF expansion.

If $\nu \geq 30$, the CF expansion to six terms (Goldberg-Levine) has an error in y_p no greater than 0.0001 when $0.1 \leq p \leq 0.9$. Only if ν or $1-p$ is very small is there much need for more than six terms of the CF expansion. See Sahai and Thompson (1974, pp. 86-88) and Johnson and Kotz (1970a, pp. 176-177). Zar (1978, pp. 280-290), using the criterion of [7.8.8], found the CF expansion to the term in $\nu^{-3/2}$ to give an approximation to y_p that is nearly as accurate as those in [7.8.11], particularly for values of p close to zero.

[7.8.14] *Bounds* on $\Pr(Y > y)$ (Wallace, 1959, p. 1127) :

$$d_\nu \exp\{1/(9\nu)\}[1 - \phi\{W_1(y)\}] < 1 - G(y;\nu) < d_\nu[1 - \Phi\{W(y)\}] \,,$$

$$W(y) = \sqrt{y - \nu - \nu \log\,(y/\nu)} \,, \quad W_1(y) = W(y) + (\sqrt{2/\nu})/3 \,,$$

$$d_\nu = \sqrt{2\pi} \exp(-\nu/2)(\tfrac{1}{2})^{(\nu-1)/2}/\Gamma(\tfrac{1}{2}\nu).$$

7.9 NONCENTRAL CHI-SQUARE

[7.9.1] The pdf $g(y; \nu, \lambda)$ of a noncentral chi-square rv with ν degrees of freedom and noncentrality parameter λ is given in *[5.3.6]*; let $G(y; \nu, \lambda)$ be the cdf of Y. The distribution of Y is that of $\sum_{i=1}^{\nu}(X_i - \mu_i)^2$, where X_1, ..., X_ν are iid N(0, 1) rvs, and $\lambda = \mu_1^2 + ... + \mu_\nu^2$.

[7.9.2] Two simple normal approximations, not very accurate, but having error $O(1/\sqrt{\lambda})$ as $\lambda \to \infty$, uniformly in y, are given (Johnson, 1959, p. 353; Johnson and Kotz, 1970b, p. 141) by
(a) $G(y; \nu, \lambda) \simeq \Phi[(y - \nu - \lambda)/\sqrt{2(\nu + 2\lambda)}]$,
(b) $G(y; \nu, \lambda) \simeq \Phi[(y - \nu - \lambda + 1)/\sqrt{2(\nu + 2\lambda)}]$.

[7.9.3] The following approximation to noncentral χ^2 involves first a central χ^2 approximation, and then a normal approximation to central χ^2. As one would expect, the first stage (direct central χ^2 approximation) is generally more accurate than the normality approximation superimposed on it; the reference sources contain further discussion (see also Johnson and Kotz, 1970b, pp. 139-142).

Abdel-Aty (1954, p. 538) applied the Wilson-Hilferty approximation of *[7.8.4]*(a) to $\{Y/(\nu+\lambda)\}^{1/3}$. Thus (Abramowitz and Stegun, 1964, Sec. 26.4.28)

$$u = \frac{\{y/(\nu + \lambda)\}^{1/3} - 1 + 2(\nu + 2\lambda)/\{9(\nu + \lambda)^2\}}{[2(\nu + 2\lambda)/\{9(\nu + \lambda)^2\}]^{1/2}}.$$

As one would expect, this approximation does better when the noncentrality parameter is small (and the distribution of Y is more like that of central chi-square), and deteriorates as λ increases; see *[7.9.6(a)]*.

[7.9.4] (a) One of three approximations of Sankaran (1959, pp. 235-236) is accurate over a wide range, even when ν is small:

$$G(y; \nu, \lambda) \simeq \Phi\{(x^h - a)/b\}, \quad x = y/(\nu + \lambda),$$

$$h = \tfrac{1}{3} + \tfrac{2}{3}\lambda^2(\nu + 2\lambda)^{-2},$$

$$a = 1 + h(h-1)\frac{\nu + 2\lambda}{(\nu + \lambda)^2} - h(h-1)(2-h)(1-3h)\frac{(\nu + 2\lambda)^2}{2(\nu + \lambda)^4},$$

$$b = h\frac{\sqrt{2(\nu + 2\lambda)}}{\nu + \lambda}\left\{1 - (1-h)(1-3h)\frac{\nu + 2\lambda}{2(\nu + \lambda)^2}\right\},$$

to terms of $O(\lambda^{-2})$. Although complicated, this approximation performs better than any so far listed. For upper and lower 5 percent points, Johnson and Kotz (1970b, p. 142) compare the absolute error for $\nu = 2, 4,$ and 7, and $\lambda = 1, 4, 16,$ and 25 with those of *[7.9.2](b)* and *[7.9.3](a)*; this error is never greater than 0.06 ($\lambda \geq 1$) or 0.03 ($\lambda \geq 4$).

(b) In the context of a more general problem (see *[7.14.1]*), and apparently unaware of the Sankaran approximation in (a) above, Jensen and Solomon (1972, pp. 899-901) gave the formula in (a) with terms truncated to $O(\lambda^{-1})$; that is, defining x and h as in (a),

$$G(y; \nu, \lambda) \simeq \Phi\{(x^h - A)/B\}, \quad A = 1 + h(h-1)\frac{\nu + 2\lambda}{(\nu + \lambda)^2},$$

$$B = h\sqrt{2(\nu + 2\lambda)} / (\nu + \lambda).$$

When $4 \leq \nu \leq 24$ and $4 \leq \lambda \leq 24$, the absolute error in approximating $G(y; \nu, \lambda)$ is less than 0.005.

[7.9.5] The asymptotic normality of Y(*[7.9.2]*) and the main result on transformations in *[6.2.2.1]* imply the asymptotic normality of $\{(Y + b)/(\nu + \lambda)\}^h$ for any choice of constants b and h, $h > 0$. Moschopoulos (1983, pp. 1873-1878) derived an improved approximation to normality when

$$h = 1 - \frac{2(\nu + \lambda)(\nu + 3\lambda)}{3(\nu + 2\lambda)^2}, \quad b = \frac{\lambda^2}{(\nu + \lambda)(\nu + 2\lambda)}.$$

The asymptotic mean and variance of $\{(Y + b)/(\nu + \lambda)\}^h$ are

$$\mu^* = 1 + \frac{h}{\nu + \lambda}\left\{(h-1)\frac{\nu + 2\lambda}{\nu + \lambda} + b\right\}, \quad \sigma^{*2} = 2h^2\frac{\nu + 2\lambda}{(\nu + \lambda)^2},$$

respectively. Then

$$G(y; \nu, \lambda) \simeq \Phi\left(\left[\{(y + b)/(\nu + h)\}^h - \mu^*\right]/\sigma^*\right).$$

A numerical evaluation indicates that this approximation is superior to those in *[7.9.3]* and *[7.9.4]*(b) over parameter values $(\nu, \lambda) = (1, 3)$, $(2, 2)$, $(4,4)$, $(4, 10)$ and $(4, 16)$.

[7.9.6] If $G(y_p; \nu, \lambda) = 1 - p = \Phi(z_p)$, approximations to percentiles are as follows:

(a) $y_p \simeq (\nu + \lambda)(z_p\sqrt{c} + 1 - c]^3, \quad c = \dfrac{2(\nu + 2\lambda)}{9(\nu + \lambda)^2}$

(Abramowitz and Stegun, 1964, Sec. 26.4.32), derived from *[7.9.3]*(a).

(b) $y_p \simeq (\nu + \lambda)(a + bz_p)^{1/h}$,

where a, b and h are as defined in *[7.9.4]*, and we may replace a and b by A and B, respectively; see Johnson and Kotz (1970b, p. 142).

7.10 STUDENT'S t DISTRIBUTION

[7.10.1] The pdf $g(y; \nu)$ of a Student t rv Y with ν degrees of freedom is given by

$$g(y; \nu) = \left[\sqrt{\nu}\, B(\tfrac{1}{2}, \tfrac{1}{2}\nu)\right]^{-1}(1 + t^2/\nu)^{-(\nu+1)/2}, \quad \nu = 1, 2, \ldots,$$

where $B(\cdot, \cdot)$ is the beta function. Let $G(y; \nu)$ be the cdf of Y. Normal approximations to t tend to improve as ν increases.

In addition to what follows in this section, approximations to the $F_{m,n}$ distribution in *[7.12]* yield approximations to t with n degrees of freedom when m = 1, since t_n^2 is distributed as $F_{1,n}$.

[7.10.2] *Simple Normal Approximations.*

(a) $G(y; \nu) \simeq \Phi(y\sqrt{\frac{\nu-2}{\nu}})$; (b) $G(y; \nu) \simeq \Phi\left[y\dfrac{1 - 1/(4\nu)}{\sqrt{1 + y^2/(2\nu)}}\right]$.

Neither approximation is recommended unless ν is fairly large; see Johnson and Kotz (1970b, p. 101) and Abramowitz and Stegun (1964, Sec. 26.7.8).

[7.10.3] *Bounds by Wallace* (1959, pp. 1124-1125). Let $G(y; \nu) = \Phi\{z(y)\}$ and $u(y) = \{\nu \log(1 + y^2/\nu)\}^{1/2}$. Suppose that $y > 0$. Then

(a) $z(y) \leq u(y)$, $\nu > 0$.
(b) $z(y) \geq u(y)\{1 - 1/(2\nu)\}^{1/2}$, $\nu > \frac{1}{2}$;
(c) $z(y) \geq u(y) - 0.368/\sqrt{\nu}$, $\nu \geq \frac{1}{2}$.

As an approximation to $z(y)$, $u(y)$ has an absolute error not exceeding $0.368/\sqrt{\nu}$. Except for very large values of y, the bound in (c) is much poorer than that in (b); the maximum of the bounds in (b) and (c) is "a good approximation." See Johnson and Kotz (1970b, pp. 108-109). As a variant of the approximation in (a), Mickey (1975, pp. 216-217) found

$$z(y) \simeq \pm\{(\nu - \tfrac{1}{2}) \log(1 + y^2/\nu)\}^{1/2}$$

to give good results if $\nu > 10$; see also *[7.10.8]*.

[7.10.4] (a) *A Recommended Approximation* (Wallace, 1959, p. 1125). If $G(y; \nu) = \Phi\{z(y)\}$ and $u(y) = \{\nu \log(1 + y^2/\nu)\}^{1/2}$, then

$$z(y) \simeq u(y)\left[1 - \frac{\{1 - \exp(-S^2)\}^{1/2}}{8\nu + 3}\right], \quad S = \frac{(0.184)(8\nu + 3)}{u(y)\sqrt{\nu}},$$

where $y > 0$. This comes within 0.02 of $z(y)$ for a wide range of y; it is better than the approximation of Peizer and Pratt (1968, p. 1428), in (c).

(b) A simpler approximation than that of (a) is given by

$$z_1(y) \simeq u(y)(8\nu + 1)/(8\nu + 3), \quad y > 0,$$

which seems to be within 0.02 of $z(y)$ when $y^2/\nu < 5$ (Wallace, 1959, p. 1125). Johnson and Kotz (1970b, pp. 108-109) give some comparisons of (a), (b) and [7.10.3].

(c) In Peizer and Pratt's approximation (op. cit.),

$$z_2(y) \simeq \pm \left(\nu - \tfrac{2}{3} + \tfrac{1}{10\nu}\right)\left\{\tfrac{1}{\nu - 5/6} \log\left(1 + \tfrac{y^2}{\nu}\right)\right\}^{1/2};$$

the sign is chosen to agree with the sign of y. While this is not as accurate as that in (a) above, it is close to that in (b) for $2 \leq \nu \leq 120$, and slightly better when $\nu \geq 11$ (Ling, 1978, pp. 276-277); see also [7.10.6].

[7.10.5] *Bounds* by Chu (1956, p. 784). If $x \geq 0$, and $\nu \geq 3$,

$$G(y; \nu) - G(-x; \nu) \leq \sqrt{\tfrac{\nu - 3/7}{\nu - 2}}\left[\Phi(y\sqrt{1 - \tfrac{2}{\nu}}) - \Phi(-x\sqrt{1 - \tfrac{2}{\nu}})\right],$$

$$G(y; \nu) - G(-x; \nu) \geq \tfrac{\nu}{\nu+1}\left[\Phi(y\sqrt{1 + \tfrac{1}{\nu}}) - \Phi(-x\sqrt{1 + \tfrac{1}{\nu}})\right].$$

The proportional error is less than $1/\nu$ for all $x \geq 0$ and $y \geq 0$ if $\nu \geq 8$.

[7.10.6] *Fisher's Expansions for the pdf and cdf of t:*

$$g(y;\nu) = \phi(y)\left[1 + \tfrac{1}{4\nu}(y^4 - 2y^2 - 1) + \tfrac{1}{96\nu^2}(3y^8 - 28y^6 + 30y^4 + 12y^2 + 3)\right.$$

$$+ \tfrac{1}{384\nu^3}(y^{12} - 22y^{10} + 113y^8 - 92y^6 - 33y^4 - 6y^2 + 15)$$

$$+ \tfrac{1}{92{,}160\nu^4}(15y^{16} - 600y^{14} + 7100y^{12} - 26616y^{10}$$

$$+ 18330y^8 + 6360y^6 + 1980y^4 - 1800y^2 - 945) + \ldots];$$

$$G(y;\nu) = \Phi(y) - y\phi(y)\left[\tfrac{1}{4\nu}(y^2 + 1) + \tfrac{1}{96\nu^2}(3y^6 - 7y^4 - 5y^2 - 3)\right.$$

$$+ \tfrac{1}{384\nu^3}(y^{10} - 11y^8 + 14y^6 + 6y^4 - 3y^2 - 15)$$

$$+ \frac{1}{92{,}160\nu^4}(15y^{14} - 375y^{12} + 2225y^{10} - 2141y^8 - 939y^6$$

$$- 213y^4 - 915y^2 + 945) + \ldots].$$

If the term in ν^{-4} is omitted, the maximum absolute error in the approximation to $G(y; \nu)$ is 5×10^{-6} (Johnson and Kotz, 1970b, pp. 101-102; Fisher, 1925, pp. 109-112); this approximation does not perform as well as those of Wallace and Peizer-Pratt in *[7.10.4]*, (b) and (c) if $\nu < 17$, but is best among these if $\nu > 17$ (Ling, 1978, pp. 276-277).

[7.10.7] An inverse hyperbolic sine approximation, suitable for smaller as well as larger values of ν (Anscombe, 1950, pp. 228-229) is

$$G(y; \nu) = \Phi(z), \quad z = \pm \sqrt{(2\nu-1)/3} \sinh^{-1}(\sqrt{3y^2/(2\nu)}).$$

[7.10.8] A *recommended accurate* approximation has been given by Hill (1970a, pp. 617-618). Let $G(y;\nu) = \Phi\{z(y)\}$ and $w = w(y) = \{(\nu - \frac{1}{2}) \log(1 + y^2/\nu)\}^{1/2}$. Then if $y > 0$,

$$z(y) \simeq w + \frac{w^3 + 3w}{b} - \frac{4w^7 + 33w^5 + 240w^3 + 855w}{10b(b + 0.8w^4 + 100)}, \quad b = 48(\nu - \tfrac{1}{2})^2.$$

The maximum |error| in $G(y; \nu)$ for all values of y is less than 10^{-1}, 10^{-3}, 10^{-5}, or 10^{-7} if $\nu \geq 1, 2, 4$, or 6, respectively.

[7.10.9] Some normalizing transformations of Moran (1966, pp. 225-230) are suitable at specific percentage points; see Johnson and Kotz (1970b, pp. 110-111), and Scott and Smith (1970, pp. 681-682).

[7.10.10] Let $G(t_{\nu,p};\nu) = \Phi(z_p) = 1 - p$. Denoting z_p by z, we have the Cornish-Fisher (CF) expansion for the *percentile* $t_{\nu,p}$ (see *[6.6.12]*):

$$t_{\nu,p} = z + \tfrac{1}{4\nu}(z^3 + z) + \tfrac{1}{96\nu^2}(5z^5 + 16z^3 + 3z) + \tfrac{1}{384\nu^3}(3z^7 + 19z^5$$

$$+ 17z^3 - 15z) + \tfrac{1}{92160\nu^4}(79z^9 + 776z^7 + 1482z^5 - 1920z^3 - 945z) + \ldots$$

(Fisher and Cornish, 1960, p. 216).

Peiser (1943, pp. 56-62) and Goldberg and Levine (1946, pp. 216-225) derived approximations to $t_{\nu,p}$ that are essentially the first two and three terms, respectively, of the above CF expansion. If $\nu \geq 6$ and $|p - \frac{1}{2}| < 0.49$, so that percentiles are not being approximated far out in the tails, then the first three terms of the expansion should be accurate enough for most purposes; but it is worthwhile to compute all five terms if ν is not large or if extreme tail probabilities are involved (Sahai and Thompson, 1974, p. 83). See also Hill (1970b, pp. 619-620) for a CF type expansion.

[7.10.11] Prescott (1974, pp. 178-180) compared inverse transformations of several discussed so far. The closest to the CF expansion above, and the one which gives the best approximation of those which he compared, is the inversion of *[7.10.4]*(b), Wallace's transformation. This gives for the percentile $t_{\nu,p}$, where $t_{\nu,p} > 0$,

$$t_{\nu,p} \simeq \sqrt{\nu} \left\{\exp(z_p^2 b^2/\nu) - 1\right\}^{1/2}, \quad b = (8\nu + 3)/(8\nu + 1).$$

The scaling factor b makes this a good approximation to $t_{\nu,p}$ over most of the distribution, including small values of ν and large values of z.

7.11 NONCENTRAL t

[7.11.1] The distribution of noncentral t is given in *[5.4.7]*. Let the pdf be $g(y; \nu, \lambda)$ and the cdf $G(y; \nu, \delta)$, for ν degrees of freedom and noncentrality parameter δ. The distribution can be represented as that of $(Z + \delta)/(\chi_\nu/\sqrt{\nu})$, where Z is a N(0, 1) rv and χ_ν a chi rv with ν df, distributed independently of Z.

[7.11.2] Let Y have a noncentral t distribution, represented as in *[7.11.1]*. Then $G(y; \nu, \delta) = \Pr(Z - y\sqrt{\nu}\,\chi_\nu \leq -\delta)$; many approximations are based on the approximation to normality of the rv $Z - y\sqrt{\nu}\,\chi_\nu$ (Johnson and Kotz, 1970b, p. 207). Thus

$$G(y; \nu, \delta) \simeq \Phi(u), \quad u = \frac{\{-\delta + y E(\chi_\nu)/\sqrt{\nu}\}}{\{1 + y^2 \operatorname{Var}(\chi_\nu)/\nu\}^{1/2}},$$

Chapter 7 217

where $E(\chi_\nu)$ and $Var(\chi_\nu)$ are the mean and variance, respectively, of χ_ν, with $Var(\chi_\nu) = \nu - \{E(\chi_\nu)\}^2$. The approximations following have been or may be used in the above.

(a) $E(\chi_\nu) \simeq \sqrt{\nu}$, $Var(\chi_\nu) \simeq 1/2$ (Johnson and Kotz, 1970b, p. 207).

(b) $E(\chi_\nu) \simeq \sqrt{\nu}\{1 - 1/(4\nu)\}$, $Var(\chi_\nu) \simeq 1/2$ (Abramowitz and Stegun, 1970, Sec. 26.7.10). Warren (1982, pp. 663-664) found this approximation to be "very good if the cumulative probability exceeds about 20%, and thus in the upper tail..." for the cases examined.

(c) $E(\chi_\nu) \simeq \sqrt{\nu}\{1 - 1/(4\nu) + 1/(32\nu^2)\}$, $Var(\chi_\nu) \simeq 1/2 + 1/(8\nu)$. Approximation (c) results from taking further terms in series expansions for $E(\chi_\nu)$; see *[5.3.5]*, or Read (1973, pp. 183-185).

(d) We have exactly that

$$E(\chi_\nu) = \sqrt{2}\Gamma\{\tfrac{1}{2}(\nu+1)\}/\Gamma(\tfrac{1}{2}\nu).$$

Tables of $E(\chi_\nu/\sqrt{\nu})$ can be used to substitute appropriate values; sources are given in *[5.3.5]*, where $E(S/\sigma)$ and $Var(S/\sigma)$ can be replaced by $E(\chi_\nu/\sqrt{\nu})$ and $Var(\chi_\nu/\sqrt{\nu})$, respectively.

While these approximations have not all been compared, it seems plausible that those which given $E(\chi_\nu)$ and $Var(\chi_\nu)$ most accurately will perform best. All improve as ν increases. For some cautionary remarks concerning percentiles, however, see *[7.11.3]*.

[7.11.3] If $G(y_p; \nu, \delta) = 1 - p = \Phi(z_p)$, we obtain the following approximation to percentiles y_p of the noncentral t distribution from the approach used in *[7.11.2]*: putting $b_\nu = E(\chi_\nu/\sqrt{\nu})$ (Johnson and Kotz, 1970b, p. 207),

$$y_p \simeq \frac{\delta b_\nu + z_p\{b_\nu^2 - z_p^2)(1 - b_\nu^2)\}^{1/2}}{b_\nu^2 - z_p^2(1 - b_\nu^2)}.$$

Various approximations to y_p arise from the substitutions for b_ν given in (a), (b), (c), and (d) of *[7.11.2]*, and the remarks given there regarding accuracy apply here also. However, for real approximations to y_p to obtain we must have $z_p^2 \leq \delta^2 + b_\nu^2/(1 - b_\nu^2)$, with analogous inequalities if (a), (b), or (c) of *[7.11.2]* is applied. The greatest range of values of p is not necessarily given by the most accurate approximation; (a) of *[7.11.2]* gives a wider range than (d), for example; see Johnson and Kotz (1970b, pp. 207-208).

[7.11.4] Expressions for percentiles y_p based on Cornish-Fisher expansions are as follows: If $z = z_p$,

$$y_p = z + \delta + \{z^3 + z + (2z^2 + 1)\delta + z\delta^2\}/(4\nu)$$

$$+ \{5z^5 + 16z^3 + 3z + 3(4z^4 + 12z^2 + 1)\delta + 6(z^3 + 4z)\delta^2$$

$$- 4(z^2 - 1)\delta^3 - 3z\delta^4\}/(96\nu^2) + O(\nu^{-3}), \quad p > 0.5;$$

$$y_p = t_{\nu,p} + \delta + \delta(1 + 2z^2 + z\delta)/(4\nu) + \delta\{3(4z^4 + 12z^2 + 1)$$

$$+ 6(z^3 + 4z)\delta - 4(z^2 - 1)\delta^2 - 3z\delta^3\}/(96\nu^2) + O(\nu^{-3}); \quad p < 0.5;$$

here $t_{\nu,p}$ is the corresponding percentile of a Student t rv having ν degrees of freedom; see Johnson and Kotz (1970b, pp. 208-209).

7.12 THE F DISTRIBUTION

[7.12.1] The F distribution is given in *[5.5.1]*. A random variable Y with an $F_{m,n}$ distribution can be represented as $(\chi_m^2/m)/(\chi_n^2/n)$, where the χ_m^2 and χ_n^2 variables are independent. Let $G(y; m, n)$ be the cdf of Y.

[7.12.2] The distribution of z, where $z = \frac{1}{2} \log Y$, is more nearly normal than that of Y; early investigations by Fisher (1924, pp. 805-813) were made of the properties of z rather than those of Y. If m and n are both large, then

$$E(z) \simeq \tfrac{1}{2}(n^{-1} - m^{-1}) + \tfrac{1}{6}(n^{-2} - m^{-2}),$$

$$\text{Var}(z) \simeq \tfrac{1}{2}(m^{-1} + n^{-1}) + \tfrac{1}{2}(m^{-2} + n^{-2}) + \tfrac{1}{3}(m^{-3} + n^{-3})$$

(Wishart, 1947, pp. 172, 174; Stuart and Ord, 1987, Sec. 16.21). For special expressions for the higher cumulants of z, see Wishart or Johnson and Kotz (1970b, pp. 78-80).

[7.12.3] Defining z as in *[7.12.2]*,

$$G(y; m, n) \simeq \Phi(u), \quad u \simeq \{\tfrac{1}{2} \log y - E(z)\}/\sqrt{\text{Var}(z)} .$$

This approximation is good when m and n are both large; the moments of z may be approximated as in *[7.12.2]*, where the first terms alone lead to the simplest approximation, but the fuller expressions give more accuracy.

[7.12.4] The Wilson-Hilferty approximation of *[7.8.4]* leads to the property that the rv

$$U = \left\{\left(1 - \tfrac{2}{9n}\right)Y^{1/3} - \left(1 - \tfrac{2}{9m}\right)\right\} \bigg/ \left\{\tfrac{2}{9n} Y^{2/3} + \tfrac{2}{9m}\right\}^{1/2}$$

has an approximate $N(0, 1)$ distribution (Paulson, 1942, pp. 233-235). This should only be used if $n \geq 3$ and for lower tail probabilities if $m \geq 3$ also; it is quite accurate if $n \geq 10$.

Let $h_1 = 2/(9m)$ and $h_2 = 2/(9n)$. Then the function u, where

$$u = [(1 - h_1) - (1 - h_2 x)]/(h_2 x^2 + h_1)^{1/2} ,$$

is monotonic increasing when $x > 0$, so that, putting $x = y^{1/3}$, $Y \leq y$ ($y > 0$) corresponds to $U_1 \leq u$, and $G(y;, m, n) \simeq \Phi(u)$. See also *[7.12.8]*, Stuart and Ord (1987, Sec. 16.21), and comments in *[7.12.6]*.

[7.12.5] The Fisher transformation of *[7.8.3]* leads to the approximation

$$G(y; m, n) \simeq \Phi(u) , \quad u = \frac{\sqrt{(2n-1)my/n} - \sqrt{2m-1}}{\sqrt{(my/n) + 1}}$$

(Laubscher, 1960, p. 1111; Abramowitz and Stegun, 1964, Sec. 26.6.14).

[7.12.6] A *recommended* accurate approximation is that of Peizer and Pratt (1968, pp. 1416-1423, 1427). When $m \neq n$, $p = n/(my+n)$ and $q = 1-p$,

$$G(y; m, n) \simeq \Phi(x), \quad x = d\left[\frac{1 + qT\left(\frac{n-1}{p(m+n-2)}\right) + pT\left(\frac{m-1}{q(m+n-2)}\right)}{pq\left(\frac{1}{2}m+\frac{1}{2}n-\frac{5}{6}\right)}\right]^{\frac{1}{2}},$$

$$d = \frac{n}{2} - \frac{1}{3} - \left(\frac{m+n}{2} - \frac{2}{3}\right)p + \epsilon\left(\frac{q}{n} - \frac{p}{m} + \frac{q-\frac{1}{2}}{m+n}\right),$$

$$T(x) = (1 - x^2 + 2x \log x)/(1-x)^2, \quad x \neq 1; \quad T(1) = 0.$$

The simpler computation is given by $\epsilon = 0$; there is a slight improvement in accuracy if $\epsilon = 0.04$. When $m = n$, the following simplification holds:

$$x = \pm\left(n - \frac{2}{3} + \frac{0.1}{n}\right)\left[\frac{-\log\{y/(y+1)^2\}}{n - 5/6}\right]^{1/2},$$

where the sign should agree with the sign of $\frac{1}{2} - 1/(y+1)$.

The absolute error when $\epsilon = 0.04$ satisfies

$$|G(y; m, n) - \Phi(x)| < \begin{cases} 0.001 & \text{if } m, n \geq 4 \\ 0.01 & \text{if } m, n \geq 2 \end{cases}.$$

In comparing approximations to $G(y; m, n)$ for $10 \leq m \leq 400$, $10 \leq n \leq 400$, Ling (1978, pp. 278-281) found that the Peizer-Pratt approximation is more accurate than that of Paulson in *[7.12.4]*, which in turn is better than the Laubscher variable U_2 in *[7.12.5]*.

[7.12.7] *Percentiles* y_p *of F*. Let $G(y_p; m, n) = 1 - p = \Phi(u_p)$. Denote by w_p the percentiles of Fisher's z transform discussed in *[7.12.2, 3]*, so that

$$w_p = \frac{1}{2} \log y_p \simeq E(z) + u_p\sqrt{\text{Var}(z)}, \quad y_p = \exp(2w_p).$$

The approximate expressions in *[7.12.2]* for $E(z)$ and $\text{Var}(z)$ may be used when m and n are large, taking the first terms only, or the full expressions for better accuracy.

Normal Approximations to Distributions

[7.12.8] Working from Cornish-Fisher-type expansions (see *[7.12.10]*), Fisher (1924, pp. 805-813) derived the simple approximation in (a) following. Successive improvements in accuracy resulted from modifications by Cochran (1940, pp. 93-95) and by Carter (1947, pp. 356-357), given in (b) and (c) following. The best of these approximations is that of Carter in (c). Here, $\Phi(u_p) = 1 - p$.

(a) $w_p = \frac{1}{2} \log y_p \simeq u_p/\sqrt{(h-1)} - (m^{-1} - n^{-1})(\lambda - \frac{1}{6})$,

$\lambda = (u_p^2 + 3)/6, \quad 2h^{-1} = m^{-1} + n^{-1}$ (Fisher) ;

(b) $w_p = \frac{1}{2} \log y_p \simeq u_p/\sqrt{h - \lambda} - (m^{-1} - n^{-1})(\lambda - \frac{1}{6})$ (Cochran) ;

(c) $w_p = \frac{1}{2} \log y_p \simeq u_p h'^{-1}\sqrt{(h'+\lambda')} - d'\left[\lambda' + \frac{5}{6} - \frac{2}{3}h'^{-1}\right]$,

$\lambda' = (u_p^2 - 3)/6, \quad 2h'^{-1} = (m-1)^{-1} + (n-1)^{-1}$,

$d' = (m-1)^{-1} - (n-1)^{-1}$ \qquad (Carter) .

This approximation compares favorably with that discussed next.

[7.12.9] From the approximation of Paulson (1942, pp. 233-235) and the monotonicity property discussed in *[7.12.4]* above,

$$y_p^{1/3} \simeq \frac{(1-h_1)(1-h_2) \pm \left[z_p^2\{(h_1 + h_2)(1 + h_1 h_2) - h_1 h_2(z_p^2 + 4)\}\right]^{1/2}}{(1-h_2)^2 - z_p^2 h_2},$$

$h_1 = 2/(9m), \quad h_2 = 2/(9n), \quad \Phi(z_p) = 1 - p$.

The sign should be chosen so that $y_p^{1/3} <$ (resp., $>$) $(1-h_1)/(1-h_2)$ if $z_p < 0$ (resp., > 0); see Ashby (1968, p. 209). This approximation should be used only if $n \geq 3$, and for lower tail probabilities if $m \geq 3$ also; it is quite accurate if $n \geq 10$; see also the sources listed in *[7.12.4]*. Ashby has given improved approximations for $n \leq 10$. If y_p is obtained as above, then the linear relation $y_p' = ky_p + c$ leads to more accuracy; Ashby has tabulated empirical values of k and c for $n = 1(1)10$ at the upper 5, 1, and 0.1 percent tail probabilities.

[7.12.10] Using approximations to the cumulants of $z = \frac{1}{2} \log Y$ when m and n are large, Fisher and Cornish (1960, pp. 209-225) gave an expansion for w_p (see *[7.12.7]*) which appears to be better than that of *[7.12.9]* if $n \leq 5$, and frequently better if $n \geq 10$, for upper tail probabilities (see Sahai and Thompson, 1974, pp. 89-91): If $b = m^{-1} + n^{-1}$, $d = m^{-1} - n^{-1}$, $u = u_p$ and $\Phi(u_p) = 1 - p$,

$$\begin{aligned} w_p = \tfrac{1}{2} \log y_p \simeq & \ u\sqrt{\tfrac{b}{2}} - \tfrac{1}{6} d(u^2 + 2) + \sqrt{\tfrac{b}{2}} \left\{ \tfrac{b}{24}(u^3 + 3u) \right. \\ & + \left. \tfrac{d^2}{72b}(u^3 + 11u) \right\} - \tfrac{db}{120}(u^4 + 9u^2 + 8) + \tfrac{d^3}{3240b}\left(3u^4 + 7u^2 - 16\right) \\ & + \sqrt{\tfrac{b}{2}} \left\{ \tfrac{b^2}{1920}(u^5 + 20u^3 + 15u) + \tfrac{d^2}{2880}(u^5 + 44u^3 + 183u) \right. \\ & + \left. \tfrac{d^4}{155{,}520b^2}\left(9u^5 - 284u^3 - 1513u\right) \right\}. \end{aligned}$$

[7.12.11] Haines (1988, pp. 95-100) incorporates the normal approximation U_1 in *[7.12.4]* in an eight-step algorithm for computing percentage points of $F_{m,n}$. This is accurate for two to three decimal places when probabilities range from 0.90 to 0.99.

7.13 NONCENTRAL F

[7.13.1] The distribution of noncentral F is given in *[5.5.4]*. Let the pdf and cdf be $g(y; m, n, \lambda)$ and $G(y; m, n, \lambda)$, respectively. A rv Y with an $F_{m,n}(\lambda)$ distribution can be represented as $\{\chi_m^2(\lambda)/m\}/(\chi_n^2/n)$, where the $\chi_m^2(\lambda)$ and χ_n^2 variables are independent.

[7.13.2] (a) Analogous to the discussion in *[7.12.4]*, $G(y; m, n, \lambda) \simeq \Phi(u_1)$, where

$$u_1 = \frac{(1-h_2)(ay)^{1/3} - (1-h_1)}{\{h_2(ay)^{2/3} + h_1\}^{1/2}}, \quad a = \tfrac{m}{m+\lambda}, \quad h_1 = \tfrac{2(m+2\lambda)}{9(m+\lambda)^2}, \quad h_2 = \tfrac{2}{9n}$$

Chapter 7

(Severo and Zelen, 1960, p. 416; Laubscher, 1960, p. 1111; Abramowitz and Stegun, 1970, Sec. 26.6.27; Johnson and Kotz, 1970b, p. 83).

(b) Analogous to the discussion in [7.12.5], $G(y; m, n, \lambda) = \Phi(u_2)$, where

$$u_2 = \frac{\sqrt{2n-1}\sqrt{my/n} - \sqrt{2(m+\lambda) - b}}{\sqrt{(my/n) + b}} \;,\; b = 1 + \frac{\lambda}{m+\lambda}$$

(Laubscher, 1960, p. 1111). Laubscher compared the approximations in (a) and (b) for a few values of λ and y, and for limited choices of m and n ($3 \leq m \leq 8$, $10 \leq n \leq 30$). Both were accurate to two decimal places; (b) performed better than (a) more often than not for the values chosen.

[7.13.3] Using the first few terms of an Edgeworth expansion ([6.4]), Mudholkar et al. (1976, pp. 353, 357) give an approximation which numerical studies indicate to be more accurate than those in [7.13.2]:

$$G(y; m, n, \lambda) \simeq \Phi(x) - \phi(x) \left[\tfrac{1}{6} \beta_1 (x^2 - 1) + \tfrac{1}{24} \beta_2 (x^3 - 3x) \right.$$

$$\left. + \tfrac{1}{72} \beta_1^2 (x^5 - 10x^3 + 15x) \right] ,$$

$x = -\kappa_1/\sqrt{\kappa_2}$, $\beta_1 = \kappa_3/\kappa_2^{3/2}$, $\beta_2 = \kappa_4/\kappa_2^2$; here κ_1, κ_2, κ_3, and κ_4 are the first four cumulants of the rv $\{\chi_m^2(\lambda)/m\}^{1/3} - y^{1/3}(\chi_n^2/n)^{1/3}$ and $\chi_m^2(\lambda)$ and χ_n^2 are independent noncentral and central chi-square rvs with m and n degrees of freedom, respectively. Mudholkar et al. give approximate expressions for κ_1, κ_2, κ_3, and κ_4.

7.14 MISCELLANEOUS CONTINUOUS DISTRIBUTIONS

[7.14.1] *Quadratic Forms.* Let $X_1, ..., X_k$ be iid $N(0, 1)$ rvs and $c_1, ..., c_k, a_1, ..., a_k$ be bounded constants such that $c > 0$, $j = 1, ..., k$; then the quadratic form defined by $Q_k = \sum_{j=1}^{k} c_j (X_j + a_j)^2$ is definite. Let $\theta_s = \sum_{j=1}^{k} c_j^s (1 + sa_j^2)$, $s = 1, 2, ...$; then the sth cumulant of the distribution of Q_k is $2^{s-1}(s-1)!\theta_s$. Using a Wilson-Hilferty-type transformation, Jensen and Solomon (1972, pp. 898-900) take the rv Z_k to be a $N(0, 1)$ variable, approximately, where

$$Z_k = \frac{\theta_1}{h\sqrt{2\theta_2}}\left[\left(\frac{Q_k}{\theta_1}\right)^h - 1 - h(h-1)\frac{\theta_2}{\theta_1^2}\right], \quad h = 1 - \frac{2}{3}\frac{\theta_1\theta_3}{\theta_2^2}.$$

This approximation compares well with its (nonnormal) competitors. For cases in which $a_1 = \ldots = a_k = 0$ and $k = 2, 3, 4, 5$, the approximation to $\Pr(Q_k \leq t)$ tends to improve as t increases, also as variation among the parameters $c_1, \ldots, c_k$ decreases (Jensen and Solomon, 1972, pp. 901-902). The case in which $c_1 = \ldots = c_k = 1$ reduces to the approximation for noncentral chi-square in [7.9.4](b), and if in addition $a_1 = \ldots = a_k = 0$, it yields the Wilson-Hilferty approximation for chi-square in [7.8.4](a).

Konishi et al. (1988, p. 279-296) extended the accuracy of the distribution of Z_k via the expansion as $\theta_1 \to \infty$ given by

$$\Pr(Z_k < x) = \Phi(x) - \phi(x)\left(\sum_{i=1}^{6} A_i \theta_1^{-i/2}\right) + O(\theta_1^{-7/2}) \; ;$$

here $A_1 = 0$ and the expressions for $A_2, \ldots, A_6$ given by the authors in terms of Hermite polynomials are fairly lengthy. The expansion assumes that $\theta_j/\theta_1 = O(1)$ for $j = 2, 3, \ldots$.

Konishi et al. also provided an improved approximation to the percentiles q_α of Q_k if z_α is the standard normal percentile; as $\theta_1 \to \infty$,

$$q_\alpha = \theta_1\left\{(2h^2\theta_2/\theta_1^2)^{1/2} x_\alpha + 1 + \theta_2 h(h-1)/\theta_1^2\right\}^{1/h} \; ;$$

$$x_\alpha = z_\alpha + \sum_{i=1}^{6} B_i \theta_1^{-i/2} + O(\theta_1^{-7/2})$$

and $B_1 = 0$ here; the authors provide expressions for $B_2, \ldots, B_6$, also fairly lengthy. If the skewness $\beta_1 = \kappa_3/\kappa_2^{3/2} \leq 1.0$, these two approximations are accurate to about four decimal places (and almost to three significant figures if terms in A_5, A_6, B_5 and B_6 are omitted), but are not very satisfactory if $\beta_1 \geq 2.0$.

Normal Approximations to Distributions

For further discussion of the distribution of quadratic forms, see Johnson and Kotz (1970b, Chap. 29), Stuart and Ord (1987, Secs. 15.10-15.21), and *[5.3.8]*.

[7.14.2] *Distance Distributions.* Let $(X_1, X_2, ..., X_p)$ and $(Y_1, Y_2, ..., Y_p)$ be two random points in a p-dimensional sphere to unit radius, and let $R = \{\sum_{i=1}^{p}(X_i - Y_i)^2\}^{1/2}$ denote the distance between them. When p is large, R has an approximate $N(\sqrt{2}, (2p)^{-1})$ distribution (Johnson and Kotz, 1970b, p. 267).

[7.14.3] The *Birnbaum-Saunders distributions* (Birnbaum and Saunders, 1969, pp. 319-327) has been described in *[5.1.6]* as that of a rv $\theta[U\sigma + \{U^2\sigma^2 + 1\}^{1/2}]^2$ where U is a N(0, 1) rv. As $\sigma \to 0$, the distribution tends to normality (Johnson and Kotz, 1970b, p. 269).

[7.14.4] Let $(n-1)^{-1} \underline{S}$ be the sample variance-covariance matrix in a sample of size n from a multivariate normal population of dimension m, with mean vector $\underline{\mu}$ and population variance-covariance matrix $\underline{V}$, assumed nonsingular; suppose that $n > m$. The determinant $|(n-1)^{-1}\underline{S}|$ is the sample *generalized variance*, and $|\underline{V}|$ the population generalized variance.

(a) Then $X = |S|/|V|$ has the distribution of $\Pi_{j=1}^{m} Y_{n-j}$, where $Y_{n-j} \sim \chi^2_{n-j}$ and $Y_{n-1}, Y_{n-2}, ..., Y_{n-m}$ are mutually independent (Anderson, 1984, p. 264). The distribution of log X can thus be approximated by treating $\sum_{j=1}^{m} \log Y_{n-j}$ as a normal rv with mean $\sum_{j=1}^{m} E(\log Y_{n-j})$ and variance $\sum_{j=1}^{m} \text{Var}(\log Y_{n-j})$; see also *[7.8.7]*. If $n-m \geq 4$,

$$E(\log X) \simeq \sum_{j=1}^{m} \log(n-j-1), \quad \text{Var}(\log X) \simeq 2\sum_{j=1}^{m}(n-j-1)^{-1}.$$

Gnanadesikan and Gupta (1970, p. 113) found that the normal approximation to log X improves with increases both in m and n.

(b) Anderson (1984, p. 266) shows that as $n \to \infty$,

$$\{(n-1)^{-[m-(1/2)]}|S|/|V|\} - \sqrt{n-1}$$

has asymptotically a N(0, 2m) distribution, leading to a second normal approximation. Note that Anderson's $\underline{S}$ is the equivalent of $(n-1)^{-1}\underline{S}$ as defined here; also that his n has the meaning of n − 1 as defined here.

[7.14.5] *Fisher Distribution.* Let θ and ϕ denote co-latitude and longitude, respectively, of a random vector on a unit sphere. In the study of directional data, a common assumption is that these directions follow the Fisher distribution with joint pdf

$$g(\theta,\phi) = [\kappa \sin \theta/(4\pi \sinh \kappa)] \exp\{\kappa[\cos \theta_0 \cos \theta + \sin \theta_0 \sin \theta \cos(\phi - \phi_0)]\},$$

$$\kappa > 0, \ 0 \leq \theta(\theta_0) \leq \pi, \ 0 \leq \phi(\phi_0) \leq \pi \, ;$$

see Fisher (1953, pp. 295-305).

If κ is large and θ_0 is not too close to 0 or to π, the deviations $\theta - \theta_0$ and $\phi - \phi_0$ will have a high probability of being small. Clark and Morrison (1983, pp. 96-104) show that θ and ϕ are approximately independent, and approximately distributed $N(\theta_0, 1/\kappa)$ and $N(\phi_0, 1/(\kappa \sin^2 \theta_0))$, respectively. The approximation breaks down if θ_0 is close to 0 or to π. It works best at the (nominal) 5 percent point when θ_0 is close to $\pi/2$.

Let $U = \phi \sqrt{\sin \theta}$. When $\theta_0 = \pi/2$ and $|\phi| \leq \pi$, U is approximately distributed $N(0, 1/\kappa)$ (Lewis and Fisher, 1982, pp. 1-13).

7.15 NORMALIZING TRANSFORMATIONS

[7.15.1] Transformations of rvs in statistics are directed toward three purposes, frequently in the context of the general linear model and regression problems:

(i) Additivity of main effects and removal of interactions.

(ii) Variance stabilizing, to make $\sigma^2(X)$ functionally free of both the mean E(X) and the sample size n.

(iii) Normality of the transformed variables.

A transformation which achieves or nearly achieves one of these aims frequently helps toward the achievement of one or both of the others, but only by not seeking optimal achievement in more than one direction (Stuart and Ord, 1991, Secs. 28.37-28.43: Hoyle, 1973, pp. 203-204). Thus, variance stabilizing may normalize approximately, but not optimally.

That a transformation leads to normality is usually inferred when it alters the skewness and kurtosis in that direction. More is required for normality than knowledge of suitable values of the first four moments, but there appear to be no important cases in which small values of $|\mu_3/\mu_2^{3/2}|$ and of $|(\mu_4/\mu_2^2) - 3|$ give a misleading picture of normal approximation (Stuart and Ord, 91, Sec. 28.42).

Some transformations are listed elsewhere in this book. Variance-stabilizing transformations that tend to normalize include angular transformations for binomial variables (see *[7.1.6]*), the logarithmic transformation for chi-square and gamma rvs of *[7.8.7]*, and the inverse tanh transformation for a bivariate normal sample correlation coefficient in *[10.8.6]*. Transformations directed toward normality do not necessarily stabilize the variance; examples include the cube root transformation of chi-square in *[7.8.4]*(a), the power transformation for general quadratic forms in *[7.14.1]*, and the Cornish-Fisher expansions (see *[6.6.12, 13]*, where z_p and x_p may be replaced by z and x, respectively). Transformations of ordered observations to normal scores in a random sample from an unknown distribution are discussed in *[8.2.11]*.

Variance-stabilizing transformations are presented in *[7.15.2]* to *[7.15.6]*, and normalizing transformations in *[7.15.6]* to *[7.15.8]*. For further discussion, see Hoyle (1973, pp. 203-223).

[7.15.2] If T_ν is a statistic such that $\sqrt{\nu}\,(T_\nu - \theta)$ has asymptotically a $N(0, \sigma^2(\theta))$ distribution as $\nu \to \infty$, then for suitable single-valued differentiable functions $g(\cdot)$, $\sqrt{\nu}\,[g(T_\nu) - g(\theta)]/[g'(\theta)\sigma(\theta)]$ has asymptotically a $N(0, 1)$ distribution (*[6.3.3]*). A form of this result by Laubscher (1960, p. 1105) shows how stabilizing the variance may sometimes lead to asymptotic normality: the rv Y_ν has a variance asymptotically stabilized at c^2, where

$$Y_\nu = c \int_k^{\sqrt{\nu}T_\nu} \{\sigma(\theta)^{-1}\} d\theta \,.$$

Further improvements can often be made (Anscombe, 1948, pp. 246-254; Hotelling, 1953, pp. 193-232) if the variance of $g(X)$ is not quite constant.

[7.15.3] *Square Root Transformations.* (a) Let X have a Poisson distribution with mean λ. For arbitrary α define

$$g(X) = \sqrt{X+\alpha}\ I_{[-\alpha,\infty)}(x), \quad h(X) = \sqrt{X} + \sqrt{X+1},$$

$I_A(x)$ being the indicator function for x lying in a set A; then $g(X) - \sqrt{\lambda+\alpha}$ converges to a N(0, 1/4) distribution as $\chi \to \infty$ (Curtiss, 1943, pp. 113-114). Anscombe (1948, pp. 246-254) found that $\alpha = 3/8$ is optimal for variance stabilizing, except for small values of λ, with variance equal to $\frac{1}{4}\{1 + 1/(16\lambda^2)\}$. Freeman and Tukey (1950, pp. 607-611) found that the *chordal transformation* h(X) stabilizes the variance over a larger range of values of λ; see Hoyle (1973, pp. 207-208).

(b) Let X have a gamma distribution with pdf proportional to $x^{(1/2)\nu-1}e^{-hx}$, $x > 0$, and define $g(\cdot)$ as in (a). Then $g(X) - \sqrt{\alpha + \nu/(2h)}$ converges to a N(0, $(4h)^{-1}$) distribution as $\nu \to \infty$ (Curtiss, 1943, p. 115). The case in which $\alpha = 0$ is related to the Fisher approximation of *[7.8.3]*.

[7.15.4] *Inverse Transformations.* The arctanh transformation is discussed in *[10.8]*; it can also arise as a variance-stabilizing device.

(a) *Angular* transformations of binomial rvs were introduced as approximations in *[7.1.6]*; these are also variance-stabilizing. Curtiss (1943, pp. 116, 117) gives some limit theorems for other transformations of binomial rvs, namely, $\sqrt{n} \log (\sqrt{Y/n} + \sqrt{(Y/n)+1})$, $\sqrt{n} \log(Y/n)$, and $\frac{1}{2}\sqrt{n} \log (Y/(n-Y))$.

(b) Let X be a negative binomial rv with pf as given in *[7.3.1]*; let

$$g(X) = \operatorname{arcsinh}\left[\{(X + \tfrac{3}{8})/(s - \tfrac{3}{4})\}^{1/2}\right].$$

This transformation stabilizes the variance at $(1/4) + O(s^{-2})$ (Anscombe, 1948; Hoyle 1973, p. 210).

(c) The transformation $g(X) = \alpha\{\operatorname{arcsinh}(\beta X) - \operatorname{arcsinh}(\beta\mu)\}$ can also be applied when X has a noncentral $t_\nu(\delta)$ distribution *[5.4.7; 7.11.1]* (Laubscher, 1960, pp. 1106, 1107). Thus

$$g(X) = \alpha\{\operatorname{arc\ sinh}(\beta X) - \operatorname{arc\ sinh}(\beta\mu)\},$$

Normal Approximations to Distributions

$$\mu = \delta\sqrt{\tfrac{1}{2}\nu}\,\frac{\Gamma(\tfrac{1}{2}(\nu-1))}{\Gamma(\tfrac{1}{2}\nu)}, \quad \alpha = \tfrac{1}{b}, \quad \beta = b\sqrt{\tfrac{\nu-2}{\nu}},$$

$$b = \left[\frac{2\{\Gamma(\tfrac{1}{2}\nu)\}^2}{(\nu-2)[\{\Gamma(\tfrac{1}{2}(\nu-1))\}^2 - 1]}\right]^{1/2}, \quad \nu \geq 4.$$

The variable g(X) approximates to a N(0, 1) rv, but the accuracy deteriorates if ν is small and δ is large simultaneously; larger values of ν improve the approximation.

See also Azorín (1953, pp. 173-198, 307-337) for simpler transformations $\sqrt{c\nu}$ arcsinh $(X/\sqrt{c\nu}) - \delta$, where c = 1 or c = 2/3.

(d) Let X have a noncentral $F_{m,n}$ (λ) distribution [5.5.4; 7.13.1]; if $\lambda \geq 4$, the transformation

$$g(X) = \sqrt{\tfrac{1}{2}n - 2} \text{ arc cosh}\left[a(X + \tfrac{n}{m})\right], \quad a = \tfrac{m}{n}\left[\tfrac{n-2}{m+n-2}\right]^{1/2},$$

stabilizes the variance, but does not give such a satisfactory normalizing approximation as those in [7.13.2] (Laubscher, 1960, pp. 1109-1111).

[7.15.5] Other Variance-Stabilizing Transformations

(a) If $E(X) = \theta$ and $Var(X) \propto \theta^4$, an appropriate transformation is given by $g(X) \propto X^{-1}$ (Hoyle, 1973, p. 207); this may reduce extreme skewness.

(b) If $E(X) = \theta$ and $Var(X) \propto \theta^2$, the logarithmic transformation $g(X) = \log(X + a)$ also reduces skewness, but to a lesser extent than in (b) (Hoyle, op. cit.). If X has a normal distribution, however, g(X) is a lognormal rv ([2.3.2]).

[7.15.6] Power Transformations. Let X be a rv; consider the class of transformations defined by $g(X) = (X + a)^b$, $b \neq 0$. (The transformation corresponding to b = 0 is logarithmic, as in [7.15.5(b)]). Box and Cox (1964, pp. 211-252) have developed a routine to estimate b so that Y = g(X) may come close to all three criteria (i), (ii), and (iii) of [7.15.1]; see also the graphic approach of Draper and Hunter (1969, pp. 23-40). Moore (1957, pp. 237-246) treated g(·) as a normalizing transformation only, where 0 < b < 1. Formally, the Box-Cox (parametric) family of

transformations is defined by

$$X^{(\lambda)} = \begin{cases} (X^\lambda - 1)/\lambda, & \lambda \neq 0, \\ \log x, & \lambda = 0. \end{cases}$$

Carroll (1980, pp. 71-78) and Hernandez and Johnson (1980, pp. 855-861) have questioned the routine use of this transformation; in particular, the latter investigate the effect of a key assumption, that there exists a λ_0 for which $X^{(\lambda_0)} \sim N(\cdot, \cdot)$ exactly. While the transformation brings X closer to normality, the maximum amount of improvement towards normality in some instances may still be unsatisfactory (Hernandez and Johnson, op. cit., p. 859).

[7.15.7] *The Johnson System of Curves.* Let X be a rv and

$$g(x) = \gamma + \delta h\{(x - \alpha)/\beta\},$$

where γ, δ, α, and β are parameters, and $h(\cdot)$ is monotonic. Johnson (1949, pp. 149-176) developed a system of frequency curves such that $Y = g(X)$ is normally distributed or approximately so; these have the property that two rvs X_1 and X_2 having the same skewness and kurtosis lead to a unique curve in the system.

Three families comprise the system: if $x' = (x - \alpha)/\beta$,
(a) Lognormal: $h(x') = \log x'$, $x' \geq 0$;
(b) The S_B system: $h(x') = \log\{x'/(1 - x')\}$, $0 \leq x' \leq 1$;
(c) The S_U system: $h(x') = \text{arcsinh } x'$, $-\infty < x' < \beta$.

The labels S_B and S_U refer to the bounded and unbounded range of values of X, respectively. There are comprehensive discussions of the Johnson system in Johnson and Kotz (1970a, pp. 22-27) and in Stuart and Ord (1987, Secs. 6.27-6.36).

Johnson (1965, pp. 547-558) gives tables of values of z and δ corresponding to pairs of values of skewness and kurtosis of X. Hill et al. (1976, pp. 180-189) fit Johnson curves by an algorithm using the moments of X; I. D. Hill (1976, pp. 190-192) gives algorithms for normal-Johnson and Johnson-normal transformations.

[7.15.8] Box and Muller (1958, pp. 610-611) suggested the following transformations of pairs of independent random numbers U_1 and U_2 to generate a pair (Z_1, Z_2) of independent standard normal variables:

$$Z_1 = (-2 \ln U_1)^{1/2} \cos(2\pi U_2), \quad Z_2 = (-2 \ln U_1)^{1/2} \sin(2\pi U_2).$$

Here U_1 and U_2 are independent random variables uniformly distributed on the unit interval (0, 1). The Box-Muller transformation can be used on small as well as large computers; it has become a standard method for generating N(0, 1) deviates: see *[3.1.4]*.

[7.15.9] Let X be a continuous rv with cdf $\Psi(x)$, and let $y = g(x) = \Phi^{-1}\{\Psi(x)\}$, the *coordinate transformation* (Hoyle, 1973, p. 211), where Φ^{-1} is the inverse of the standard normal cdf Φ. If $Y = g(X)$, then Y has a N(0, 1) distribution, since $\Phi\{g(x)\} = \Psi(x)$.

The function $\Phi^{-1}(\cdot)$ cannot be expressed in closed form. Any of the approximations to normal quantiles given in *[3.8]* may be used, however, noting that, if $\Phi(z) = 1 - p$, then $z = \Phi^{-1}(1-p)$; in *[3.8]*, replace z_p by z.

[7.15.10] *The Probit Transformation* (Bliss, 1935, pp. 134-167; Finney, 1971, p. 19). This is defined by $p = \Phi(y-5)$, where p is an observable proportion, estimating the probability with which the equivalent normal deviate $y-5$ is exceeded in a standard normal distribution. The equivalent normal deviate is reduced by 5 to make the probability of negative values of y negligibly small. The quantity y is the *probit* of the proportion p; the corresponding rv is Y, where $Y = (X - \mu)/\sigma$, and in bioassay problems, X is the logarithm of "dose" (Finney, 1971, p. 11). For an interesting history of the subject, see Finney (1971, pp. 38-42). Fisher and Yates (1964, Table IX) give probits corresponding to values of p.

[7.15.11] Let X be a standardized rv with cdf $G(\cdot)$, and $Z \sim N(0, 1)$. The quantiles x_p of X and z_p of Z satisfy $G(x_p) = \Phi(z_p)$. Suppose that X depends on n(n = 1, 2, ...) such that X converges in distribution to Z as n $\to \infty$. Let κ_3 be the third cumulant of X. Shore (1986, pp. 242-246) gives an approximation $\hat{x}_p$ to x_p via

$$\hat{x}_p = \begin{cases} (1 - 0.41781)z_p - \frac{1}{3}\kappa_3, & z_p < 0, \\ (1 + 0.41781)z_p - \frac{1}{3}\kappa_3, & z_p \geq 0. \end{cases}$$

Here $\hat{x}_p$ is the quantile of a rv $\hat{X}$ such that

$$E(\hat{X}) = E(X), \quad \sigma^2(\hat{X}) = \sigma^2(X) + O(n^{-1}), \quad \kappa_3(X) = \kappa_3(\hat{X}) + O(n^{-3/2}).$$

If X is discrete add a continuity correction $-(2\sigma(X))^{-1}$ to the expression for $\hat{x}_p$. For further discussion and an application of this approximation to the Central Limit Theorem see [6.5.1, 2].

REFERENCES

Abdel-Aty, S. H. (1954). Approximate formulae for the percentage points and the probability integral of the non-central χ^2 distribution, *Biometrika* 41, 538-540. [7.9.3]

Abramowitz, M. and Stegun, I. A. (eds.) (1970). *Handbook of Mathematical Functions*. Washington, D.C.: National Bureau of Standards. [Intr.; 7.6.7; 7.8.4; 7.9.3, 6; 7.10.2; 7.11.2; 7.12.5; 7.13.2]

Alfers, D. and Dinges, H. (1984). A normal approximation for beta and gamma tail probabilities, *Zeitschrift für Wahrscheinlichkeitstheorie und verwandte Gebiete* 65, 399-419. [Intr.; 7.1.9; 7.6.5]

Anderson, T. W. (1984). *Introduction to Multivariate Statistical Analysis* (2nd edn.), New York: Wiley. [7.14.4]

Anscombe, F. J. (1948). The transformation of Poisson, binomial, and negative binomial data. *Biometrika* 35, 246-254. [7.1.6.1, 2; 7.15.2, 3, 4]

Anscombe, F. J. (1950). Table of the hyperbolic transformation $\sinh^{-1}\sqrt{x}$, *Journal of the Royal Statistical Society* A113, 228-229. [7.10.7]

Ashby, T. (1968). A modification to Paulson's approximation to the variance ratio distribution, *The Computer Journal* 11, 209-210. [7.12.9]

Azorín, P. F. (1953). Sobre la distribución t no central, I, II, *Trabajos de Estadística* 4, 173-198, 307-337. [7.15.4]

Bartko, J. J. (1966). Approximating the negative binomial, *Technometrics* 8, 345-350; (1967): erratum, *Technometrics* 9, 498. [7.3.5]

Birnbaum, Z. W. and Saunders, S. C. (1969). A new family of life

Birnbaum, Z. W. and Saunders, S. C. (1969). A new family of life distributions, *Journal of Applied Probability 6*, 319-327. *[7.14.3]*

Bliss, C. I. (1935). The calculation of the dosage mortality curve, *Annals of Applied Biology 22*, 134-167. *[7.15.8]*

Bohman, H. (1963). Two inequalities for Poisson distributions, *Skandinavisk Aktuarietidskrift 46*, 47-52. *[7.2.11]*

Borenius, G. (1953). On the statistical distribution of mine explosions, *Skandinavisk Aktuarietidskrift 36*, 151-157. *[7.5.2]*

Borges, R. (1970). Eine Approximation der Binomialverteilung durch die Normalverteilung der Ordnung 1/n, *Zeitschrift für Wahrscheinlichkeitstheorie und verwandte Gebiete 14*, 189-199. *[7.1.8]*

Box, G. E. P. and Cox, D. R. (1964). An analysis of transformations, *Journal of the Royal Statistical Society B26*, 211-252. *[7.15.6]*

Box, G.E.P. and Muller, M. E. (1958). A note on the generation of random normal deviates, *Annals of Mathematical Statistics, 29*, 610-611. *[7.15.8]*

Camp, B. H. (1951). Approximation to the point binomial, *Annals of Mathematical Statistics 22*, 130-131. *[7.1.7]*

Carroll, R. J. (1980). A robust method for testing transformations to achieve approximate normality, *Journal of the Royal Statistical Society B 42*, 71-78. *[7.15.6]*

Carter, A. H. (1947). Approximation to percentage points of the z distribution, *Biometrika 34*, 352-358. *[7.12.8]*

Cheng, T. T. (1949). The normal approximation to the Poisson distribution and a proof of a conjecture of Ramanujan, *Bulletin of the American Mathematical Society 55*, 396-401. *[7.2.5]*

Chu, J. T. (1956). Errors in normal approximations to t, τ and similar types of distribution, *Annals of Mathematical Statistics 27*, 780-789. *[7.10.5]*

Clark, R. M. and Morrison, B. J. (1983). A normal approximation to the Fisher distribution, *Australian Journal of Statistics 25*, 96-104. *[7.14.5]*

Cochran, W. G. (1940). Note on an approximate formula for the significance levels of z, *Annals of Mathematical Statistics 11*, 93-95. *[7.12.8]*

Curtiss, J. H. (1943). On transformations used in the analysis of variance, *Annals of Mathematical Statistics 14*, 107-122. *[7.1.6.1; 7.15.3, 4]*

Dixon, W. J. and Massey, F. J. (1983). *Introduction to Statistical Analysis* (4th ed.), pp. 372-374, 614, New York: McGraw Hill. *[7.1.6]*

Draper, N. R. and Hunter, W. G. (1969). Transformations: Some examples revisited, *Technometrics 11*, 23-40. *[7.15.6]*

Feller, W. (1968). *Introduction to Probability Theory and Its Applications*, Vol. 1 (3rd ed.), New York: Wiley. *[7.4.4]*

Finney, D. J. (1971). *Probit Analysis* (3rd ed.), Cambridge: Cambridge University Press. *[7.15.8]*

Fisher, R. A. (1924). On a distribution yielding the error functions of several well-known statistics, *Proceedings of the International Mathematical Congress*, Toronto, 805-813. *[7.12.2, 8]*

Fisher, R. A. (1925). Expansion of Student's integral in powers of n^{-1}, *Metron 5(3)*, 109-112. *[7.10.6]*

Fisher, R. A. (1953). Dispersion on a sphere, *Proceedings of the Royal Society of London A 217*, 295-305. *[7.14.5]*

Fisher, R. A. and Cornish, E. A. (1960). The percentile points of distributions having known cumulants, *Technometrics 2*, 205-226. *[7.8.8; 7.10.10; 7.12.10]*

Fisher R. A. and Yates, F. (1964). *Statistical Tables for Biological, Agricultural, and Medical Research*, London and Edinburgh: Oliver & Boyd. *[7.15.8]*

Foster, F. G. and Stuart, A. (1954). Distribution-free tests in time series based on the breaking of records, *Jouranl of the Royal Statistical Society B 16*, 1-16. *[7.5.4]*

Freeman, M. F. and Tukey, J. W. (1950). Transformations related to the angular and the square root, *Annals of Mathematical Statistics 21*, 607-611. *[7.1.4; 7.2.3; 7.15.3]*

Gebhardt, F. (1969). Some numerical comparisons of several approximations to the binomial distribution, *Journal of the American Statistical Association 64*, 1638-1646. *[7.1.5.1, 7.1.6.1; 7.1.8]*

Gebhardt, F. (1971). Incomplete Beta-integral B(x; 2/3, 2/3) and $[p(1-p)]^{-1/6}$ for use with Borges' approximation of the binomial distribution, *Journal of the American Statistical Association 66*, 189-191. *[7.1.8]*

Ghosh, B. K. (1980). Two normal approximations to the binomial distribution, *Communications in Statistics A 9*, 427-438. *[7.1.2.2; 7.1.5.2; 7.1.7]*

Gnanadesikan, M. and Gupta, S. S. (1970). A selection procedure for multivariate normal distributions in terms of the generalized variances, *Technometrics* 12, 103-117. *[7.14.4]*

Goldberg, G. and Levine, H. (1946). Approximate formulas for the percentage points and normalization of t and χ^2, *Annals of Mathematical Statistics* 17, 216-225. *[7.8.13; 7.10.10]*

Goldstein, R. B. (1973). Chi-square quantiles. Algorithm 451, *Communications of the Association for Computing Machinery* 16, 483-485. *[7.8.11]*

Goria, M. N. (1992). On the fourth-root transformation of chi-square, *Australian Journal of Statistics* 34, 55-64. *[7.8.5]*

Govindarajulu, Z. (1965). Normal approximations to the classical discrete distributions, *Sankhyā A27*, 143-172. *[7.1.3; 7.2.5]*

Haines, P. D. (1988). A closed form approximation for calculating the percentage points of the F and t distributions, *Applied Statistics* 37, 95-100. *[7.12.11]*

Hawkins, D. M. and Wixley, R. A. J. (1986). A note on the transformation of chi-squared variables to normality, *American Statistician* 40, 296-298. *[7.8.5, 7]*

Hemelrijk, J. (1967). The hypergeometric, the normal and chi-squared, *Statistica Neerlandica* 21, 225-229. *[7.4.2]*

Hernandez, F. and Johnson, R. A. (1980). The large-sample behavior of transformations to normality, *Journal of the American Statistical Association* 75, 855-861. *[7.15.6]*

Hill, G. W. (1970a). Algorithm 395: Student's t-distribution, *Communications of the Association for Computing Machinery* 13, 617-619. *[7.10.8]*

Hill, G. W. (1970b). Algorithm 396: Student's t-quantiles, *Communications of the Association for Computing Machinery* 13, 619-620. *[7.10.10]*

Hill, G. W. (1976). New approximations to the von Mises distribution, *Biometrika* 63, 673-676. *[7.7.3, 4, 5]*

Hill, I. D. (1976). Algorithm AS 100: Normal-Johnson and Johnson-normal transformations, *Applied Statistics* 25, 190-192. *[7.15.7]*

Hill, I. D., Hill, R. and Holder, R. L. (1976). Algorithm AS 99: Fitting Johnson curves by moments, *Applied Statistics* 25, 180-189. *[7.15.7]*

Hotelling, H. (1953). New light on the correlation coefficient and its transforms, *Journal of the Royal Statistical Society B15*, 193-232. *[7.15.2]*

Hoyle, M. H. (1973). Transformations--An introduction and a bibliography, *International Statistical Review 41*, 203-223; erratum (1976); *ibid 44*, 368. *[7.1.6.2; 7.15.1, 3]*

Jensen, D. R. and Solomon, H. (1972). A Gaussian approximation to the distribution of a definite quadratic form, *Journal of the American Statistical Association 67*, 898-902. *[7.9.4; 7.14.1]*

Johnson, N. L. (1949). Systems of frequency curves, generated by methods of translation, *Biometrika 36*, 149-176. *[7.15.7]*

Johnson, N. L. (1959). On an extension of the connextion between Poisson and χ^2-distributions, *Biometrika 46*, 352-363. *[7.9.2, 3]*

Johnson, N. L. (1965). Tables to facilitate fitting S_U frequency curves, *Biometrika 52*, 547-558. *[7.15.7]*

Johnson, N. L. and Kotz, S. (1970a). *Distributions in Statistics: Continuous Univariate Distributions, Vol. 1*, New York: Wiley. *[Intr., 7.8.4, 7, 13; 7.15.7]*

Johnson, N. L. and Kotz, S. (1970b). *Distributions in Statistics: Continuous Univariate Distributions, Vol. 2*, New York: Wiley. *[Intr., 7.9.2, 3, 4, 5; 7.10.2, 3, 4, 6; 7.11.2, 3, 4, 7.12.2; 7.13.2; 7.14.1, 2, 3]*

Johnson, N. L., Kotz, S. and Kemp, A. W. (1992). *Univariate Discrete Distributions* (2nd edn.), New York: Wiley. *[Intr.; 7.1.3; 7.3.1, 4; 7.4.7]*

Konishi, S., Niki, N. and Gupta, A. K. (1988). Asymptotic expansions for the distribution of quadratic forms in normal variables, *Annals of the Institute of Statistical Mathematics, 40*, 279-296. *[7.14.1]*

Laubscher, N. F. (1960). Normalizing the noncentral t and F distributions, *Annals of Mathematical Statistics 31*, 1105-1112. *[7.12.5; 7.13.2; 7.15.2, 4]*

Lewis, T. and Fisher, N. I. (1982). Graphical methods for estimating the fit of a Fisher distribution to spherical data, *Geophysical Journal of the Royal Astronomical Society 69*, 1-13. *[7.14.5]*

Ling, R. F. (1978). A study of the accuracy of some approximations for t, χ^2, and F tail probabilities, *Journal of the American Statistical Association 73*, 274-283. *[Intr.; 7.8.1, 6; 7.10.4, 6; 7.12.6]*

Ling, R. F. and Pratt, J. W. (1984). The accuracy of Peizer approximations to the hypergeometric distribution, with comparisons to some other approximations, *Journal of the American Statistical Association* 79, 49-60. *[7.4.6, 7]*

Makabe, H. and Morimura, H. (1955). A normal approximation to the Poisson distribution, *Reports on Statistical Applications Research* (Union of Japanese Scientists and Engineers) 4, 37-46. *[7.2.5]*

Martin, D. C. and Katti, S. K. (1962). Approximations to the Neyman Type A distribution for practical problems, *Biometrics* 18, 354-364. *[7.5.1]*

Mathur, R. K. (1961). A note on Wilson-Hilferty transformation, *Calcutta Statistical Association Bulletin* 10, 103-105. *[7.8.4]*

Mickey, M. R. (1975). Approximate test probabilities for Student's t distribution, *Biometrika* 62, 216-217. *[7.10.3]*

Molenaar, I. W. (1970). *Approximations to the Poisson, Binomial, and Hypergeometric Distribution Functions*, Mathematical Centre Tracts, Vol. 31, Amsterdam: Mathematisch Centrum. *[Intr., 7.1.4; 7.1.5.1; 7.1.6.1; 7.1.7, 8, 9, 10, 11, 12; 7.2.2, 3, 5, 6, 7, 8, 9, 10; 7.4.1, 2, 3, 5, 8; 7.6.3, 6]*

Molenaar, I. W. (1985). Normal approximations to the Poisson, binomial, negative binomial and hypergeometric distributions, *Encyclopedia of Statistical Sciences, Vol. 6* (S. Kotz, N. L. Johnson and C. B. Read, eds.), 340-347, New York: Wiley. *[Intr., 7.4.6]*

Moore, P. G. (1957). Transformations to normality using fractional powers of the variables, *Journal of the American Statistical Association* 52, 237-246. *[7.15.6]*

Moran, P. A. P. (1966). Accurate approximations for t-tests, in *Research Papers in Statistics: Festschrift for J. Neyman* (F. N. David, ed.), 225-230. *[7.10.9]*

Moschopoulos, P. G. (1983). On a new transformation to normality, *Communications in Statistics A* 12, 1873-1878. *[7.9.5]*

Mudholkar, G. S., Chaubey, Y. P. and Lin, C.-C. (1976). Some approximations for the noncentral F-distribution, *Technometrics* 18, 351-358. *[7.13.3]*

Neyman, J. (1939). On a new class of "contagious" distributions, applicable in entomology and bacteriology, *Annals of Mathematical Statistics* 10, 35-57. *[7.5.1]*

Nicholson, W. L. (1956). On the normal approximation to the hypergeometric distribution, *Annals of Mathematical Statistics* 27, 471-483. *[7.4.5]*

Patil, G. P. (1960). On the evaluation of the negative binomial distribution with examples, *Technometrics* 2, 501-505. *[7.3.2]*

Paulson, E. (1942). An approximate normalization of the analysis of variance distribution, *Annals of Mathematical Statistics* 13, 233-235. *[7.12.4, 9]*

Peiser, A. M. (1943). Asymptotic formulas for significance levels of certain distributions, *Annals of Mathematical Statistics* 14, 56-62. *[7.8.13; 7.10.10]*

Peizer, D. B. and Pratt, J. W. (1968). A normal approximation for binomial, F, beta, and other common, related tail probabilities, I, *Journal of the American Statistical Association* 63, 1416-1456. *[Intr., 7.1.9; 7.2.8; 7.3.6; 7.6.5; 7.8.6; 7.10.4; 7.12.6]*

Pratt, J. (1968). A normal approximation for binomial, F, beta, and other common, related tail probabilities, II, *Journal of the American Statistical Association* 63, 1457-1483. *[7.1.11]*

Prescott, P. (1974). Normalizing transformations of Student's t distribution, *Biometrika* 61, 177-180. *[7.10.11]*

Raff, M. S. (1956). On approximating the point binomial, *Journal of the American Statistical Association* 51, 293-303. *[7.1.2.2; 7.1.6.1; 7.1.7]*

Read, C. B. (1973). An application of a result of Watson to estimation of the normal standard deviation, *Communications in Statistics* 1, 183-185. *[7.11.2]*

Riordan, J. (1949). Inversion formulas in normal variable mapping, *Annals of Mathematical Statistics* 20, 417-425. *[7.2.9]*

Sahai, H. and Thompson, W. O. (1974). Comparisons of approximations to the percentiles of the t, χ^2, and F distributions, *Journal of Statistical Computation and Simulation* 3, 81-93. *[7.8.11, 13; 7.10.10; 7.12.10]*

Sankaran, M. (1959). On the noncentral chi-square distribution, *Biometrika* 46, 235-237. *[7.9.4]*

Scott, A. and Smith, T. M. F. (1970). A note on Moran's approximation to Student's t, *Biometrika* 57, 681-682. *[7.10.9]*

Severo, N. C. and Zelen M. (1960). Normal approximation to the chi-square and noncentral F probability functions, *Biometrika* 47, 411-416. *[7.8.3, 4, 12; 7.13.2]*

Shore, H. (1986). An approximation for the error of the normal approximation to a linear combination of independently distributed random errors, *Transactions of the Institute of Industrial Engineers* 20, 242-246. *[7.15.11]*

Slud, E. V. (1977). Distribution inequalities for the binomial law, *Annals of Probability* 5, 404-412. *[7.1.13]*

Stuart, A. and Ord, J. K. (1987). *Kendall's Advanced Theory of Statistics, Vol. 1* (5th edn.), New York: Oxford University Press. *[7.8.3, 4; 7.12.2; 7.14.1]*

Stuart, A. and Ord, J. K. (1991). *Kendall's Advanced Theory of Statistics, Vol. 2* (5th edn.), New York: Oxford University Press. *[7.15.1]*

Upton, G. J. C. (1974). New approximations to the distribution of certain angular statistics, *Biometrika* 61, 369-373. *[7.7.2]*

Wallace, D. L. (1959). Bounds on normal approximations to Student's and the chi-square distributions, *Annals of Mathematical Statistics* 30, 1121-1130; correction (1960); *ibidm* 31, 810. *[7.8.14; 7.10.3, 4, 6]*

Wallis, W. A. and Moore, G. H. (1941). A significance test for time series analysis, *Journal of the American Statistical Association* 36, 401-412. *[7.5.3]*

Warren, W. G. (1982). On the adequacy of the chi-squared approximation for the coefficient of variation, *Communications in Statistics B* 11, 659-666. *[7.11.2]*

Wilson, E. B. and Hilferty, M. M. (1931). The distribution of chi-square, *Proceedings of the National Academy of Science* 17, 684-688. *[7.2.6; 7.8.4; 7.8.10]*

Wishart, J. (1947). The cumulants of the z and of the logarithmic χ^2 and t distributions, *Biometrika* 34, 170-178, 374. *[7.12.2]*

Zar, J. H. (1978). Approximations for the percentage points of the chi-squared distribution, *Applied Statistics* 27, 280-290. *[7.8.8, 11, 12, 13]*

Chapter 8

ORDER STATISTICS FROM NORMAL SAMPLES

In this chapter a number of results are listed which relate to order statistics from normally distributed parent populations. Asymptotic distributions are included along with discussion of each statistic of interest. Limiting normal distributions of linear combinations and other functions of order statistics are also listed.

The notation is defined in *[8.1.1]* and the warning given there about conflicting notation in the literature should be noted. Basic properties of independence and distributions of order statistics appear in Section 8.1, while moments are given in Section 8.2 and in Tables 8.2 (a) to 8.2 (d) which collect together results for the first four moments of order statistics in samples not larger than 5. Properties of deviates of order statistics from the sample mean are given in Section 8.3, as well as properties of the ratio of such deviates to an estimate of the standard deviation. In Section 8.4 appear properties of the sample range and of the ratio of the range to an estimate of standard deviation. Quasi-ranges are discussed in Section 8.5, and several results relating to the sample median and to the sample midrange are given in Section 8.6. Asymptotic properties and limiting distributions involving normal order statistics are covered in Section 8.7. Sample quantiles other than the median are briefly covered in Section 8.8; properties of Gini's mean difference, of the trimmed mean, and of some miscellaneous statistics are given in Section 8.9.

A good source book with proofs and extensive discussion is David (1981); see also chapter 14 of Stuart and Ord (1994); Galambos (1978) has a detailed exposition of the asymptotic theory of extreme order statistics, which updates Gumbel's classic work of 1958.

8.1 ORDER STATISTICS: BASIC RESULTS

[8.1.1] Let $X_1, X_2, ..., X_n$ be a random sample of size n from a normal distribution. If these rvs are arranged in ascending order of magnitude and written

$$X_{(1;n)} \leq X_{(2;n)} \leq \cdots \leq X_{(n;n)} ,$$

then $X_{(r;n)}$ is the *rth order statistic*. When no confusion arises, we write more simply

$$X_{(1)} \leq X_{(2)} \leq \cdots \leq X_{(n)} .$$

In some sources, the labeling is made in reverse order, with $X_{(1)}$ as the largest rather than as the smallest order statistic, and it is wise to check in each case, particularly when tables are consulted.

[8.1.2.1] Let $X_1, X_2, ..., X_n$ be a random sample from a $N(\mu, \sigma^2)$ distribution, and let $X_{(1)}, ..., X_{(n)}$ be the order statistics. Let

$$U = \sum_{i=1}^{n} c_i X_{(i)} , \quad \sum_{i=1}^{n} c_i = 0 .$$

Then if $S^2 = (n-1)^{-1} \sum_{i=1}^{n}(X_i - \overline{X})^2$ is the sample variance, U/S, $\overline{X}$, and S are mutually independent (David, 1981, pp. 89, 111). This property applies to several linear combinations of order statistics discussed in this chapter.

Deviates from the mean: $(X_{(i)} - \overline{X})/S$, $\overline{X}$ and S are mutually independent (*[8.3.1]*); $X_{(i)} - \overline{X}$ and $\overline{X}$ are independent.

Range: $(X_{(n)} - X_{(1)})/S$, $\overline{X}$, and S are mutually independent, and $X_{(n)} - X_{(1)}$ and $\overline{X}$ are independent (Section [8.4]).

Quasi-ranges: $(X_{(n-r+1)} - X_{(r)})/S$, $\overline{X}$, and S are mutually independent, and $X_{(n-r+1)} - X_{(r)}$ and $\overline{X}$ are independent (Section [8.5]).

[8.1.2.2] Let $\overline{X}_k = \sum_{i=n-k+1}^{n} X_{(i)}/k$, so that $\overline{X}_n = \overline{X}$. The *selection differential* or *reach statistic* Δ_k is defined by

$$\Delta_k = \overline{X}_k - \overline{X}_n .$$

Order Statistics From Normal Samples

Then Δ_k/S, $\bar{X}$ and S are mutually independent; and are independent (see also [8.2.2.3] and Arnold and Balakrishnan, 1989, pp. 18-20).

[8.1.3] *Distributions.* Let the parent distribution be $N(0, 1)$, and let $F_r(x)$ and $f_r(x)$ denote the cdf and pdf of $Z_{(r;n)}$, respectively. Then, when $1 \leq r \leq n$,

(a) $F_r(z) = \sum_{i=r}^{n} \binom{n}{i} p^i (1-p)^{n-i} = I_p(r, n-r+1),$

(b) $f_r(z) = r\binom{n}{r} p^{r-1}(1-p)^{n-r} \phi(z).$

(c) $F_r(-z) = 1 - F_{n-r+1}(z),$

(d) $f_r(-z) = f_{n-r+1}(z),$

where $p = \Phi(z)$, $I_p(a, b)$ is the incomplete beta function, and $B(a, b)$ is the beta function (David, 1981, p. 9). Special cases when $r = 1$ or $r = n$ are given by

$$\Pr(Z_{(1)} \leq z) = 1 - (1-p)^n, \quad \Pr(Z_{(n)} \leq z) = p^n,$$

$$f_n(z) = np^{n-1}\phi(z), \quad f_1(z) = n(1-p)^{n-1}\phi(z).$$

For a discussion of the extremes $Z_{(1)}$ and $Z_{(n)}$, see Gumbel (1958, pp. 129-140).

[8.1.4] *Percentage Points of Order Statistics with a $N(\mu, \sigma^2)$ Parent.*
Let $\Pr\left[X_{(r)} < x_{(r), p}\right] = 1-p$, and let Y have a F distribution with $2(n-r+1)$ and $2r$ degrees of freedom, such that

$$\Pr\left(Y < y_{2(n-r+1), 2r; \beta}\right) = 1 - \beta$$

(see [7.12.7]). Then

$$\Phi\left(\frac{x_{(r), 1-\alpha} - \mu}{\sigma}\right) = \frac{r}{r + (n-r+1) y_{2(n-r+1), 2r; \alpha}} = q,$$

say (Guenther, 1977, pp. 319-320). To obtain $x_{(r), 1-\alpha}$, first derive the percentile $y_{2(n-r+1), 2r; \alpha}$ of F from tables of F or by means of a desk calculator (Guenther, 1977, p. 319; see *[7.12.7]* to *[7.12.10]* and Johnson and Kotz, 1970, chap. 26); then solve $\Phi(y) = q$ for y using standard normal tables (see also Section 3.8). For extreme-order statistics, solve

$$\Phi\big(\{x_{(1),\alpha} - \mu\}/\sigma\big) = 1 - \alpha^{1/n},$$

$$\Phi\big(\{x_{(n),\alpha} - \mu\}/\sigma\big) = (1 - \alpha)^{1/n}$$

(Gupta, 1961, p. 889).

[8.1.5] Tables of the cdf and percentage points of $X_{(1;n)}$ and $X_{(n;n)}$ are given in several standard sources, listed in Table 8.1. In addition, Govindarajulu and Hubacker (1964, pp. 77-78) tabulate percentage points of $X_{(r;n)}$ corresponding to α, $1 - \alpha = 0.50, 0.25, 0.10, 0.025, 0.01$; $n = 1(1)30$; $r = 1(1)(1 + [n/2])$, when the parent distribution is $N(0, 1)$; Gupta (1961, pp. 890-891) tabulates similar percentage points corresponding to $1 - \alpha = 0.50, 0.75, 0.90, 0.95, 0.99$; $n = 1(1)10$, $r = 1(1)n$, and values of r corresponding to the extreme and central order statistics when $n = 11(1)20$, to four decimal places. See also David (1981, pp. 287-298) for references to other tables.

[8.1.6] *Joint distributions.* Let the parent distribution be $N(0, 1)$, and let $f(z_1, z_2, ..., z_k; r_1, r_2, ..., r_k)$ be the joint pdf of $Z_{(r_1)}, Z_{(r_2)}, ..., Z_{(r_k)}$; $1 \leq k \leq n$; $1 \leq r_1 \leq r_2 \leq \cdots \leq r_k \leq n$. Then (David, 1981, p. 10) provided $Z_1 \leq Z_2 \leq \cdots \leq Z_k$,

$f(z_1, ..., z_k; r_1, ..., r_k)$

$$= c\{\Phi(z_1)\}^{r_1-1} \prod_{i=1}^{k-1} \left[\{\Phi(z_{i+1}) - \Phi(z_i)\}^{r_{i+1}-r_i-1} \right] \{1 - \Phi(z_k)\}^{n-r_k} \prod_{i=1}^{k} \phi(z_i),$$

where $c = n! \Big/ \Big[(r_1 - 1)! \big(\prod_{i=1}^{k-1}\{(r_{i+1} - r_i - 1)!\}\big)(n - r_k)!\Big]$. The joint pdf of all n order statistics is $n!\phi(z_1)...\phi(z_k)$ if $z_1 \leq z_2 \leq \cdots \leq z_k$, and is zero otherwise.

If the parent distribution is $N(\mu, \sigma^2)$, the joint pdf is

$$\sigma^{-k} f\left(\frac{x_1 - \mu}{\sigma}, ..., \frac{x_k - \mu}{\sigma}; r_1, ..., r_k\right).$$

TABLE 8.1

Normal Order Statistics: Tables and Coverages in Some Standard Sources

Function	Source[a]	Coverage	Accuracy[b]
$EX_{(i;n)}$	PH1, 190	$n=2(1)26(2)50,\ n-i+1=1(1)[\tfrac{1}{2}n]$	3, 2
	OW, 151-154	$n=2(1)50,\ n-i+1=1(1)[\tfrac{1}{2}n]$	4
	RMM, 89-90	$n=2(1)40,\ n-i+1=1(1)[\tfrac{1}{2}n]$	2
	PH2, 27,205-210	$n=2(1)100(25)200,\ n-i+1=1(1)[\tfrac{1}{2}n]$	5
	Hart 2, 425-455	$n=2(1)100$; values to 400,	5
	KS, 888-891	$n=2(1)79,\ n-i+1=1(1)[\tfrac{1}{2}n]$	5
	SG, 193	$n=2(1)20,\ n-i+1=1(1)[\tfrac{1}{2}n]$	10
	Y, 33-37	$n=2(1)50,\ n-i+1=1(1)[\tfrac{1}{2}n]$	20
$E\{X_{(i;n)}X_{(j;n)}\}$:			
	SG, 191,194-199	$n=1(1)20,\ j=1(1)n;\ n-i+1=1(1)[\tfrac{1}{2}n]$	10
	Y, 38-49	$n=2(1)30,\ j=1(1)n;\ n-i+1=1(1)[\tfrac{1}{2}n]$	8
$Cov\{X_{(i;n)}X_{(j;n)}\}$:			
	OW, 163-169	$n=2(1)20,\ j=1(1)n;\ n-i+1=1(1)[\tfrac{1}{2}n]$	4
	PH2, 211-213	$n=2(1)20,\ j=1(1)n;\ n-i+1=1(1)[\tfrac{1}{2}n]$	6
	SG, 191,200-205	$n=2(1)20,\ j=1(1)n;\ n-i+1=1(1)[\tfrac{1}{2}n]$	10
	KS, 892-897	$n=2(1)20,\ i=1(1)[n/2];\ i\le j\le n-1$	10
$\sqrt{\operatorname{Var} X_{(i;n)}}$:			
	Y, 33-37	$n=2(1)50,\ n-i+1=1(1)[\tfrac{1}{2}n]$	20

$(X_{n;n}-\mu)/\sigma,\ (\mu-X_{(1;n)})/\sigma$:

(a) Percent Points

	RMM, 94	$n=1(1)30,\ \alpha=0.05,0.01,0.001$	3
	PH1, 18 SG, 322	$\begin{cases} n=1(1)30;\ \alpha,\ 1-\alpha=0.10,0.05,0.025,\\ \qquad\qquad\qquad\qquad 0.01,0.005,0.001 \end{cases}$	3
(b) cdf	PH2, 184-187	$n=1(1)25(5)60,100,\ -2.6\le x\le 6.1$	7

TABLE 8.1 (continued)

Function	Source[a]	Coverage	Accuracy[b]
$X_{(1;n)}$, $X_{(n;n)}$ to μ_2, μ_2, μ_4, ratios:			
	PH2, 216	n=1(1)50	
	SG, 188		
$E\{X^r_{(n;n)}\}$	SG, 186	n=1(1)50, r=1(1)10	(10)

$(X_{(n;n)} - \bar{X})/\sigma$, $(\bar{X} - X_{(1;n)})/\sigma$:

(a) Percent Points

	PH1, 184	n=3(1)9; α, $1-\alpha$=0.10,	
	SG, 322	0.05,0.025,0.01,0.005, 0.0001	3
(b) cdf	PH2, 189-199	n=3(1)25; $0.00 \leq x \leq 4.90$	4,5,6

$(X_{(n;n)} - \bar{X})/S_\nu$; S_ν from indpt. sample:

| percent points | PH1, 185-186 | n=3(1)10,12; α=0.10,0.05,0.025,0.01, 0.005,0.001 ν=10(1)20,24,30,40,60,120,∞ Also for α=0.05,0.01; ν=5(1)9 | 2 |
| | SG, 326 | n=3(1)10,12; α=0.05,0.01; ν=10(1)20,24,30,40,60,120,∞ | 2 |

$(X_{(n;n)} - \bar{X})/S$, S from same sample:

| percent points | SG, 324 | n=3(1)25, α=0.10,0.05,0.025,0.01 | 3 |
| | Y, 90 | n=3(1)74; α=0.10,0.05,0.01 | 3 |

Range W

percent points	OW, 139	n=2(1)20(2)40(10)100; α, $1-\alpha$=0.10,0.05,0.025,0.01,0.001	3
	PH1, 177	n=2(1)20; α, $1-\alpha$=0.10,0.05,0.025, 0.01,0.005,0.0001	2
	Hart 1, 372-374	n=2(1)20(2)40(10)100; α=0.10(0.1) 0.90; $\alpha, 1-\alpha$=0.05,0.025,0.01,0.005, 0.001,0.0005,0.0001	6
	SG, 327	n=2(1)20; $\alpha, 1-\alpha$=0.10,0.05, 0.025,0.01,0.005,0.001	2

TABLE 8.1 (continued)

Function	Source[a]	Coverage	Accuracy[b]
percent points	Y, 61	n=2(1)40(5)50(10)100;α,$1-\alpha$= 0.5,0.25,0.10,0.05,0.025,0.01,0.005	3
cdf	PH1, 178-183	n=2(1)20; x=0.00(0.05)7.30	4
	Hart 1, 240-370	n=2(1)20(2)40(10)100; x=0.00(0.01)10.47	8
pdf	Hart 1, 38-97	n=2(1)16; x=0.00(0.01)9.99	8
	Y, 54-59	n=3(1)20; x=0.00(0.05)7.65	4
moments	OW, 140	μ, σ^2, $\sqrt{\beta_1}$, β_2, n=2(1)20(2)40(10)100	3
	PH1, 176	μ, σ, σ^2, β_1, β_2, μ/σ^2, μ^2/σ^2; n=2(1)20	5,4,3,2
	Hart 1, 376,377	μ, σ_2, $\sqrt{\beta_1}$, β_2; n=2(1)100	10,8,7
	Y, 60	μ, σ^2, $\sqrt{\beta_1}$, β_2; n=2(1)100	7,6,5

Studentized range W/S_ν (S_ν from indpt. sample)

Function	Source[a]	Coverage	Accuracy[b]
percent points	OW, 144-148	n=2(1)20,24,30,40,60,100; α, $1-\alpha$=0.10,0.05,0.025,0.01,0.005; $\nu=1,3,5(5)20,60,\infty$	3
	PH1, 191-193	n=2(1)20; α=0.10,0.05,0.01; ν=1(1)20,24,30,40,60,120,∞	2
	Hart 1, 624-661	n=2(1)20(2)40(10)100; α, $1-\alpha$= 0.001,0.005,0.01,0.025,0.05,0.1(0.1) 0.50; ν=1(1)20,24,30,40,60,120,∞	3
	SG, 114-115	n=2(10)20; α=0.05,0.01; ν=1(1)20,24,30,40,60,120,∞	2
	Y, 63	n=2(1)6,8,10,15,20,30; α=0.05,0.01; ν=1(1)20,24,30,40,60,120,∞	4
cdf	Hart 1, 382-622	n=2(1)20(2)40(10)100; ν=1(1)20, 24,30,40,60,120; $0 \leq x \leq 2000$	6

TABLE 8.1 (continued)

Function	Source[a]	Coverage	Accuracy[b]
W/S (S from same sample)			
percent points	PH1, 200	n=3(1)20(5)100,150,200,500,1000; α, $1-\alpha$=0.10,0.05,0.25,0.01,0.005,0.001	3,2
percent points	SG, 328-329	As for PH1, p. 200; but lower % points missing for n=3(1)9	3,2
$(\bar{X}-\mu)/W$, $\bar{X}$ and W from same sample:			
percent points, $|\bar{X}-\mu|/W$	OW, 142	n=2(1)20; α=0.05,0.025,0.01, 0.005,0.001,0.0005	3,2
	SG, 120	n=2(1)12; α=0.05,0.01	3,2
Indpt. ranges W_1, W_2			
W_1/W_2: percent points	PH1, 196-199	n_1,n_2=2(1)15; α=0.5,0.25,0.10,0.05, 0.025,0.01,0.005,0.001	(4)
	Hart 1, 224-227	As for PH1	(4)
W_1/W_2: cdf	Hart 1, 172-221	$n_1, n_2 = 2(1)15$; $1 \leq x \leq 600$	5
W_1/W_2: pdf	Hart 1, 100-169	n_1, n_2=2(1)15; $0 \leq x \leq 600$	6
$|\bar{X}_1 - \bar{X}_2|/(W_1+W_2)$			
percent points	PH1, 194-195	n_1, n_2=2(1)20; $\alpha = 0.10,0.05,0.02,0.01$	3
$|\bar{X}_1-\bar{X}_2|/\{\frac{1}{2}(W_1+W_2)\}$ percent points	SG, 120	$n_1=n_2$=2(1)12; α=0.05,0.01	3
Quasi range: $W_r = X_{(n-r;n)} - X_{(r+1;n)}$			
EW_r	Hart 2, 136-139	n=2(1)100; r=0(1)8	6
Var W_r	Hart 2, 142-144	n=2(1)100; r=0(1)8	5
s.d. W_r	Hart 2, 146-148	n=2(1)100; r=0(1)8	5
percent points	Hart 2, 296-319	n=2(1)20(2)40(10)100; r=0(1)8; α=0.1(0.1)0.9; $\alpha, 1-\alpha$=0.05,0.025,0.01, 0.005,0.001,0.0005,0.0001	6
cdf	Hart 2, 160-294	n=2(1)20(2)40(10)100; r=0(1)8; x=0.05(0.05)10.0	8

TABLE 8.1 (continued)

Function	Source[a]	Coverage	Accuracy[b]
k indpt. samples of size n			
S^2_{max}/S^2_{min}: percent points	OW, 101	n=3(1)11,13,16,21,31,61,∞; k=2(1)12; α=0.05,0.01	2,1,0
	PH1, 202	n=3(1)11,13,16,21,31,61,∞; k=2(1)20; α=0.05,0.01	2,1,0
	Y, 72-75	n=3(1)31,41,61,121,∞, k=2(1)20; α=0.05,0.01	2,1,0
$S^2_{max}/\sum S^2_j$ percent points	PH1, 203	n=2(1)11,17,37,145,∞; k=2(1)10,12,15,20; α = 0.05,0.01	4
	Y, 76-79	n=2(1)31,41,61,121,∞; k=2(1)20; α=0.05,0.01	4
S^2_{max}/S^2_0 from indpt. sample:			
percent points	PH1, 176	n=2; k=1(1)10; ν=10,12,15, 20,30,60,∞; α=0.05,0.01	2
	Y, 68-71	n=2; k=1(1)10(2)30; ν=9(1)30(2)50, 60,80,120,240, ∞; α, n=0.05,0.01	2
Ranges:			
$W_{max}/\Sigma W_j$: percent points	PH1, p. 205	n=2(1)10; k=2(1)10,12,15,20; α=0.05	3
Y half-normal:			
$EY_{(i;n)}$	PH2, p. 226	n=1(1)30; i=1(1)n	4

[a]OW=Owen (1962)
RMM= Rao, Mitra, and Matthai (1966)
PH1 = Pearson and Hartley (1966), vol. 1
PH2 = Pearson and Hartley (1971), vol. 2
Hart 1 = Harter (1969a)
Hart 2 = Harter (1969b)
SG = Sarhan and Greenberg (1962)
Y = Yamauti (1972)
KS = Krishnaiah and Sen (1984)

[b]Accuracy in no. of decimal places or significant figures (in parens.)

8.2 MOMENTS

[8.2.1] *Notation.* Let

$$\mu_{(r;n)} = E(X_{(r;n)}), \quad \mu_{(r,s;n)} = E(X_{(r;n)}X_{(s;n)}),$$

$$\sigma_{(r,s;n)} = \text{Cov}(X_{(r;n)}X_{(s;n)}) = E\left[\{X_{(r;n)} - \mu_{(r;n)}\} \times \{X_{(s;n)} - \mu_{(s;n)}\}\right],$$

$$\mu_{(r;n)}^{(k)} = E(X_{(r;n)}^k), \; 1 \leq r \leq n, \; 1 \leq s \leq n.$$

[8.2.2.1] The following *moment relations* hold when the order statistics have a N(0, 1) parent:

$$\mu_{(r;n)} = -\mu_{(n-r+1;n)},$$

$$\sigma_{(r,s;n)} = \sigma_{(s,r;n)} = \sigma_{(n-r+1,n-s+1;n)} = \sigma_{(n-s+1,n-r+1;n)},$$

$$= \mu_{(r,s;n)} - \mu_{(r;n)}\mu_{(s;n)},$$

$$\sum_{r=1}^{n}\sum_{s=1}^{n} \sigma_{(r,s;n)} = n, \quad 1 \leq r \leq s \leq n$$

(Sarhan and Greenberg, 1962, p. 191; Govindarajulu, 1963, p. 636; Jones, 1948, pp. 271-273). Harter (1969b, p. 26) gives the following:

$$\mu_{(r+1;n)} = \{n\mu_{(r;n-1)} - (n-r)\mu_{(r;n)}\}/n,$$

$$\mu_{(r+1;n)} = \{r\mu_{(r+1;n)} + (n-r)\mu_{(r;n)}\}/n, \quad 1 \leq r \leq s \leq n-1.$$

Using these recursive relations, the values of $\mu_{(1;n)}$ $(1 \leq n \leq N)$ are sufficient to compute the set of values of $\mu_{(r;n)}$ $(2 \leq r \leq n, 2 \leq n \leq N)$. See *[8.2.3]* below.

If n is even, then

$$\mu_{(n/2;n-1)}^{(k)} = \begin{cases} \mu_{(n/2;n)}^{(k)}, & k \text{ even} \\ 0, & k \text{ odd} \end{cases}$$

Order Statistics From Normal Samples

(David, 1981, p. 47). This result holds for any parent distribution symmetric about the origin, and includes a correction by David.

[8.2.2.2] Arnold and Balakrishnan (1980), p.18, give the following relations for a standard normal distribution for $1 \leq i \leq n$:

$$\sum_{j=1}^{n} \mu_{(i,j;n)} = 1, \quad \sum_{j=1}^{n} \sigma_{(i,j;n)} = 1,$$

$$\mu_{(i;n)}^{(2)} = 1 + n \binom{n-1}{i-1} \sum_{j=0}^{n-i} (-1)^j \binom{n-i}{j} \left(\frac{1}{i+j}\right) \mu_{(1,2;i+j)}.$$

(See also Balakrishnan and Malik 1988, pp. 2657-2694).

[8.2.2.3] Joshi and Balakrishnan (1981, pp. 203-213) established the following relations both for standard normal and half-normal distributions for $1 \leq r \leq s \leq n$:

$$\sum_{s=r}^{n} \mu_{(r,s;n)} = 1 + \sum_{s=r}^{n} \mu_{(r-1,s;n)}, \quad r = 1, 2, ..., n,$$

$$\sum_{s=r+1}^{n} \mu_{(r,s;n)} = \sum_{s=r+1}^{n} \mu_{(s,s;n)} - (n-r), \quad r = 1, 2, ..., n-1,$$

$$\sum_{s=1}^{n} \mu_{(r,s;n)} = 1 + n\mu_{(1;1)}\mu_{(r-1;n-1)},$$

$$\sum_{s=r}^{n} \sigma_{(r,s;n)} = 1 - (n-r+1)\mu_{(1;1)}(\mu_{(r;n)} - \mu_{(r-1;n)}),$$

where $\mu_{(0,t;n)} = \mu_{(0;t)} = 0$ for $t \geq 1$.

These relations are useful in finding the variance of the selection differential or reach statistic defined by $\Delta_k = \bar{X}_k - \bar{X}_n$, where $\bar{X}_k = \frac{1}{k}\sum_{i=n-k+1}^{n} X_{(i)}$. For a selected fraction k/n, $v_k = k \text{ var } \bar{X}_k$ remains almost constant. Joshi and Balakrishnan (1981) have tabulated the mean and variance of $\bar{X}_k$ for n up to 50. See Barnett and Lewis (1978) and Hawkins (1979, pp. 227-236) on the internally studentized version of Δ_k, viz, $D_k = k\Delta_k/S$, where S^2 is the usual sample variance. Then D_k and S are independent. Based on a N(0,1) parent ,

$$E(\Delta_k) = E(\overline{X}_k), \quad E(\Delta_k^2) = \text{var}(\overline{X}_k) + \left[E(\overline{X}_k)\right]^2 - 1/n \;.$$

For k=1, Borenius (1966, pp. 1-15) has tabulated these quantities for $n \leq 120$.

[8.2.3] *Moment Values.*

(a) Define $F_r(x)$ as in *[8.1.3]*. Then for a $N(0, \sigma^2)$ parent,

$$\mu_{(r;n)} = \int_0^\infty \{F_{n-r+1}(x) - F_r(x)\}dx$$

(David, 1981, p. 38).

(b) Bose and Gupta (1959, p. 437) give expressions for $\mu_{(r;n)}^{(k)}$, $k = 1$, 2, 3, 4; these are in terms of certain integral functions. If $r = n$ and

$$I_n(k) = \int_{-\infty}^\infty \{\Phi(x)\}^{n-k} \exp(-kx^2/2)dx \;,$$

then for a $N(0, 1)$ parent,

$$\mu_{(n;n)} = 2\binom{n}{2}(2\pi)^{-1} I_n(2) \;,$$

$$\mu_{(n;n)}^{(2)} = 1 + 3\binom{n}{3}(2\pi)^{-3/2} I_n(3) \;,$$

$$\mu_{(n;n)}^{(3)} = (\tfrac{5}{2}\mu_{(n;n)} + 4\binom{n}{4}(2\pi)^{-2} I_n(4) \;,$$

$$\mu_{(n;n)}^{(4)} = -\tfrac{4}{3} + \tfrac{13}{3}\mu_{(r;n)}^{(2)} + 5\binom{n}{5}(2\pi)^{-5/2} I_n(5) \;.$$

(c) From Jones (1948, p. 270) and Godwin (1949, p. 284) we can derive *exact values* for $\mu_{(r;n)}$ and $\mu_{(r;n)}^{(2)}$ for n = 2, 3, 4, and 5. These appear in Tables 8.2A and B, respectively. From Bose and Gupta (1959, pp. 438-439) we can derive third and fourth moments; these appear in Tables 8.2C and D. These references also include numerical approximations that are accurate to several decimal places; the tables clearly show how these moments differ from the corresponding first four moments (0, 1, 0, and 3, respectively) of unordered random variables.

Order Statistics From Normal Samples

TABLE 8.2
Standard Normal Order Statistics: Moments, n=2,3,4,5

A. Exact Values of $EZ_{(r;n)}$

r \ n:	2	3	4	5
1	$-\dfrac{1}{\sqrt{\pi}}$	$-\dfrac{3}{2\sqrt{\pi}}$	$-\dfrac{3}{2\sqrt{\pi}}\left(1+\dfrac{2a}{\pi}\right)$	$-\dfrac{5}{4\sqrt{\pi}}\left(1+\dfrac{6a}{\pi}\right)$
2	$\dfrac{1}{\sqrt{\pi}}$	0	$-\dfrac{3}{2\sqrt{\pi}}\left(1-\dfrac{6a}{\pi}\right)$	$-\dfrac{5}{2\sqrt{\pi}}\left(1-\dfrac{6a}{\pi}\right)$
3	--	$\dfrac{3}{2\sqrt{\pi}}$	$\dfrac{3}{2\sqrt{\pi}}\left(1-\dfrac{6a}{\pi}\right)$	0
4	--	--	$\dfrac{3}{2\sqrt{\pi}}\left(1+\dfrac{2a}{\pi}\right)$	$\dfrac{5}{2\sqrt{\pi}}\left(1-\dfrac{6a}{\pi}\right)$
5	--	--	--	$\dfrac{5}{4\sqrt{\pi}}\left(1+\dfrac{6a}{\pi}\right)$

B. Exact Values of $EZ^2_{(r;n)}$

r \ n:	2	3	4	5
1	1	$1+\dfrac{\sqrt{3}}{2\pi}$	$1+\dfrac{\sqrt{3}}{\pi}$	$1+\dfrac{5\sqrt{3}}{4\pi}+\dfrac{5\sqrt{3}}{2\pi^2}b$
2	1	$1-\dfrac{\sqrt{3}}{\pi}$	$1-\dfrac{\sqrt{3}}{\pi}$	$1-\dfrac{10\sqrt{3}}{\pi^2}b$
3	--	$1+\dfrac{\sqrt{3}}{2\pi}$	$1-\dfrac{\sqrt{3}}{\pi}$	$1-\dfrac{5\sqrt{3}}{2\pi}+\dfrac{15\sqrt{3}}{\pi^2}b$
4	--	--	$1+\dfrac{\sqrt{3}}{\pi}$	$1-\dfrac{10\sqrt{3}}{\pi^2}b$
5	--	--	--	$1+\dfrac{5\sqrt{3}}{4\pi}+\dfrac{5\sqrt{3}}{2\pi^2}b$

Note: $a = \arcsin(\tfrac{1}{3}) \simeq 0.33983\ 69094$
$b = \arcsin(\tfrac{1}{4}) \simeq 0.25268\ 02552$
$\pi = 3.14159\ 2654$; $\sqrt{\pi} \simeq 1.77245\ 3851$

TABLE 8.2 (continued)

C. Exact Values of $EZ^3_{(r;n)}$

r \ n:	2	3	4	5
1	$-\dfrac{5}{2\sqrt{\pi}}$	$-\dfrac{15}{4\sqrt{\pi}}$	$-\dfrac{1}{\pi^{3/2}}\left(\dfrac{1}{2\sqrt{2}}+15c\right)$	$-\dfrac{25}{4\pi^{3/2}}\left(-\pi+\dfrac{1}{5\sqrt{2}}+6c\right)$
2	$\dfrac{5}{2\sqrt{\pi}}$	0	$-\dfrac{15}{\pi^{3/2}}\left(\pi-\dfrac{1}{10\sqrt{2}}-3c\right)$	$-\dfrac{25}{\pi^{3/2}}\left(\pi-\dfrac{1}{10\sqrt{2}}-3c\right)$
3	--	$\dfrac{15}{4\sqrt{\pi}}$	$\dfrac{15}{\pi^{3/2}}\left(\pi-\dfrac{1}{10\sqrt{2}}-3c\right)$	0
4	--	--	$\dfrac{1}{\pi^{3/2}}\left(\dfrac{1}{2\sqrt{2}}+15c\right)$	$\dfrac{25}{\pi^{3/2}}\left(\pi-\dfrac{1}{10\sqrt{2}}-3c\right)$
5	--	--	--	$\dfrac{25}{4\pi^{3/2}}\left(-\pi+\dfrac{1}{5\sqrt{2}}+6c\right)$

D. Exact Values of $EZ^4_{(r;n)}$

r \ n:	2	3	4	5
1	3	$3+\dfrac{13}{2\pi\sqrt{3}}$	$3+\dfrac{13}{\pi\sqrt{3}}$	$3+\dfrac{\sqrt{5}}{4\pi^2}+d$
2	3	$3-\dfrac{13}{2\pi\sqrt{3}}$	$3-\dfrac{13}{\pi\sqrt{3}}$	$3+\dfrac{65}{\pi\sqrt{3}}-\dfrac{\sqrt{5}}{\pi^2}-4d$
3	--	$3+\dfrac{13}{2\pi\sqrt{3}}$	$3-\dfrac{13}{\pi\sqrt{3}}$	$3-\dfrac{130}{\pi\sqrt{3}}+\dfrac{3\sqrt{5}}{2\pi^2}+6d$
4	--	--	$3+\dfrac{13}{\pi\sqrt{3}}$	$3+\dfrac{65}{\pi\sqrt{3}}-\dfrac{\sqrt{5}}{\pi^2}-4d$
5	--	--	--	$3+\dfrac{\sqrt{5}}{4\pi^2}+d$

Note: $c = \arctan(\sqrt{2}) \simeq 0.95531\ 6618$
$d = 65(\pi^2\sqrt{3})^{-1}\arctan(\sqrt{5/3}) \simeq 3.46675\ 52225\ 38$
$\pi = 3.14159\ 2654;\quad \sqrt{\pi} \simeq 1.77245\ 3851$
Sources: See [8.2.3]

Order Statistics From Normal Samples

(d) Parrish (1992a, pp. 57-70), uses a Gauss-Legendre quadrature method to compute expected values and variances of standard normal order statistics for sample sizes n = 2(1)50(10)200(25)500. Values are given to 25 decimal places, taken from final values accurate to at least 27 places. His results support those of Harter (5 places) and Yamauti (10 places, n ≤ 50). Tietjen et al. obtain values differing from Harter and Yamauti, sometimes at the fourth decimal place.

An abbreviated table of expected values for is published in the paper. Parrish (1992b, pp. 71-101), presents tables of variances and covariances to 25 decimal places for pairs of standard normal order statistics for sample sizes n = 2(1)20; also tables of product moments $E(X_{i;n}X_{j;n})$ to 25 decimal places for n = 20, to 20 decimal places for n = 30, to 15 decimal places for n = 40 and to 10 decimal places for n = 50.

[8.2.4] Expansions. By inverting the probability integral transformation, David and Johnson (1954, pp. 228-240) develop a series in powers of $(n + 2)^{-1}$, giving moments and joint cumulants of N(0, 1) order statistics. Convergence may be slow or may not hold at all when r/n is close to zero or one; however, a series by Plackett (1958, pp. 131-142) may converge a little more rapidly, although it has less computational advantage (David, 1981, p. 81). See also Saw (1960, pp. 79-86), who developed bounds for the remainder of the David-Johnson series after an even number of terms, and compared these with bounds derived by Plackett (op. cit.) for his series.

[8.2.5.1] David (1981, pp. 77-78) gives the following *bounds* for $\mu_{(r;n)}$, of which the lower bound is rather poor as an approximation: if r ≥ (n+1)/2,

$$\Phi^{-1}\{(r-1)/n\} \leq \mu_{(r;n)} \leq \min\left[\Phi^{-1}\{r/(n+\tfrac{1}{2})\}, \Phi^{-1}\{(r-\tfrac{1}{2})/n\}\right],$$

$$\mu_{(r;n)} \leq \min\left[\Phi^{-1}\{1 - \exp\left(-\sum_{i=n-r+1}^{n} i^{-1}\right)\}, \Phi^{-1}\{1 + \exp\left(-\sum_{i=n-r+1}^{n} i^{-1}\right)^{-1}\}\right].$$

The author has communicated that an improved lower bound is $\Phi^{-1}\{r/(n+1)\}$.

[8.2.5.2] Gascual and Caraux (1992, pp. 143-148), give the following bounds on expectations of order statistics via extremal dependencies. For

an iid sample from a N(0, 1) parent,

$$L_{n,r} = -\frac{n}{r}\phi\left[\Phi^{-1}\left(\frac{r}{n}\right)\right] \leq \mu_{(r)} \leq \frac{n}{n-r+1}\phi\left[\Phi^{-1}\left(\frac{n-r+1}{n}\right)\right] = U_{n,r};$$

$$-\left[2\log\tfrac{n}{r} - \log\log\tfrac{n}{r}\right]^{\frac{1}{2}} \leq L_{n,r}, \quad n/r \geq \sqrt{e},$$

$$U_{n,r} \leq \left[2\log\tfrac{n}{n-r+1} - \log\log\tfrac{n}{n-r+1}\right], \quad n/(n-r+1) \geq \sqrt{e}.$$

The lower and upper bounds $L_{n,r}$ and $U_{n,r}$ may be approximated via these last two inequalities, which become sharp for $L_{n,r}$ when $n/r \to \infty$, and for $U_{n,r}$ likewise when $n/(n-r+1) \to \infty$.

[8.2.6] Approximations of the form

$$\mu_{(r;n)} \simeq \Phi^{-1}\{(r - \alpha_{r,n})/(n - 2\alpha_{r,n} + 1)\}$$

due to Blom (1958) are discussed in Harter (1961, pp. 153-156) and in Harter (1969b, p. 456), where values of $\alpha_{r,n}$ are tabulated that make the approximation accurate; these are for n = 25, 50, 100, 200, 400. Approximate algorithms for $\alpha_{r,n}$ for intermediate values of n are also given. If $n \leq 20$, then $0.33 \leq \alpha_{r,n} \leq 0.39$ for all r; so that taking $\alpha_{r,n} = \frac{3}{8}$ (David, 1981, pp. 80-82),

$$\mu_{r,b} \simeq \Phi^{-1}\{(r - \tfrac{3}{8})/(n + 1 - \tfrac{3}{4})\}, \quad n \leq 20.$$

[8.2.7] *Normal scores.* Consider the order statistics for a combined sample from one or more unknown parent distributions, not necessarily normal. Nonparametric approaches often replace these order statistics by their ranks, or by their expected values $E\{X_{(1,n)}\}, \ldots, E\{X_{(n;n)}\}$ for a N(0, 1) parent, that is, by the *normal scores* of the ordered sample. This happens, for example, for a combined sample from two populations; the *normal scores statistic* is then the sum of the *normal scores* in the combined sample of the measurements belonging to the second population. See *[7.15.1]*, *[8.8.6]* and Lehmann (1975, pp. 96-97) for further discussion and references.

Order Statistics From Normal Samples

Tables of normal scores for different (combined) sample sizes n are referenced in *[8.2.8]* and Table 8.1. David et al. (1968, Table III) tabulate normal scores $E\{Z_{(r;n)}\}$ to 5 decimal places for $n = 2(1)100(25)250(50)400$ and $r = \{n + 1 - [n/2]\}(1)n$. Royston (1982, pp. 161-165), along with Königer (1983, pp. 223-224), provides an algorithm, NSCOR1, for computing exact normal scores; Royston also gives an algorithm, NSCOR2, for approximating expected values of normal order statistics. The latter is suitable for small computers, is fast, requires little storage, and is capable of accuracy within 0.0001.

Balakrishnan (1984, pp. 242-245) gives an algorithm to approximate the sum of squares S_n of normal scores:

$$S_n = \sum_{r=1}^{n} \{\mu_{(r;n)}\}^2.$$

The author provides other references and also gives bounds for the error involved.

[8.2.8] Tables of moments of N(0,1) order statistics appear in general standard sources, listed in Table 8.1. Others are referenced in David (1981, pp. 287-298); Harter (1961, pp. 158-165) tabulates values of $\mu_{(r;n)}$ to 5 decimal places for n=2(1)100(25)250(50)400, and Teichroew (1956, pp. 416-422) tabulates values of $\mu_{(r;n)}$ and of $E\{X_{(i;n)}X_{(j;n)}\}$ to 10 decimal places for $n = 2(1)20$. Ruben (1954, pp. 224-226) tabulates values of $\mu_{(n;n)}^{(k)}$ to 10 significant figures for $n = 1(1)50$ and $k = 1(1)10$; he also tabulates values of the variance, standard deviation, third and fourth central and standardized moments of $X_{(n;n)}$ to 8 and 7 decimal places for $n = 1(1)50$. All of these are based on N(0,1) parents. Tables to 10 decimal places of $\sigma_{r,s;n}$ appear in Owen et al. (1977, Table 1) for $s \leq r$; $r = 1(1)([n/2]+1)$; $n = 2(1)50$.

[8.2.9] Some exact values of joint moments $\mu_{(r,s;n)}$, $2 \leq n \leq 5$, were obtained by Jones (1948), p. 270) and Godwin (1949, pp. 284-285) for a N(0,1) parent; these are as follows:

n = 2: $\mu_{(1,2;2)} = 0$, $\sigma_{(1,2;2)} = 1/\pi$,

n = 3: $\mu_{(1,2;3)} = \mu_{(2,3;3)} = \sqrt{3}/(2\pi)$, $\mu_{(1,3;3)} = -\sqrt{3}/\pi$,

$$\sigma_{(1,2;3)} = \sqrt{3}/(2\pi),\ \sigma_{(2,3;3)} = \sqrt{3}/(2\pi),\ \sigma_{(1,3;3)} = (9 - 4\sqrt{3})/(4\pi).$$

n = 4: $\mu_{(1,2;4)} = \mu_{(3,4;4)} = \sqrt{3}/\pi,\ \mu_{(1,3;4)} = \mu_{(2,4;4)} = (2\sqrt{3} - 3)/\pi$.

$$\mu_{(1,4;4)} = -3/\pi,\ \mu_{(2,3;4)} = (2\sqrt{3} - 3)/\pi.$$

n = 5: $\mu_{(1,2;5)} = \mu_{(4,5;5)} = 5\sqrt{3}/(4\pi)$,

$$\mu_{(1,3;5)} = \mu_{(3,5;5)} = 5(3 - \sqrt{3})/(2\pi) - (5\sqrt{3}/\pi^2)\alpha - (15/\pi^2)\beta,$$

$$\mu_{(1,4;5)} = \mu_{(2,5;5)} = -15/(2\pi) + (45/\pi^2)\beta,\ \mu_{(1,5;5)} = (30/\pi^2)\beta.$$

$\pi = 3.14159\ 2654,\ \alpha = \arcsin(\tfrac{1}{4}) \simeq 0.25268\ 02552$,
$\beta = \arcsin(\tfrac{1}{\sqrt{6}}) \simeq 0.42053\ 4335$.

[8.2.10] Bounds and approximations to means, variances and covariances of order statistics can be obtained using results in David (1981, pp. 66-70). General bounds for $E\{X_{(s;n)} - X_{(r;n)}\}$ and for $|EX_{(r;n)}|$ are given in David (1981, pp. 62-68, 83-84); some of these apply to general classes of parent distributions, such as those which are symmetrical about zero. For example, for independent samples from any parent distribution, $\sigma_{(r,s;n)} \geq 0$ for all r, s, and n (Bickel, 1967, p. 575).

Davis and Stephens (1978, pp. 206-212) give a Fortran computer program for approximating the covariance matrix of order statistics from a N(0,1) parent. If n ≤ 20, the maximum error is less than 0.00005.

8.3 ORDERED DEVIATES FROM THE SAMPLE MEAN

[8.3.1] The statistics $X_{(i;n)} - \bar{X}$, based on a sample of size n from a $N(\mu, \sigma^2)$ population, are *ordered deviates from the mean*. Let S or S' be an estimate of σ with ν degrees of freedom; then $\{X_{(i;n)} - \bar{X}\}/S'$ is a *Studentized deviate* from $\bar{X}$ if S' is based on a second independent sample $X_1^*, X_2^*, ..., X_{\nu+1}^*$ and

Order Statistics From Normal Samples

$$S'^2 = \nu^{-1} \sum_{i=1}^{\nu+1}(X_i^* - \bar{X}^*)^2, \quad \bar{X}^* = (\nu+1)^{-1} \sum_{i=1}^{\nu+1} X_i^*.$$

Frequently, however, S is based on the same sample $X_1, \ldots, X_n$, and then $\{X_{(i;n)} - \bar{X}\}/S$ is not studentized, since $X_{(i;n)} - \bar{X}$ and S are not then independent. However $\{X_{(i;n)} - \bar{X}\}/S$, $\bar{X}$ and S are mutually independent in this situation (see [8.1.2.1]). Some writers (e.g., Berman, 1962, p. 154) have used the term Studentized in the latter sense.

[8.3.2.1] In this subsection, $\bar{X}$ is the sample mean, whatever the sample size may be, and the order statistics come from a $N(\mu,1)$ parent.

If $n = 2$, then $X_{(2;2)} - \bar{X} = \{X_{(2;2)} - X_{(1;2)}\}/2$, with pdf $g_2(x)$ given by

$$g_2(x) = 2\pi^{-1/2}\exp(-x^2), \quad x > 0,$$

a half-normal distribution; if $n = 3$, then $X_{(3;3)} - \bar{X}$ has pdf $g_3(x)$ given by

$$g_3(x) = 3\left(\tfrac{3}{\pi}\right)^{1/2} \exp\left(-\tfrac{3x^2}{4}\right)\{\Phi\left(\tfrac{3x}{\sqrt{2}}\right) - \tfrac{1}{2}\}, \quad x > 0$$

(McKay, 1935, p. 469). Recursively, if $X_{(n;n)} - \bar{X}$ has pdf $g_n(x)$ and cdf $G_n(x)$, then

$$g_n(x) = \tfrac{n}{\sqrt{2\pi}} \exp\left(-\tfrac{1}{2}\tfrac{n}{n-1}x^2\right) \sqrt{\tfrac{n}{n-1}} \, G_{n-1}\left(\tfrac{nx}{n-1}\right), \quad n = 2, 3, \ldots.$$

See also David (1981, pp. 90-91) and Grubbs (1950, pp. 41-44).

[8.3.2.2] For samples of size n from a $N(\mu, 1)$ parent, let $X_{(n;n)} - \bar{X}$ have cdf $G_n(x)$. Then when x is very small

$$1 - G_n(x) \simeq n\left[1 - \Phi\left(x\sqrt{\tfrac{n}{n-1}}\right)\right]$$

(McKay, 1935, p. 469). This leads to a good approximation to the upper percent points, based on the usual range of probabilities. Thus, if $G_n(x_\alpha) = 1 - \alpha$, tables of $N(0,1)$ probabilities can be used to solve for x_α from the relation

$$\tfrac{\alpha}{n} \simeq 1 - \Phi\left[x_\alpha \sqrt{\tfrac{n}{n-1}}\right].$$

If this approximation is x'_α, then

$$\alpha - \tfrac{1}{2}(n-1)\,\alpha^2/n < 1 - G_n(x'_\alpha) < \alpha$$

(David, 1956a, p. 86). In particular, for all n,

$$0.04875 < 1 - G_n(x'_{0.05}) < 0.05, \qquad 0.00995 < 1 - G_n(x'_{0.01}) < 0.01.$$

A "generally very accurate second approximation" $\tilde{x}_\alpha$ follows from the relation

$$\alpha + \tfrac{1}{2}(n-1)\,\alpha^2/n \simeq n - n\Phi\!\left[\tilde{x}_\alpha\sqrt{\tfrac{n}{n-1}}\right].$$

[8.3.3] Based on a $N(\mu,\sigma^2)$ parent, let the cumulants of $X_{(i;n)} - \overline{X}$ be $\kappa'^{(k)}_{(i)}$ and let those of $X_{(i;n)}$ be $\kappa^{(k)}_{(i)}$; $k = 1, 2, \ldots$. Then (David, 1981, p. 111)

$$\kappa'^{(1)}_{(i)} = E\{X_{(i;n)} - \overline{X}\} = \kappa^{(1)}_{(i)} - \mu,$$

$$\kappa'^{(2)}_{(i)} = \mathrm{Var}\{X_{(i;n)} - \overline{X}\} = \kappa^{(2)}_{(i)} - \sigma^2/n,$$

$$\vdots$$

$$\kappa'^{(k)}_{(i)} = \kappa^{(k)}_{(i;n)},\; k = 3, 4, \ldots .$$

[8.3.4] Grubbs (1950, pp. 31-37) gives a table of the cdf $G_n(x)$ of $(X_{(n;n)} - \overline{X})$ for an $N(\mu,\sigma^2)$ parent, to 5 decimal places, for $x = 0.00(0.05)4.90$ and $n = 2(1)25$. He also tabulates (p. 45) upper percent points of the distribution for $n = 2(1)25$ and $\alpha = 0.10, 0.05, 0.01$, and 0.005, to 3 decimal places; and (p. 46) the mean, standard deviation, skewness and kurtosis for $n = 2(1)15, 20, 60, 100, 200, 500, 1000$, also to 3 decimal places. See Table 8.1 for coverage in some standard sources.

[8.3.5] Let $Y_{(i)} = \{X_{(i;n)} - \overline{X}\}/S$, where $S^2 = \Sigma(X_i - \overline{X})^2/(n-1)$, so that $Y_{(i)}$ is not externally Studentized; the parent population is $N(\mu,\sigma^2)$.

Then $Y_{(i)}$, $\overline{X}$, and S are mutually independent (*[8.1.2.1]*) from which it follows for the moments about zero of $Y_{(i)}$ (see, for example, Quesenberry and David, 1961, p. 381) that

$$E\{Y_{(i)}^r\} = E\{(X_{(i;n)} - \bar{X})^r\} \Big/ E(S^r).$$

[8.3.6] With the notation and conditions in [8.3.5], let $Y = (X_i - \bar{X})/S$, one of the unordered Y terms (see [5.4.3]). Then (David, 1981, pp. 111-112)

$$Y_{(n-1)} \leq \sqrt{\{(n-1)(n-2)/(2n)\}}$$

$$\Pr\{Y_{(n)} > y \geq \sqrt{(n-1)(n-2)/(2n)}\} = n \Pr(Y > y)$$

$$\Pr\{Y_{(1)} < y \leq -\sqrt{(n-1)(n-2)/(2n)}\} = n \Pr(Y < y).$$

[8.3.7] Upper 10, 5, 2.5, and 1 percent points for $Y_{(n)}$ as defined in [8.3.5] are tabulated for $n = 3(1)25$ to 3 decimal places by Grubbs (1950, p. 29). Pearson and Chandra Sekar (1936, p. 318) tabulate some corresponding percent points for $n = 3(1)19$, but for $\sqrt{n/(n-1)}\, Y_{(n)}$. See also Table 8.1.

[8.3.8] Consider the *absolute difference ratios* $|X_1 - \bar{X}|/S, \ldots, |X_n - \bar{X}|/S$, where S is defined in [8.3.5], and let the ordered set of these statistics be

$$0 \leq U_{(1)} \leq U_{(2)} \leq \cdots \leq U_{(n)}.$$

Then (Pearson and Chandra Sekar, 1936, pp. 313-315) the $U_{(i)}$ terms are bounded almost surely as follows: If $n - i$ is even and $i > 1$,

$$U_{(i)} \leq \left\{\frac{n-1}{n-i+1+(i-1)^{-1}}\right\}^{1/2},$$

$$U_{(1)} \leq \left\{\frac{(n-1)^2}{n^2+n}\right\}^{1/2}, \quad n \text{ odd},$$

$$U_{(i)} \leq \sqrt{\frac{n-1}{n-i+1}}, \quad n-i \text{ odd}.$$

Borenius (1958, pp. 152-156) gives formulas for the pdf of $\sqrt{n/(n-1)}\, U_{(n)}$.

[8.3.9] The *Studentized extreme deviate* is $\{X_{(n;n)} - \bar{X}\}/S'$ or $\{\bar{X} - X_{(1;n)}\}/S'$, where S' is defined in *[8.3.1]* from an independent sample; the parent population is $N(\mu, \sigma^2)$ and S' has ν degrees of freedom. An approximation to the upper 100α percent point is given by $(\sqrt{(n-1)/n})t_{\nu;\alpha/n}$ where $t_{\nu;\alpha/n}$ is the upper $100\alpha/n$ percent point of a Student t rv with ν degrees of freedom (David, 1981, p. 111).

Tables of upper percent points of the Studentized extreme deviate are listed in Table 8.1 from standard sources. They also appear in David (1956b, p. 450) to 2 decimal places for $\alpha = 0.10, 0.05, 0.025, 0.01$ and 0.005, n = 3(1)10, 12, and $\nu = 10(1)20, 24, 30, 40, 60, 120, \infty$. Corresponding values appear to 1 decimal place for $\alpha = 0.001$. The use of the preceding approximation leads to errors no greater than 0.07 for the values covered in David's tables. The third edition of *Biometrika Tables for Statisticians*, Vol. 1 (Pearson and Hartley, 1966, Table 26) contains corrections by David (op. cit.) to some earlier approximate values when $\nu > 20$.

Pillai (1959, p. 473) gives upper 5 and 1 percent points to 2 places for n = 2(1)10, 12, and ν = 3(1)10; also to 1 decimal place for ν = 2 and the nearest integer for ν = 1; his values are reproduced in Pearson and Hartley (1966), except for ν = 1(1)4 and for n = 2.

[8.3.10] Halperin et al. (1955, pp. 187-188) give tables to two decimal places of upper and lower bounds for the upper 5 and 1 percent points of the *maximum absolute Studentized deviate*, i.e., of

$$\max\left\{\frac{X_{(n;n)} - \bar{X}}{S'}, \frac{\bar{X} - X_{(1;n)}}{S'}\right\}$$

where S' is defined in *[8.3.1]*. These tables are for n = 3(1)10, 15, 20, 30, 40, 60, and ν = 3(1)10, 15, 20, 30, 40, 60, 120, ∞.

[8.3.11] Quesenberry and David (1961, pp. 379-390) considered statistics $\{X_{(i;n)} - \bar{X}\}/S^*$, where for a $N(\mu, \sigma^2)$ parent,

$$S^{*2} = (n-1)S^2 + \nu S_\nu^2 .$$

S^2 is the sample variance and S_ν^2 is an independent mean-squared estimate of σ^2 based on ν degrees of freedom. Let

Order Statistics From Normal Samples

$$V = \max\left[\{X_{(i;n)} - \overline{X}\}\right]/S^*, \quad V^* = \max\left[|X_{(i;n)} - \overline{X}|/S^*\right], i = 1, 2, \ldots, n.$$

V is a pooling of information in $Y_{(n)}$ and $U_{(n)}$, defined above; but note that S^{*2} is not divided by its degrees of freedom, $n + \nu - 1$.

Then $E(V^r) = E(\{X_{(n;n)} - \overline{X}\}^r/E(S^{*r})$. Tables of percent points of V (Quesenberry and David, 1961, pp. 388-390; Pearson and Hartley, 1966, pp. 187-188) for $\alpha = 0.05$ and $n = 3(1)10, 12, 15, 20$, $\nu = 0(1)10, 12, 15, 20, 24, 30, 40, 50$ are given to 3 decimal places, and for $\alpha = 0.01$ to 4 decimal places. Analogous tables of percent points of V^* are given, but to 3 decimal places; in Quesenberry and David (1961) these appear as bounds, with a condensed version in Pearson and Hartley (1966).

8.4 THE SAMPLE RANGE

[8.4.1.1] The range W of a sample of size n is $X_{(n;n)} - X_{(1;n)}$. For a $N(0, 1)$ parent, the pdf $g(w;n)$ and cdf $G(w;n)$ are given by

$$g(w;n) = n(n-1) \int_{-\infty}^{\infty} [\Phi(x+w) - \Phi(x)]^{n-2} \phi(x)\phi(x+w)dx, \quad w > 0,$$

$$G(w;n) = n \int_{-\infty}^{\infty} [\Phi(x+w) - \Phi(x)]^{n-1} \phi(x)dx$$

$$= n \int_{0}^{\infty} \left[\{\Phi(x+w) - \Phi(x)\}^{n-1} + \{\Phi(w-x) + \Phi(x) - 1\}^{n-1}\right] \phi(x)dx, \quad w > 0$$

(David, 1981, pp. 11-12; Harter, 1969a, pp. 4, 13). Also (Hartley, 1942, p. 342)

$$G(2x;n) = \{2\Phi(x) - 1\}^n + 2n \int_{0}^{\infty} \{\Phi(x+w) - \Phi(x-w)\}^{n-1} \phi(x+w) \, dx.$$

[8.4.1.2] Special cases of the pdf and cdf in *[8.4.1.1]* are given by

$$g(0;2) = 1\sqrt{\pi} \quad ; \quad g(0;n) = 0, \quad n > 2$$

(Harter, 1969a, p. 4). If $w > 0$, then (Lord, 1947, p. 44)

$$g(w;2) = \sqrt{2}\phi(2/\sqrt{2}), \quad G(w;2) = 2\Phi(w/\sqrt{2}) - 1 .$$

$$g(w;3) = 3\sqrt{2}\left[2\Phi(w/\sqrt{6}) - 1\right]\phi(w/\sqrt{2})$$

(McKay and Pearson 1933, p. 417).

$$G(w;3) = 1 - 12T(w/\sqrt{2}, 1/\sqrt{3})$$

(Bland et al. 1966, p. 246), where $T(h,a) = \int_0^a \{\phi(h)\phi(hx)/(1 + x^2)\}dx$, and $T(\cdot,\cdot)$ has been tabulated by Owen (1956, pp. 1080-1087); see also [9.2.2.2].

When n = 3, W has the same distribution as $3\sqrt{3}$ M/2 or $3M_1$, where

$$M = \sum_{i=1}^{3} |X_i - \bar{X}|/3, \quad M_1 = \sum_{i=1}^{3} |X_i - X_{(2;3)}|/3 .$$

When n = 4,

$$G(w;4) = 12\Phi(w/\sqrt{2}) - 1 - 48S(w/\sqrt{2},1/\sqrt{3},\sqrt{2}) - 48S(w/\sqrt{2},1/\sqrt{3},1/\sqrt{2}),$$

where $S(y, a, b) = \int_{-\infty}^{y} T(ax,b)\phi(x)dx$ (Bland et al.), and $S(\cdot,\cdot,\cdot)$ is tabulated by Steck (1958, pp. 790-799). Bland et al. also give expressions for G(w;5) and g(w;5).

[8.4.2.1] The *moments* of the range can be derived from those of $X_{(n)}$ and $X_{(1)}$ discussed above in Section 8.2. However, for some small sample sizes, we give the following results for a N(0, 1) parent:

n = 2: $\quad E(W) = 2/\sqrt{\pi}, \ E(W^2) = 2, \ E(W^3) = 8/\sqrt{\pi}$,

n = 4: $\quad E(W) = 6 \text{ arccos}(-1/3)/\pi^{3/2}$

$$= 6\{\tfrac{1}{2}\pi + \arcsin\tfrac{1}{3}\}/\pi^{3/2},$$

$$E(W^2) = 2 + 6\{1 + (1/\sqrt{3})\}/\pi .$$

Order Statistics From Normal Samples 265

From *[8.2.9]*, Tables 8.2A and 8.2B, we have

$$n = 3: \quad E(W) = 3/\sqrt{\pi}, \quad E(W^2) = 2 + (3\sqrt{3}/\pi),$$

$$n = 5: \quad E(W) = \frac{5}{2\sqrt{\pi}} \left\{ 1 + \frac{6 \arcsin(1/3)}{\pi} \right\},$$

$$E(W^2) = 2 + \frac{5\sqrt{3}}{2\pi} + \frac{5\sqrt{3}}{\pi^2} \arcsin(1.4) + \frac{60}{\pi^2} \arcsin \frac{1}{\sqrt{6}}$$

(Ruben, 1956, p. 460; Sarhan and Greenberg, 1962, p. 185). Generally, when n = 3,

$$E(W^k) = \frac{3}{\pi} 2^{k+1} \Gamma(\tfrac{1}{2}k + 1) \int_0^{\pi/6} \cos^k \theta \, d\theta$$

(McKay and Pearson, 1933, p. 417).

[8.4.2.2] With the notation of *[8.4.2.1]* and of *[8.2.1]*, for a N(0, 1) parent (David, 1981, p. 37)

$$E(W) = 2\mu_{(n;n)}, \quad Var(W) = 2\{\sigma_{(n,n;n)} - \sigma_{(n,1;n)}\}.$$

[8.4.3] Tables giving the cdf, percentage points, and moments of the range are listed from several standard sources in Table 8.1. Barnard (1978, pp. 197-198) gives a Fortran computer program for the cdf of W, from a N(0, 1) parent. Harter (1960, pp. 1125-1131) gives tables of percent points and of moments of W; the coverage is that given in Table 8.1 for Harter's later tables (Harter, 1969a, pp. 372-377).

[8.4.4.1] David (1981, pp. 185-188) discusses three approximations to the distribution of W/σ, based on chi or chi-square. The most accurate is given by

$$W/\sigma \simeq (\chi_\nu^2/c)^a$$

(Cadwell, 1953b, pp. 337-338, 344), where ν, c and α are appropriate constants. Cadwell's Table 1 gives values of ν, $1/\alpha$, and log c for n = 2(1)20.

[8.4.4.2] Cadwell (1954, pp. 803-805) gives an approximation to the cdf of the range W from N(0, 1) samples. This is G(w;n), where

$$G(2y; n) \simeq \{2\Phi(y)-1\}^n + 2na\{2\Phi(y)-1\}^{n-1}$$
$$\times \exp\{-(y^2/2)(1-a^2)\Phi\{1-\Phi(ay)\}\}$$
$$a^{-2} = 1 + (n-1)y\phi(y)/\{\Phi(y)-1/2\}.$$

When n = 20, 60, or 100, the maximum error of this approximation is 0.0031, 0.0040 or 0.0043, respectively. An improved approximation to G(w;n) results if the factor $1 - \Phi(ay)$ is replaced by

$$\{1 - \Phi(ay) - (n-1)a^4 P(y)Q(ay)\},$$
$$P(x) = \frac{x^2}{8}\left\{\frac{\phi(x)}{0.5-\Phi(x)}\right\}^2 + \frac{x^3-3x}{24}\left\{\frac{\phi(x)}{0.5-\Phi(x)}\right\},$$
$$Q(x) = (x^4 + 6x^2 + 3)\{1-\Phi(x)\} - (x^3 + 5x)\phi(x).$$

If n = 20, 60 or 100, the maximum error is only -0.00052, -0.00070, or -0.00075, respectively. Cadwell notes that errors in each of these approximations will initially increase with n (from n = 2) and then fall asymptotically to zero.

[8.4.4.3] Johnson (1952, p. 418) gives approximations to the cdf G(y;n) that are useful when $n \leq 15$ and $y \leq 2$. He also gives approximations to the quantiles y_α, where $G(y_\alpha;n) = 1 - \alpha$, but these hold only when $n \leq 5$ and for the lower quantiles $(1-\alpha \leq 0.025)$. The simplest is given by

$$y_a \simeq \sqrt{2\pi}\,(a/\sqrt{n})^{1/(n-1)}.$$

[8.4.5] As in *[8.3.1]*, let S or S' be an estimate of σ with ν degrees of freedom. Then W/S' is the *Studentized range* if S' is based on a second independent sample of size $\nu + 1$; frequently S is based on the same sample, and W/S is not "Studentized", in the sense that W ans S are not independent. Sometimes W/S is referred to as *internally Studentized*, and W/S' as *externally Studentized*. In both cases, however, the statistic has a distribution free both of μ and of σ^2, for a general $N(\mu, \sigma^2)$ parent.

Order Statistics From Normal Samples 267

[8.4.6] The distribution of W/S was studied by David et al. (1954, pp. 482-493). The rvs W/S, $\bar{X}$ and S are mutually independent (*[8.1.2.2]*). The authors constructed a table of percent points of W/S, extended along with moments by Pearson and Stephens (1964, pp. 484-487). The latter quote exact results when n = 3: if H is the cdf of W/S, then

$$x = 2\cos\{(1 - H(x))\pi/6\}, \quad \sqrt{3} \le x \le 2.$$

[8.4.7] The following bounds hold for samples from any distribution:

$$W/S \ge \begin{cases} \sqrt{2(n-1)/n}, & \text{n even,} \\ \sqrt{2n/(n+1)}, & \text{n odd} \end{cases}$$

(Thomson, 1955, p. 268).

$$\{X_{(n-1;n)} - X_{(1;n)}\}/S \le \sqrt{3(n-1)/2}$$

(David et al., 1954, p. 492).

[8.4.8] The moments of W/S satisfy

$$E\{(W/S)^r\} = E(W^r)/E(S^r),$$

(David et al., 1954, p. 483). These moments may thus be determined from those of W (see *[8.4.2.2]* and S (see *[5.3.5]*). For a $N(\mu, \sigma^2)$ parent (David, 1981, p. 89)

$$E\left[\left(\frac{W}{S}\right)^r\right] = \{\tfrac{1}{2}(n-1)\}^{r/2} \frac{\Gamma[(n-1)/2]}{\Gamma[(n+r-1)/2]} E\left[\left(\frac{W}{\sigma}\right)^r\right].$$

[8.4.9] The cdf of the Studentized range Q, where Q = W/S', as defined in *[8.4.6]*, is

$$G(q;\nu,n) = 2\{(\nu/2)^{\nu/2}/\Gamma(\nu/2)\} \int_0^\infty x^{\nu-1} e^{-\nu x^2/2} G_0(qx;n) dx;$$

$G_0(\cdot; n)$ is the cdf of the range W (see *[8.4.1]*) (Harter, 1969a, p. 18). Pillai (1952, p. 195) gives the pdf of Q in the form of an infinite series.

[8.4.10] The moments of the Studentized range Q satisfy

$$E(Q^r) = E(W^r)E(S'^{-r}) = (\tfrac{1}{2}\nu)^{r/2} \frac{\Gamma[(\nu-r)/2]}{\Gamma(\nu/2)}\; E\!\left[(\tfrac{W}{\sigma})^r\right],\; 0 \leq r < \nu$$

(David et al., 1954, p. 483; David, 1981, p. 88).

[8.4.11] Tables of the cdf and percentage points of Q and of W/S are listed in Table 8.1 for standard sources. See also Pearson and Stephens (1964, pp. 485-486) for tables of percent points and moments of Q and Harter (1960, pp. 1132-1143) for percent points of Q [n = 2(1) 20(2) 40(10) 100; α = 0.10, 0.05, 0.025, 0.01, 0.005, 0.001; ν = 1(1)20, 24, 30, 40, 60, 120, ∞; to 3 decimal places or 4 significant figures.]

[8.4.12] The distribution of the statistic $(\overline{X} - \mu)/W$ is discussed by Daly (1946, p. 71-74). Tables of percent points are listed in Table 8.1.

8.5 QUASI-RANGES

[8.5.1] The *rth quasi-range* of a random sample of size is W'_r, where $W'_r = X_{(n-r;n)} - X_{(r+1;n)}$ and is used to estimate the standard deviation σ. The sample range is W'_0 and the standardized quasi-range is W'_r/σ denoted by W_r.

[8.5.2] The pdf and cdf of W_r for a $N(\mu, \sigma^2)$ parent are given by $g_r(w;n)$ and $G_r(w;n)$, where

$$g_r(w;n) = \frac{n!}{(n-2r-2)!(r!)^2} \int_{-\infty}^{\infty} \left[\Phi(x)\right]^r \left[1 - \Phi(x+w)\right]^r$$
$$\cdot \left[\Phi(x+w) - \Phi(x)\right]^{n-2r-2} \phi(x+w)dx$$

(Cadwell, 1953a, p. 604; Harter, 1969b, p. 12),

$$G_r(w;n) = \int_{-\infty}^{\infty} \sum_{k=0}^{r} \frac{n(n-1)\ldots(n-2r+k)}{r!(r-k)!} \left[1 - \Phi(x+w)\right]^{r-k}$$
$$\cdot \left[\Phi(x+w) - \Phi(x)\right]^{n-2r+k-1} \left[\Phi(x)\right]^r \phi(x)dx$$

(Harter, 1969b, p. 12). Cadwell (op cit., p. 605) gives an asymptotic series for the pdf.

[8.5.3] Moments.

$$E(W_r) = 2E\{X_{(n-r)}\} = 2(r+1)\binom{n}{r+1}\int_{-\infty}^{\infty} x[1-\Phi(x)]^r[\Phi(x)]^{n-r-1}\phi(x)dx,$$

$$E(W_r^2) = \frac{n!}{(n-2r-2)!(r!)^2}\int_{-\infty}^{\infty}\left\{\int_0^{\infty} H(x,y)\,dy\right\}[\Phi(x)]^r\phi(x)dx,$$

$$H(x,y) = y^2\left[1-\Phi(x+y)\right]^r[\Phi(x+y)-\Phi(x)]^{n-2r-2}\phi(x+y)$$

(Harter, 1969b, pp. 3-4).

[8.5.4] Extensive tables of the cdf, percent points and first two moments appearing in Harter (1969b, pp. 136-319) are listed in Table 8.1. Harter (1959, pp. 982-987) tabulates $E(W_r)$ to 6 decimal places and $Var(W_r)$ to 5 places for n = 2(1)100, r = 0(1)8. Cadwell (1953a, p. 610) tabulates the first four moments of the first quasi-range W_1, as well as upper and lower 5, 2.5, 1, and 0.1 percent points for n = 10(1)30, to 2 (percent points) and 3 or 4 (moments) decimal places.

8.6 MEDIAN AND MIDRANGE

[8.6.1] If n = 2m + 1, where m is an integer, the *sample median* is $X_{(m+1;n)}$, i.e., $X_{((n+1)/2;n)}$; if n = 2m, it is $\frac{1}{2}\{X_{(m;n)} + X_{(m+1;n)}\}$, i.e., $\frac{1}{2}\{X_{(n/2;n)} + X_{((n/2)+1;n)}\}$. The *sample midrange* is $\frac{1}{2}\{X_{(n;n)} + X_{(1;n)}\}$.

[8.6.2.1] If n is odd, the distribution of the median $\tilde{X}$ is easier to express. Let the pdf of $\tilde{X}$ for a sample from a N(0,1) parent be g(y;n).

(a) Then

$$g(y; 2m+1) = \frac{(2m+1)!}{m!m!}[\Phi(y)\{1-\Phi(y)\}]^m\phi(y)$$

(Stuart and Ord, 1994, p. 477); g(y: 2m+1) is approximately proportional to

$$\{1 + 2(\pi-3)my^4/(3\pi^2)\}\exp\{-(\tfrac{1}{2})y^2 - (2my^2/\pi)\}$$

(Cadwell, 1952, p. 208).

(b) If $n = 2m$, $g(y; 2m)$ is proportional to

$$\exp(-y^2) \int_0^\infty [\Phi(y-x)\{1-\Phi(y+x)\}]^{m-1} \exp(-x^2) dx$$

(Cadwell, 1952, p. 209).

[8.6.2.2] Let the parent population be $N(\mu, 1)$ and let $n = 2m + 1$. The cdf $H(y;n)$ of the rv $\{X_{(m+1;n)} - \mu\}/\sigma_m$ satisfies the following inequalities, where $\sigma_m^2 = \pi/(2n)$:

$$0.9929(1 + m/8)\sqrt{1 - 1/(2m+2)} \{\phi(y) - \phi(-x)\}$$

$$\leq H(y;n) - H(-x;n)$$

$$\leq (1 + m/8)\sqrt{1 + 1/(2m)} \{\phi(y) - \phi(-x)\}, \quad y > 0, x > 0$$

(Chu, 1955, pp. 114-115). Note that σ_m^2 is the asymptotic variance of the median, so that H is the cdf of an "asymptotically standardized" rv. See [8.7.2.2] below.

[8.6.3] The mean of the sample median is zero for a $N(0, 1)$ parent, as is the third moment. If $n = 2m + 1$,

$$\text{Var}\{X_{(m+1;n)}\} = 1 + \frac{(2m+1)!}{4\pi m! m!} \sum_{j=0}^m (-1)^j \binom{m}{j}(m+j)(m+j-1)$$

$$\cdot \frac{1}{\sqrt{2\pi}} \int_{-\infty}^\infty \{\Phi(x)\}^{m+j-2} \exp(-\tfrac{3}{2}x^2)$$

$$= \frac{\pi}{2(n+2)} + \frac{\pi^2}{4(n+2)(n+4)} + O(n^{-3})$$

when n is large (Stuart and Ord, 1994, pp. 478, 480). If σ_m^2 is defined as in [8.6.2.2], then for $n \geq 7$,

$$\text{Var}\{X_{(m+1;n)}\} = \sigma_m^2 \left[1 - (2 - \tfrac{1}{2}\pi)n^{-1} - (3\pi - 4 - 13\pi^2/24)n^{-2}\right] + O(n^{-4}),$$

(Chu and Hotelling, 1955, p. 601). Further,

Order Statistics From Normal Samples

$$B_m\{1 - 1/(n+1)\}^{3/2} \leq \text{Var}\{X_{(m+1;n)}\}/\sigma_m^2 \leq B_m\{1 + 1/(n-1)\}^{3/2},$$

$$B_m = \frac{(2m+1)!}{m!m!} 2^{-n} \sqrt{\frac{2\pi}{n}}$$

(ibid., p. 602). In the latter inequalities, if $n \geq 4$,

$$1 + \frac{1}{8m} - \frac{7m+3}{24m^2(2m+1)} < B_m < 1 + \frac{1}{8m} + \frac{1}{16(8m-1)}.$$

See also Chu and Hotelling (1955, pp. 601-604), in particular their expressions (49) and (56), and remarks 1 and 2.

[8.6.4] The ratio c_n is of interest, where

$$c_n = \frac{\text{standard deviation of sample median } \tilde{X} \text{ of n values}}{\text{standard deviation of sample mean } \overline{X} \text{ of n values}}.$$

Asymptotically, $c_n \to \sqrt{\pi/2}$ as $n \to \infty$. If $n = 2m + 1$,

$$c_n \simeq \left\{\frac{\pi(2m+1)}{\pi + 4m}\right\}^{1/2} \left\{1 + \frac{4(\pi - 3)m}{(\pi + 4m)^2}\right\} = \sqrt{\frac{\pi}{2}} - \frac{0.2690}{n} - \frac{0.0782}{n^2} + \cdots$$

(Cadwell, 1952, p. 208; Pearson, 1931, p. 363). If $n = 2m$,

$$c_n \simeq \left(\frac{\pi m}{\pi + 2m - 2}\right)^{1/2} \left\{1 - \frac{4 - \pi}{4(\pi + 2m - 2)} + \frac{(\pi - 3)(m - 1)}{(\pi + 2m - 2)^2}\right\}$$

$$= \sqrt{\frac{\pi}{2}} - \frac{0.8941}{n} + 0(n^{-2})$$

(Cadwell, 1952, pp. 209-210). Stuart and Ord (1994, p. 478) give a table of values of c_n for $n = 2(1)10, 15, 20, \infty$ to 3 decimal places, showing that c_n increases from 1.000 to 1.214 as n increases from 2 to 20, and that $c_n \to 1.253$ as $n \to \infty$.

[8.6.5] The fourth moment ratio $\beta_2 = \mu_4/\mu_2^2$ of $\tilde{X}$ when n is large satisfies (Cadwell, 1952, pp. 208, 209)

$$\beta_2 \simeq 3 + \frac{16(\pi - 3)m}{(\pi + 4m)^3}, \quad n = 2m + 1,$$

$$\beta_2 \simeq 3 + \frac{4(\pi - 3)(m - 1)}{(\pi + 2m - 2)^2}, \quad n = 2m.$$

[8.6.6] The cdf $H(y;n)$ of the midrange $(\frac{1}{2})\{X_{(1)} + X_{(n)}\}$ is given by

$$H(y;n) = n \int_{-\infty}^{y} \left[\Phi(2y-x) - \Phi(x)\right]^{n-1} \phi(x) dx$$

(Gumbel, 1958, p. 109). Gumbel refers to $X_{(1)} + X_{(n)}$ as the "midrange". The mean of the midrange of an iid $N(\mu,\sigma^2)$ sample is μ.

[8.6.7] The pdf $h(y;n)$ of the midrange is given by

$$h(y;n) = n(n-1)\pi^{-1}\exp\left(-(ny^2/2)\right) \sum_{i=0}^{\infty} B_i\, y^{2i},$$

where Pillai (1950, pp. 101-102) gives expressions for B_0, B_1, and B_2, and tabulates the first five B coefficients when $3 \leq n \leq 10$. When $n = 2$,

$$h(y;2) = \pi^{-1/2}\exp(-y^2) = \phi(\sqrt{2y}).$$

[8.6.8] The *rth midrange* is $\frac{1}{2}\{X_{(n-r+1;n)} + X_{(r;n)}\}$; $r = 1, 2, ..., \left[\frac{1}{2}n\right]$. Leslie and Culpin (1970, pp. 317-322) have tabulated the standard deviations for $N(0, 1)$ samples; $n = 2(1)21$; also the cdfs, $n = 3(2)15$, 18 and $r = 1, 2$. For $n = 5, 11, 13$, and 15, the cdf is given for $r = 3$, and then for $n = 18$ and $r = 4$. All tables are to 4 decimal places. See also *[8.9.1]*.

When $r = 1$ and $r = \left[\frac{1}{2}n\right]$, this statistic becomes the midrange and median, respectively. As r increases from 1 to $\left[\frac{1}{2}n\right]$, the standard deviation of the rth midrange decreases, and then increases again. Leslie and Culpin's Table 1 (1970, p. 317) indicates that the approximate values of r giving the least variance are given for $n = 3(1)21$ by $r \simeq [n/3]$. Mosteller (1946, pp. 387-389) found that for large n, the value of r that minimizes the variance of the rth midrange is given by $r/n \simeq 0.2702$. Dixon (1957, p. 807) tabulates the efficiencies of these optimum r^{th} midranges with respect to the sample mean $\bar{X}$ for $n = 2(1)20, \infty$, when the data come from a $N(\mu, \sigma^2)$ parent. As n increases from 3 to 20, the efficiency decreases from 0.92 to 0.824, with a limiting efficiency of 0.810 as $n \to \infty$.

The sequence of rth midrange statistics is equivalent to the *quasi-medians* of Hodges and Lehmann (1967, p. 928). These are given by

Order Statistics From Normal Samples 273

$$\tfrac{1}{2}\{X_{(m+1-r;2m+1)} + X_{(m+1+r;2m+1)}\}, \quad \tfrac{1}{2}\{X_{(m-r;2m)} + X_{(m+1+r;2m)}\};$$

$r = 0, 1, 2, \ldots, \left[\tfrac{1}{2}n\right]$. Hodges and Lehmann (1967, pp. 928-931) give an approximation to the pdf of these statistics; for a $N(\mu, \sigma^2)$ parent, the variance is approximately

$$\frac{\pi\sigma^2}{4m}\left[1 - \frac{4r + 6 - \pi}{4m}\right]$$

with $n = 2m$ or $n = 2m + 1$, when r is fixed and n is large; see *[8.7.2.2]*.

8.7 ASYMPTOTIC PROPERTIES

[8.7.1] Throughout this sub-section $Z_{(1;n)}, \ldots, Z_{(n;n)}$ are order statistics from an iid sample of size n from a $N(0,1)$ parent.

Gnedenko (1943, pp. 423-453) proved that

$$\Pr\{|Z_{(n;n)} - \sqrt{2\log n}| < \epsilon\} \to 1$$

as $n \to \infty$, for all $\epsilon > 0$. Galambos (1978, pp. 52, 65-67) shows that

$$\lim_{n\to\infty} \Pr\{\sqrt{2\log n}\,(Z_{(n;n)} - \alpha_n) < x\} = \exp(-e^{-x}), \quad |x| < \infty,$$

$$\lim_{n\to\infty} \Pr\{\sqrt{2\log n}\,(Z_{(1;n)} - \alpha_n) < x\} = \exp(-e^{-x}), \quad |x| < \infty,$$

where

$$\alpha_n = \sqrt{2\log n} - \tfrac{1}{2}(\log\log n + \log(4\pi))/\sqrt{2\log n}.$$

For the same values of α_n and for fixed $r \geq 1$,

$$\lim_{n\to\infty} \Pr\{\sqrt{2\log n}\,(Z_{(n-r+1;n)} - \alpha_n) < x\} = \exp(-e^{-x})\sum_{t=0}^{r-1}\frac{e^{-tx}}{t!},$$

$$\lim_{n\to\infty} \Pr\{\sqrt{2\log n}\,(Z_{(r;n)} - \alpha_n) < x\} = 1 - \exp(-e^{-x})\sum_{t=0}^{r-1}\frac{e^{-tx}}{t!}, \quad |x| < \infty$$

(Galambos, 1978, p. 105). The choice of α_n goes back to Cramer (1946, pp. 374-378). More generally, suppose that one looks for norming constants a_n and b_n such that

$$\lim_{n\to\infty} \Pr\left[\{X_{(n;n)} - \alpha_n\}/b_n < x\right] = \exp(-e^{-x}).$$

Hall (1979, pp. 433-439) showed that, if a_n is the solution of the equation

$$2\pi a_n^2 \exp(a_n^2) = n^2$$

and if $b_n = 1/a_n$, then

$$\sup_{|x|<\infty} \left|\{\Phi(b_n x + a_n)\}^n - \exp(-e^{-x})\right| < 3/\log n.$$

This rate of convergence cannot be improved by choosing a different sequence (a_n, b_n). Suitable norming constants (α_n, β_n) are provided by α_n as defined above, and $\beta_n = 1/\alpha_n$, since as $n \to \infty$, $\beta_n/b_n \to 1$ and $(\alpha_n - a_n)/b_n \to 0$.

[8.7.2.1] The asymptotic distribution of $Z_{(n-r+1;n)}$, suitably normed, has pdf $g(y;r,n)$, where

$$g(y;r,n) = r^r\{(r-1)!\}^{-1} \exp(-ry - re^{-y}), \quad -\infty < y < \infty.$$

If $Z_{(n-r+1;n)}$ has a mode at m, then $g(y; r, n)$ is the asymptotic distribution of

$$(r/n)\phi(m)\{Z_{(n-r+1;n)} - m\}$$

(Stuart and Ord, 1994, pp. 488-491, 500; Woodroofe, 1975, p. 259). As for the case in which $m = 1$, convergence to the limiting form is slow.

Let α_n be defined as in *[8.7.1]* and S^2 as in *[8.3.5]* be the sample variance. Then the normed extreme deviate for a sample from a $N(\mu, \sigma^2)$ parent, i.e.,

$$\sqrt{2 \log n} \left\{\frac{X_{(n;n)} - \overline{X}}{S} - \alpha_n\right\},$$

Order Statistics From Normal Samples

has the limiting distribution as n → ∞ with cdf $\exp(-e^{-x})$, $-\infty < x < \infty$ (David, 1981, p. 270).

Fisher and Tippett (1928, p. 183) give a "penultimate form" of limiting distribution for a suitably normed linear function of $X_{(n;n)}$; this has pdf of the form $k(-x)^k \exp\{-(-x)^k\}$, for suitable choice of k. This is Gumbel's "third asymptote" (Gumbel, 1958, Chaps. 5, 7).

[8.7.2.2] For a sample of size n = 2m + 1 from a $N(\mu, \sigma^2)$ parent, the distribution of the *sample median* $\tilde{X}$ (see *[8.3.2]*) is asymptotically normal with mean μ and variance $\pi\sigma^2/(2n)$ (Chu, 1955, p. 114).

Hodges and Lehmann (1967, pp. 927-928) have given a large-sample approximation to the variance of the sample median. Whether n = 2m or n = 2m + 1, this is given for a $N(\mu, \sigma^2)$ parent by

$$\text{Var}(\tilde{X}) \simeq \frac{\pi\sigma^2}{4m}\left[1 - \frac{(6-\pi)\pi\sigma^2}{2m}\right].$$

Insofar as this formula is accurate, it pays to base the median on an even number (n = 2m) of observations, rather than on the next highest odd number (2m + 1). Dixon (1957, p. 807) tabulates the efficiency of $\tilde{X}$ with respect to the sample mean $\overline{X}$, and the exact variance of $\tilde{X}$, for n = 2(1)20, ∞.

[8.7.3.1] The *midrange* of a sample of size n from a $N(\mu, \sigma^2)$ parent converges in probability to μ as n → ∞; $\pi^2\sigma^2/(24\log n)$ is the asymptotic variance (Stuart and Ord, 1994, p. 498); the latter tends to zero more slowly than σ^2/n, the variance of $\overline{X}$. The midrange becomes increasingly sensitive to outlying values in the tails of the parent distribution, as the sample size becomes larger, at least when the parent distribution is not restricted to a finite interval. Thus, for a $N(\mu, \sigma^2)$ parent, let

$$d_n = \frac{\text{standard deviation of sample midrange of n values}}{\text{standard deviation of sample mean of n values}}.$$

Then, as n increases from two to 20, d_n increases from 1.000 to 1.691, and $d_n \to \infty$ as n → ∞ (Stuart and Ord, 1994, p. 478). Dixon (1957, p. 807) tabulates the exact variance of the midrange and its efficiency with respect to the sample mean $\overline{X}$, for n = 2(1) 20, ∞.

[8.7.3.2] The asymptotic distribution of Gumbel's "reduced" midrange for a N(0,1) parent has the logistic form $\exp(-y)\{1 + \exp(-y)\}^{-2}$ for its pdf and

$$\lim_{n\to\infty} \Pr\left[\left(\sqrt{2\log n}\right)\{X_{(1;n)} + X_{(n;n)}\} < x\right] = (1 - e^{-x})^{-1}$$

(Galambos, 1978, p. 109; Gumbel, 1958, p. 311). This result is an exception to the more common phenomenon that statistical measures of location, at least for symmetric populations, have an asymptotic normal distribution.

[8.7.4.1] An approximation to the pdf of the *range* W for a $N(\mu, 1)$ parent is the asymptotic distribution given by Cadwell (1953a, p. 607); this is g(y;n), where

$$g(2y;n) \simeq \frac{n(n-1)\sqrt{\pi}\{\phi(y)\}^2\{2\Phi(y)-1\}^{n-3/2}}{\{2\Phi(y)-1-(n-2)\phi'(y)\}^{1/2}}.$$

Accuracy, as measured by the lower moments, shows this to improve upon earlier approximations; a further improvement is indicated. See also David (1981, pp. 267-269).

[8.7.4.2] For the constants α_n defined in *[8.7.1]*, i.e.,

$$\alpha_n = \sqrt{2\log n} - \tfrac{1}{2}(\log\log n + \log(4\pi))/\sqrt{2\log n},$$

and with a $N(\mu, 1)$ parent, the limiting distribution of W is given by

$$\lim_{n\to\infty} \Pr\left[\sqrt{2\log n}\,\{W - 2\alpha_n\} < x\right] = \int_{-\infty}^{\infty} e^{-y}\exp(-e^{-y} - e^{y-x})dy.$$

The range W and the extreme value $X_{(n;n)}$ are asymptotically independent (Galambos, 1978, p. 109).

[8.7.5] Let $\overline{X}$ and $\tilde{X}$ be the sample mean and median, respectively, from a sample of size n from a $N(\mu, \sigma^2)$ population. Then $(\overline{X}, \tilde{X})$ have a joint bivariate normal distribution asymptotically, as $n \to \infty$, where

$$E(\overline{X}) = \mu, \quad E(\tilde{X}) = \mu, \quad \text{Cov}(\overline{X}, \tilde{X}) = \sigma^2/n,$$
$$\text{Var}(\overline{X}) = \sigma^2/n, \quad \text{Var}(\tilde{X}) \simeq \pi^2\sigma^2/(2n)$$

(Wilks, 1962, p. 275).

[8.7.6] *Independence of functions of order statistics.* The order statistics are not mutually independent. However, for fixed values of r and s, and as $n \to \infty$ (David, 1981, pp. 267, 269-270),

(a) $X_{(r;n)}$ and $X_{(n-s+1;n)}$ are asymptotically independent,
(b) both are asymptotically independent of the central order statistics,
(c) both are asymptotically independent of the sample mean $\overline{X}$.

Further, the order statistics of a random sample from any population have a Markov property; for a $N(\mu, \sigma^2)$ parent with cdf $F(x)$, this is given for $1 \le r \le n-1$ by

$$\Pr\{X_{(r+1;n)} \le x | X_{(r;n)} = y\} = 1 - \{1-F(x)\}^{n-r}\{1-F(y)\}^{-(n-r)}, \; x > y$$

(Pyke, 1965, p. 399). See *[8.9.1]* for certain functions of order statistics that are uncorrelated.

[8.7.6.1] Haldane and Jayakar (1963, pp. 89-94) exhibit a linear combination of $X^2_{(n;n)}$ and of $X^2_{(1;n)}$ that converges in distribution to the extreme-value form with cdf $\exp(-e^{-y})$, $|y| < \infty$, and also exhibit a limiting cdf for $X^2_{(m;n)}$, $2 \le m \le n-1$. See also David (1981, pp. 265-266).

[8.7.7] Hodges and Lehmann (1967, pp. 926-927) have tabulated the exact and approximate efficiency $e(n)$ of $\tilde{X}$ for $n = 1(1)20, \infty$, where $e(n) = \text{Var}(\overline{X})/\text{Var}(\tilde{X})$, for samples of size n from a $N(\mu, \sigma^2)$ parent, $\overline{X}$ being the sample mean. For large n, $e(n) = 2/\pi + a/n$, approximately, where a $= 4/\pi - 1$ if n is odd, and a $= 6/\pi - 1$ if n is even. The efficiency of $\tilde{X}$ decreases from 1.0 to 0.637 as n increases from one to ∞, but does so through persistently higher values when n is odd. See also *[8.6.8]*.

[8.7.7.1] The following is a central limit theorem for order statistics. Let $X_1, X_2, \ldots$ be an iid sequence of rvs with an absolutely continuous cdf

$F(x)$ and a density function $f(x)$ continuous and positive if $a \leq x < b$. Let $X_{(1;n)} \leq X_{(2;n)} \leq \ldots \leq X_{(n;n)}$ be the order statistics based on X_1, $X_2, \ldots, X_n$ and q a number such that $0 < F(a) < q < F(b)$. If $\{k_n\}$ is a sequence of integers such that

$$\lim_{n \to \infty}\left\{\sqrt{n}|(k_n/n) - q|\right\} = 0 ,$$

and if $q = F(Q)$, $D = \sqrt{q(1-q)}/f(Q)$, then (Reńyi, 1970, p. 490)

$$\lim_{n \to \infty}\left\{\frac{\sqrt{n}(X_{(k_n;n)} - Q)}{D} < x\right\} = \Phi(x) .$$

[8.7.7.2] Bjerve (1977, p. 365) gives a Berry-Esseen bound $Kn^{-1/2}$ for $|\bar{H}_n(x) - \Phi(x)|$, where $\bar{H}_n(\cdot)$ is the cdf of trimmed linear functions $\sum_{i=1}^{n} c_{in} X_{(i;n)}$ of order statistic. This allows fairly general weights on those observations which are not trimmed, but requires a smoothness condition for the cdf. See also Helmers (1977, p. 941).

8.8 QUANTILES

[8.8.1] If Z is an $N(0,1)$ random variable, then with the notation used throughout this book, the p-quantile of Z is z_{1-p}, where

$$\Pr(Z \leq z_{1-p}) = \Phi(z_{1-p}) = p.$$

In a sample of size n, and if np is not an integer, the order statistic $X_{([np]+1;n)}$ is the unique sample p-quantile, where [np] is the greatest integer less than or equal to p.

If np is an integer, then any quantity in the interval between (and including) $X_{([np];n)}$ and $X_{([np+1];n)}$, can be used to define the sample p-quantile. In what follows we assume that np is not an integer.

[8.8.2] Let $Y = X_{([np+1];n)}$, where the sample has a $N(\mu, \sigma^2)$ parent having cdf $F(\cdot, \mu, \sigma^2)$ and pdf $f(\cdot, \mu, \sigma^2)$. The pdf of Y is $g(y, \mu, \sigma^2)$, where (Cramér, 1946, p. 368)

$$g(y, \mu, \sigma^2) = \binom{n}{[np]} (n - [np])\{F(y,\mu,\sigma^2)\}^{[np]}$$
$$\cdot \{1 - F(y,\mu,\sigma^2)\}^{n-[np]-1} f(y,\mu,\sigma^2), \quad -\infty < y < \infty$$

[8.8.3] The asymptotic distribution of Y as $n \to \infty$ is normal, with mean $(z_{1-p} - \mu)/\sigma$ and variance $[\sigma^2/\{\phi(z_{1-p})\}^2]p(1-p)/n$ (Cramér, 1946, p. 369; Wilks, 1962, p. 273; Cadwell, 1952, pp. 210-211).

[8.8.4] Mosteller (1946, pp. 383-384) gives the asymptotic joint distribution of the sample quantiles corresponding to $p_1, p_2, ..., p_k, 0 < p_1 < p_2 < ... < p_k < 1$. For a $N(\mu, \sigma^2)$ parent, this asymptotic joint distribution (as $n \to \infty$) is multivariate normal, with mean vector $(z_{1-p_1}, z_{1-p_2}, ..., z_{1-p_k})$ and

$$\text{Cov}\left(X_{([np_i]+1;n)}, X_{([np_j]+1;n)}\right) \simeq \frac{p_i(1-p_j)\sigma^2}{n\phi(z_{1-p_i})\phi(z_{1-p_j})}, \quad 1 \leq i \leq j \leq k.$$

See also Cramér (1946, pp. 369-370).

[8.8.5] The median is the 0.5 quartile, and is discussed in *[8.6]*. The semi-interquartile range in a $N(0, 1)$ distribution is

$$(\tfrac{1}{2})\{Z_{([3n/4]+1;n)} - Z_{([n/4]+1;n)}\},$$

and for a $N(\mu, \sigma^2)$ parent has an asymptotic normal distribution as $n \to \infty$ with mean $\tfrac{1}{2}\sigma(z_{1/4} - z_{3/4})$, or $(0.6744898)\sigma$, and asymptotic variance (Cramér, 1946, p. 370)

$$\frac{\sigma^2}{16n} \frac{1}{\{\phi(z_{1/4})\}^2} = \left\{(0.786716) \frac{\sigma}{\sqrt{n}}\right\}^2.$$

[8.8.6] Let $X_{(1;n)}, X_{(2;n)}, ..., X_{(n;n)}$ be order statistics in a combined sample from one or more unknown parent distributions, replaced by rank statistics as in *[8.2.7]*. As an alternative to normal scores, replace $X_{(r;n)}$ by $\Phi^{-1}[r/(n+1)]$; in the notation used throughout this book, if $p(r) = r/(n+1)$, then

$$\Phi(z_{1-p(r)}) = p(r), \quad \Phi^{-1}(p(r)) = z_{1-p(r)}.$$

The *Van der Waerden statistic* is the sum of the scores $\Phi^{-1}(p(r))$ over those values r for measurements belonging to the second of two populations involved in a combined sample. See Lehmann (1975, p. 97) for further discussion and references.

8.9 MISCELLANEOUS RESULTS

[8.9.1] Certain statistics discussed in this chapter and in Chapter 5 are uncorrelated. The sample mean, sample median, sample midrange, and sample rth midranges are all *odd location statistics* $T(X_1,...,X_n)$ such that, for all h

$$T(x_1 + h, ..., x_n + h) = T(x_1, ..., x_n) + h$$

$$T(-x_1, ..., -x_n) = -T(x_1, ..., x_n).$$

The sample variance, sample range, quasi-ranges, sample mean deviation from the mean or median, and the sample interquartile range are all *even location-free statistics* $T(X_1,...,X_n)$ such that, for all h

$$T(x_1 + h, ..., x_n + h) = T(x_1, ..., x_n)$$

$$T(-x_1,...,-x_n) = T(x_1,...,x_n).$$

Hogg (1960, pp. 265-267) shows that for $N(\mu, \sigma^2)$ samples, or for any symmetric distribution, the correlation coefficient, if any, of an odd location statistic and an even location-free statistic is zero.

One other even location-free statistic is Gini's mean difference, discussed next.

[8.9.2.1] *Gini's mean difference* G is defined by

$$G = \{n(n-1)\}^{-1} \sum_{i=1}^{n} \sum_{j=1}^{n} |X_i - X_j|$$

(David, 1981, p. 192). Computationally, however, one of two other forms, functions of the order statistics, may be more suitable:

$$G = \{n(n-1)\}^{-1} \sum_{i=1}^{[n/2]} (n-2i+1)W'_r$$

$$= 4\{n(n-1)\}^{-1} \sum_{i=1}^{n} \{i - \tfrac{1}{2}(n+1)\}X_{(i;n)},$$

where W'_r, the rth quasi-range, is defined in *[8.5.1]*. See David (1981), pp. 191, 192, 216. For a sample of size n from a $N(\mu, \sigma^2)$ parent,

$$E(G) = 2\sigma/\sqrt{\pi} = 4\int_{-\infty}^{\infty} x\{F(x;\mu,\sigma^2) - \tfrac{1}{2}\}f(x;\mu,\sigma^2)dx,$$

$$\text{Var}(G) = 4\sigma^2\{\pi n(n-1)\}^{-1}\{2(n-2)\sqrt{3} - 2(2n-3) + (n+1)\pi/3\}$$

$$\simeq (0.8068)^2 \sigma^2/n$$

when n is large (David, 1981, p. 216). Further,

$$E(G) = E|X_i - X_j|, \; i \neq j$$

in the unordered sample, and $E(\sqrt{\pi}G/2) = \sigma$.

[8.9.2.2] The limiting distribution of $\{G - E(G)\}/\sqrt{\text{Var}(G)}$ as $n \to \infty$ is standard normal for independent and identically distributed samples, whenever the parent distribution has a finite second moment (Stigler, 1974, pp. 683, 690-691).

[8.9.3.1] For a random sample of size n from a normal population with mean μ and variance σ^2, and for fixed k, $\{nF(X_{(k;n)};\mu, \sigma^2)\}$ is a sequence of rvs converging in distribution as $n \to \infty$ ($n \geq k$) to the gamma distribution with cdf (Wilks, 1962, p. 269)

$$G(y;k) = \int_0^y (1/\Gamma(k))x^{k-1}e^{-x}dx, \; y > 0.$$

[8.9.3.2] With the conditions of *[8.9.3.1]*, let np^* be an integer such that $p^* = p + O(1/n)$, $0 < p < 1$. Then, for large n, $F(X_{(np^*;n)};\mu, \sigma^2)$ is asymptotically distributed as a normal rv with mean p and variance $p(1-p)/n$ (Wilks, 1962, p. 271).

[8.9.4] Let π_n be the proportion of observations that are greater than the sample mean $\bar{X}$ in a normally distributed sample of size n. Then $\sqrt{n}(\pi_n - 1/2)$ has an asymptotic $N(0, (1/4) - 1/(2\pi))$ distribution (David, 1962, p. 1161). Let

$$P(n;k) = \Pr\{X_{(k;n)} \leq \bar{X} \leq X_{(k+1;n)}\} \; .$$

Kendall (1954, pp. 560-564) gives an Edgeworth series as an approximation to $P(n;k)$; David (1963, pp. 49-54) gives some exact expressions and bounds for n = 4, 5, and arbitrary n. Specifically, if

$$C(n) = n\sqrt{n-1} \; \Gamma\{(n-1)/2\} \big/ \{2(n-2)! \pi(n-1)/2\} \; ,$$

$$C(n)(1/2)^{(n-2)/2} \leq P(n,1) \leq C(n)(1 - n^{-1})^{(n-2)/2} \; .$$

The exact values of P(4,1), P(5,1) and P(6,1) are 0.175, 0.049, and 0.011, respectively.

The asymptotic value of $P(n,k)$ as $n \to \infty$ and for fixed values of k is

$$(k^{n-k-1} e^{n-2k})^{1/2} \big/ \{2^{n-k+1}(k!)^2 n^{n-3k-1} \pi^{n-k}\}^{1/2}$$

(David, 1963, p. 53).

[8.9.5.1] Let $X_1, X_2, ..., X_n$ be a random sample from a $N(\mu, \sigma^2)$ distribution. If a proportion α of the ordered sample is "trimmed" off the lower end, and a proportion $1 - \beta$ is "trimmed" off the upper end, then the average of the remaining observations is the trimmed mean, T_n, say, where, if $0 < \alpha < \beta < 1$, and [x] is the largest integer less than or equal to x,

$$T_n = \{[\beta n] - [\alpha n]\}^{-1} \sum_{i=[\alpha n]+1}^{[\beta n]} X_{(i;n)} \; .$$

For given α, β, and n, the mean and variance of T_n can be computed from results given in [8.2.3], [8.2.4], and [8.2.6]; the trimmed mean is less sensitive to outliers in the sample than is the sample mean $\bar{X}$. Stigler (1973, p. 473) gives the asymptotic distribution of T_n for a $N(\mu, \sigma^2)$ parent, as follows:

Let μ_t and σ_t^2 be the mean and variance, respectively, of the doubly truncated $N(\mu, \sigma^2)$ distribution, truncated below at a and above at b, where

$$\Pr(X_i \leq a) = \alpha, \quad \Pr(X_i \geq b) = 1 - \beta .$$

Expressions for μ_t and σ_t^2 appears in [2.4.5]; note that, if for a $N(0, 1)$ random variable Z, $\Pr(Z \leq z_{1-\alpha}) = \Phi(z_{1-\alpha}) = \alpha$ as in [2.1.2], then

$$a = \mu + z_{1-\alpha}\sigma, \quad b = \mu + z_{1-\beta}\sigma .$$

The asymptotic distribution of $\sqrt{n}(T_n - \mu_t)$ as $n \to \infty$ is normal with mean zero and variance σ^{*2}, where

$$\sigma^{*2} = (\beta - \alpha)^{-2}\{(\beta - \alpha)\sigma_t^2 + \beta(1 - \beta)(b - \mu_t)^2$$

$$+ \alpha(1 - \alpha)(\mu_t - a)^2 + 2a(1 - \beta)(b - \mu_t)(\mu_t - a)\} .$$

If $\alpha = 1 - \beta$, so that the trimming is symmetrical, then $\mu_t = \mu$ and

$$\sigma^{*2} = \{1 - 2\alpha + 2z_{1-a}\phi(z_{1-a}) + 2\alpha z_{1-a}^2\}\sigma^2 \big/ (1 - 2a)^2, \; 0 < a < 1/2.$$

[8.9.5.2] We now consider the asymptotic distribution of the trimmed mean in a random sample from a general distribution. Let $G(x)$ be the cdf of the underlying population. If G is continuous and strictly increasing in $\{x: 0 < G(x) < 1\}$, then the asymptotic distribution of T_n is normal, with mean and variance as given in [8.9.5.1], μ_t and σ_t^2 being the mean and variance of the distribution $G(\cdot)$ truncated below and above at a and b, respectively.

For other cases, $G(\cdot)$ is truncated to include $a \leq x < b$ only where $a = \sup\{x: G(x) \geq \alpha\}, b = \inf\{x: G(x) \geq \beta\}$; μ_t and σ_t^2 are the mean and variance of the truncated distribution. Let

$$A = a - \inf\{x: G(x) \geq \alpha\}, \quad B = \sup\{x: G(x) \leq \beta\} - b .$$

Then the limiting distribution of $\sqrt{n}(T_n - \mu_t)$ is that of a rv Z, where

$$Z = (\beta - \alpha)^{-1}\{Y_1 + (b - \mu_t)Y_2 + (a - \mu_t)Y_3 + B\max(0, Y_2) - A\max(0, Y_3)\},$$

$$E(Z) = \left[B\sqrt{\beta(1-\beta)} - A\sqrt{\alpha(1-\alpha)}\right]/\{\sqrt{2\pi}(\beta - \alpha)\},$$

where $Y_1 \sim N(0, (\beta - \alpha)\sigma_t^2)$, (Y_2, Y_3) and Y_1 are independent, and (Y_2, Y_3) is bivariate normal with mean vector zero and variance-covariance matrix

$$\begin{pmatrix} \beta(1-\beta) & -\alpha(1-\beta) \\ -\alpha(1-\beta) & \alpha(1-\alpha) \end{pmatrix}$$

(Stigler, 1973, p. 473).

Notice that, for a non-normal limiting distribution to occur, trimming must take place when α or β corresponds to a non-unique percentile of the underlying distribution $G(\cdot)$. Stigler (1973, p. 477) suggests a more smoothly trimmed mean which is asymptotically normal for any distribution $G(\cdot)$.

[8.9.6.1] *The Ratio of Two Ranges.* Let W_1 and W_2 be the sample ranges from two independent random samples of sizes n_1 and n_2, respectively, from normal populations having the same variance, σ^2. Consider the ratio $R = W_1/W_2$; the statistic R is sometimes used in place of the usual F statistic. The pdf and cdf of R are, respectively,

$$h(r; n_1, n_2) = \int_0^\infty x g_2(x; n_2) g_1(xr; n_1) dx,$$

$$H(r; n_1, n_2) = \int_0^r h(x; n_1, n_2) dx = 1 - H(1/r; n_2, n_1),$$

where g_i is the pdf of R_i/σ (i = 1,2) (Harter, 1969a, pp. 5-6).

[8.9.6.2] Some particular cases of $H(r; n_1, n_2)$ are given by $H(r; 2, 2) = (2/\pi)\arctan(r)$, $H(r; 2, 3) = (6/\pi)\arctan\{r/(4+3r^2)^{1/2}\}$, $H(r; 3, 2) = (6/\pi)\arctan\{(3+4r^2)^{1/2}\} - 2$ (Link, 1950, p. 113).

[8.9.6.3] Sources for tables of the pdf, cdf and percent points of R are listed in Table 8.1. Also listed are sources for tables of percent points of

Order Statistics From Normal Samples

$|\bar{X}_1 - \bar{X}_2|/(W_1 + W_2)$, where $\bar{X}_1$ and $\bar{X}_2$ are the sample means of the two samples.

[8.9.7.1] *Mean Range.* The mean range is used as an estimator of standard deviation in a one-way layout and the associated analysis of variance (David, 1981, pp. 205-211). Suppose m samples, all of size n, are taken from the same normal distribution with $\sigma = 1$. Then the *mean range* $\bar{W}_{n;m}$ is the mean of the m sample ranges. Let g(y;n,m) be the pdf of $\bar{W}_{n,m}$, and G(y;n,m) the cdf. If $T(h,\lambda)$ is the function related to bivariate normal probabilities and tabulated by Owen (1956, pp. 1080-1087) (see *[9.2.2.2]*), then

$$G(y;2,2) = \Pr(\bar{W}_{2,2} \leq y) = [2\Phi(y) - 1]^2, \quad y > 0,$$

$$g(y;2,3) = 4\sqrt{6}\phi(y\sqrt{3/2})\{1 - 6T(y\sqrt{3}/2, \sqrt{3})\}, \quad y > 0$$

(Bland et al., 1966, pp. 246-247). $\bar{W}_{3/2}$ has mean $3/\sqrt{\pi}$ and variance $1 - 3(3 - \sqrt{3})/(2\pi)$; Bland et al. give expressions for g(y;3,2) and g(y;2,4). For a discussion of some approximations to the distribution of $\bar{W}_{n,m}$ by chi-square, see David (1981, pp. 185-188) and Cadwell (1953b, pp. 336-346).

[8.9.7.2] Bliss et al. (1956, p. 420) have tabulated upper 5 percent points of the distribution of $T = W_{max}/\Sigma_{j=1}^{k} W_j$, where W_{max} is the largest among k mutually independent sample ranges $W_1, ..., W_k$, each based upon a random sample of size n from a common $N(\mu, \sigma^2)$ parent. The statistic T has been suggested to test for the presence of outliers. The percent points are based upon the use of two approximations (Bliss et al., 1956, p. 421). See also Table 8.1 for other sources of tables.

[8.9.8.1] Let $S_1^2, ..., S_k^2$ be sample variances of k mutually independent random samples, each of size n, based on a common $N(\mu, \sigma^2)$ distribution, and let

$$U = S_{max}^2 \bigg/ \sum_{j=1}^{k} S_j^2,$$

where S_{max}^2 is the largest of $S_1^2, ..., S_k^2$. Cochran (1941, p. 50) has tabulated the upper 5 percent points of U for $k = 3(1)10$ and $n = 2(1)11$. When n = 3, and each S_j^2 has two degrees of freedom,

$$\Pr(U > y) = k(1-y)^{k-1} - \binom{k}{2}(1-2y)^{k-1} + \ldots$$

$$+ (-1)^{h-1}\binom{k}{h}(1-hy)^{k-1},$$

where h is the greatest integer less than 1/y (ibid., p. 47).

If the number k of samples is large, then approximately,

$$\Pr(U \leq y) \simeq \left(1 - e^{-k(n-1)y/2}\right)^k,$$

and U has an approximate pdf

$$g(u;k,n) \simeq (\tfrac{1}{2})k^2(n-1)e^{-k(n-1)y/2} \cdot \left(1 - e^{-k(n-1)y/2}\right)^{k-1}, \; y > 0,$$

with (approximate) mean value $2(\gamma + \log k)/\{(n-1)/k\}$ (ibid., 1941, p. 51); γ is Euler's constant. See also Table 8.1 for other sources with tables.

[8.9.8.2] Hartley (1950, p. 308) introduced the ratio $F_{max} = S_{max}^2/S_{min}^2$ to test for heterogeneity of variance, where S_{max}^2 is defined in [8.9.8.1] above, and S_{min}^2 is the smallest of $S_1^2, \ldots, S_k^2$. David (1952, p. 424) gives a table of upper 5 and 1 percent points of the distribution of F_{max}, for the degrees of freedom $n - 1 = 2(1)10, 12, 15, 20, 30, 60, \infty$, and $k = 2(1)12$. See Table 8.1 for other sources of tables.

Let $g(y; \nu)$ and $G(y; \nu)$ be the pdf and cdf, respectively, of a chi-square rv with ν degrees of freedom. Then (David, 1952, p. 423)

$$\Pr(F_{max} \leq y) = k \int_0^\infty g(x;n-1)[G(xy;n-1) - G(x;n-1)]^{k-1} dx.$$

[8.9.9] If $X_1, X_2,\ldots$ are equicorrelated rvs with $EX_i = 0$, $EX_i^2 = 1$, $EX_iX_{i+N} = \rho$; $i = 1,2,\ldots$; $N = 1,2,\ldots$, and if $X_{(n,n)} = \max(X_1,\ldots,X_n)$, then the limiting distribution of $X_{(n,n)} - \sqrt{2(1-\rho)\log n}$, when $\rho \geq 0$ is $N(0,\rho)$ (David, 1981, p. 273).

[8.9.10] Let $X_1, \ldots, X_n$ be an iid sample from a distribution with cdf G, and let

$$S_n = \sum_{i=1}^{n} c_{in} X_{(i;n)},$$

a linear function of the order statistics, with "weights" c_{in}. Considerable research, particularly in the mid-1960s to mid-1970s, addressed the question of what conditions lead to the asymptotic normality of S_n. For a good readable discussion and some references, see Stigler (1969, pp. 770-772; 1974, pp. 676-677, 687-688, 690-692). There is generally a trade-off between the amount of weight to be permitted to the extreme order statistics and smoothness restrictions on $G(\cdot)$, particularly in the tails, and on the density of G in the support of G.

[8.9.11] Robison-Cox (1992, pp. 3497-3520) has considered the following linear trend order statistics. Let $Z_i \sim N(0,1)$, iid for $i = 1, \ldots, n$. Let

$$X_i(\tau) = Z_i + (i - \tfrac{1}{2}(n+1))\tau; \quad i = 1, \ldots, n.$$

Denote the ordered $X_i(\tau)$ by $X_{(1)}(\tau) \leq X_{(2)}(\tau) \leq \ldots \leq X_{(n)}(\tau)$. Then $X_{(r)}(\tau)$ and $X_{(r)}(-\tau)$ have the same distribution, while $X_{(r)}(\tau)$ and $-X_{(n-r+1)}(\tau)$ also have the same distribution. The author tabulates to four decimal places the means and variances of these order statistics $X_{(r)}(\tau)$ for $n \leq 10$, and $\tau = 0(0.1)0.6, 0.8, 1.0(0.25)2.0, 2.5, 3.0$. The covariances of $X_r(\tau)$ and $X_s(\tau)$ are tabulated similarly for $n \geq 10$.

REFERENCES

Arnold, B. C. and Balakrishnan, N. (1989). *Relations, Bounds and Approximations for Order Statistics*, New York, Springer-Verlag. *[8.1.2.2, 8.2.2.2]*

Balakrishnan, N. (1984). Approximating the sum of squares of normal scores, *Applied Statistics* 33, 242-245. *[8.2.7]*

Balakrishnan, N. and Malik, H. J. (1988). Recurrence relations and identities for moments of order statistics, II: Specific continuous distributions, *Communications in Statistics - A17*, 2657-2694. *[8.2.2.2]*

Barnard, J. (1978). Algorithm AS126: Probability integral of the normal range, *Applied Statistics* 27, 197-198. *[8.4.3]*

Barnett, V. and Lewis, T. (1978). *Outliers in Statistical Data*, New York: John Wiley. *[8.2.2.3]*

Berman, S. (1962). Limiting distribution of the studentized largest observation, *Skandinavisk Aktuarietidskrift 45*, 154-161. *[8.3.1.1]*

Bickel, P. J. (1967). Some contributions to the theory of order statistics, *Proceedings of the Fifth Berkeley Symposium on Mathematical Statistics and Probability 1*, 575-591. *[8.2.10]*

Bjerve, S. (1977). Error bounds for linear combinations of order statistics, *Annals of Statistics 5*, 357-360. *[8.7.7.2]*

Bland, R. P., Gilbert, R. D., Kapadia, C. H. and Owen, D. B. (1966). On the distributions of the range and mean range for samples from a normal distribution, *Biometrika 53*, 245-248. *[8.4.1.2; 8.9.7.1]*

Bliss, C. I., Cochran, W. G. and Tukey, J. W. (1956). A rejection criterion based upon the range, *Biometrika 43*, 418-422. *[8.9.7.2]*

Blom, G. (1958). *Statistical Estimates and Transformed Beta-Variables*, Uppsala, Sweden: Almqvist & Wiksell; New York: Wiley. *[8.2.6]*

Borenius, G., (1958). On the distribution of the extreme values in a sample from a normal distribution, *Skandinavisk Aktuarietidskrift 41*, 131-166. *[8.3.8]*

Borenius, G. (1966). On the limit distribution of an extreme value in a normal distribution, *Skandinavisk Aktuarietidskrift* 1965, 1-15. *[8.2.2.3]*

Bose, R. C. and Gupta, S. S. (1959). Moments of order statistics from a normal population, *Biometrika 46*, 433-440. *[8.2.3]*

Cadwell, J. H. (1952). The distribution of quantiles of small samples, *Biometrika 39*, 207-211. *[8.6.2.1; 8.6.4; 8.6.4, 5; 8.8.3]*

Cadwell, J. H. (1953a). The distribution of quasi-ranges in samples from a normal population, *Annals of Mathematical Statistics 24*, 603-613. *[8.5.2,4; 8.7.4.1]*

Cadwell, J. H. (1953b). Approximating to the distributions of measures of dispersion by a power of χ^2, *Biometrika 40*, 336-346. *[8.4.4.1; 8.9.7.1]*

Cadwell, J. H. (1954). The probability integral of range for samples from a symmetrical unimodal population, *Annals of Mathematical Statistics 25*, 803-806. *[8.4.4.2]*

Chu, J. T. (1955). On the distribution of the sample median, *Annals of Mathematical Statistics 26*, 112-116. *[8.6.2.2; 8.7.2.2]*

Chu, J. T. and Hotelling, H. (1955). The moments of the sample median, *Annals of Mathematical Statistics* 26, 593-606. *[8.6.3]*

Cochran, W. G. (1941). The distribution of the largest of a set of estimated variances as a fraction of their total, *Annals of Human Genetics* 11, 47-52. *[8.9.8.1]*

Cramér, H. (1946). *Mathematical Methods of Statistics*, Princeton, N.J.: Princeton University Press. *[8.7.1; 8.8.2; 3, 4, 5]*

Daly, J. D. (1946). On the use of the sample range in an analogue of Student's t-test, *Annals of Mathematical Statistics* 17, 71-74. *[8.4.12]*

David, F. N., Barton, D. E., et al. (1968). *Normal Centroids, Medians, and Scores for Ordinal Data, Tracts for Computers*, Vol. XXIX, Cambridge: Cambridge University Press. *[8.2.7]*

David, F. N., and Johnson, N. L. (1954). Statistical treatment of censored data, *Biometrika, 41*, 228-240. *[8.2.4]*

David, H. A., (1952). Upper 5% and 1% points of the maximum F-ratio, *Biometrika 39*, 422-424. *[8.9.8.2]*

David, H. A., (1956a). On the application to statistics of an elementary theorem in probability, *Biometrika 43*, 85-91. *[8.3.2.2]*

David, H. A., (1981). Order Statistics (2nd Ed.), New York: Wiley. *[8.1.2.1, 2; 8.1.3, 5, 6; 8.2.2.1; 8.2.3, 4; 8.2.5.1, 8.2.6; 8.2.10; 8.3.2.1, 8.3.3, 6, 9; 8.4.1.1; 8.4.2.2; 8.4.4.1; 8.4.8; 8.7.2.1; 8.7.4.1; 8.7.6; 8.7.6.1; 8.9.2.1; 8.9.7.1; 8.9.9]*

David, H. A., Hartley, H. O. and Pearson, E. S. (1954). The distribution of the ratio, in a single normal sample, of range to standard deviation, *Biometrika 41*, 482-493. *[8.4.6; 8.4.7, 8, 10]*

David, H. T. (1962). The sample mean among the moderate order statistics, *Annals of Mathematical Statistics* 33, 1160-1166. *[8.9.4]*

David, H. T. (1963). The sample mean among the extreme normal order statistics, *Annals of Mathematical Statistics* 34, 33-55. *[8.9.4]*

Davis, C. S., and Stephens, M. A. (1978). Approximating the covariance matrix of normal order statistics, *Applied Statistics* 27, 206-212. *[8.2.10]*

Dixon, W. J. (1957). Estimates of the mean and standard deviation of a normal population, *Annals of Mathematical Statistics* 28, 806-809. *[8.6.8; 8.7.2.2; 8.7.3.1]*

Fisher, R. A., and Tippett, L. H. C. (1928). Limiting forms of the frequency distribution of the largest or smallest member of a sample, *Proceedings of the Cambridge Philosophical Society 24*, 180-190. *[8.1.7.2; 8.7.2.1]*

Galambos, J. (1978). *The Asymptotic Theory of Extreme Order Statistics*, New York: Wiley. *[8.7.1; 8.7.3.2; 8.7.4.2]*

Gascual, O. and Caraux, G. (1992). Bounds on expectations of order statistics via extremal dependencies, *Statistics and Probability Letters 15*, 143-148. *[8.2.5.2]*

Gnedenko, B. (1943). Sur la distribution limite du terme maximum d'une série aléatoire, *Annals of Mathematics 44*, 423-453. *[8.7.1]*

Godwin, H. J. (1949). Some low moments of order statistics, *Annals of Mathematical Statistics 20*, 279-285. *[8.2.3, 9]*

Govindarajulu, Z. (1963). On moments of order statistics and quasi-ranges from normal populations, *Annals of Mathematical Statistics 34*, 633-651. *[8.2.2]*

Govindarajulu, Z. and Hubacker, N. W. (1964). Percentiles of order statistics in samples from uniform, normal, chi (1d.f) and Weibull populations, *Reports on Statistical Applications Research, JUSE 11*, 64-90. *[8.1.5]*

Grubbs, F. E. (1950). Sample criteria for testing outlying observations, *Annals of Mathematical Statistics 32*, 27-58. *[8.3.2.1; 8.3.4, 7]*

Guenther, W. C. (1977). An easy method for obtaining percentage points of order statistics, *Technometrics 19*, 319-322. *[8.1.4]*

Gumbel, E. J. (1958). *Statistics of Extremes*, New York: Columbia University Press. *[8.1.3; 8.6.6; 8.7.2.1.; 8.7.3.2]*

Gupta, S. S. (1961). Percentage points and modes of order statistics from the normal distribution, *Annals of Mathematical Statistics 32*, 888-893. *[8.1.4, 5]*

Haldane, J. B. S. and Jayakar, S. D. (1963). The distribution of extremal and nearly extremal values in samples from a normal distribution, *Biometrika 50*, 89-94. *[8.7.6.1]*

Hall, P. (1979). On the rate of convergence of normal extremes, *Journal of Applied Probability 16*, 433-439. *[8.7.1]*

Halperin, M., Greenhouse, S. W., Cornfield, J. and Zalokar, J. (1955). Tables of percentage points for the Studentized maximum absolute deviate in normal samples, *Journal of the American Statistical Association 50*, 185-195. *[8.3.10]*

Harter, H. L. (1959). The use of sample quasi-range in estimating population standard deviation, *Annals of Mathematical Statistics 30*, 980-999. *[8.5.4]*

Harter, H. L. (1960). Tables of range and Studentized range, *Annals of Mathematical Statistics 31*, 1122-1147. *[8.4.3, 11]*

Harter, H. L. (1961). Expected values of normal order statistics, Biometrika 48, 151-165; correction: 476. *[8.2.6, 8]*

Harter, H. L. (1969a). *Order Statistics and Their Use in Testing and Estimation*, Vol. 1, Aerospace Research Laboratories, USAF. *[8.4.1.1, 2; 8.4.9; 8.9.6.1; Table 8.1]*

Harter, H. L. (1969b). *Order Statistics and Their Use in Testing and Estimation*, Vol. 2, Aerospace Research Laboratories, USAF. *[8.2.2, 6; 8.5.2, 3, 4; Table 8.1]*

Hartley, H. O. (1942). The range in normal samples, *Biometrika 32*, 334-348. *[8.4.1.1]*

Hartley, H. O. (1950). The maximum F ratio as a short-cut test for heterogeneity of variance, *Biometrika 37*, 308-312.

Hawkins, D. M. (1979). Fractiles of an extended multiple outlier test, *Journal of Statistical Computation and Simulation* 8, 227-236. *[8.2.2.3]*

Helmers, R. (1977). The order of the normal approximation for linear combinations of order statistics with smooth weight functions, *Annals of Probability 5*, 940-953. *[8.7.7.2]*

Hodges, J. L., and Lehmann, E. L. (1967). On medians and quasi medians, *Journal of the American Statistical Association 62*, 926-931. *[8.6.8; 8.7.2.2; 8.7.7]*

Hogg, R. V. (1960). Certain uncorrelated statistics, *Journal of the American Statistical Association 55*, 265-267. *[8.9.1]*

Johnson, N. L. and Kotz, S. (1970). *Distributions in Statistics, Continuous Univariate Distributions*, Vol. 2, New York: Wiley. *[8.1.4]*

Jones, H. L. (1948). Exact lower moments of order statistics in small samples from a normal distribution, *Annals of Mathematical Statistics 19*, 270-273. *[8.2.2, 3, 9]*

Joshi, P. C. and Balakrishnan, N. (1981). An identity for the moments of normal order statistics with applications, *Scandinavian Actuarial Journal*, 203-213. *[8.2.2.2]*

Kendall, M. G. (1954). Two problems in sets of measurements, *Biometrika 41*, 560-564. *[8.9.4]*

Königer, W. (1983). A remark on AS177. Expected normal order statistics (exact and approximate), *Applied Statistics 32*, 223-224. *[8.2.7]*

Krishnaiah, P. R., and Sen, P. K. (1984). *Tables for order statistics, Handbook of Statistics 4: Nonparametric Methods* (Krishnaiah, P. R. and Sen, P. K., eds.), 873-935, New York: North-Holland. *[Table 8.1]*

Lehmann, E. L. (1975). *Nonparametrics: Statistical Methods Based on Ranks*, San Francisco: Holden-Day. *[8.2.7; 8.8.6]*

Leslie, R. T., and Culpin, D. (1970). Distribution of quasi-midranges and associated mixtures, *Technometrics 12*, 311-325. *[8.6.8]*

Link, R. F. (1950). The sampling distribution of the ratio of two ranges from independent samples, *Annals of Mathematical Statistics 21*, 112-116. *[8.9.6.2]*

Lord, E. (1947). The use of range in place of standard deviation in the t-test, *Biometrika 34*, 41-67. *[8.4.1.2]*

McKay, A. T. (1935). The distribution of the difference between the extreme observation and the sample mean in samples of n from a normal universe, *Biometrika 27*, 466-471. *[8.3.2.1, 2]*

McKay, A. T. and Pearson, E. S. (1933). A note on the distribution of range in samples of n, *Biometrika 25*, 415-420. *[8.4.1.2; 8.4.2.1]*

Mosteller, F. (1946). On some useful "inefficient" statistics, *Annals of Mathematical Statistics 17*, 377-408. *[8.6.8; 8.8.4]*

Owen, D. B. (1956). Tables for computing bivariate normal probabilities, *Annals of Mathematical Statistics 27*, 1075-1090. *[8.4.1.2; 8.9.7.1]*

Owen, D. B. (1962). *Handbook of Statistical Tables*, Reading, MA: Addison-Wesley. *[Table 8.1]*

Owen, D. B., Odeh, R. E. and Davenport, J. M. (1977). *Selected Tables in Mathematical Statistics*, Vol. V. Providence, RI: American Mathematical Society. *[8.2.8]*

Parrish, R. S. (1992a). Computing expected values of normal order statistics, *Communications in Statistics - B21*, 57-70. *[8.2.3]*

Parrish, R. S. (1992b). Computing variances and covariances of normal order statistics, *Communications in Statistics - B21*, 71-101. *[8.2.3]*

Pearson, E. S. and Chandra Sekar, C. (1936). The efficiency of statistical tools and a criterion for the rejection of outlying observations, *Biometrika 28*, 308-320. *[8.3.7, 8, 9]*

Pearson, E. S. and Hartley, H. O. (1966). *Biometrika Tables for Statisticians*, Vol. 1 (3rd ed.), London: Cambridge University Press. *[8.3.11, Table 8.1]*

Pearson, E. S. and Hartley, H. O. (1972). *Biometrika Tables for Statisticians*, Vol. 2, London: Cambridge University Press. *[Table 8.1]*

Pearson, E. S. and Stephens, M. A. (1964). The ratio of range to standard deviation in the same normal sample, *Biometrika 51*, 484-4876. *[8.4.6, 11]*

Pillai, K. C. S. (1950). On the distribution of midrange and semi-range in samples from a normal population, *Annals of Mathematical Statistics 21*, 100-105. *[8.6.7]*

Pillai, K. C. S. (1952). On the distribution of "Studentized" range, *Biometrika 39*, 194-195. *[8.4.9]*

Pillai, K. C. S. (1959). Upper percentage points of the extreme Studentized deviate from the sample mean, *Biometrika 46*, 473-474. *[8.3.9]*

Plackett, R. L. (1958). Linear estimation from censored data, *Annals of Mathematical Statistics 29*, 131-142. *[8.2.4]*

Pyke, R. (1965). Spacings, *Journal of the Royal Statistical Society B 27*, 395-449 (with discussion). *[8.7.6]*

Quesenberry, C. P. and David, H. A. (1961). Some tests for outliers, *Biometrika 48*, 379-390. *[8.3.5, 11]*

Rao, C. R., Mitra, S. K. and Matthai, A. (eds.) (1966). *Formulae and Tables for Statistical Work*, Calcutta: Statistical Publishing Society. *[Table 8.1]*

Renyi, A. (1970). Probability Theory. Budapest: Akadémiai Kiadó. *[8.7.7.1]*

Robison-Cox, J. (1992). Tables of order statistics of normal random variables under linear trend, *Communications in Statistics - A 21*, 3497-3520. *[8.9.11]*

Royston, J. P. (1982). Expected normal order statistics (exact and approximate), *Applied Statistics 31*, 161-165. *[8.2.7]*

Ruben, H. (1954). On moments of order statistics in samples from normal populations, *Biometrika 41*, 200-227. *[8.2.8]*

Ruben, H. (1956). On the moments of the range and product moments of extreme order statistics in normal samples, *Biometrika* 43, 458-460. *[8.4.2.1]*

Sarhan, A. E. and Greenberg, B. G. (eds.) (1962). *Contributions to Order Statistics*, New York: Wiley. *[8.2.2; 8.4.2.1; Table 8.1]*

Saw, J. G. (1960). A note on the error after a number of terms of the David-Johnson series for the expected values of normal order statistics, *Biometrika* 47, 79-86. *[8.2.4]*

Steck, G. P. (1958). A table for computing trivariate normal probabilities, *Annals of Mathematical Statistics* 29, 780-800. *[8.4.1.2]*

Stigler, S. M. (1974). Linear functions of order statistics with smooth weight functions, *Annals of Statistics* 2, 676-693. *[8.9.2.2; 8.9.5.1, 2; 8.9.10]*

Stuart, A. and Ord, J. K. (1994). *Kendall's Advanced Theory of Statistics, Vol. 1, Distribution Theory* (6th ed.). New York: Oxford University Press. *[8.6.2.1; 8.6.3, 4; 8.7.2.1; 8.7.3.1]*

Teichroew, D. (1956). Tables of expected values of order statistics and products of order statistics for samples of size twenty and less from the normal distribution, *Annals of Mathematical Statistics* 27, 410-426. *[8.2.8]*

Tietjen, G. L., Kahaner, D. K. and Beckman, R. J. (1977). Variances and covariances of the normal order statistics for sample sizes 2 to 50, *Selected Tables in Mathematical Statistics*, 5, 1-73. *[8.2.3]*

Thomson, G. W. (1955). Bounds for the ratio of range to standard deviation, *Biometrika* 42, 268-269. *[8.4.7]*

Wilks, S. S. (1962). *Mathematical Statistics*, New York: Wiley. *[8.7.5; 8.8.3; 8.9.3.1, 2]*

Woodroofe, M. (1975). *Probability with Applications*, New York: McGraw-Hill. *[8.7.2.1]*

Yamauti, Z. (ed.) (1972). *Statistical Tables and Formulas with Computer Applications*, Tokyo: Japanese Standards Association. *[Table 8.1, 8.2.3]*

Chapter 9
THE BIVARIATE NORMAL DISTRIBUTION

Although the main focus of this book is on univariate normality, we shall present a detailed discussion in this chapter of the bivariate normal distribution. Section *9.1* presents definitions, basic properties of the distribution and its moments; algorithms for computational purposes and sources for tables are given in Section *9.2*. Approximations are discussed in Section *9.3*; some characterizations of and sufficient conditions for bivariate normality appear in Section *9.4*. Certain distributions associated with the bivariate normal, such as those arising from truncation of one or both variables, are described in Section *9.5*. Properties of circles and of ellipses offset from the center of a bivariate normal distribution are given in Section *9.6*.

Sampling distributions are covered in Chapter 10.

9.1 DEFINITIONS AND BASIC PROPERTIES

[9.1.1.1] Let $\underline{\theta}' = (\mu_1, \mu_2; \sigma_1^2, \sigma_2^2, \rho)$ be a vector of parameters.

The random vector $\underline{X}' = (X, Y)$ has a bivariate normal (BVN) distribution if it has the joint pdf

$$f(\underline{x}; \underline{\theta}) = f(x, y; \mu_1, \mu_2; \sigma_1^2, \sigma_2^2, \rho)$$

$$= \frac{1}{2\pi\sigma_1\sigma_2\sqrt{(1-\rho^2)}} \exp\left[\frac{-1}{2(1-\rho^2)}\left\{\frac{(x-\mu_1)^2}{\sigma_1^2} - 2\rho\frac{(x-\mu_1)(y-\mu_2)}{\sigma_1\sigma_2} + \frac{(y-\mu_2)^2}{\sigma_2^2}\right\}\right]$$

We denote the joint cdf of (X, Y) by $F(\underline{x}; \underline{\theta})$. The marginal distributions of X and of Y are $N(\mu_1, \sigma_1^2)$ and $N(\mu_2, \sigma_2^2)$, respectively. $\underline{X}$ has variance-covariance matrix

$$\begin{bmatrix} \sigma_1^2 & \rho\sigma_1\sigma_2 \\ \rho\sigma_1\sigma_2 & \sigma_2^2 \end{bmatrix}$$

and ρ is the *correlation coefficient* between X and Y. We shall say that $\underline{X}$ has a BVN($\underline{\theta}$) distribution.

If $\rho = \pm 1$, there is perfect correlation and $\Pr(Y = \alpha + \beta X) = 1$ for constants α and β, with $\beta > 0$ if $\rho = +1$ and $\beta < 0$ if $\rho = -1$.

[9.1.1.2] If we standardize X and Y via

$$Z_1 = \frac{X - \mu_1}{\sigma_1}, \quad Z_2 = \frac{Y - \mu_2}{\sigma_2}$$

then $\underline{Z}' = (Z_1, Z_2)$ has a standardized bivariate normal (SBVN) distribution with joint cdf $\Psi(z_1, z_2; \rho)$ where

$$\Psi(z_1, z_2; \rho) = F(z_1, z_2\,;\,0, 0, 1, 1, \rho) = \Psi(z_2, z_1; \rho) .$$

Then

$$\Psi(h, k; \rho) = \Phi(k) - \Psi(-h, k; -\rho),$$

$\Phi(\cdot)$ being the standard normal cdf.

The joint pdf of (Z_1, Z_2) is given by

$$\psi(z_1, z_2; \rho) = \frac{1}{2\pi\sqrt{(1-\rho^2)}} \exp\left[-\frac{1}{2(1-\rho^2)}(z_1^2 - 2\rho z_1 z_2 + z_2^2)\right] .$$

We shall say that $\underline{Z}$ has a SBVN(ρ) distribution. Then

$$\psi(z_1, z_2; \rho) = \frac{1}{\sqrt{(1-\rho^2)}} \phi(z_1) \phi\left[\frac{z_2 - \rho z_1}{\sqrt{1-\rho^2}}\right].$$

Note also Mehler's identity (Mehler, 1866, pp. 161-176); if $\phi_n(x) = d^n\phi(x)/dx^n$, then

$$\psi(z_1, z_2; \rho) = \sum_{n=0}^{\infty} \rho^n \, \phi_n(z_1) \, \phi_n(z_2)/n!$$

$$= \phi(z_1)\phi(z_2)\left[1 + \sum_{n=1}^{\infty} \rho^n H_n(z_1) H_n(z_2)/n!\right],$$

where $H_n(\cdot)$ is the Tchebyshev-Hermite polynomial of degree n satisfying

$$\int_{-\infty}^{\infty} H_n(x)H_m(x)\phi(x)dx = \begin{cases} n! \,, & n = m\,; \\ 0 \,; & n \neq m\,; \end{cases}$$

see Stuart and Ord (1987, Secs. 6.14-15), and [2.1.8].

[9.1.1.3] If $\rho = 0$, (X, Y) has an *elliptical normal* distribution, with joint pdf

$$\frac{1}{2\pi\sigma_1\sigma_2} \exp\left[-\frac{1}{2}\left\{\frac{(x-\mu_1)^2}{\sigma_1^2} + \frac{(y-\mu_2)^2}{\sigma_2^2}\right\}\right]$$

If in addition, $\sigma_1 = \sigma_2$, (X, Y) has a *circular normal* distribution. These names arise because the contours of constant density in the (x, y)–plane are ellipses and circles, respectively, centered at (μ_1, μ_2).

[9.1.2] One needs to be careful in stating implications arising between the independence of variates X and Y, zero correlation between X and Y, and bivariate normality of (X, Y). If we assume only that the pertinent moments exist, the following results hold:

i) Given that (X, Y) is BVN, X and Y are independent if and only if $\rho = 0$.

ii) For any random vector (X, Y), X and Y are independent only if $\rho = 0$.

iii) If for some bivariate distribution zero correlation implies independence of X and Y, it does not follow that (X, Y) is BVN. An example is the generalized Farlie-Gumbel-Morgenstern distribution (Johnson and Kotz, 1975, pp. 415-416) with joint pdf

$$h(x, y) = g_1(x)g_2(y)[1 + \alpha\{1 - 2\,G_1(x)\}\{1 - 2G_2(y)\}], \;|\alpha| < 1,$$

and joint cdf

$$H(x, y) = G_1(x)G_2(x)[1 + \alpha\{1 - G_1(x)\}\{1 - G_2(y)\}]\,.$$

Here the marginal cdfs of X and Y are $G_1(\cdot)$ and $G_2(\cdot)$, respectively. If $G_2 \equiv \Phi$ and $G_2 = \Phi$, the same example demonstrates that (X, Y) may have normally distributed marginals for each of X and Y and yet not have a joint BVN distribution.

[9.1.3] If (X, Y) has a BVN($\underline{\theta}$) distribution, aX + bY has a N(aμ_1 + bμ_2, $a^2\sigma_1^2 + 2\rho ab\sigma_1\sigma_2 + b^2\sigma_2^2$) distribution, where a and b are constants. The vector (aX + bY, cX + dY) has a BVN distribution with mean vector (aμ_1 + bμ_2, cμ_1 + dμ_2) and covariance matrix

$$\begin{bmatrix} a^2\sigma_1^2 + 2\rho ab\sigma_1\sigma_2 + b^2\sigma_2^2 & ac\sigma_1^2 + \rho(ac+bd)\sigma_1\sigma_2 + bd\sigma_2^2 \\ ac\sigma_1^2 + \rho(ac+bd)\sigma_1\sigma_2 + bd\sigma_2^2 & c^2\sigma_1^2 + 2\rho cd\sigma_1\sigma_2 + d^2\sigma_2^2 \end{bmatrix}$$

[9.1.4] <u>Conditional</u> <u>Distributions</u> and <u>Regression</u>
If (X, Y) ~ BVN($\underline{\theta}$), the conditional pdf of Y, given X = x, is given by

$$f(y|x) = \frac{1}{\sqrt{(2\pi\sigma_2^2(1-\rho^2))}} \exp\left[-\frac{1}{2\sigma_2^2(1-\rho^2)}\left\{y - \mu_2 - \rho\frac{\sigma_2}{\sigma_1}(x-\mu_1)\right\}^2\right]$$

That is, given X = x, Y has a N($\mu_2 + \rho(\sigma_2/\sigma_1)(x-\mu_1)$, $\sigma_2^2(1-\rho^2)$) distribution. The <u>conditional</u> <u>variance</u> of each subpopulation of Y's at fixed values x of X is $\sigma_2^2(1-\rho^2)$ and is free of x; this is the homoscedastic property of regression.

The <u>regression</u> <u>equation</u> is the conditional expectation of Y given x, i.e.,

$$E(Y|x) = \mu_2 + \rho(\sigma_2/\sigma_1)(x-\mu_1) ;$$

the regression coefficient β is the slope of this line, $\rho\sigma_2/\sigma_1$.

Two properties hold that are of interest:
(i) The variables X and Y − E(Y|X) are independent.
(ii) The <u>regression</u> <u>function</u> $\mu_2 + \rho(\sigma_2/\sigma_1)(x-\mu_1)$ is not only linear in x, it is the linear combination $\alpha + \beta x$ that minimizes $E\{(Y - \alpha' - \beta' x)^2 | X=x\}$ among all choices of α' and β'. The corresponding minimum expected value is the conditional variance $\sigma_2^2(1-\rho^2)$, also known as the <u>residual</u> or <u>error</u> <u>variance</u> of Y given x. It is zero if and only if $\rho = \pm 1$.

Property (ii) holds for bivariate distributions in general, in the sense that, for real single-valued functions u(x), $E[(Y-u(x))^2|Y=x]$ is minimized by u(x) = E(Y|x), and by u(x) = $\mu_2 + \rho(\sigma_2/\sigma_1)(x-\mu_1)$ if u(x) is restricted to

The Bivariate Normal Distribution

be a linear function of x. Wilks (1962, pp. 83-88, 163) discusses this in detail. He calls the line

$$y = \mu_2 + \rho(\sigma_2/\sigma_1)(x - \mu_1)$$

in the (x, y) − plane the least squares regression line of Y on x, although some would argue that the term least squares belongs more properly with the fitting of a straight line through a scatterplot of data.

[9.1.5] Lancaster (1957, p. 290) proves a theorem with useful applications in contingency table and correspondence analysis.

Let (X, Y) have a BVN($\underline{\theta}$) distribution; if separate transformations g(·) on X and h(·) on Y lead to the bivariate distribution of (g(X), h(Y)) having correlation coefficient ρ', then $|\rho'| \leq |\rho|$ (see also Kendall and Stuart (1967, p. 569)). Thus, if we have bivariate grouped data or an r × s contingency table with qualitative categories, and if we then seek separate scoring systems for each classification which will maximize the magnitude of the product-moment correlation coefficient between them, then effectively we are trying to fit the data to the closest to a BVN distribution that the data will allow.

[9.1.6] For a BVN($\underline{\theta}$) random vector (X, Y) denote the (r, s)th joint moment $E(X^r Y^s)$ of (X, Y) by μ'_{rs}. Denote the corresponding joint central moment $E[(X-\mu_1)^r / (Y-\mu_2)^s]$ by μ_{rs}. The covariance of X and Y is μ_{11}, where $\mu_{11} = \rho \sigma_1 \sigma_2$. The BVN($\underline{\theta}$) distribution is uniquely determined by parameters $\rho, \mu_{10}, \mu_{01}, \mu_{20}, \mu_{02}$, i.e. by $\rho, \mu_1, \mu_2, \sigma_1^2$ and σ_2^2. The joint moment-generating function $M(t_1, t_2)$ is $E[\exp(t_1 X + t_2 Y)]$;

$$M(t_1, t_2) = \exp[t_1\mu_1 + t_2\mu_2 + \tfrac{1}{2}(t_1^2\sigma_1^2 + 2\rho t_1 t_2 \sigma_1 \sigma_2 + t_2^2\sigma_2^2)].$$

The logarithm of $M(t_1, t_2)$ is the joint cumulant generating function

$$K(t_1, t_2) = t_1\mu_1 + t_2\mu_2 + \tfrac{1}{2}(t_1^2\sigma_1^2 + 2\rho t_1 t_2 \sigma_1 \sigma_2 + t_2^2\sigma_2^2)$$

$$= \sum_{r=0}^{\infty} \sum_{s=0}^{\infty} \kappa_{rs} t_1^r t_2^s,$$

say. The cumulants κ_{rs} vanish when r + s > 2.

[9.1.7] Let $\lambda_{rs} = \mu_{rs}/(\sigma_1^r \sigma_2^s)$, the standardized joint central moment of order (r, s) of (X, Y). Then $\lambda_{rs} = 0$ whenever r + s is odd. Further, $\lambda_{sr} = \lambda_{rs}$, and listing some of the latter for r < s,

$$\lambda_{11} = \rho, \qquad \lambda_{13} = 3\rho, \qquad \lambda_{15} = 15\rho,$$

$$\lambda_{17} = 105\rho, \qquad \lambda_{22} = 1 + 2\rho^2, \qquad \lambda_{24} = 3(1 + 4\rho^2),$$

$$\lambda_{26} = 15(1+6\rho^2), \qquad \lambda_{33} = 3\rho(3+2\rho^2), \qquad \lambda_{35} = 15\rho(3 + 4\rho^2),$$

$$\lambda_{44} = 3(3 + 24\rho^2 + 8\rho^4).$$

Stuart and Ord (1987, p. 114) give the following recursion formulas and expansions for $\{\lambda_{rs}\}$, where t = min(r, s):

$$\lambda_{rs} = (r + s - 1)\rho\lambda_{r-1,s-1} + (r-1)(s-1)(1-\rho^2)\lambda_{r-2,s-2},$$

$$\lambda_{2r,2s} = \frac{(2r)!(2s)!}{2^{r+s}} \sum_{j=0}^{t} \frac{(2\rho)^{2j}}{(r-j)!(s-j)!(2j)!},$$

$$\lambda_{2r+1,2s+1} = \frac{(2r+1)!(2s+1)!}{2^{r+s}} \rho \sum_{j=0}^{t} \frac{(2\rho)^{2j}}{(r-j!)(s-j)!(2j+1)!}.$$

[9.1.8] If (X, Y) has a joint distribution, then $E|X^r Y^s|$ is the *joint absolute moment* of order (r, s), and $E|(X-\mu_1)^r(Y-\mu_2)^s|$ is the *joint central moment* of order (r, s). Let

$$\nu_{rs} = E\left|(X-\mu_1)^r(Y-\mu_2)^s \Big/ (\sigma_1^r \sigma_2^s)\right|,$$

the *standardized joint absolute moment* of order (r, s). If (Z_1, Z_2) has an SBVN(ρ) distribution, then $\nu_{rs} = E|Z_1^r Z_2^s| = \nu_{sr}$. If $\tau = \sqrt{(1-\rho^2)}$, then (Kamat, 1953, pp. 26-27)

$$\nu_{11} = 2\{\tau + \rho \arcsin \rho\}/\pi, \qquad \nu_{12} = (1+\rho^2)\sqrt{(2/\pi)},$$
$$\nu_{13} = 2\{\tau(2+\rho^2) + 3\rho \arcsin \rho\}/\pi, \qquad \nu_{22} = 1 + 2\rho^2,$$
$$\nu_{14} = (3 + 6\rho^2 - \rho^4)\sqrt{(2/\pi)}, \qquad \nu_{23} = 2(1+3\rho^2)\sqrt{(2/\pi)},$$

$$\nu_{24} = 3(1 + 4\rho^2),$$
$$\nu_{33} = 2\{(4 + 11\rho^2)\tau + 3\rho(3 + 2\rho^2) \arcsin \rho\}/\pi.$$

In series form,

$$\nu_{m,n} = \pi^{-1} 2^{(m+n)/2} \tau^{m+n+1} \sum_{k=0}^{\infty} \Gamma\{\tfrac{1}{2}(m+1) + k\}$$
$$\cdot \Gamma\{\tfrac{1}{2}(n+1) + k\}(2\rho)^{2k}/(2k)!$$

9.2 PROBABILITY FUNCTIONS AND TABLES

[9.2.1] *Probabilities.* In the historical development of the study of the bivariate normal distribution, interest centered around algorithms for computing $\Pr(Z_1 \geq h, Z_2 \geq k)$, denoted $L(h, k, \rho)$, where (Z_1, Z_2) has a SBVN(ρ) distribution, that is,

$$L(h, k, \rho) = \int_h^\infty \int_k^\infty \psi(x, y; \rho) \, dy \, dx.$$

Then
$$\Psi(h, k; \rho) = \Psi(k, h; \rho) = L(-h, -k, \rho) = L(-k, -h, \rho).$$

In particular, let $(X, Y) \sim \text{BVN}(0, 0; \sigma_1^2, \sigma_2^2, \rho)$. Then (Sheppard, 1898, pp. 101-167)

$$\Pr(X \leq 0, Y \leq 0) = \Pr(Z_1 \leq 0, Z_2 \leq 0) = L(0, 0, \rho)$$
$$= \tfrac{1}{4} + (2\pi)^{-1} \arcsin \rho.$$

Kepner et al. (1989, pp. 48-49) give two concise proofs of this result. For a direct approximation to $L(h, k, \rho)$ see *[9.3.4]*.

Available tables of $L(h, k, \rho)$ are given solely for nonnegative values of h and k, and so the following relations need to be used (Abramowitz and Stegun, 1970, Sec. 26.3; National Bureau of Standards, 1959, pp. vi-viii):

$$\Psi(h, k; \rho) = L(-h, -k, \rho) = L(h, k, \rho) + \Phi(h) + \Phi(k) - 1 ,$$
$$\Psi(-h, k; \rho) = L(h, -k, \rho) = 1 - \Phi(h) - L(h, k, -\rho) ,$$
$$\Psi(h, -k; \rho) = L(-h, k, \rho) = 1 - \Phi(k) - L(h, k, -\rho) ,$$
$$\Pr(|Z_1| \leq h, |Z_2| \leq k) = 2[L(h, k, \rho) + L(h, k, -\rho) + \Phi(h) + \Phi(k) - 1] ,$$
$$\Psi(h, k, 0) = \Phi(h)\Phi(k) \quad \text{from independence} ,$$
$$\Psi(h, k, -1) = \begin{cases} 0, & h + k \leq 0, \\ \Phi(h) + \Phi(k) - 1, & h + k \geq 0. \end{cases}$$

Note that $L(h, k, -1) = 1 - \Phi(h) - \Phi(k)$ if $h + k \leq 0$ and that

$$\Psi(h, k, 1) = \begin{cases} \Phi(k), & k \leq h, \\ \Phi(h), & k \geq h. \end{cases}$$

Further,

$$L(h, k, \rho) = \int_h^\infty \phi(x)[1 - \Phi\{(k - \rho x)/\sqrt{(1-\rho^2)}\}] \, dx$$
$$= \int_k^\infty \phi(y)[1 - \Phi\{(h - \rho y)/\sqrt{(1-\rho^2)}\}] \, dy .$$

Polya (1949, pp. 70-73) proved that, if $\rho h > k$, and $0 < \rho < 1$,

$$1 - \Phi(h) - (1-\rho^2)(\rho h - k)^{-1}\phi(k)[1 - \Phi\{(h - \rho k)/\sqrt{(1-\rho^2)}\}]$$
$$< L(h, k, \rho) < 1 - \Phi(h) .$$

[9.2.2.1] Another function that has attracted some attention is $V(h, k)$, given by

$$V(h, k) = \frac{1}{2\pi} \int_0^h \int_0^{kx/h} \exp[-(x^2+y^2)/2] \, dy \, dx ,$$

the integral of the standard circular normal density over the triangle with vertices $(0, 0)$, $(h, 0)$, and (h, k). Then $V(0, k) = 0 = V(h, 0)$,

$$L(h, k, \rho) = L(0, 0, \rho) - \tfrac{1}{2}[\Phi(h) + \Phi(k) - 1]$$
$$+ V\left(h, (k - \rho h)/\sqrt{(1-\rho^2)}\right) + V\left(k, (h - \rho k)/\sqrt{(1-\rho^2)}\right) ,$$

The Bivariate Normal Distribution

and
$$V(h, k) = L(h, 0, \tau) - L(0, 0, \tau) + \Phi(h) - \tfrac{1}{4},$$

where $\tau = -k\,\mathrm{sgn}(h)/\sqrt{h^2 + k^2}$ and $L(0, 0, \rho) = \tfrac{1}{4} + \arcsin\{\rho/(2\pi)\}$ (see [9.2.1]). Further,

$$V(h, k) = V(-h, -k) = -V(-h, k) = V(h, -k),$$
$$V(h, k) + V(k, h) = \{\Phi(h) - \tfrac{1}{2}\}\{\Phi(k) - \tfrac{1}{2}\}.$$

[9.2.2.2] For some computational purposes it is convenient to work with the integral $T(h, \lambda)$ (Owen, 1956, p. 1078; 1980, pp. 391, 414-416), where

$$T(h, \lambda) = \int_0^\lambda \frac{\phi(h)\,\phi(hx)}{1 + x^2}\,dx, \quad |h| < \infty, \ \lambda > 0,$$
$$= (2\pi)^{-1} \arctan \lambda - V(h, \lambda h)$$
$$= (2\pi)^{-1} \arctan \lambda - \int_0^h \phi(x)\Phi(\lambda x)dx + \tfrac{1}{2}\Phi(h) - \tfrac{1}{4}.$$

Here $(2\pi)^{-1} \arctan \lambda$ is the integral of the circular normal density $\psi(x, y; 0)$ over the sector in the positive quadrant of the (x, y)-plane bounded by the lines $y = 0$, $y = \lambda x$. Then

$$T(h, \lambda) = T(-h, \lambda) = -T(h, -\lambda),$$
$$T(h, 0) = 0\,;\ T(0, \lambda) = (2\pi)^{-1} \arctan \lambda,$$
$$T(h, 1) = \tfrac{1}{2}\Phi(h)\{1 - \Phi(h)\},$$
$$T(h, \infty) = \tfrac{1}{2}\left[1 - \Phi(|h|)\right];$$

further,
$$T(h, \lambda) + T(\lambda h, \lambda^{-1}) = \tfrac{1}{2}\Phi(\lambda h) + \tfrac{1}{2}\Phi(h) - \Phi(h)\Phi(\lambda h) - \tfrac{1}{2} I_{(-\infty, 0)}(\lambda),$$

where $I_{(a,b)}(\lambda) = 1$ if $a < \lambda < b$ and equals zero otherwise.

[9.2.3] Working with the properties listed in the preceding paragraphs, Owen (1956, p. 1076) provided the following formula for cumulative BVN probabilities:

$$\Psi(h, k; \rho) = \tfrac{1}{2}\{\Phi(h) + \Phi(k)\} - T(h, a_h) - T(k, a_k) - c(h, k),$$

where

$$a_n = \frac{k - \rho h}{h\sqrt{1-\rho^2}}, \quad a_k = \frac{h - \rho k}{k\sqrt{1-\rho^2}}, \quad c(h, k) = \begin{cases} 0, & hk > 0, \\ \tfrac{1}{2}, & \text{otherwise.} \end{cases}$$

This formula or some variant of it is the cornerstone of algorithms and approximations to $\Psi(\cdot, \cdot; \cdot)$. These are discussed in [9.3] and in [9.3.3.1] in particular.

[9.2.4] *Tables.* The results given above in [9.2.1] to [9.2.3] are frequently used in conjunction with available tables to evaluate $\Psi(h, k; \rho)$, $L(h, k, \rho)$, $V(h, k)$, and $T(h, \lambda)$. We list some of the most accessible tables in Table 9.1, along with the coverage and number of decimal places given. Linear interpolation in these and other tables gives varying degrees of accuracy, and the recommended interpolation procedures given in the appropriate sources should be followed if a high degree of accuracy is important.

The first extensive set of tables of bivariate normal probabilities was for $L(h, k, \rho)$, edited by Karl Pearson (1931), and containing a number of earlier sets of tables. Nicholson (1943, pp. 59-72) and Owen (1956, pp. 1075-1090) were the first to tabulate values of $V(h, k)$ and $T(h, \lambda)$, respectively. Zelen and Severo (1960, pp. 621-623) (see Abramowitz and Stegun, 1970, Sec. 26.3) provide charts to read off values of $L(h, 0, \rho)$, and show how to use the charts to obtain values of $L(h, k, \rho)$.

[9.2.5] To two decimal places, Fieller et al. (1955) list 3000 random pairs from each of nine SBVN distributions with correlation coefficients 0.1(0.1)0.9. Each page of 50 entries includes sums, sums of squares and products, and sample correlation coefficients. Wold (1948, p. xii) suggested using $(X, \rho X + \sqrt{(1-\rho^2)} \cdot Y)$ for such pairs, where X and Y are independent $N(0, 1)$ rvs, and chosen from tables of random normal deviates.

The Bivariate Normal Distribution

TABLE 9.1 Tables of Bivariate Normal Probabilities

Source	Function	Coverage	Decimal places
Pearson (1931)	$L(h, k, \rho)$	$h, k = 0.0(0.1)2.6$ $\rho = 0(0.05)1$	6
		$h, k = 0.0(0.1)2.6$ $-\rho = 0.0(0.05)1$	7
Nicholson (1943)	$V(h, k)$	$h, k = 0.1(0.1)3.0;\ k = \infty$	6
National Bureau of Standards (1959)	$L(h, k, \rho)$	$h, k = 0.0(0.1)4.0$ $\rho = 0.0(0.05)0.95(0.01)1$	6
		$h, k = 0.0(0.1)4.0$ $-\rho = 0.0(0.05)0.95(0.01)1$	7
	$V(h, \lambda h)$	$\lambda = 0.1(0.1)1.0$ $h = 0.0(0.01)4.0(0.02)$ $\quad 4.6(0.1)5.6, \infty$	7
	$V(\lambda h, h)$	$\lambda = 0.1(0.1)1.0$ $h = 0.0(0.01)4.0(0.02)5.6, \infty$	7
Yamauti (1972)	$L(h, k, \rho)$	$\rho = 1.0(-0.01)0.95(-0.05)$ $\quad 0.60,-0.60(-0.05)-0.95$ $\quad (-0.01)-0.99$ $h, k = 0.0(0.1)4.0$	7
		$\rho = 0.55, 0.50(-0.10)-0.50,-0.55$ $h, k = 0.0(0.1)2.6$	
	$V(h, \lambda h)$	$\lambda = 0.0(0.1)1.0$ $h = 0.0(0.01)4.0(0.02)4.6$ $\quad (0.1)5.0, 6.0, \infty$	7
	$(2\pi)^{-1} \arctan \lambda,$ $(2\pi)^{-1} \arcsin \lambda$	$\lambda = 0.01, (0.01)1.0$	10
Owen (1956)	$T(h, \lambda)$	$h = 0.0(0.01)2.0(0.02)3.0$ $\lambda = 0.25(0.25)1.0$	
		$h = 0.0(0.25)3.0$ $\lambda = 0.0(0.01)1.0, \infty$	6
		$h = 3.0(0.05)3.5(0.1)4.7, 4.76$ $\lambda = 0.1, 0.2(0.05)0.5(0.1)0.8,$ $\quad 1.0, \infty$	

(Table 9.1 continued)

	T(h, λ)	h = 0.0(0.01)3.14 λ = 0.25(0.25)1.0	
Owen (1962) Table 8.5		h = 0.0(0.25)3.25 λ = 0.0(0.01)1.0,1.25, 1.50,2.00,∞	6
		h = 3.0(0.05)3.50(0.1)4.0 (0.2)4.6,4.76 λ = 0.1,0.2(0.05)0.5(0.1) 0.8, 1.0, ∞	
	T(h, λ)	h = 0(0.01)3.0; λ=0(0.01)1.0 h = 3.0(0.05)4.0; λ = 0.05(0.05)1.0	
Smirnov and Bol'shev (1962)	T(h, 1)	h = 4.0(0.1)5.2; λ = 0.1(0.1)1.0	7
	T(h, 1)	h = 0(0.001)3.0(0.005)4.0 (0.01)5.0(0.1)6.0	
	$(2\pi)^{-1}\arctan \lambda$	λ = 0(0.001)1.0	

[9.2.6] Owen et al. (1975, pp. 127-138) have tabulated values of β such that, when (X, Y) has an SBVN (ρ) distribution,

$$\Pr(Y \geq z_\gamma \mid X \geq z_\beta) = \delta$$

for given values of γ, δ, and ρ, listed below. These appear in the context of a screening variable X, which is used to indicate the acceptability of a performance variable Y. Coverage in the table is to 4 decimal places, for $\rho = 0.3(0.05)1.0$; $\gamma = 0.75(0.01)0.94$ with $\delta = 0.95, 0.99, 0.999$; also for $\gamma = 0.95(0.01)0.98$ with $\delta = 0.99, 0.999$; and for $\gamma = 0.99$ with $\delta = 0.999$.

[9.2.7] Let $x - \mu_1 = r \cos \alpha$, $y - \mu_2 = r \sin \alpha$; if (X, Y) has pdf f(x, y; $\underline{\theta}$) as in *[9.1.1]*, then this polar coordinate transformation is centered at the mean (μ_1, μ_2). In (r, α) coordinates the BVN pdf is

$$f(r, \alpha) = \frac{1}{2\pi\sigma_1\sigma_2 \sqrt{(1-\rho^2)}} \exp\left[-r^2 \, Q(\alpha)\right],$$

The Bivariate Normal Distribution

$$Q(\alpha) = \frac{1}{2(1-\rho^2)}\left[\frac{\cos^2\alpha}{\sigma_1^2} - \frac{2\rho\cos\alpha\sin\alpha}{\sigma_1\sigma_2} + \frac{\sin^2\alpha}{\sigma_2^2}\right].$$

The equation

$$r^2 Q(\alpha) = \text{constant}$$

represents an ellipse centered at the mean (μ_1, μ_2). The probability that (X, Y) lies in a region between two such ellipses can be expressed in closed form. Define the region R as that bounded by the two ellipses $r^2Q(\alpha) = k_1$ and $r^2Q(\alpha) = k_2$, $k_1 < k_2$, and by the rays $\alpha = \alpha_1$ and $\alpha = \alpha_2$, $\alpha_1 < \alpha_2$. Then (Lowerre, 1983, pp. 235-236)

$$\Pr((X, Y) \in R) = (e^{-k_1} - e^{-k_2})\left\{(2\pi)^{-1}(\arctan[H(\alpha_1)] - \arctan[H(\alpha_2)])\right.$$

$$\left. + \tfrac{1}{2}\delta(\alpha_1, \alpha_2)\right\},$$

$$H(\alpha) = \frac{\sigma_2}{\sqrt{(1-\rho^2)}}\left(\frac{\cot\alpha}{\sigma_1} - \frac{\rho}{\sigma_2}\right),$$

$$\delta(\alpha_1, \alpha_2) = \begin{cases} 1 & \text{if } 0 < \alpha_1 < \pi < \alpha_2, \\ 0 & \text{if } 0 < \alpha_1 < \alpha_2 < \pi \text{ or } \pi < \alpha_1 < \alpha_2 < 2\pi. \end{cases}$$

In particular, with $k_1 = 0$, $k_2 = \infty$, $\alpha_1 = \pi/4$ and $\alpha_2 = 3\pi/4$, and if $\sigma_2 > \sigma_1$, this reduces to

$$\Pr[|X - \mu_1| < |Y - \mu_2|] = 1 - \tfrac{1}{\pi}\arctan\left[\frac{2\sigma_1\sigma_2\sqrt{(1-\rho^2)}}{\sigma_2^2 - \sigma_1^2}\right].$$

9.3 ALGORITHMS AND APPROXIMATIONS

[9.3.1.1] Mee and Owen (1983, pp. 72-75) have approximated the joint cdf $\Psi(h, k, \rho)$ of a SBVN (ρ) distribution via

$$\Psi(h, k, \rho) \doteq \Phi(h)\,\Phi[(k-\mu)/\sigma] = B(h, k, \rho),$$

where

$$\mu = -\rho\phi(h)/\Phi(h),$$
$$\sigma^2 = 1 + \rho h\mu - \mu^2.$$

The error of approximation can be minimized as follows, using properties of $\Psi(\cdot, \cdot, \rho)$ given in *[9.1.1]*. Choose h and k so that $|h| \geq |k|$, by switching them if necessary. Then, if $h \leq 0$, approximate $\Psi(h, k, \rho)$ by $B(h, k, \rho)$, and if $h > 0$, by $\Phi(k) - B(-h, k, -\rho)$. Using this rule, the maximum absolute error of approximation when $|\rho| = 0.1, 0.3, 0.5, 0.7$, and 0.9 is $0.00001, 0.00019, 0.00079, 0.00164$, and 0.00765, respectively. For a polynomial approximation with decreasing errors as $|\rho|$ increases, see Moskowitz and Tsai (1989, pp. 1424-1426).

[9.3.1.2] Several algorithms based on Gaussian or Gauss-Legendre quadrature, and capable of achieving 14-digit accuracy, are compared by Terza and Welland (1991, pp. 115-127). They include Drezner (1978, pp. 277-279), Bouver and Bargmann (1979, pp. 344-348), and Parrish and Bargmann (1981, pp. 241-257). Each of these, however, is outperformed by Divgi (1979, pp. 903-910); see the discussion in *[9.3.4]*.

[9.3.1.3] Drezner and Wesolowsky (1990, pp. 101-107) showed that

$$L(h, k, \rho) = \frac{1}{2\pi} \int_0^\rho \frac{1}{\sqrt{(1-r^2)}} \exp\left\{-\frac{h^2 - 2rhk + k^2}{2(1-r^2)}\right\} dr + \Phi(-h)\Phi(-k),$$

and used Gaussian quadrature based on Legendre polynomials to evaluate $L(h, k, \rho)$ in this form. The algorithm gives some improvement over the Mee-Owen approximation in *[9.3.1.1]*, even with a quadrature based on three or five points.

[9.3.1.4] Moskowitz and Tsai (1989, pp. 1421-1438) have developed a polynomial approximation to $\Psi(h, k, \rho)$, based on the univariate standard normal distribution. In Mee-Owen *[9.1.3.1]* the errors decrease monotonically as $|\rho|$ decreases; in Moskowitz-Tsai they decrease as $|\rho|$ increases, so that this approximation is more accurate than Mee-Owen if $|\rho| \geq 0.6$. The algorithms given are error-bounded and confined to quadratics and cubics. The authors use the same approach to approximate values of $P(Z_1 \leq c | Z_2 \geq d)$ when $(Z_1, Z_2) \sim \text{SBVN}(\rho)$.

[9.3.1.5] Cox and Wermuth (1991, p. 263-269) provide an explicit approximation to $L(a, b, \rho)$, viz.,

$$L(a, b, \rho) \simeq \Phi(-a)\Phi\left\{\frac{\rho\,\mu(a) - b}{\sqrt{(1-\rho^2)}}\right\} = \Phi(-a)\Phi\{\xi(a, b; \rho)\},$$

say, where
$$\mu(a) = E(Z|Z > a) = \phi(a)/\Phi(-a),$$

$Z \sim N(0, 1)$. A refinement of this result is given by

$$L(a, b, \rho) \simeq \Phi(-a)[\Phi\{\xi(a, b; \rho)\} - \tfrac{1}{2}\rho^2(1-\rho^2)^{-1}\xi(a, b; \rho)$$
$$\cdot \phi\{\xi(a, b; \rho)\} \sigma^2(a)],$$

$$\sigma^2(a) = \text{Var}(Z|Z > a) = 1 + \mu(a) - \mu^2(a).$$

The least accurate results are obtained for $a = b$ when ρ is large, $\rho > 0$; the error in the simpler version is about 10% when $\rho = 0.9$, but is reduced in the refined version.

Clearly a and b are interchangeable. Provided a or b is positive, the simpler version should be used, taking a as the larger of the arguments. If $a < 0$ and $b < 0$, with $-a$ as the larger of $-a$ and $-b$, compute

$$L(a, b, \rho) = 1 - \Phi(-a) - \Phi(-b) + L(-a, -b, \rho).$$

If $\rho < 0$, compute $L(a, -b, -\rho)$ noting that

$$L(a, b, \rho) = \Phi(-a) - L(a, -b, -\rho).$$

[9.3.2] Gideon and Gurland (1978, pp. 681-684), give a polynomial approximation to the ratio $D(r, \theta)/D(r, \pi/2)$, where

$D(r, \theta) = \Pr((Z_1, Z_2)$ lies in a certain half-infinite triangle),

and $(Z_1, Z_2) \sim \text{SBVN}(\rho)$. The author's equation (3) relates $D(r, \theta)$ to the joint cdf $\Psi(\cdot, \cdot; \cdot)$ of (Z_1, Z_2), and a table of coefficients for the polynomial approximation is provided. The error in approximating $D(r, \theta)$ is no greater than 5×10^{-6}.

[9.3.3.1] Owen (1956, p. 1079) showed that

$$T(h, a) = (2\pi)^{-1}\left(\arctan a - \sum_{j=0}^{\infty} c_j a^{2j+1}\right),$$

$$c_j = \left[(-1)^j/(2j+1)\right]\left[1 - \exp(-\tfrac{1}{2}h^2)\sum_{i=0}^{j} h^{2i}/(2^i i!)\right];$$

this converges rapidly when a and h are small, and converts readily into a series in V(h, ah) via

$$T(h, a) = (2\pi)^{-1} \arctan a - V(h, ah).$$

T(h, a) is defined in *[9.2.2.2]* and V(h, k) in *[9.2.2.1]*. The IMSL subroutine DBNRDF essentially computes the BVN probability as given in *[9.2.3]* via this expansion of T(h, a); see IMSL (1987).

[9.3.3.2] Sowden and Ashford (1969, pp. 169-180) suggest a composite method of computing L(h, k, ρ), defined in *[9.2.1]*, incorporating Owen's algorithm of *[9.3.3.1]* for T(h, λ) when h and λ are small, and introducing Hermite-Gauss quadrature or Simpson's rule otherwise.

[9.3.3.3] Daley (1974, pp. 435-438) asserts that integration via Simpson's rule alone, based on the normal distribution of X, given Y = y, is better than the rule in *[9.3.3.2]*; the level of accuracy is 3 × 10^{-7}.

Daley also gives an approximation to T(h, λ), based on that of Cadwell (1951), pp. 475-479:

$$T(h, \lambda) \doteq (2\pi)^{-1}(\arctan \lambda)\exp(-\tfrac{1}{2}h^2/\arctan \lambda)\{1 + (0.00868)h^4\lambda^4\},$$

with a maximum error for all h when $|\lambda| \leq 1$ of 5.1 × 10^{-5}.

[9.3.3.4] If h $\geq$ 5.6 and 0.1 $\leq \lambda \leq$ 1, V(h, λh) is equal (to 7 decimal places) to V(∞, $\lambda \cdot \infty$), that is, to $(2\pi)^{-1}$ arctan λ; but for small values of λ, V(λh, h) differs considerably at h = 5.6 from V($\lambda\cdot\infty$, ∞), that is, from $(2\pi)^{-1}$ arc cot λ; if h $\geq$ 5.6 in this region, a better approximation is given (National Bureau of Standards, 1959, pp. vii-viii) by

$$V(\lambda h, h) \doteq (2\pi)^{-1} \operatorname{arc\,cot} \lambda - \tfrac{1}{2}\{1 - \Phi(\lambda h)\}.$$

The error in this approximation, for small values of λ, is less than $\tfrac{1}{2} \cdot 10^{-7}$. If h $\leq$ 0.8 and $\lambda \leq$ 1,

The Bivariate Normal Distribution

$$V(h, \lambda h) \doteq (4\pi)^{-1} \lambda h^2 \{1 - \tfrac{1}{4} \lambda^2 (1 + \tfrac{1}{3}\lambda^2)\}$$

(Johnson & Kotz, 1972, p. 97).

[9.3.3.5] Young and Minder (1974, pp. 455-457) published an algorithm for computing T(h, a) that the authors claimed is accurate to at least 6 decimal places. Their procedure uses 10-point Gaussian quadrature, without incorporating an algorithm for the normal cdf $\Phi(\cdot)$. Corrections and/or improvements to the computing procedure were published by the authors (ibid., 1979, p. 113) and by Thomas (1986, pp. 310-312). The latter provides a complete listing of the algorithm, incorporating some earlier corrections.

Hill (1978, p. 379) and Chou (1985, pp. 100-101) extended the Young and Minder algorithm to cover $-\infty < a < 0$ and cases $h = +\infty$, $h = -\infty$ and/or $a = +\infty$, $a = -\infty$, respectively. Claiming that, even with these modifications the algorithm fails to provide sufficient accuracy for T(h, a) when h is small and a is large, Boys (1989, pp. 580-582) gave the approximation

$$T(h, \lambda/h) \doteq \tfrac{1}{4} - \frac{1}{2\pi} \left[\frac{e^{-\lambda^2/2}}{\lambda} + \sqrt{2\pi} \, \{\Phi(\lambda) - \tfrac{1}{2}\} \right] h$$

$$+ \frac{1}{12\pi} \left[\frac{\lambda^2 + 2}{\lambda^3} e^{-\lambda^2/2} + \sqrt{2\pi} \, \{\Phi(\lambda) - \tfrac{1}{2}\} \right] h^3 ,$$

which is accurate to at least five decimal places in the range $0 < h < 0.3$, $a = \lambda/h > 7$.

[9.3.4] Divgi (1979, pp. 903-910) has developed an accurate and fast approximation to L(h, k, ρ) that avoids computation of T(h, a), but is based on an approach involving polar coordinates first introduced by Ruben (1961, pp. 171-186). Let

$$R^2 = (h^2 + k^2 - 2hk\rho)/(1 - \rho^2), \quad \pi/2 - \theta = \arcsin(h/R) ,$$

$$\theta - \eta = \arcsin(k/R).$$

Then
$$L(h, k, \rho) = W(R, \pi/2 - \theta) + W(R, \theta - \eta) + C,$$

where $C = 1$ or $C = 0$ according as h and k are both negative or not, where $W(R, \psi) = -W(R, |\psi|)$ if $\psi < 0$, and where

$$W(R, \psi) \doteq (2\pi)^{-1} \exp(-\tfrac{1}{2}R^2) \left[\psi - \sum_{k=0}^{J} d_k\, R^{k+1} \int_0^{\psi} (\cos u)^{k+1} du \right],$$

$$|\psi| \leq \pi/2.$$

In this expression the constants d_k are as follows when $J = 10$:

$d_0 = 1.2532\ 98042$ $d_1 = -0.99973\ 16607$

$d_2 = 0.62501\ 92459$ $d_3 = -0.32819\ 15667$

$d_4 = 0.14703\ 31965$ $d_5 = -5.4948\ 56177 \times 10^{-2}$

$d_6 = 1.6298\ 27794 \times 10^{-2}$ $d_7 = -3.5912\ 57830 \times 10^{-3}$

$d_8 = 5.4066\ 19903 \times 10^{-4}$ $d_9 = -4.8902\ 54061 \times 10^{-5}$

$d_{10} = 1.9847\ 41031 \times 10^{-6}$

This provides an accuracy of the order of 10^{-7}. Higher orders of accuracy result when J is chosen to be larger than 10. For details on deriving the coefficients d_k when $J \neq 10$, see Divgi's paper.

It turns out that Divgi's method "decisively outperforms" several others, "achieving 14 digit accuracy ten times faster than its nearest competitor." This was the conclusion indicated from a comparison of several BVN algorithms by Terza and Welland (1991, pp. 115-127). Attention was restricted to algorithms having the potential for such a high degree of accuracy. They included Owen (1956, pp. 1075-1090) discussed in *[9.2.3]* and *[9.3.3.1]*, Young and Minder (1974, pp. 455-457) discussed in *[9.3.3.5]*, Daley (1974, pp. 435-438), and the four methods involving Gaussian quadrature listed in *[9.3.1.2]*.

[9.3.5] Wang (1987, pp. 185-190) describes an approach that gives a "good" approximation to integrals of BVN distributions over rectangular regions, and that "can be easily programmed and executed on a personal computer". The (x, y)-plane is partitioned into rectangular regions, over

The Bivariate Normal Distribution

which integrals of the BVN joint p.d.f. can be considered as a population contingency table, with discretized univariate normal distributions in the margins.

9.4 CHARACTERIZATIONS

Some of the characterizations listed in Chapter 4 generalize to the multivariate normal distribution, and to the BVN distribution in particular. Some of these are listed here (all in *[9.4.1]*, for example), as well as some specific to the bivariate case only. Hamedani (1992), pp. 2665-2688, provides a compendium of bivariate and multivariate normal characterizations. In what follows, (X, Y) has some bivariate distribution, with mean μ_x, μ_y, variances σ_x^2, σ_y^2, and correlation ρ.

[9.4.1] (X, Y) have a BVN distribution if and only if:
(a) For any constants a and b, not both zero, aX + bY has a normal distribution (Johnson and Kotz, 1972, p. 59); or
(b) (X, Y) have a joint moment generating function $M(t_1, t_2)$ of the form given in *[9.1.6]*; or
(c) The sample mean vector and the sample covariance matrix, exhibited in *[11.1]*, are independent (Stuart and Ord, 1987, Sec. 15.24).

[9.4.2] We now give some characterizations based on regression and conditional distribution properties; see Hamedani (1992), pp. 2665-2688; also Khatri (1979), pp. 589-598, and Bhattacharyya (1943), pp. 399-406.

[9.4.2.1] (X, Y) is BVN if and only if $X \sim N(\mu, \sigma^2)$, and for all real x, the conditional distribution of Y|X = x, i.e., of Y given that X = x, is $N(ax + b, \sigma_1^2)$. Ahsanullah (1985, pp. 215-218) spells out the corresponding joint distribution:

$$(X, Y) \sim BVN(\mu, a + b\mu; \sigma^2, \sigma_1^2 + b^2\sigma^2, b\sigma(\sigma_1^2 + b^2\sigma^2)^{-1/2}).$$

[9.4.2.2] (X, Y) is BVN if and only if the conditional distribution of Y|X = x is normal for all real x, and the equiprobable density contours f(x, y) = constant are similar concentric ellipses.

[9.4.2.3.] (X, Y) is BVN if and only if both of the conditional

distributions of $X|Y = y$ and of $Y|X = x$ are normal for any real x and y and any one of the following three conditions holds:
   (a) The conditional variance of $X|Y = y$ or of $Y|X = x$ is constant;
   (b) The marginal distribution of X or of Y is normal;
   (c) The regression of X on Y or of Y on X is linear with nonzero slope.

[9.4.2.4] Let (X, Y) have a bivariate distribution. If the conditional variables
   (i) $X|Y = y \sim N(a + by, 1)$, for all real y, and
   (ii) $Y|X = x \sim N(cx + d, 1)$, for all real x,
then $b = c$ with $|c| < 1$, and (X, Y) has a BVN distribution with respective mean vector and variance-covariance matrix

$$\frac{1}{1-c^2}\begin{pmatrix} a + cd \\ d + ac \end{pmatrix}, \quad \frac{1}{1-c^2}\begin{pmatrix} 1 & c \\ c & 1 \end{pmatrix}$$

(Brucker, 1979, p. 175). By incorporating a scale change, these sufficient conditions can be stated as a characterization.

Fraser and Streit (1980, p. 1098) weakened condition (i) and Ahsanullah (1985, op. cit.) generalized the result further: If X and Y are identically distributed, and if for all x the conditional distribution of $Y|X = x$ is $N(ax + b, \sigma^2)$ with $|a| < 1$, then

$$(X, Y) \sim BVN(\frac{b}{1-a}, \frac{b}{1-a}; \frac{\sigma^2}{1-a^2}, \frac{\sigma^2}{1-a^2}, a).$$

[9.4.2.5] (X, Y) is BVN if and only if for some constants c, c', d and d',
(a) the regression of Y on X is linear; given $X = x$, $Y = cx + d + \epsilon$,
(b) the regression of X on Y is linear; given $Y = y$, $X = c'y + d' + \epsilon'$,
(c) the distributions of the rvs ϵ and ϵ' do not depend on x or on y
   (Stuart and Ord, 1991, Sec. 26.44).
This characterization holds unless X and Y are independent or are functionally related.

[9.4.2.6] Castillo and Galambos (1989, pp. 209-214) and Hamedani (1991, pp. 255-258) proved the following:

Let (X, Y) have a joint continuous density, with marginal normal

The Bivariate Normal Distribution

densities for each of X and Y. Let the conditional distribution of Y given $X = x$ have mean $E(Y|x)$ and variance $\sigma^2(Y|x)$, and that of X given $Y = y$ have mean $E(X|y)$ and variance $\sigma^2(X|y)$. Then (X, Y) is BVN if and only if one of the following properties holds:

(i) $\sigma(Y|x)$ or $\sigma(X|y)$ is constant (i.e., free of x or y, respectively).
(ii) As $y \to +\infty$, $y^2\sigma^2(X|y) \to +\infty$
or as $x \to +\infty$, $x^2\sigma^2(Y|x) \to \infty$.
(iii) As $y \to +\infty$, $\liminf \sigma(X|y) \neq 0$
or as $x \to +\infty$, $\liminf \sigma(Y|x) \neq 0$.

Under any one of i), ii) or iii), $E(Y|x)$ and $E(X|y)$ are linear.

[9.4.2.7] Let (X, Y) be a random vector in which the conditional distributions of X, given $Y = y$ and of Y, given $X = x$ are normal for all x and for all y. Then (X, Y) does not necessarily have a BVN distribution (Gelman and Meng, 1991, pp. 125-126). After standardizing, the most general joint pdf is

$$g(x, y) \propto \exp\left[-\tfrac{1}{2}(Ax^2y^2 + x^2 + y^2 - 2Bxy - 2Cx - 2Dy)\right].$$

(For this to be a pdf, one requires that $A > 0$, or $A = 0$ and $|B| < 1$.) Then, for conditional distributions,

$$X|Y = y \sim N\left(\frac{By + C}{Ay^2 + 1}, \frac{1}{Ay^2 + 1}\right), \quad Y|X = x \sim N\left(\frac{Bx + D}{Ax^2 + 1}, \frac{1}{Ax^2 + 1}\right).$$

[9.4.3] We now give some characterizations based on independence.

[9.4.3.1] Let X and Y be rvs and a and b be constants such that (i) $X - aY$ and Y are independent, and (ii) $Y - bX$ and X are independent. Then $(X,Y) \sim \text{BVN}(\cdot)$ if $ab \neq 0$ and $ab \neq 1$ (Rao, 1975, pp. 1-13).

[9.4.3.2] Let X, Y, U_1 and U_2 be rvs and a and b be constants such that (i) $Z_1 = X + aY + U_1$ and (Y, U_1, U_2) are independent, and (ii) $Z_2 = bX + Y + U_2$ and (X, U_1, U_2) are independent. Then (Z_1, Z_2) has a BVN distribution if $a \neq 0$ and $b \neq 0$; further (Z_1, Z_2) and (U_1, U_2) then are independent (Khatri and Rao, 1976, pp. 83-84).

[9.4.3.3] Let X and Y be identically distributed with zero means, variances and correlation coefficient ρ, $0 < |\rho| < 1$. Then (X, Y) is BVN if and only if X and $(Y - \rho X)/\sqrt{1-\rho^2}$ are independent and identically distributed (Fieger, 1977, pp. 135-140).

9.5 ASSOCIATED DISTRIBUTIONS

[9.5.1.1] Let $(Z_1, Z_2) \sim$ SBVN(ρ). Fieller (1932, pp. 428-440) showed that the ratio Z_1/Z_2 has a Cauchy distribution, with pdf

$$\sqrt{(1-\rho^2)} \,/\, \{\pi(1 - 2\rho u + u^2)\} \;, \quad -\infty < u < \infty .$$

He also derived the distribution of X/Y when $(X, Y) \sim$ BVN(μ_1, μ_2; σ_1^2, σ_2^2, ρ). Let

$$a(u) = \left(\frac{u^2}{\sigma_1^2} - \frac{2\rho u}{\sigma_1 \sigma_2} + \frac{1}{\sigma_2^2}\right)^{\frac{1}{2}}, \quad b(u) = \frac{\mu_1 u}{\sigma_1^2} - \frac{\rho(\mu_1 + \mu_2 u)}{\sigma_2 \sigma_2} + \frac{\mu_2}{\sigma_2^2},$$

$$c = \mu_2^2 \, a^2(\mu_1/\mu_2), \qquad d(u) = \exp\left[\frac{b^2(u) - ca^2(u)}{2(1-\rho^2)a^2(u)}\right].$$

Then the pdf of X/Y is h(u), where

$$h(u) = \frac{b(u)\, d(u)}{\sqrt{(2\pi)}\, \sigma_1 \sigma_2 a^3(u)} \left[\Phi\left\{\frac{b(u)}{\sqrt{(1-\rho^2)}\, a(u)}\right\} - \Phi\left\{-\frac{b(u)}{\sqrt{(1-\rho^2)}\, a(u)}\right\}\right]$$

$$+ \frac{\sqrt{(1-\rho^2)}}{\pi \sigma_1 \sigma_2 a^2(u)} \exp\left\{-\frac{c}{2(1-\rho^2)}\right\}.$$

Hinkley (1969, pp. 635-639) showed that as $\theta_2/\sigma_2 \to \infty$ (i.e., as Pr(Y > 0) $\to 1$), the cdf H(u) of X/Y tends to H*(u), where

$$H^*(u) = \Phi\left[\frac{\mu_2 u - \mu_1}{\sigma_1 \sigma_2 a(u)}\right].$$

This approximation is useful when θ_2 is much larger than σ_2; also

The Bivariate Normal Distribution

$|H^*(u) - H(u)| \leq \Phi(-\theta_2/\sigma_2)$, a bound that is attained at $u = \pm\infty$. See also Johnson and Kotz (1972, pp. 123-124).

[9.5.1.2] Aroian (1986, pp. 612-613) has also given expressions for the distribution of X/Y in the general BVN set up. If $U = X/Y$, define parameters a and b, and the variable T:

$$a = \left(\frac{\mu_1}{\sigma_1} - \rho\frac{\mu_2}{\sigma_2}\right) / \sqrt{(1-\rho^2)}, \quad b = \mu_2/\sigma_2,$$

$$T = \left(\frac{\sigma_2}{\sigma_1}U - \rho\right) / \sqrt{(1-\rho^2)}.$$

Then T has cdf $G(t)$, where

$$G(t) = \tfrac{1}{2} + (1/\pi)\arctan t - 2V(b, a)$$
$$+ 2V\{(bt-a)(1+t^2)^{-1/2},\ (at+b)(1+t^2)^{-1/2}\}$$
$$= L\{(a-bt)(1+t^2)^{-1/2},\ -b,\ t(1+t^2)^{-1/2}\}$$
$$+ L\{(-a+bt)(1+t_2)^{-1/2},\ b,\ t(1+t^2)^{-1/2}\};$$

$L(\cdot, \cdot, \cdot)$ is defined in *[9.2.1]* and $V(\cdot,\cdot)$ in *[9.2.2]*. T has pdf

$$g(t) = \tfrac{1}{\pi} \cdot \frac{1}{1+t^2} \exp\left\{-\tfrac{1}{2}(a^2+b^2)\right\}\left[1 + \frac{q}{\phi(q)}\int_0^q \phi(z)dz\right],$$

where $q = (b+at)/(1+t^2)$. The pdf of U is then

$$h(u) = \frac{\sigma_2}{\sigma_1} \cdot \frac{1}{\sqrt{(1-\rho^2)}} g(t).$$

Öksoy and Aroian (1986, pp. 660-663) give some numerical tabulations.

[9.5.2] If (X, Y) has a circular normal distribution with zero means, and $\sigma_x^2 = \sigma_y^2 = \sigma^2$, the radial error R, where $R^2 = X^2 + Y^2$, has a Rayleigh distribution with p.d.f.

$$g(r) = r\sigma^{-2}\exp(-\tfrac{1}{2}r^2/\sigma^2) \quad (r > 0).$$

The circular probable error is the 50-percentile radial error 0.774σ. For the distribution of R when (X, Y) have a general BVN distribution but are uncorrelated, see Weil (1954, pp. 168-170) and Laurent (1957, pp. 75-89).

[9.5.3] Let $(X, Y) \sim BVN(\mu_1, \mu_2; \sigma_1^2, \sigma_2^2, \rho)$ as in *[10.1.1]*. Craig (1936, pp. 1-15) and Aroian et al. (1978, pp. 165-172) express the distribution of the product XY in series form and integral form, respectively. Define

$$Z = \frac{XY}{\sigma_1 \sigma_2}, \quad \delta_j = \frac{\mu_j}{\sigma_j}, \quad j = 1, 2.$$

Craig showed that Z has mean μ_Z and variance σ_Z^2, where

$$\mu_Z = \delta_1 \delta_2 + \rho, \quad \sigma_Z^2 = \delta_1^2 + \delta_2^2 + 2\rho\delta_1\delta_2 + 1 + \rho^2.$$

The pdf of Z has a singularity at $z = 0$. Springer (1983, p. 211) points out that if $\delta_1 = \delta_2 = 0$ and $|\rho| < 1$, the pdf of the product Z is given by

$$h(z) = \frac{e^{\rho z/(1-\rho^2)}}{\pi\sqrt{1-\rho^2}} K_0\left(\frac{z}{1-\rho^2}\right), \quad |z| < \infty,$$

where $K_0(\cdot)$ is a Bessel function of the second kind with an imaginary argument of order zero.

Meeker et al. (1981) have tabulated the fractiles of W, where $W = (Z - \mu_Z)/\sigma_Z$, for 39 values of α between 0.0005 and 0.9995; the fractiles w are solutions to the equation $\Pr(W \leq w) = \alpha$. These are given to five decimal places, for $\delta_1 \leq \delta_2$, $\delta_j = 0(0.4)4, 6, 12, \infty$ (j = 1,2), w = $-6(0.4) - 3.2(0.2)3.2(0.4)6$, and for $\rho = -1, -0.95, -0.9, -0.8(0.2)0.8$, 0.9, 0.95, 1. The case in which X and Y are independent (see *[5.1.3]*) is given when $\rho = 0$.

[9.5.4] The conditional distribution of Y, given $X \leq h$, is not normal; if (X, Y) is distributed SBVN(ρ), the pdf is

$$f(y|X \leq h) = \phi(y)\Phi\left[(h - \rho y)/(1 - \rho^2)^{1/2}\right] \Big/ \Phi(h),$$

where $\phi(\cdot)$ and $\Phi(\cdot)$ are the standard normal pdf and cdf, respectively.

The Bivariate Normal Distribution

The mean and variance are $-\rho\phi(h)/\Phi(h)$ and $1 + \rho h\mu - \mu^2$, respectively. Higher moments are given by Johnson and Kotz (1972, pp. 112, 114). If $|\rho|$ is small, the distribution can be approximated by that of a normal variable having the same mean and variance (Mee and Owen, 1983, pp. 72-75).

[9.5.5] If $(Z_1, Z_2) \sim \text{SBVN}(\rho)$, then (David, 1981, pp. 51-52)

$$E[\max(Z_1, Z_2)] = \sqrt{[(1-\rho)/\pi]},$$

$$\text{Var}[\max(Z_1, Z_2)] = 1 - (1-\rho)/\pi.$$

More generally, if $(X, Y) \sim \text{BVN}(\mu_1, \mu_2; \sigma_1^2, \sigma_2^2, \rho)$ with cdf $F(\cdot)$ and if $W = \max(X, Y)$,

$$\Pr(W \leq w) = F(w, w).$$

Kella (1986, pp. 3265-3276) derived expressions for the mean and variance of W when $|\rho| < 1$:

$$E(Z) = \mu_1 \Phi + \mu_2(1-\Phi) + \xi\phi,$$

$$\text{Var}(Z) = \sigma_1^2 \Phi + \sigma_2^2(1-\Phi) + (\Delta\Phi + \xi\phi)[\Delta(1-\Phi) - \xi\phi],$$

where

$$\Phi \equiv \Phi(\Delta/\xi), \quad \phi = \phi(\Delta/\xi), \quad \Delta = \mu_1 - \mu_2,$$

$$\xi^2 = \sigma_1^2 - 2\rho\sigma_1\sigma_2 + \sigma_2^2.$$

[9.5.6] In this section we consider properties of a <u>singly truncated</u> SBVN(ρ) distribution; (X, Y) is truncated to include values of X greater than a only. In *[9.5.7-8]* cases in which each of X and Y is singly truncated below or doubly truncated are discussed. For a general discussion see Johnson and Kotz (1972, pp. 112-117).

The joint pdf of $(X, Y)|X > a$ is $g(x, y)$, where if $x > a$,

$$g(x, y) = \psi(x, y; \rho)/[1 - \Phi(a)]$$

and $\psi(\cdot, \cdot; \rho)$ is the SBVN(ρ) pdf, defined in *[9.1.1.2]*.

[9.5.6.1] Let $E_T(X)$, $\text{Var}_T(X)$ and ρ_T denote the mean and variance

of X, and the correlation coefficient between X and Y in a SBVN(ρ) distribution singly truncated so that X > a. Also, let

$$q(a) = \phi(a)/\{1 - \Phi(a)\} = 1/R(x) ,$$

where R(x) is Mills' ratio (see [3.4] – [3.7]). Then

$$E_T(X) = q(a) , \qquad E_T(Y) = \rho q(a) ,$$
$$E_T(X^2) = 1 + aq(a) , \qquad E_T(Y^2) = 1 + \rho^2 aq(a) ,$$
$$\text{Var}_T(X) = 1 - q(a)\{q(a) - a\}, \qquad \text{Var}_T(Y) = 1 - \rho^2 q(a)\{q(a) - a\},$$
$$\rho_t = \rho\sqrt{\{\text{Var}_T(X)/\text{Var}_T(Y)\}}$$

Since $\text{Var}_t(X) \leq \text{Var}_t(Y)$, it follows that $|\rho_t| \leq |\rho|$. See Rao et al. (1968, pp. 433-436) for these results and for other references; see also Weiler (1959, pp. 73-81).

In screening problems an aptitude test is given to all applicants for employment in a large company, giving rise to an aptitude score X. Later, a performance score Y is available for all who are hired (i.e., all for which X > a, say). Suppose that (X, Y) has a BVN(μ_x, μ_y; σ_x^2, σ_y^2, ρ) distribution over all applicants, and a singly truncated BVN distribution over those hired, for which X > a, with parameters μ_x^*, μ_y^*, σ_x^{*2}, σ_y^{*2}, and ρ^*. Birnbaum et al. (1950, pp. 191-204) obtained the following relationships:

$$\mu_y = \mu_y^* + \rho^*(\sigma_y^*/\sigma_x^*)(\mu_x - \mu_x^*) ,$$
$$\sigma_y = \sigma_y^*(1 - \rho^{*2} + \rho^{*2}\sigma_x^2/\sigma_x^{*2})^{1/2} ,$$
$$\rho = \rho^*\sigma_x\sigma_y^*/(\sigma_x^*\sigma_y) .$$

These relationships would be useful if μ_x, σ_x, μ_x^*, μ_y^*, σ_x^*, σ_y^* and ρ^* are known, which would hold approximately using estimates if a large number of records is available from the past; μ_y, σ_y and ρ would, always remain unknown, however, because Y is only observed on applicants who are hired.

See Chou and Owen (1984, pp. 2535-2537).

[9.5.6.2] Tallis (1961, p. 225) derived the moment generating function of the SBVN (ρ) distribution with X truncated below a;

$$E(e^{t_1 X + t_2 Y}) = \frac{1 - \Phi(a - t_1 - \rho t_2)}{1 - \Phi(a)} \exp\{\tfrac{1}{2}(t_1^2 + 2\rho t_1 t_2 + t_2^2)\}.$$

The cumulant generating function is $\log\left[E(e^{t_1 X + t_2 Y})\right]$. Chou and Owen (1984, pp. 2540-2545) derived the following expressions for the cumulants κ_{ij} of the SBVN(ρ) distribution with X truncated below a, where q(a) is defined in [9.5.6.1]:

$$\kappa_{10} = q(a),$$
$$\kappa_{01} = \rho q(a) = \rho \kappa_{10}$$
$$\kappa_{20} = 1 - q(a)\{q(a) - a\},$$
$$\kappa_{02} = 1 - \rho^2 q(a)\{q(a) - a\} = \rho^2(\kappa_{20} - 1) + 1,$$
$$\kappa_{11} = \rho \kappa_{20};$$

$$\kappa_{ij} = \rho^j \, \frac{\partial^{i+j-1}}{\partial x^{i+j-1}} \left\{\frac{\phi(x)}{\Phi(x)}\right\} \bigg|_{x=-a}, \quad i+j \neq 2;$$

$$= \rho^j \, \kappa_{i+j,0}, \quad i + j > 2.$$

See also Cook (1951a, pp. 179-195).

Chou and Owen make use of these cumulants and of an expansion of Cornish-Fisher type to obtain an approximation to percentile points of one variable of a BVN distribution when the other is truncated.

Gajjar and Subrahmaniam (1978, p. 456) give the following recursive relation between the moments of $(X, Y) | X > a$:

$$\mu'_{r+1,s} - a\mu'_{r,s} = r(1 - \rho^2)\mu'_{r-1,s} + \rho\mu'_{r,s+1} - a(r - 1)(1 - \rho^2)\mu'_{r-2,s}$$

$$- a\rho\mu'_{r-1,s-1}, \quad r \geq 2, \, s \geq 0.$$

See also Johnson and Kotz (1972), p. 114.

[9.5.6.3] Truncation of a SBVN(ρ) distribution so that X > a does not affect the regression of Y on X (Rao et al., 1968, p. 435); see *[9.1.4]*. With the notation of *[9.5.6.1]* above, however, the regression of X on Y is the expected value of X, given Y = y, i.e. (Johnson and Kotz, 1972, p. 113),

$$E(X|Y = y) = \rho y + \sqrt{(1-\rho^2)} \cdot q((a-\rho y)/\sqrt{(1-\rho^2)}).$$

[9.5.7] We next discuss a SBVN(ρ) distribution truncated so that observations are available only in the region

$$D = \{(x, y): x \geq a, y \geq b\}.$$

Denote the truncated vector $(X, Y)|X \geq a, Y \geq b$ by (X', Y'), where $(X, Y) \sim$ SBVN(ρ), and let

$$P = \int\int_D \Psi(x, y; \rho) \, dy dx = L(a, b; \rho).$$

Then the pdf of the truncated distribution is $\psi(x, y; \rho)/P$ inside D and zero elsewhere. The truncation is single or it is one-sided in both variables. Weiler (1959, pp. 73-81), Rosenbaum (1961, pp. 405-408) and Regier and Hamdan (1971, pp. 77-82) obtained the following moments:

$$E(X') = [\phi(a)Q(A) + \rho\phi(b)Q(B)]/P,$$

$$E(Y') = [\phi(b)Q(B) + \rho\phi(a)Q(A)]/P,$$

where

$$A = (b - \rho a)/(1 - \rho^2)^{1/2},$$
$$B = (a - \rho b)/(1 - \rho^2)^{1/2},$$
$$Q(a) = 1 - \Phi(a);$$
$$E(X'^2) = [a\phi(a)Q(A) + \rho^2 b\phi(b)Q(B) + P + \rho(1-\rho^2)\psi(a,b;\rho)] / P,$$

$$E(Y'^2) = [b\phi(b)Q(B) + \rho^2 a\phi(a)Q(A) + P + \rho(1-\rho^2)\psi(a, b; \rho)]/P,$$

The Bivariate Normal Distribution

$$E(X'Y') = \left[\rho a\phi(a)Q(A) + \rho b\phi(b)Q(B) + \rho P + (1-\rho^2)\psi(a, b; \rho)\right]/P.$$

From these results the variances $\sigma^2(X')$ and $\sigma^2(Y')$, and the correlation coefficient $\rho(X', Y')$ follow. Regier and Hamdan have tabulated values of $\rho' = \rho(X', Y')$ for $b = a$, $-2.5 \le a \le 2.5$, and $0.05 \le \rho \le 0.95$, Nath (1973, pp. 563-587) has tabulated values of ρ' for $a = -2.5(0.5)2.0$, $-2.5 \le b \le 2.0$ and $0.05 \le \rho \le 0.99$. See also Gajjar and Subrahmaniam (1978, pp. 456-458).

[9.5.8] Suppose that X and Y are both doubly truncated in a SBVN (ρ) distribution, so that only values in the rectangular region $\{(x, y); a_1 < x < a_2, b_1 < y < b_2\}$ are available. Denote the truncated vector by (X', Y'), so that inside this region the pdf of (X', Y') is $\psi(x, y; \rho)/P$, where

$$P = \int_{a_1}^{a_2} \int_{b_1}^{b_2} \psi(x, y; \rho)\, dy\, dx.$$

Gupta and Tracy (1980, pp. 593-613) have obtained the first two moments of (X', Y') as follows (we have tried to simplify the notation): Let

$$\lambda = (1 - \rho^2)^{-1/2},$$

$$A_i = \lambda(b_i - \rho a_i), \quad B_i = \lambda(a_i - \rho b_i), \quad i = 1, 2,$$

$$A_3 = \lambda(b_1 - \rho a_2), \quad B_3 = \lambda(a_1 - \rho b_2),$$

$$A_4 = \lambda(b_2 - \rho a_1), \quad B_4 = \lambda(a_2 - \rho b_1);$$

further, define

$$\alpha_1 = \phi(a_1)[\Phi(A_4) - \Phi(A_1)], \quad \beta_1 = \phi(b_1)[\Phi(B_4) - \Phi(B_1)],$$

$$\alpha_2 = \phi(a_2)[\Phi(A_3) - \Phi(A_2)], \quad \beta_2 = \phi(b_2)[\Phi(B_3) - \Phi(B_2)],$$

$$C(\rho) = \psi(a_1, b_1; \rho) + \psi(a_2, b_2; \rho) - \psi(a_1, b_2; \rho) - \psi(a_2, b_1; \rho).$$

Then

$$\text{PE}(X') = \alpha_1 + \alpha_2 + \rho(\beta_1 + \beta_2),$$

$$\text{PE}(Y') = \beta_1 + \beta_2 + \rho(\alpha_1 + \alpha_2),$$

$$\text{PE}(X'^2) = a_1\alpha_1 + a_2\alpha_2 + \rho^2(b_1\beta_1 + b_2\beta_2) + \rho(1-\rho^2)C(\rho) + P,$$

$$\text{PE}(Y'^2) = b_1\beta_1 + b_2\beta_2 + \rho^2(a_1\alpha_1 + a_2\alpha_2) + \rho(1-\rho^2)C(\rho) + P,$$

$$\text{PE}(X'Y') = \rho(a_1\alpha_1 + a_2\alpha_2 + b_1\beta_1 + b_2\beta_2) + (1-\rho^2)C(\rho) + \rho P.$$

The variances $\sigma^2(X')$ and $\sigma^2(Y')$ and the correlation coefficient $\rho(X', Y')$ can be obtained from these expressions.

Muthén (1990, pp. 131-143) gives alternative expressions for these moments (and for those of analogous censored BVN distributions). One can obtain these from those of Gupta and Tracy by noting that

$$\psi(x, y; \rho) = \lambda\phi(y)\phi\{\lambda(x - \rho y)\}$$
$$= \lambda\phi(x)\phi\{\lambda(y - \rho x)\}$$

(see [9.1.1.2]), so that now, for example,

$$C(\rho) = \lambda\phi(b_1)\{\phi[\lambda(a_1 - \rho b_1)] - \phi[\lambda(a_2 - \rho b_1)]\}$$
$$+ \lambda\phi(b_2)\{\phi[\lambda(a_2 - \rho b_1)] - \phi[\lambda(a_1 - \rho b_2)]\}.$$

9.6 OFFSET CIRCLES AND ELLIPSES

[9.6.1] The circular coverage function $p(R, d)$ is the probability mass of that portion of a standard circular normal, i.e., SBVN(0) distribution that lies in a circle Γ of radius R, with center of Γ offset by a distance d from the mean $(0, 0)$ of the distribution.

Formally $p(R, d)$ is defined via

$$p(R, d) = (2\pi)^{-1} \int\int_\Gamma \exp\{-\tfrac{1}{2}(x^2 + y^2)\}\,dy\,dx,$$

$$\Gamma = \{(x, y): (x - d)^2 + y^2 \leq R^2\}.$$

The Bivariate Normal Distribution

The function $p(R, \cdot)$ arises as the probability of destruction of a point target at the origin, when damage is complete for a distance $r \leq R$ from the point of impact and damage is zero otherwise (the "cookie cutter" model) and the point of impact has a circular normal distribution about a mean impact (or aiming) point. However, $p(R, \cdot)$ is also relevant in certain problems involving areal targets; see the surveys by Eckler (1969, pp. 561-589; 1988, pp. 180-183). There is also some discussion in Johnson and Kotz (1970, pp. 144, 164-165, 179-180).

In polar coordinates,

$$p(R, d) = (2\pi)^{-1} \int \int_\Gamma r \exp(-r^2/2) d\theta \, dr \, ,$$

$$p(R, 0) = \Gamma - \exp(-R^2/2) \, ,$$

where $x = r \cos \theta$, $y = r \sin \theta$, and $\Gamma = \{(r, \theta): r^2 - 2rd \cos \theta + d^2 \leq R^2\}$;

$$p(R, d) = \exp(-d^2/2) \int_0^R r \exp(-r^2/2) \, I_0(rd) \, dr,$$

where $I_0(\cdot)$ is a modified Bessel function of the first kind of order zero, viz.

$$I_0(z) = \pi^{-1} \int_0^\pi \exp(-z \cos \theta) \, d\theta \quad \text{(Di Donato and Jarnagin, 1962, p. 348)}.$$

[9.6.2] The circular coverage function has the following reproducing property. Let (X, Y) have a BVN $(a, b; \sigma^2, \sigma^2, 0)$ distribution, let $d^2 = a^2 + b^2$, $\sigma_2^2 = \sigma^2 + \sigma_1^2$ and $r^2 = x^2 + y^2$. Then (Read, 1971, p. 1733)

$$p(R/\sigma_2, d/\sigma_2) = (2\pi \sigma^2)^{-1} \int_{-\infty}^\infty \int_{-\infty}^\infty p(R/\sigma_1, r/\sigma_1)$$

$$\cdot \exp[-\{(x-a)^2 + (y-b)^2\}/(2\pi \sigma^2)] \, dx \, dy.$$

[9.6.3] Let (Z_1, Z_2) have a standard circular normal [i.e., SBVN (0)] distribution. Then (Owen, 1962, p. 172)

$$\Pr(a_1 Z_1^2 + a_2 Z_2^2 \leq t) = p(A, B) - p(B, A) \, ,$$

where

$$A = \tfrac{1}{2}[\sqrt{t/a_1} + \sqrt{t/a_2}] \text{ and } B = \tfrac{1}{2}|\sqrt{t/a_1} - \sqrt{t/a_2}|.$$

This gives the c.d.f. of a quadratic form in two dimensions; further,

$$\Pr(W \leq t) = p(t, \sqrt{\lambda}),$$

where W has a noncentral chi-square distribution with two degrees of freedom and noncentrality parameter λ.

[9.6.4] The circular coverage function arises as the probability of destruction of a point target at the origin, when damage is complete for a distance $r \leq R$ from the point of impact (the "cookie-cutter") and zero otherwise, and the point of impact has a circular normal distribution about a mean impact point; p(R, d) is also relevant in certain problems involving areal targets. Two surveys covering material in this and the next subsection in detail are those by Guenther and Terragno (1964, pp. 232-260) and by Eckler (1969, pp. 561-589); these list tables, some of which appear only in technical reports with limited circulation. The latter tables generally have the most detailed coverage, but unfortunately are the least easily obtained. We cite only sources which are readily accessible or are not referenced in the above surveys. Table 9.2 lists some of the available tables and charts, with coverages, of p(R, d). The algorithms used to generate the tables are described in the relevant sources.

[9.6.5] The following recursive procedure for the computation of $1 - p(R, d)$ gets round the problem of interpolating from values in the tables (Brennan and Reed, 1965, pp. 312-313). It may also be less complicated than other methods described in the literature.

Replace the zero-order Bessel function by its expansion,

$$I_0(rd) = \sum_{n=0}^{\infty} (\tfrac{1}{2} rd)^{2n}/(n!)^2,$$

and substitute in the integral of *[9.6.1]* to obtain

$$1 - p(R, d) = 1 - \sum_{n=0}^{\infty} g_n k_n,$$

TABLE 9.2 Tables and Charts of p(R, d) with Coverage

Source	Function	Coverage	Decimal places
Solomon (1953, Fig. 1)	p(R, r)	p = 0.05(0.05)0.95 0 ≤ d ≤ 10 (horizontally) 0 ≤ R ≤ 8 (vertically)	Graphed
Burington and May (1970, Tables 11.9.1, 11.11.1)	R	p(R, 0) = 0.0(0.05)1.0	4
	p(R,d)	d=0(0.1)3.0(0.2)6 R=0.1(0.1)1.0(0.2)3	3 or 4
Groves and Smith (1957)	R	R = 0.1, 0.5, 1(1)12 0 ≤ d ≤ 9 (horizontally) $0.0^3 1 \le 1 - p(R,d) \le 0.9^4$	Graphed
Owen (1962, Tables 8.1, 8.2)	R	d=0 p=0.01(0.01)0.99(0.001) $0.9^3(0.0^31)0.9^4$ to 0.9^9	4
	1 − p(R,d)	R − d = − 3.9(0.1)4.0 d = 0.1(0.1)6(0.5)10(1)20,∞	3
DiDonato and Jarnagin (1962)	R	p(R,d)=0.01(0.05)0.95,0.97, 0.99, 0.995, 0.9^3 to 0.9^6 d = 0.1,0.5(0.5)2(1)6,8,10	6
		d = 20,30,50,80,120	4-5
Bark et al. (1964, Table I)	1 − p(R,d)	R = 0.0(0.02)7.84 d = 0.0(0.02)3.00 See Sec. 2 of source for wider coverage	6

where

$$g_0 = 1 - \exp(-\tfrac{1}{2}R^2) \ , \qquad g_n = g_{n-1} - (\tfrac{1}{2}R^2)^n \exp(-\tfrac{1}{2}R^2)/n! \ ,$$

$$k_n = (\tfrac{1}{2}d^2)^n \exp(-\tfrac{1}{2}d^2)/n! = k_{n-1} (\tfrac{1}{2}d^2)/n.$$

After N iterations, the remainder is R_N, where $R_N = \sum_{n=N}^{\infty} g_n k_n$. Then

$$R_N < k_N \, g_N \{1 - (\tfrac{1}{2} Rd/N)^2\}^{-1} \quad , \quad N > \tfrac{1}{2} Rd$$

$$R_{N+1} \leq k_N \, g_N \qquad\qquad\qquad , \quad N > Rd/\sqrt{2} .$$

For large enough values of N, the summation can be stopped at any desired accuracy level; the second bound on R_N avoids computation of R_N at each iteration. Convergence is rapid once the series approaches its limit, and an initial error in g_0 or in k_0 will not grow, but will retain the same percentage error in $1 - r(R, d)$. The series $\Sigma g_n k_n$ for $p(R, d)$ is essentially that derived by Gilliland (1962, pp. 758-767).

[9.6.6.1] Approximations to $p(R, d)$ are also available; the first terms of the series in [9.6.5] provide one. Grubbs (1964, pp. 52-55) suggested the following:

$$p(R, d) \simeq \Pr(\chi_\nu^2 \leq mR^2/v) ,$$

where χ_ν^2 is a central chi-square r.v., $\nu = 2m^2/v$ d.f., $m = 1 + d^2/2$, and $v = 1 + d^2$. Hence $p(R, d) \simeq I(\tfrac{1}{2}R^2/\sqrt{v}, (m^2/v) - 1)$, where $I(\cdot, \cdot)$ is Pearson's Incomplete Gamma Function (Pearson, 1922). The Wilson-Hilferty approximation (see [9.2.6]) then gives

$$p(R, d) \simeq \Phi\left[\{(R^2/m)^{1/3} - (1 - vm^{-2}/9)\}/\sqrt{vm^{-2}/9}\right],$$

for which the greatest observed discrepancy was 0.023.

[9.6.6.2] Wegner's (1963) approximation (Read, 1971, p. 1731)

$$p(R, d) \simeq \begin{cases} 1 - \Phi\{d - (R^2 - 1)^{\frac{1}{2}}\} , & R \geq 3; \; d > 1.5, \; R > 1.8, \\ R^2(2 + \tfrac{1}{2}R^2)^{-1} \exp\{-d^2(2 + \tfrac{1}{2}R^2)^{-1}\}, & d \leq 1.5, \; R \leq 1; d \geq 1.5, R \geq 1.8, \\ 1 - \exp[-R^2\{1.416 + (0.397 - 0.0159R^2)d^2\}^{-2}], & \text{otherwise.} \end{cases}$$

The Bivariate Normal Distribution

has a maximum error 0.02 in the neighborhood of $(R, d) = (2, 1.5)$.

[9.6.6.3] If d is large, then (DiDonato and Jarnagin, 1962, p. 354)

$$p(R, d) \simeq \Phi(R - d - \tfrac{1}{2}d^{-1}) + D,$$

where $D = O(d^{-2})$. The error D is given in Table II of Bark et al. (1964). DiDonato and Jarnagin (1962, p. 353) tabulate values of R inversely as a function of $p(\cdot, \cdot)$ and d; for coverage, see Table 9.2.

[9.6.7] Offset circles and ellipses: *elliptical normal variables*.

Let (X, Y) have a BVN$(\mu_1, \mu_2; \sigma_1^2, \sigma_2^2, 0)$ distribution; we wish to evaluate the probability P defined by

$$P = \int\int_{x^2+y^2 \leq R^2} f(x, y; \mu_1, \mu_2; \sigma_1^2, \sigma_2^2, 0) \, dy \, dx \,,$$

where $f(\cdot)$ is the BVN joint density of *[9.1.1.1]*, the offset circle is centered at the origin, and $\sigma_1^2 \neq \sigma_2^2$.

We discuss the case in which $\mu_1 = \mu_2 = 0$ in sections *[9.6.7.1-9.6.12]*, and that for which $\mu_1 \neq \mu_2 \neq 0$ in sections *[9.6.13.1-2]*.

[9.6.7.1] Case I: $\mu_1 = \mu_2 = 0$. Without loss of generality $\sigma_1^2 \geq \sigma_2^2$. Let $R' = R/\sigma_1$ and $c = \sigma_2/\sigma_1$, so that $0 \leq c \leq 1$. The probability in *[9.6.7]* then depends only on R' and c, and is denoted by $P(R', c)$; other sources use the form $P(a_1, a_2, t)$ or $P(a_2, a_1, t)$, where $a_1 \leq a_2$, $c^2 = a_1/a_2$, and $R'^2 = t/a_2$. Then (Guenther and Terragno, 1964, p. 240)

$$P(R', c) = P(a_1, a_2, t) = \Pr(Z_1^2 + c^2 Z_2^2 \leq R'^2)$$
$$= \Pr(a_1 Z_1^2 + a_2 Z_2^2 \leq t) \,,$$

the c.d.f. of a quadratic form in i.i.d. N(0, 1) variables Z_1 and Z_2.

[9.6.7.2] Suppose we require $\Pr(b_1 X^2 + b_2 Y^2 \leq r)$, the probability of falling inside an ellipse centered at the origin, where $b_1 > 0$ and $b_2 > 0$. If $c' = b_1/\sigma_1^2 + b_2/\sigma_2^2$; $a_i' = b_1/(c\sigma_i^2)$, where $i = 1, 2$; and $t' = r/c$. Then (Owen, 1962, p. 181)

$$\Pr(b_1 X^2 + b_2 Y^2 \leq r) = P(a_1', a_2', t)$$

and we have reduced the problem to that of *[9.6.7]*.

TABLE 9.3 Tables of $P(R', c)$ and $P(a_1, a_2, t)$ with Coverage

Source	Function	Coverage	Decimal places
Grad and Solomon (1955)	$P(a_1, a_2, t)$	$(a_2, a_1) = (.5,.5),(.6,.4),(.7,.3),$ $(.8,.2),(.9,.1),(.95,.05),$ $(.99, .01), (1,0)$	4
		$t = 0.1(0.1)1(0.5)2(1)5$	
Owen (1962, Tables 8.3, 8.4)	$P(a_1, a_2, t)$	$(a_2, a_1) = (.5,.5),(.6,.4), (2/3,1/3),$ $(.7,.3),(.75,.25),(.8,.2)(.875,.125),$ $(.9,.1),(.95,.05),(.99,.01);$ $t = 0.1((0.1)1,1.5,2.0(1)5$	5
	t	$P(a_1, a_2, t) = 0.05(0.05)0.30(0.1)0.7$ $(0.05)0.95$	
Harter (1960)	$P(R', c)$	$c = 0(0.1)1.0$ $R' = 0.1(0.1)6.0$	7
	R'	$P = .5, .75, .95, .975,$ $.99, .995, .9975, .999$	5
Weingarten and DiDonato (1961)	R'	$c = 0.05(.05)1$ $P(R', c) = .05,(.05).95(.01).99$	5
Beyer (1966)	$P(R', c)$	$c = 0(0.1)1.0$ $R' = 0.1(0.1)5.8$	7

[9.6.8] Table 9.3 describes some available tabulations relating to $P(R', c)$; for others, see Eckler (1969, p. 564). Harter (1960, p. 724), used the following form for numerical integration by the trapezoidal rule, with some advice concerning chopping and truncation errors:

$$P(R', c) = \frac{2c}{\pi} \int_0^\pi \frac{1 - \exp\{-R'^2/(4c^2) \cdot [(1 + c^2) - (1 - c^2)\cos\phi]\}}{(1+c^2) - (1-c^2)\cos\phi} d\phi.$$

The Bivariate Normal Distribution

[9.6.9] Guenther and Terragno (1964, p. 241) show that

$$P(R', c) = p[\tfrac{1}{2}R'(c^{-1} + 1), \tfrac{1}{2}R'(c^{-1} - 1)]$$
$$- p[\tfrac{1}{2}R'(c^{-1} - 1), \tfrac{1}{2}R'(c^{-1} + 1)],$$

so that $P(R', c)$ may be obtained from the circular coverage function.

[9.6.10] The approximation to the circular coverage function $p(R, d)$ in *[9.6.6.1]* by Grubbs (1964, pp. 52-55) applies also to $P(R', c)$, being based on fitting the first two moments of a weighted sum of non-central chi-square variables to those of a central chi-square rv. Thus (see *[9.6.8.1]*)

$$P(R', c) \simeq \Pr(\chi_\nu^2 \leq R'^2/\{v(1 + c^2)\}),$$

where $\nu = 2/v$ and $v = 2(1 + c^4)/(1 + c^2)^2$; and

$$P(R', c) \simeq \Phi[\{R'^2/(1 + c^2)\}^{\tfrac{1}{3}} - (1 - v/9)\}/\sqrt{v/9}\,].$$

[9.6.11] Let $P_n(\lambda) = \Pr(Y > n)$, where Y has a Poisson distribution with mean λ. Then (Gilliland, 1962, pp. 758-767; Eckler, 1969, p. 565)

$$P(R', c) \simeq \frac{2c}{1 + c^2} \sum_{n=0}^{\infty} \left[\frac{(2n)!}{2^{2n}(n!)^2} \left(\frac{1-c^2}{1+c^2}\right)^{2n} P_{2n}\left\{\frac{(1+c^2)R'^2}{4c^2}\right\} \right].$$

[9.6.12] Lilliefors (1957, pp. 416-421) gives an approximation requiring no tabulated functions. Let $A = 1 + \tfrac{1}{4}R'^2$ and $B = \tfrac{1}{2}(c^2 - 1)$. Then

$$P(R', c) \simeq \sum_{n=1}^{\infty} \frac{(-1)^{n+1}R'^{2n}}{2^n n! c^{2n-1}} a_n;$$

$$a_1 = 1, \quad a_2 = A, \quad a_3 = A^2 + \tfrac{1}{2}B^2, \quad a_4 = A^3 + 3B^2 A/2,$$

$$a_5 = A^4 + 3B^2 A^2 + 3B^4/8.$$

For algorithms to derive higher factors a_n, see the source paper, or Eckler (1969, p. 565).

[9.6.13.1] <u>Case</u> <u>II</u>: $\mu_1 \neq 0$, $\mu_2 \neq 0$. We return to the general discussion at the beginning of *[9.6.7]*. The probability P has been tabulated to three decimal places by Lowe (1960, pp. 177-187). Four parameter values are involved: Lowe's tables cover $\sigma_1/\sigma_2 = 1, 2, 4, 8$; $R/\sigma_2 = 1, 2, 4, 8, 16, 32, 64$; and a range of values of μ_1/σ_1 and μ_2/σ_2 chosen to cover that region with substantial variation in the values of P.

Some extensive tables by Groenewoud et al. (1967), which give values of P to 5 decimal places, were derived using a recursive scheme operating on a series expansion; the authors describe the technique. The coverage is as follows:

$$\mu_1/\sigma_1 = \{0.0(0.5)5.0\}(\sigma_2/\sigma_1); \quad \mu_2/\sigma_2 = 0.0(0.5)5.0,$$

$$\sigma_2/\sigma_1 = 0.1(0.1)1.0; \quad R/\sigma_1 = 2A, 3A, 4A;$$

values of A are tabulated to 2 or 3 decimal places, and are chosen so that the probability range is reasonably covered.

[9.6.13.2] Gilliland (1962, pp. 759-761) gives a series expansion for P; see Eckler (1969, p. 565). The approximation of Grubbs (1964, pp. 51-62) given in *[9.6.6.1]* holds for P if

$$m = 1 + (\mu_1^2 + \mu_2^2)/(\sigma_1^2 + \sigma_2^2),$$
$$V = 2(\sigma_1^4 + \sigma_2^4 + 2\mu_1^2\sigma_1^2 + 2\mu_2^2\sigma_2^2)/(\sigma_1^2 + \sigma_2^2)^2.$$

Then (see also Eckler, 1969, pp. 566-567), if χ_ν^2 is a central chi-square rv, where $\nu = 2m^2/v$ d.f., and $\sigma^2 = \sigma_1^2 + \sigma_2^2$,

$$P \simeq \Pr(\chi_\nu^2 \leq mR^2/v),$$
$$\simeq \Phi[\{(R^2/(m\sigma^2))^{1/3} - (1 - vm^{-2}/9)\}\big/\{(vm^{-2}/9)^{1/2}\}].$$

REFERENCES

Abramowitz, M. and Stegun, I. A. (eds.) (1970). *Handbook of Mathematical Functions*, Washington, D.C. : National Bureau of Standards. *[9.2.1, 4; 9.8]*

Ahsanullah, M. (1985). Some characterizations of the bivariate normal distribution, *Metrika 32*, 215-218. *[9.4.2.1,4]*

Aroian, L. A. (1986). The distribution of the quotient of two correlated normal random variables, *A.S.A., Proceedings of the Business and Economics Section*, 612-613. *[9.5.1.2]*

Aroian, L. A., Taneja, V. S. and Cornwell, L. W. (1978). Mathematical forms of the distribution of the product of two normal variables, *Communications in Statistics A 7*, 165-172. *[9.5.3]*

Bark, L. S., Bol'shev, L. N., Kuznetsov, P. E. and Cherenkov, A. P. (1964). *Tables of the Rayleigh-Rice Distributions*, Moscow: Voychislityelnoy Tsentr. *[9.6.6.3; Table 9.2]*

Beyer, W. H. (ed.) (1966). *Handbook of Tables for Probability and Statistics*, Cleveland, Ohio: Chemical Rubber Co. *[Table 9.3]*

Bhattacharyya, A. (1943). On some sets of sufficient conditions leading to the normal bivariate distribution, *Sankhyā 6*, 399-406. *[9.4.2]*

Birnbaum, Z. W., Paulson, E. and Andrews, F. C. (1950). On the effect of selection performed on some coordinates of a multidimensional population, *Psychometrika 15*, 191-204. *[9.5.6.1]*

Bouver, H. and Bargmann, R. E. (1979). Comparison of computational algorithms for the evaluation of the univariate and bivariate normal distribution, *Proceedings of the Computer Science and Statistics 12th Annual Symposium on the Interface*, 344-348. *[9.3.1.2]*

Boys, R. J. (1989). A remark on Algorithm A576: An integral useful in calculating noncentral t and bivariate normal probabilities, *Applied Statistics 38*, 580-582. *[9.3.3.5]*

Brennan, L. E. and Reed, I. S. (1965). A recursive method of computing the Q Function, *IEEE Transactions on Information Theory*, 312-313. *[9.6.5]*

Brucker, J. (1979). A note on the bivariate normal distribution, *Communications in Statistics A 8*, 175-177. *[9.4.2.4]*

Burington, R. S. and May, D. C. (1970). *Handbook of Probability and Statistics with Tables*. Sandusky, Ohio: Handbook Publishers. *[Table 9.2]*

Cadwell, J. H. (1951). The bivariate normal integral, *Biometrika 38*, 475-479. *[9.3.3.2]*

Castillo, E. and Galambos, J. (1989). Conditional distributions and the bivariate normal distribution, *Metrika 36*, 209-214. *[9.4.2.6]*

Chou, Y-M (1985). A remark on Algorithm AS76: An integral useful in calculating noncentral T and bivariate normal probabilities, *Applied Statistics* 34, 100-101. *[9.3.3.5]*

Chou, Y.-M., and Owen, D. B. (1984). An approximation to percentiles of a variable of the bivariate normal distribution when the other variable is truncated, with applications, *Communications in Statistics A* 13, 2535-2547. *[9.5.6.1,2]*

Cook, M. B. (1951a). Bivariate k-statistics and cumulants of their joint sampling distribution, *Biometrika* 38, 179-195. *[9.5.6.2]*

Cox, D. R. and Wermuth, N. (1991). A simple approximation for bivariate and trivariate normal integrals, *International Statistical Review* 59, 263-269. *[9.3.1.5]*

Craig, C. C. (1936). On the frequency function of xy, *Annals of Mathematical Statistics* 7, 1-15. *[9.5.3]*

Daley, D. J. (1974). Computation of bi- and tri-variate normal integrals, *Applied Statistics* 23, 435-438. *[9.3.3.2]*

David, H. A. (1981). *Order Statistics* (2nd ed.), New York: Wiley. *[9.5.5]*

Di Donato, A. R. and Jarnagin, M. P. (1962). A method for computing the circular coverage function, *Mathematics of Computation* 16, 347-355. *[9.6.1; 9.6.6.3; Table 9.2]*

Divgi, D. R. (1979). Calculation of univariate and bivariate normal probability functions, *Annals of Statistics* 7, 903-910 *[9.3.4]*

Drezner, Z. (1978). Computation of the bivariate normal integral, *Mathematics of Computation* 32, 277-279. *[9.3.1.2]*

Drezner, Z. and Wesolowsky, G. O. (1990). On the computation of the bivariate normal integral, *Journal of Statistical Computation and Simulation* 35, 101-107. *[9.3.1.3]*

Eckler, A. R. (1969). A survey of coverage problems associated with point and area targets, *Technometrics* 11, 561-589. *[9.6.4, 8, 11, 12; 9.6.13.2]*

Eckler, A. R. (1988). Target coverage, *Encyclopedia of Statistical Sciences*, Vol. 9 (Kotz, S., Johnson, N. L. and Read, C. B., eds.) , 180-183, New York: Wiley. *[9.6.1]*

Fieger, W. (1977). A characterization of the multivariate normal distribution, *Communications in Statistics A* 6, 135-140. *[9.4.3.3]*

Fieller, E. C. (1932). The distribution of the index in a normal bivariate population, *Biometrika* 24, 428-440. *[9.5.1.1]*

Fieller, E. C., Lewis, T. and Pearson, E. S. (1955). *Correlated Random Normal Deviates, Tracts for Computers 26*, Cambridge University Press; Corrigenda (1956): *Biometrika* 43, 496. *[9.2.5]*

Fraser, D.A.S. and Streit, F. (1980). A further note on the bivariate normal distribution, *Communications in Statistics A* 9, 1097-1099. *[9.4.2.4]*

Gajjar, A. V. and Subrahmaniam, K. (1978). On the sample correlation coefficient in the truncated bivariate normal population, *Communications in Statistics B* 7, 455-478. *[9.5.6.2; 9.5.7]*

Gelman, A. and Meng, X-L. (1991). A note on bivariate distributions that are conditionally normal, *American Statistician* 45, 125-126. *[9.4.2.7]*

Gideon, R. A. and Gurland, J. (1978). A polynomial type approximation for bivariate normal variates, *S.I.A.M. Journal on Applied Mathematics* 34, 681-684. *[9.3.2]*

Gilliland, D. C. (1962). Integral of the bivariate normal distribution over an offset circle, *Journal of the American Statistical Association* 57, 758-768. *[9.6.5, 11; 9.6.13.2]*

Grad, A. and Solomon, H. (1955). Distribution of quadratic forms and some applications, *Annals of Mathematical Statistics* 26, 464-477. *[Table 9.3]*

Groenewoud, C., Hoaglin, D. C. and Vitalis, J. A. (1967). *Bivariate Normal Offset Circle Probability Tables*, Buffalo, NY: Cornell Aeronautical Laboratory, Inc. (2 vols.). *[9.6.13.1]*

Groves, A. D. and Smith, E. S. (1957). Salvo hit probabilities for offset circular targets, *Operations Research* 5, 222-228. *[Table 9.2]*

Grubbs, F. E. (1964). Approximate circular and noncircular offset probabilities of hitting, *Operations Research* 12, 51-62. *[9.6.6.1; 9.6.10; 9.6.13.2]*

Guenther, W. C. and Terragno, P. J. (1964). A review of the literature on a class of coverage problems, *Annals of Mathematical Statistics* 35, 232-260. *[9.6.4; 9.6.7.1; 9.6.9]*

Gupta, A. K. and Tracy, D. S. (1980). A study of the effects of truncation in bivariate normal distribution, *Biometrical Journal* 22, 593-613. *[9.5.8]*

Hamedani, M. (1991). On a recent characterization of the bivariate normal distribution, *Metrika* 38, 255-258. *[9.4.2.6]*

Hamedani, M. (1992). Bivariate and multivariate characterizations: A brief survey, *Communications in Statistics A 21*, 2665-2688. *[9.4.2]*

Harter, H. L. (1960). Circular error probabilities, *Journal of the American Statistical Association 55*, 723-731. *[9.6.8; Table 9.3]*

Hill, I. D. (1978). A remark on Algorithm AS76: An integral useful in calculating noncentral t and bivariate normal probabilities, *Applied Statistics 27*, 379. *[9.3.3.5]*

Hinkley, D. V. (1969). On the ratio of two correlated normal random variables, *Biometrika 56*, 635-639. *[9.5.1.1]*

IMSL (1987). *IMSL Subroutine Library Documentation*, Houston: IMSL. *[9.3.3.1]*

Johnson, N. L. and Kotz, S. (1975). On some generalized Farlie-Gumbel-Morgenstern distributions, *Communications in Statistics 4*, 415-427. *[9.1.2]*

Johnson, N. L. and Kotz, S. (1970). *Distributions in Statistics. Continuous Univariate Distributions-2*, New York: Wiley. *[9.6.1]*

Johnson, N. L. and Kotz, S. (1972). *Distributions in Statistics: Continuous Multivariate Distributions*, New York: Wiley. *[9.3.3.4; 9.4.1; 9.5.1.1, 9.5.4,6; 9.5.6.2,3; 9.6.1]*

Kamat, A. R. (1953). Incomplete and absolute moments of the multivariate normal distribution with some applications, *Biometrika 40*, 20-34. *[9.1.8]*

Kella, O. (1986). On the distribution of the maximum of bivariate normal random variables with general means and variances, *Communications in Statistics A 15*, 3265-3276. *[9.5.5]*

Kendall, M. G. and Stuart, A. (1967). *Advanced Theory of Statistics, Vol. 2* (3rd edn.), New York: Macmillan. *[9.1.5]*

Kepner, J. L., Harper, J. D. and Keith, S. Z. (1989). A note on evaluating a certain orthant probability, *The American Statistician 43*, 48-49. *[9.2.1]*

Khatri, C. G. (1979). Characterizations of multivariate normality. II: Through linear regressions, *Journal of Multivariate Analysis 9*, 589-598. *[9.4.2]*

Khatri, C. G. and Rao, C. R. (1976). Characterizations of multivariate normality. I: Through independence of some statistics, *Journal of Multivariate Analysis 6*, 81-94. *[9.4.3.2]*

Lancaster, H. O. (1957). Some properties of the bivariate normal distribution considered in the form of a contingency table, *Biometrika* **44**, 289-292. [9.1.5]

Laurent, A. G. (1957). Bombing problems – a statistical approach, *Operations Research* **5**, 75-89. [9.5.2]

Lilliefors, H. W. (1957). A hand-computation determination of kill probability for weapons having spherical lethal volume, *Operations Research* **5**, 416-421. [9.6.12]

Lowe, J. R. (1960). A table of the integral of the bivariate normal distribution over an offset circle, *Journal of the Royal Statistical Society B* **22**, 177-187. [9.6.13.1]

Lowerre, J. M. (1983). An integral of the bivariate normal and an application, *American Statistician* **37**, 235-236. [9.2.7]

Mee, R. W. and Owen, D. B. (1983). A simple approximation for bivariate normal probabilities, *Journal of Quality Technology* **15**, 72-75. [9.3.1.1, 3, 4; 9.5.4]

Meeker, W. Q., Jr., Cornwell, L. W. and Aroian, L. A. (1983). *Selected Tables in Mathematical Statistics* (Vol. VII): *The Product of Two Normally Distributed Random Variables*, Providence, RI: American Mathematical Society. [9.5.3]

Mehler, G. (1866). Reihenentwicklungen und Laplacesche Funktionen, *J. reine Angew. Math.* **66**, 161-176. [9.1.1.2]

Moskowitz, H. and Tsai, H.-T. (1989). An error-bounded polynomial approximation for bivariate normal probabilities, *Communications in Statistics B* **18**, 1421-1438. [9.3.1.4]

Muthén, B. (1990). Moments of the censored and truncated bivariate normal distribution, *British Journal of Mathematical and Statistical Psychology* **43**, 131-143. [9.5.8]

Nath, G. B. (1973). Correlation in a truncated bivariate normal distribution, *Statistica* **33**, 563-587. [9.5.7]

National Bureau of Standards (1959). *Tables of the Bivariate Normal Distribution Function and Related Functions*, Applied Mathematics Series 50, U.S. Government Printing Office, Washington 25, D.C. [9.2.1; 9.3.3.4]

Nicholson, C. (1943). The probability integral for two variables, *Biometrika* **33**, 59-72. [9.2.4]

Öksoy, D. and Aroian, L. A. (1986). Computational techniques and examples of the density and the distribution of the quotient of two correlated normal variables, *A.S.A., Proceedings of the Business and*

Economics Section, 660-663. *[9.5.1.2]*

Owen, D. B. (1956). Tables for computing bivariate normal probabilities, *Annals of Mathematical Statistics* 27, 1075-1090. *[9.2.2.2; 9.2.3; 9.3.3.1]*

Owen, D. B. (1962). *Handbook of Statistical Tables*, Reading, Mass: Addison-Wesley. *[9.6.3; 9.6.7.2; Table 9.2]*

Owen, D. B. (1980). A table of normal integrals, *Communications in Statistics B* 9, 389-419. *[9.2.3]*

Owen, D. B., McIntire, D. and Seymour, E. (1975). Tables using one or two screening variables to increase acceptable product under one-sided specifications, *Journal of Quality Technology* 7, 127-138. *[9.2.6]*

Parrish, R. S. and Bargmann, R. E. (1981). A method for the evaluation of cumulative probabilities of bivariate distributions using the Pearson family, *Statistical Distributions in Scientific Work 5*, (C. Tallie, G. P. Patil and B. A. Baldessari, eds.), 241-257. D. Reidel. *[9.3.1.2]*

Pearson, K. (ed.) (1922). *Tables of the Incomplete* Γ*-Function*, London: H. M. Stationery Office. *[9.6.6.1]*

Pearson, K. (1931). *Tables for Statisticians and Biometricians 2*, Cambridge University Press, England. *[9.2.4]*

Pólya, G. (1949). Remarks on computing the probability integral in one and two dimensions, *Proceedings of the First Berkeley Symposium on Mathematical Statistics and Probability*, 63-78. Berkeley, Calif: University of California Press. *[9.2.1]*

Rao, B. R., Garg, M. L. and Li, C. C. (1968). Correlation between the sample variances in a singly truncated bivariate normal distribution, *Biometrika* 55, 433-436. *[9.5.6.1, 3]*

Rao, C. R. (1975). Some problems in characterization of the multivariate normal distribution, *A Modern Course on Statistical Distributions in Scientific Work 3, Characterizations and Applications* (G. P. Patil et al, eds.), New York: D. Reidel. *[9.4.3.1]*

Read, C. B. (1971). Two casualty-estimation problems associated with area targets and the circular coverage function, *Operations Research* 19, 1730-1741. *[9.6.2; 9.6.6.2]*

Regier, M. H. and Hamdan, M. A. (1971). Correlation in a bivariate normal distribution with truncation in both variables, *Australian Journal of Statistics* 13, 77-82. *[9.5.7]*

Rosenbaum, S. (1961). Moments of a truncated bivariate normal distribution, *Journal of the Royal Statistical Society B* 23, 405-408.

[9.5.7]

Ruben, H. (1961). Probability contents of regions under spherical normal distributions, III: The bivariate normal integral, *Annals of Mathematical Statistics* 32, 171-186. [9.3.4]

Sheppard, W. F. (1898). On the application of the theory of error to cases of normal distributions and normal correlation, *Philosophical Transactions of the Royal Society of London* 192, 101-167. [9.2.1]

Smirnov, N. V. and Bol'shev, L. N. (1962). *Tables for Evaluating a Function of a Bivariate Normal Distribution*, Is'datel'stov Akademii Nauk SSR, Moscow. [Table 9.1]

Solomon, H. (1953). Distribution of the measure of a random two-dimensional set, *Annals of Mathematical Statistics* 24, 650-656. [Table 9.2]

Springer, M. D. (1983). Review of *Selected Tables in Mathematical Statistics* (Vol. VII): *The Product of Two Normally Distributed Random Variables* (Meeker, W. Q., Jr., Cornwell, L. W. and Aroian, L. A., 1983), *Technometrics* 25, 211-212. [9.5.3]

Sowden, R. R. and Ashford, J. R. (1969). Computation of the bi-variate normal integral, *Applied Statistics* 18, 169-180. [9.3.3.2]

Stuart, A. and Ord, J. K. (1987), *Kendall's Advanced Theory of Statistics, Vol. 1* (5th edn.) New York: Oxford. [9.1.1.2; 9.1.7; 9.4.1]

Stuart, A. and Ord, J. K. (1991). *Kendall's Advanced Theory of Statistics, Vol. 2* (5th edn.). New York: Oxford. [9.4.2.5]

Tallis, G. M. (1961). The moment generating function of the truncated multinormal distribution, *Journal of the Royal Statistical Society B* 23, 223-229. [9.5.6.2]

Terza, J. V. and Welland, U. (1991). A comparison of bivariate normal algorithms, *Journal of Statistical Computation and Simulation* 39, 115-127. [9.3.1.2; 9.3.4]

Thomas, G. E. (1986). A remark on Algorithm AS76: An integral useful in calculating non-central t and bivariate normal probabilities, *Applied Statistics,* 35, 310-312. [9.3.3.5]

Wang, Y. J. (1987). The probability integrals of bivariate normal distributions: a contingency table approach, *Biometrika* 74, 185-190. [9.3.5]

Wegner, L. H. (1963). Quick-count, a general war casualty estimation model, *RM 3811-PR*, The Rand Corporation. [9.3.6.2]

Weil, H. (1954). The distribution of radial error, *Annals of Mathematical Statistics* 25, 168-170. *[9.5.2]*

Weiler, H. (1959). Mean and standard deviations of a truncated normal bivariate distribution, *Australian Journal of Statistics 1*, 73-81. *[9.5.6.1; 9.5.7]*

Weingarten, H. and DiDonato, A. R. (1961). A table of generalized circular error, *Mathematics of Computation 15*, 169-173. *[Table 9.3]*

Wilks, S. S. (1962). *Mathematical Statistics*, New York: Wiley. *[9.1.4]*

Wold, H. (1948). *Random Normal Deviates*, Tracts for Computers 25, Cambridge, England: Cambridge University Press. *[9.2.5]*

Yamauti, Z. (ed.) (1972). *Statistical Tables and Formulas with Computer Applications*, Tokyo: Japanese Standards Association. *[Table 9.1]*

Young, J. C. and Minder, C. E. (1974). Algorithm AS 76. An integral useful in calculating non-central t and bivariate normal probabilities, *Applied Statistics 23*, 455-457. Correction (1979)*28*, 113. *[9.3.3.5]*

Zelen, M. and Severo, N. C. (1960). Graphs for bivariate normal probabilities, *Annals of Mathematical Statistics 31*, 619-624. *[9.2.4]*

Chapter 10
BIVARIATE NORMAL SAMPLING DISTRIBUTIONS

In Chapter 9 the structure and properties of the bivariate normal (BVN) distribution and of associated distributions were set out. In this chapter we discuss distributions based on simple random samples (X_1, Y_1), $(X_2, Y_2),...,(X_n, Y_n)$ from a BVN parent. In Section *10.1*, properties of the sample mean vector, variances and covariances, sample regression statistics, concomitants of order statistics of $X_1,...,X_n$, and statistics based on truncated BVN distributions are presented, along with a brief discussion of bivariate normal tolerance regions.

Of prime interest is the distribution of the sample correlation coefficient R. Exact results involving R and transformations of R appear in Section *10.2*, while those of approximations to the distribution of R are given in Section *10.3*.

10.1 BVN STATISTICS – GENERAL

[10.1.1] The <u>joint pdf of an independent sample</u> (X_1, Y_1), (X_2, Y_2), ..., (X_n, Y_n) of size n from a BVN distribution is given by

$$f(\underline{x}_n, \underline{y}_n; \mu_1, \mu_2; \sigma_1^2, \sigma_2^2, \rho) = \left[(2\pi\sigma_1\sigma_2)^n(1-\rho^2)^{n/2}\right]^{-1} \exp[-\tfrac{1}{2}(1-\rho^2)^{-1}$$

$$\cdot \{\sum(x_i-\mu_1)^2/\sigma_1^2 - 2\rho\sum\left[(x_i-\mu_1)(y_i-\mu_2)/(\sigma_1\sigma_2)\right] + \sum(y_i-\mu_2)^2/\sigma_2^2\}]$$

$$= f(\overline{x}, \overline{y}; \mu_1, \mu_2, \sigma_1^2/n, \sigma_2^2/n, \rho) \ h(s_1^2, s_2^2, r; \sigma_1^2, \sigma_2^2, \rho),$$

the product of the joint p.d.f. of $(\overline{X}, \overline{Y})$ and the joint p.d.f. of (S_1^2, S_2^2, R), given in *[10.1.2]* below, where

$$S_1^2 = \sum (X_i - \bar{X})^2/(n-1), \quad S_2^2 = \sum (Y_i - \bar{Y})^2/(n-1),$$

$$R = \frac{\sum (X_i - \bar{X})(Y_i - \bar{Y})}{\left[\sum (X_i - \bar{X})^2 \sum (Y_i - \bar{Y})^2\right]^{1/2}}.$$

R is the sample or product-moment correlation coefficient. Thus $(\bar{X}, \bar{Y}, S_1^2, S_2^2, R)$ is jointly sufficient for $\underline{\theta}$, and $(\bar{X}, \bar{Y})$ is jointly independent of (S_1^2, S_2^2, R).

[10.1.2.1] The distribution of $(\bar{X}, \bar{Y})$ is $BVN(\mu_1, \mu_2; \sigma_1^2/n, \sigma_2^2/n, \rho)$. The marginal distributions of $\bar{X}$ and $\bar{Y}$ are $N(\mu_1, \sigma_1^2)$ and $N(\mu_2, \sigma_2^2)$, respectively.

The distribution of (S_1^2, S_2^2, R) has joint p.d.f. given by

$$h(s_1^2, s_2^2, r; \sigma_1^2, \sigma_2^2, \rho) = [n\{2\pi \sigma_1\sigma_2 \sqrt{(1-\rho^2)}\}^{n-1}]^{-1}(s_1 s_2)^{n-2}(1-r^2)^{(n-4)/2}$$

$$\cdot \exp\left[-\frac{n}{2(1-\rho^2)}\left\{\frac{s_1^2}{\sigma_1^2} - 2\frac{\rho r s_1 s_2}{\sigma_1 \sigma_2} + \frac{s_2^2}{\sigma_2^2}\right\}\right],$$

$s_1 > 0$, $s_2 > 0$, $|r| \le 1$. See also Fisher (1915, pp. 507-521).

If in addition $\rho = 0$, then R, S_1^2 and S_2^2 are mutually independent (Kshirsagar, 1972, pp. 32-34).

The marginal distributions of $(n-1)S_1^2/\sigma_1^2$ and of $(n-1)S_2^2/\sigma_2^2$ are χ^2_{n-1}; see *[5.3.2]*. The sampling distribution of R is treated separately in *[10.2]* and *[10.3]*.

[10.1.2.2] From Prabhakaran et al. (1991, pp. 95-102) we can list some of the joint moments of (S_1^2, S_2^2, R); see also *[5.3.4]*:

$$\text{Var}(S_1^2) = 2\sigma_1^4/(n-1),$$

$$\text{Cov}(S_1^2, S_2^2) = 2\rho\sigma_1\sigma_2/(n-1),$$

$$E(R^2 S_1^2) = \rho^2\sigma_1^2 + \frac{(1-\rho^2)}{n-1}\sigma_1^2,$$

The Bivariate Normal Distribution

$$\text{Cov}(S_1^2, R^2 S_1^2) = \text{Var}(R^2 S_1^2) = \frac{2\sigma_1^4}{n-1}\rho^2(2-\rho^2) + \frac{2\sigma_1^4}{(n-1)^2}(1-\rho^2)^2,$$

$$E(R^2 S_1^4) = \rho^2 \sigma_1^4 + \frac{1 + 3\rho^2 - 2\rho^4}{n-1}\sigma_1^4 + \frac{2(1-\rho^2)^2}{(n-1)^2}\sigma_1^4,$$

$$E(R^4 S_1^4) = \rho^4 \sigma_1^4 + \frac{2\rho^2(3 - 2\rho^2)}{n-1}\sigma_1^4 + \frac{3(1-\rho^2)^2}{(n-1)^2}\sigma_1^4.$$

Also, when n is large,

$$\text{Var}(R^2 S_2^2/S_1^2) \doteq \text{Cov}(R^2 S_2^2/S_1^2, S_2^2/S_1^2) \doteq \frac{4\rho^2(1-\rho^2)}{n-3}\frac{\sigma_2^4}{\sigma_1^4},$$

$$\text{Var}\left[(1-R^2)S_2^2/S_1^2\right] \doteq \frac{4(1-\rho^2)}{n-3}\frac{\sigma_2^4}{\sigma_1^4}.$$

[10.1.3] If n pairs of observations from a BVN distribution are available, the <u>sample regression coefficient</u> of Y on X is b, where $b = RS_2/S_1$; see *[10.1.1]*. Let $\beta = \rho\sigma_2/\sigma_1$, the population regression coefficient of Y on X. Then (Stuart and Ord, 1987, Secs. 16.35-37)

$$(n-1)^{\frac{1}{2}}(b-\beta)\sigma_1/\{\sigma_2(1-\rho^2)^{\frac{1}{2}}\} \sim \text{'t'}_{n-1},$$

$$(n-2)^{\frac{1}{2}}(b-\beta)S_1/\{S_2(1-R^2)^{\frac{1}{2}}\} \sim \text{'t'}_{n-2};$$

$$n\, S_2^2(1-R^2)/(n-2) \sim \chi_{n-2}^2,$$

and is independent of b. The second of these results is more useful than the first, since ρ, σ_1 and σ_2 are generally unknown. When $\beta = 0$, and hence $\rho = 0$, this is the variable described in *[10.1.6]*.

The sampling distribution of b (Stuart and Ord, 1991, p. 1005) has mean β;

$$\text{Var}(b) = (\sigma_2^2/\sigma_1^2)(1-\rho^2)/(n-3), \quad n \geq 4,$$

$$\mu_3(b)/\sigma^3(b) = 0 \quad \text{(skewness)}, \quad n \geq 5,$$

$$\mu_4(b)/\sigma^4(b) = 6/(n-5) \text{ (kurtosis)}, \quad n \geq 6.$$

[10.1.4] If $\sigma_1^2 = \sigma_2^2$, the sample *Intraclass Correlation Coefficient* U may be of interest; for example, when it is not clear which variable should be labeled X and which Y. Thus, X and Y might denote the heights of twin brothers, and a measure of association between them is desired. The statistic U is defined by

$$U = 2S_{12}/(S_1^2 + S_2^2),$$

where S_1^2 and S_2^2 are defined in *[10.1.1]* and

$$S_{12} = \sum_{i=1}^{n} \{(X_i - \bar{X})(Y_i - \bar{Y})\}/(n-1),$$

the sample covariance. The p.d.f. of U (DeLury, 1938, pp. 149-151) is given by

$$g(u) = \begin{cases} \dfrac{\Gamma\{\frac{1}{2}(n+1)\}}{\sqrt{\pi}\,\Gamma(\frac{1}{2}n)} (1-\rho^2)^{\frac{1}{2}n}(1-\rho u)^{-n}(1-u^2)^{\frac{1}{2}(n-1)} &, |u| \leq 1; \\ 0 &, |u| > 1. \end{cases}$$

If $\rho=0$, the distribution of U can be compared with that of R in *[10.1.4.1]*.

[10.1.5] Let (X_i, Y_i), $i = 1,...,n$ be a random sample from a BVN(θ) distribution as defined in *[9.1.1]*. Let

$$X_{(1)} \leq X_{(2)} \leq \cdots \leq X_{(n)}$$

define the order statistics of $(X_1, ..., X_n)$, and let $Y_{[1]}, Y_{[2]}, ..., Y_{[n]}$ be the Y's corresponding to $X_{(1)}, ..., X_{(n)}$. These are the *concomitants* of the order statistics or *induced order statistics*, and they will not in general be ordered.

For general surveys on research into the properties of concomitants see David (1982, pp. 89-100), Bhattacharya (1984, pp. 383-403) and David (1993, pp. 507-518).

The Bivariate Normal Distribution

[10.1.5.1] From the regression equation in *[9.1.4]* David (1981, p. 110) derives moments of $Y_{[r]}$, $1 \leq r \leq n$. Let $Z_{(1)} \leq Z_{(2)} \leq \ldots \leq Z_{(n)}$ be order statistics of a random sample $Z_1, \ldots, Z_n$ from a $N(0,1)$ distribution, and define for $1 \leq r, s \leq n$,

$$\alpha_{r:n} = E(Z_{(r)}) \,, \quad \beta_{rs:n} = \text{Cov}(Z_{(r)}, Z_{(s)}) \,, \quad \beta_{rr:n} = \text{Var}(Z_{(r)}) \,.$$

Then

$$E(Y_{[r]}) = \mu_2 + \rho\sigma_2 \alpha_{r:n} \,,$$
$$\text{Var}(Y_{[r]}) = (1-\rho^2)\sigma_2^2 + \rho^2\sigma_2^2 \beta_{rr:n} \,,$$
$$\text{Cov}(X_{(r)}, Y_{[s]}) = \rho\sigma_1\sigma_2 \beta_{rs:n} \,,$$
$$\text{Cov}(Y_{[r]}, Y_{[s]}) = \rho^2\sigma_2^2 \beta_{rs:n} \,, \quad r \neq s \,.$$

See also Watterson (1959, pp. 814-824) and David (1973, pp. 295-300).

[10.1.5.2] Now write $Y_{[r:n]}$ for $Y_{[r]}$ and consider the case in which $(X, Y) \sim \text{SBVN}(\rho)$. If $\lim_{n\to\infty}(r/n) = \lambda$, where $0 < \lambda < 1$, then as n increases,

$$E(Y_{[r:n]}) \sim \rho Z_\lambda \,,$$

where the $N(0, 1)$ quantile Z_λ is defined in *[2.1.2]*. Then the rv $Y_{[r:n]} - \rho Z_\lambda$ has asymptotically a $N(0,(1-\rho^2)\sigma_2^2)$ distribution (David, 1981, pp. 282-283). Further,

$$E(Y_{[r:n]}) \sim \begin{cases} \rho\sqrt{(2\log n)} \,, & \lambda = 1, \\ -\rho\sqrt{(2\log n)} \,, & \lambda = 0 \,. \end{cases}$$

Let $R_{t,n}$ be the rank of $Y_{[t:n]}$ among the Y_i's, i.e., the number of integers j such that $Y_{[t:n]} > Y_{[j:n]}$. David and Galambos (1974, pp. 762-770) have determined the asymptotic distribution and moments of $R_{t,n}$ when $t/n \to \lambda$, $0 < \lambda < 1$, as $n \to \infty$. They show that, if $t/n \to \lambda$ and $r/n \to \eta$, then

$$\lim_{n\to\infty} \Pr(Y_{[t:n]} \geq Y_{[r:n]}) = \Phi\{\rho[2(1-\rho^2)]^{-\frac{1}{2}}(Z_\lambda - Z_\eta)\}$$

for $0 \leq \lambda \leq 1$ and $0 \leq \eta \leq 1$, but excepting $\lambda = \eta = 1$ or $\lambda = \eta = 0$. In these cases the asymptotic probability is $\frac{1}{2}$. David et al. (1977, pp. 216-223) provide an illustrative table of the distribution of $R_{t:n}$ for different values of ρ and of the ratio $r/(n+1)$ when $n = 9$; they provide a similar table leading to values of $E(R_{t:n})/(n+1)$ when $n = 9, 19, 39, \infty$. Each table runs to four decimal places.

[10.1.5.3] In the set up of this section, for any $1 \leq r_1 < ... < r_k \leq n$,

$$\lim_{n \to \infty} \Pr\left[Y_{[r_1]} \leq y_1, ..., Y_{[r_k]} \leq y_k\right] = \prod_{i=1}^{k} \Phi(y_i/\sigma),$$

where $\sigma^2 = \sigma_2^2(1 - \rho^2)$. This describes the asymptotic independence of the concomitants (Bhattacharya, 1984, pp. 392-393).

[10.1.6] Rao et al. (1968, pp. 434-436) show that, if ρ_T^* is the correlation coefficient between the sample variances s_1^2 and s_2^2 of X and Y, respectively, in samples of size n from a SBVN(ρ) distribution truncated as in *[9.5.6]*, and if μ_{rs} is the (r,s)th central moment $E\{(X - E_T(X))^r(Y - E_T(Y))^s\}$, then to $O(n^{-1})$,

$$\rho_T^* = (\mu_{22} - \mu_{20}\,\mu_{02})/\{(\mu_{40} - \mu_{20}^2)(\mu_{04} - \mu_{02}^2)\}^{1/2}.$$

Further,

$$\rho_T^* \geq \rho_T^2 \text{ and } |\rho_T^*| \leq |\rho_T|$$

for all choices of a and of ρ; Gajjar and Subrahmanian (1978, p. 458) give expressions for the moments of the sample correlation coefficient r_T; see also Cook (1951, pp. 368-376).

[10.1.7] Let $X_1, X_2,...,X_n$ be a random sample from a $N(\mu, \sigma^2)$ population and $a_1,...,a_n, b_1,...,b_n$ be constants. Then $(\Sigma_{i=1}^n a_i X_i, \Sigma_{i=1}^n b_i X_i)$ has a BVN($\underline{\theta}$) distribution, where in the notation of *[9.1.1.1]*,

$$\underline{\theta}' = \left[(\Sigma a_i)\mu, (\Sigma b_i)\mu; (\Sigma a_i^2)\sigma^2, (\Sigma b_i^2)\sigma^2, (\Sigma a_i b_i)/\sqrt{(\Sigma a_i^2)(\Sigma b_i^2)}\right]$$

$$= \left[(\Sigma a_i)\mu, (\Sigma b_i)\mu; \sigma^2, \sigma^2, \Sigma a_i b_i\right]$$

The Bivariate Normal Distribution

if $\sum a_i^2 = \sum b_i^2 = 1$, and

$$\underline{\theta}' = \left[0, 0; \sigma^2, \sigma^2, \sum a_i b_i \right]$$

if, in addition, $\sum a_i = \sum b_i = 0$.

Hence $\sum a_i X_i$ and $\sum b_i X_i$ are independent if and only if $\sum a_i b_i = 0$, that is, these linear transformations are orthogonal. In the design of experiments, $X_1, \ldots, X_n$ may represent treatment means under equal group sizes; $\sum a_i X_i$ is termed a contrast whenever $\sum a_i = 0$, the normalizing condition $\sum a_i^2 = 1$ being added for convenience.

[10.1.8] Mason et al. (1987, pp. 448-451) have provided a technique for generating samples from a BVN($\underline{\theta}$) distribution with known parameters. The technique is "simpler and more efficient" than earlier procedures, which are referenced in their paper.

[10.1.9] One application of the sampling distributions discussed in this section is to the construction of <u>bivariate normal tolerance regions</u>, that contain at least a stated proportion δ of a BVN distribution with a given level of confidence γ. This requires defining a region A such that

$$\Pr\left[\int\int_A f(x, y: \mu_1, \mu_2; \sigma_1^2, \sigma_2^2, \rho)\, dxdy > \delta \right] = \gamma,$$

where A is specified in terms of sample statistics such as $\overline{X}$, $\overline{Y}$, S_1^2, S_2^2 and R based on a sample of size n from the distribution. Hall and Sheldon (1979, pp. 13-19) give procedures and tables that "are more accurate than those previously published", along with a bibliography of other work on this problem and several illustrative applications. The regions they define are circular and elliptical.

<u>Case I.</u> (μ_1, μ_2) known, $\sigma_1 = \sigma_2$ unknown, $\rho = 0$.

$$A = \{(x, y): (x - \mu_1)^2 + (y - \mu_2)^2 < Ks^2\},$$
$$s^2 = (2n)^{-1}\left[\sum_{i=1}^n (x_i - \mu_1)^2 + \sum_{i=1}^n (y_i - \mu_2)^2\right].$$

Then

$$K = -4n \ln(1 - \delta)/\chi^2_{1-\gamma}(2n),$$

where $\chi^2_\alpha(\nu)$ is the (100α) percentile of chi-square with ν degrees of freedom.

Cases II, III and IV are handled via Monte Carlo simulation, and with K defined as below, the authors tabulate values of K to two decimal places for $\delta = 0.50$, 0.80, 0.90 and 0.95, $\gamma = 0.75$, 0.90 and 0.95, and $n = 10(1)20(2)30(5)50$.

<u>Case II.</u> (μ_1, μ_2) known, (σ_1, σ_2) unknown, $\rho = 0$.

$$A = \{(x, y): \left(\frac{x-\mu_1}{S_1^*}\right)^2 + \left(\frac{y-\mu_2}{S_2^*}\right)^2 < K\},$$

where here $S_1^{*2} = [(n-1)/n] S_1^2$ and $S_2^{*2} = [(n-1)/n] S_2^2$.

<u>Case III.</u> (μ_1, μ_2) unknown, (σ_1, σ_2) unknown, $\rho = 0$.

$$A = \left\{(x, y): \left(\frac{x-\bar{x}}{S_1}\right)^2 + \left(\frac{y-\bar{y}}{S_2}\right) < K\right\}$$

<u>Case IV.</u> $(\mu_1, \mu_2, \sigma_1, \sigma_2, \rho)$ unknown.

$$A = \left\{(x, y): \left(\frac{x-\bar{x}}{S_1}\right)^2 - 2r\left(\frac{x-\bar{x}}{S_1}\right)\left(\frac{y-\bar{y}}{S_2}\right)^2 + \left(\frac{y-\bar{y}}{S_2}\right)^2 < K(1-r^2)\right\},$$

$$r = \sum_{i=1}^{n} \frac{(x_i - \bar{x})(y_i - \bar{y})}{(n-1)s_1 s_2}.$$

10.2 THE SAMPLE CORRELATION COEFFICIENT R

The distribution of R in the set up of *[10.1.1-2]* has received considerable attention in the literature. We shall have occasion in what follows to refer to the *hypergeometric function* (Abramowitz and Stegun, 1970, Chap. 15)

$$F(a, b; c; d) = \frac{\Gamma(c)}{\Gamma(a)\Gamma(b)} \sum_{j=0}^{\infty} \left[\frac{\Gamma(a+j)\Gamma(b+j)}{\Gamma(c+j)} \frac{d^j}{j!}\right].$$

The Bivariate Normal Distribution

[10.2.1.1] Fisher (1915, pp. 507-521) derived the pdf of R, given by

$$g_n(r, \rho) = \frac{(1-\rho^2)^{(n-1)/2}}{\pi \, \Gamma(n-2)} (1-r^2)^{(n-4)/2} \frac{d^{n-2}}{d(r\rho)^{n-2}} \left[\frac{\arccos(-r\rho)}{(1-r^2\rho^2)^{\frac{1}{2}}} \right],$$

$$|\rho| \leq 1, \ |r| \leq 1.$$

[10.2.1.2] Hotelling (1953, pp. 193-232) expressed the pdf of R in terms of the hypergeometric function and beta function $B(\cdot, \cdot)$;

$$g_n(r, \rho) = \frac{(n-2)}{\sqrt{2}\,(n-1)B(\frac{1}{2}, n-\frac{1}{2})} (1-\rho^2)^{(n-1)/2}(1-r^2)^{(n-4)/2}(1-\rho r)^{\frac{3}{2}-n}$$

$$\cdot \, F(\tfrac{1}{2}, \tfrac{1}{2}; n-2; \tfrac{1}{2}(1+\rho r)).$$

This leads to an asymptotic representation for $\Pr(R > r)$, i.e.,

$$\frac{(n-2)}{\sqrt{2}(n-1)^2 B(\frac{1}{2}, n-\frac{1}{2})} \cdot \frac{(1-\rho^2)^{(n-1)/2}(1-r^2)^{(n-2)/2}}{(r-\rho)(1-\rho r)^{n-5/2}} \{1 + O(n^{-1})\}.$$

See also Stuart and Ord (1987, Sec. 16.31).

[10.2.1.3] Cramér (1946, p. 398) derives the infinite series representation

$$g_n(r, \rho) = \frac{2^{n-3}}{\pi(n-3)!} (1-\rho^2)^{(n-1)/2}(1-r^2)^{(n-4)/2}$$

$$\cdot \sum_{i=0}^{\infty} \{\Gamma(\tfrac{1}{2}(n+i-1))\}^2 (2\rho r)^i/i!, \quad |\rho| \leq 1, |r| \leq 1.$$

Integration term by term leads to rapid convergence of $\Pr(0 < R < r)$, with a good bound on error; see Guenther (1977, pp. 45-58), who also gives computational procedures for calculators programmed to compute probabilities based on the F-distribution.

Johnson and Kotz (1970, pp. 221-223) and Anderson (1984, p. 113) give some other expressions for $g_n(r, \rho)$.

[10.2.2.1] David (1938, 1954) has tabulated cumulative probabilities $G_n(r; \rho)$ and ordinates $g_n(r; \rho)$ of the distribution of R as follows:

n = 3(1)25
$\rho = 0.0(0.1)0.4$;   r = $-1.00(0.05)1.00$
$\rho = 0.5(0.1)0.9$;   r = $-1.00(0.05)0.60(0.025)1.00$ Ordinate:
$\rho = 0.90$;   r = $0.80(0.01)0.95(0.005)1.00$ 2 dec. places

n = 50
$\rho = 0.0(0.1)0.4$;   r = $-0.75(0.05)1.00$
$\rho = 0.5(0.1)0.7$;   r = $-0.25(0.05)1.00$ Ordinate:
$\rho = 0.80$;   r = $0.22(0.02)1.00$ 2 dec. places
$\rho = 0.90$;   r = $0.61(0.01)1.00$ Ordinate: 1 dec. place.

Further tabulations appear, for n = 100, 200 and 400, and for $\rho = 0.0(0.1)0.9$. Coverage for r is over a range outside of which the ordinates effectively are zero. This range varies; for example, if n = 200, and $\rho = 0.5$, then r = $0.15(0.01)0.79$.

Probabilities are given to five decimal places; the 1954 printing contains a few corrections to errors in the 1938 edition, on pp. viii and xxxii.

[10.2.2.2] Öksoy and Aroian (1982, pp. 302-305) have published tables of the cdf and of percentage points of R. The cdf $G_n(r, \rho)$ is tabulated for n = 3(1)10(2)24, 25(5)40(10)100(100)500; for $\rho = 0(0.05)0.90$, 0.92, 0.94, 0.95, 0.96, 0.98, and for choices of r that vary with the choice of ρ and n.

Percentage points r_α satisfying $G_n(r_\alpha, \rho) = 1 - \alpha$ are tabulated for thirteen values of α in each of the intervals [0.0005, 0.1000] and [0.9000, 0.9995] and for $\alpha = 0.50$. See also Öksoy and Aroian (1981, pp. 328-332).

For $3 \leq n \leq 10$, the areas are computed from formulas of Garwood (1933, pp. 71-78); these formulas have been updated by Aroian and Öksoy (1982, pp. 306-308).

The Bivariate Normal Distribution

[10.2.2.3] Odeh (1982, pp. 1-26) has tabulated percentage points r_α of R, satisfying $G_n(r_\alpha, \rho) = 1 - \alpha$. Values of r_α are given to five decimal places for

$$\rho = 0(0.10)0.90,\ 0.95$$
$$n = 4(1)30(2)40(5)50(10)100(20)200(100)1000;$$
$$\alpha = 0.005,\ 0.01,\ 0.025,\ 0.05,\ 0.10,\ 0.25,\ 0.75,\ 0.90,\ 0.95,$$
$$0.975,\ 0.99,\ 0.995\ .$$

Percentage points for $\rho < 0$ can be obtained via the relation

$$G_n(-r, -\rho) = 1 - G_n(r, \rho)\ ;$$

i.e., if r_α works for positive ρ, $-r_{1-\alpha}$ works for corresponding negative ρ.

[10.2.3.1] Anderson (1984, pp. 151-152) gives the *moments* of R; for integer values of h,

$$E(R^{2h+1}) = \frac{(1-\rho^2)^{(n-1)/2}}{\sqrt{\pi}\ \Gamma(\tfrac{1}{2}(n-1))} \sum_{i=0}^{\infty} \frac{(2\rho)^{2i+1}}{(2i+1)!}\ \frac{\Gamma^2(\tfrac{1}{2}n+i)\ \Gamma(h+i+\tfrac{3}{2})}{\Gamma(\tfrac{1}{2}(n+1)+h+i)},$$

$$E(R^{2h}) = \frac{(1-\rho^2)^{(n-1)/2}}{\sqrt{\pi}\ \Gamma(\tfrac{1}{2}(n-1))} \sum_{i=0}^{\infty} \frac{(2\rho)^{2i}}{(2i)!}\ \frac{\Gamma^2(\tfrac{1}{2}(n-1)+i)\ \Gamma(h+i+\tfrac{1}{2})}{\Gamma(\tfrac{1}{2}(n-1)+h+i)}.$$

A simplified series obtains when $h = 0$, i.e.,

$$E(R) = \frac{(1-\rho^2)^{\tfrac{1}{2}(n-1)}}{\Gamma(\tfrac{1}{2}(n-1))} \sum_{i=0}^{\infty} \frac{\rho^{2i+1}}{i!}\ \frac{\Gamma^2(\tfrac{1}{2}n+i)}{\Gamma(\tfrac{1}{2}(n+1)+i)}.$$

[10.2.3.2] Ghosh (1966, pp. 258-262) gives the first two moments of R:

$$E(R) = \frac{2}{n-1} \left[\frac{\Gamma(\tfrac{1}{2}n)}{\Gamma\{\tfrac{1}{2}(n-1)\}}\right]^2 \rho\ F(\tfrac{1}{2}, \tfrac{1}{2}; \tfrac{1}{2}(n+1); \rho^2)\ ,$$

$$E(R^2) = 1 - (n-2)(1-\rho^2)(n-1)^{-1}\ F(1, 1; \tfrac{1}{2}(n+1); \rho^2)\ .$$

If $m = n + 6$, then

$$E(R) = \rho - \frac{\rho(1-\rho^2)}{2m}\left\{1 + \frac{9}{4m}(3+\rho^2) + \frac{3(121 + 70\rho^2 + 25\rho^4)}{8m^2}\right\} + O(m^{-4}),$$

$$\text{Var}(R) = \frac{(1-\rho^2)^2}{m}\left\{1 + \frac{14 + 11\rho^2}{2m} + \frac{98 + 130\rho^2 + 75\rho^4}{2m}\right\} + O(m^{-4}).$$

[10.2.3.3] Hotelling (1953, p. 212) shows that

$$E(R) = \rho + (1-\rho^2)\left\{-\frac{\rho}{2(n-1)} + \frac{\rho - 9\rho^3}{8(n-1)} + \frac{\rho + 42\rho^3 - 75\rho^5}{16(n-1)}\right\}$$
$$+ O(n^{-4}),$$

$$\text{Var}(R) = (1-\rho^2)^2\left\{\frac{1}{n-1} + \frac{11\rho^2}{2(n-1)^2} + \frac{-24\rho^2 + 75\rho^4}{2(n-1)^3}\right\} + O(n^{-4}),$$

$$\mu_3(R) = (1-\rho^2)^3\left\{-\frac{6\rho}{(n-1)^2} + \frac{15\rho - 88\rho^3}{(n-1)^3}\right\} + O(n^{-4}),$$

$$\mu_4(R) = (1-\rho^2)^4\left\{\frac{3}{(n-1)^2} + \frac{-6 + 105\rho^2}{(n-1)^3}\right\} + O(n^{-4}).$$

Thus if n is large, the bias in $E(R)$ is of order (n^{-1}), the variance is of order (n^{-1}), while the skewness and kurtosis differ from normality by order $(n^{-1/2})$ and (n^{-1}) respectively, i.e.,

$$\mu_3(R)/\sigma^3(R) = -6\rho\, n^{-1/2} + O(n^{-3/2}),$$

$$\mu_4(R)/\sigma^4(R) = 3 - 6(1 - 12\rho^2)n^{-1} + O(n^{-2}).$$

[10.2.4.1] *Distribution of R when $\rho = 0$.* We have

$$g_n(r; 0) = (1 - r^2)^{(n-4)/2} / B(\tfrac{1}{2}, \tfrac{1}{2}(n-2)),\ 0 < |r| < 1$$

(Stuart and Ord, 1987, Sec. 16.28; Student, 1908, pp. 302–310).

If $T = \{(n-2)R^2/(1-R^2)\}^{1/2}$, then when $\rho = 0$, T has a Student t-distribution with $n-2$ degrees of freedom. This property is very useful for tests of independence of X and Y in a BVN population. See also *[10.1.3]*.

The Bivariate Normal Distribution

Thus the pdf $g_n(r; 0)$ is symmetric about zero with mean zero and variance $1/(n-1)$. The kurtosis is given by

$$\mu_4(R|0)/\sigma^4(R|0) = 3 - 6/(n+1).$$

[10.2.4.2] More generally, if R is based on a sample of size n from a BVN ($\underline{\theta}$) distribution, define T by

$$T = \frac{kR - \rho w}{\ell\sqrt{\{(1-\rho^2)(1-R^2)\}}} \sqrt{(n-2)},$$

where

$$k = 2\sigma_1\sigma_2 + (\sigma_1^2 + \sigma_2^2)\sqrt{(1-\rho^2)},$$

$$\ell = 2\sigma_1\sigma_2\sqrt{(1-\rho^2)} + (\sigma_1^2 + \sigma_2^2),$$

$$w = \frac{\sigma_1^2 s_2^2 + \sigma_2^2 s_1^2 + \sigma_1^2\sigma_2^2(s_1^2 + s_2^2)\sqrt{(1-\rho^2)}}{\sqrt{(s_1^2 s_2^2)}}.$$

Then (Muddapur, 1988, pp. 99 – 108) T has an exact Student t distribution with n − 2 degrees of freedom.

[10.2.5] *Asymptotic normality of R.* As n → ∞, the limiting distribution of $\sqrt{n}(R-\rho)/(1-\rho^2)$ is N(0, 1). A proof of this result is given by Anderson (1984, pp. 120-122). A drawback is that the convergence to normality is rather slow. This, together with the mathematical difficulties inherent in $g_n(r; \rho)$ led to a search for a suitable transformation.

10.3 APPROXIMATIONS TO THE DISTRIBUTION OF R

[10.3.1] The most well-known and the simplest transformation of R is that of Fisher (1921, pp. 3-32), viz.,

$$z = \tfrac{1}{2}\ln\left[(1+R)/(1-R)\right];$$

see also Anderson (1984, pp. 122-123) and Stuart and Ord (1987, Sec. 16.33). Fisher expanded the pdf of z in inverse powers of $n-1$ and in powers of ζ, where

$$\zeta = \tfrac{1}{2} \ln\left[(1+\rho)/(1-\rho)\right].$$

Then $\sqrt{n}\,(z-\zeta)$ has a limiting $N(0,1)$ distribution. Fisher's z tends to normality much faster than R, but also has a variance that is (almost) free of ρ.

Fisher expanded the moments of z as follows, where corrections of Gayen noted by Hotelling (1953, p. 216) are included (Stuart and Ord, 1987, Sec. 16.33):

$$E(z) = \zeta + \frac{\rho}{2(n-1)}\left\{1 + \frac{5+\rho^2}{4(n-1)} + \frac{11+2\rho^2+3\rho^4}{8(n-1)^2}\right\} + O(n^{-4}),$$

$$\mathrm{Var}(z) = \frac{1}{n-1}\left\{1 + \frac{4-\rho^2}{2(n-1)} + \frac{22-6\rho^2-3\rho^4}{6(n-1)^2}\right\} + O(n^{-4}),$$

$$\frac{\mu_3}{\sigma^3} = \frac{\rho^3}{(n-1)^{3/2}} + O(n^{-2}) \quad \text{(skewness)},$$

$$\frac{\mu_4}{\sigma^4} = 3 + \frac{2}{n-1} + \frac{4+2\rho^2-3\rho^4}{(n-1)^2} + O(n^{-3}) \quad \text{(kurtosis)}.$$

The commonest approximation treats z as a $N(\zeta, 1/(n-3))$ variable, and others treat z as normal, taking $\mathrm{Var}(z)$ as $(n-3)^{-1}$ or as above to terms of order $(n-1)^{-3}$, and taking $E(z)$ to terms of order $(n-1)^{-1}$ or $(n-1)^{-2}$. David (1954) and Kraemer (1973, pp. 1004-1008) state that great accuracy is achieved with higher order approximations, even when $n \leq 10$, provided that $|\rho|$ is not too close to unity. When $n = 11$ and $\rho = 0.90$, the agreement with the exact sampling distribution of R is reasonably good.

Hotelling (1953, pp. 223-224) suggested two modifications of z to stabilize the variance further and give a closer approximation to normality. These are given by

$$z* = z - (3z + R)/\{4(n-1)\},$$

and
$$z^{**} = z^* - (23z + 33R - 5R^3)/\{96(n-1)^2\} \ .$$
Then
$$E(z^*) = \zeta - \frac{3\zeta + \rho}{4(n-1)} + \frac{\rho}{2(n-1)} + \frac{3\rho}{8(n-1)^2} + O(n^{-3}) \ ,$$
and $E(z^{**})$ differs from $E(z^*)$ only by terms of order n^{-3}. However,
$$\text{Var}(z^*) = (n-1)^{-1} + O(n^{-3}) \ ,$$
while
$$\text{Var}(z^{**}) = (n-1)^{-1} + O(n^{-4}) \ .$$

[10.3.1.2] Harley (1954, pp. 278-280) gave Edgeworth and Cornish-Fisher expansions for the cdf and percentiles of Fisher's z; these incorporate corrections by Winterbottom (1980, pp. 599-609), and depend upon the cumulants κ_j of $z - \zeta$ (see also Gayen, 1951, pp. 219-247). In what follows n denotes the sample size minus one. The cumulants have the form

$$\kappa_1 = (1/n) \sum_{s=0}^{\infty} \kappa_{1,s+1}/n^s \ , \qquad \kappa_j = n^{1-j} \sum_{s=0}^{\infty} \kappa_{j,j+s-1}/n^s \ ;$$

$\kappa_{11} = \rho/2 \ ,$	$\kappa_{12} = \rho(5 + \rho^2)/8 \ ,$	$\kappa_{21} = 1 \ ,$
$\kappa_{22} = (4 - \rho^2)/2 \ ,$	$\kappa_{23} = (22 - 6\rho^2 - 3\rho^4)/6 \ ,$	$\kappa_{32} = 0 \ ,$
$\kappa_{33} = \rho^3 \ ,$	$\kappa_{43} = 2 \ ,$	$\kappa_{44} = 3(4 - \rho^4),$
$\kappa_{54} = 0 \ ,$	$\kappa_{65} = 24.$	

If $x = \sqrt{n}(z - \zeta)$, the standard normal deviate, and $H_m(\cdot)$ is the mth Tchebyshev-Hermite polynomial as set out in *[2.1.9]*, then the cdf $G(z, \zeta)$ of Fisher's z is given by

$$G(z, \zeta) = \Phi(x) - \phi(x) \left[\frac{\kappa_{11}}{n^{1/2}} + \frac{1}{24n} \left\{ 12(\kappa_{11}^2 + \kappa_{22}) H_1(x) + \kappa_{43} H_3(x) \right\} \right.$$

$$+ \frac{1}{120n^{3/2}} \left\{ 120\kappa_{12} + 20(\kappa_{33} + 3\kappa_{11}\kappa_{22} + \kappa_{11}^3) H_2(x) \right.$$

$$+ 5\kappa_{11}\kappa_{43}H_4(x) \right\} + \frac{1}{5760n^2} \left\{ 2880(\kappa_{23} + 2\kappa_{11}\kappa_{12}) H_1(x) \right.$$

$$+ 240 \left(\kappa_{44} + 4\kappa_{11}\kappa_{33} + \kappa_{11}^4 + 3\kappa_{22}^2 + 6\kappa_{11}^2 \kappa_{22} \right) H_3(x)$$

$$+ 8(\kappa_{65} + 15\kappa_{43}\kappa_{22} + 15\kappa_{22}\kappa_{11}^2) H_5(x)$$

$$\left. \left. + 5\kappa_{43}^2 H_7(x) \right\} \right] + O(n^{-\frac{5}{2}}) .$$

For a more concise expansion of this type see *[10.3.5]*. If u is a specified percentile of the standard normal distribution, the corresponding percentile z(u) of Fisher's z is given by

$$z(u) = \zeta + \frac{u}{n^{1/2}} + \frac{\rho}{2n} + \frac{1}{12n^{3/2}} \left\{ u^3 + 3(3 - \rho^2)u \right\}$$

$$+ \frac{1}{24n^2} \left\{ 4\rho^3 u^2 + (15\rho - \rho^3) \right\} + \frac{1}{480n^{5/2}} \left\{ u^5 + 10(8 + 3\rho^2 - 6\rho^4)u^3 \right.$$

$$\left. + 3(125 - 7\rho^2 + 15\rho^4)u \right\} + O(n^{-3}) .$$

These results can be used to approximate the cdf and percentiles of R by inverting Fisher's z-transformation;

$$R = \frac{e^{2z} - 1}{e^{2z} + 1} = \tanh z.$$

Winterbottom (1980, pp. 599-609) provides some numerical comparisons with exact results and with approximations given in *[10.2.2-4]*. He concludes that the Harley-Winterbottom formula for the cdf of R is "very accurate overall", and that the formula for the percentiles "provides the best approximations ... and maintains reasonable accuracy for sample sizes as small as eleven."

The Bivariate Normal Distribution 357

[10.3.1.3] Niki and Konishi (1984, pp. 169-182) provide asymptotic expansions of G(z, ζ) and of z(u) along the lines of *[10.3.1.2]*, but to terms in n^{-4}, inclusive. The expressions involved in the additional terms, however, are somewhat involved. If $|r - 1| < 0.005$, their expansion of G(z, ζ) "guarantees accuracy to five decimal places for a sample of size eleven or more...". Asymptotic expansions for the first ten cumulants of z are provided to terms in n^{-4} inclusive. Once again the sample size in these expansions is n + 1.

[10.3.2] Ruben (1966, pp. 518-519) approximates the distribution of R by the relation

$$\Pr(R \leq r) \simeq \Phi\left[\frac{\sqrt{(n - 5/2)}\,\tilde{r} - \sqrt{(n - 3/2)}\,\tilde{\rho}}{\{1 + \tfrac{1}{2}(\tilde{r}^2 + \tilde{\rho}^2)\}^{1/2}}\right],$$

where $\tilde{r} = r/\sqrt{(1 - r^2)}$, $\tilde{\rho} = \rho/\sqrt{(1 - \rho^2)}$, and ϕ is the standard normal c.d.f. This approximation improves upon Fisher's z-transform, but lacks the variance-stabilizing property of z. See also Kshirsagar (1972, pp. 88-92).

[10.3.3] Kraemer (1973, pp. 1004-1008) developed another approximation. If $\rho' = \rho'(\rho, n)$, such that $|\rho'| \geq \rho$, $\rho' = \rho$ whenever ρ is 0, 1 or -1, $\rho'(-\rho, n) = -\rho'(\rho, n)$ and as $n \to \infty$ we have that $\rho' \to \rho$, then

$$(R - \rho')\sqrt{\frac{(n - 2)}{(1 - R^2)(1 - \rho'^2)}}$$

is approximately distributed as Student's t with $n - 2$ degrees of freedom. If $n > 10$, the optimum choice of ρ' is the median of the distribution of R, given ρ and n, but if $n > 25$, one may take ρ' equal to ρ. Kraemer compared her approximation with those for z discussed in *[10.8.6.1]* and found it to give more accurate probabilities for R if $|\rho| < 0.8$.

[10.3.4] Mudholkar and Chaubey (1976, pp. 163-172) approximate the distribution of Fisher's z by adjusting the kurtosis with a mixture of normal and logistic distributions. Using the notation of *[10.8.6.1]* let $y = [z - E(z)]/\sqrt{\text{Var}(z)}$, where $E(z)$ and $\text{Var}(z)$ are approximated as in *[10.8.6.1]*. Then

$$\Pr(R \leq r) \simeq \lambda \Phi(y) + (1-\lambda)\{1 + \exp(-\pi y/\sqrt{3})\}^{-1},$$

where Φ is the standard normal c.d.f. and

$$1 - \lambda = (\mu_4(z)/\sigma^4(z) - 3)/1.2 \simeq \left[\frac{2}{n-1} + \frac{4 + 2\rho^2 - 3\rho^4}{(n-1)^2}\right] / 1.2.$$

The $100(\alpha)$th percentile of z is approximated by $\lambda z_{1-\alpha} - (1-\lambda)\sqrt{3} \cdot \{\ell n(\alpha^{-1} - 1)\}/\pi$, where $\phi(z_{1-\alpha}) = \alpha$. The authors compare exact and approximate probabilities based on their mixture distribution with those of Fisher's z-transformation (see [10.3.1.1]), of Ruben (1966, pp. 513-515) and of Kraemer (1973, pp. 1004-1008) for a range of values of R when $\rho = 0.0(0.1)(0.90)$; and $n = 11, 21$; and they compare percentiles of R computed by these methods with exact values, for $\rho = 0.5, 0.9$; $n = 11, 25, 50$; $\alpha > 0.95$. The approximations of Kraemer and of Mudholkar and Chaubey are uniformly accurate to two or three decimal places; those of Ruben are almost as good, while Fisher's z approximation sacrifices accuracy for simplicity, unless n is large.

[10.3.5] Let $m = n - (5/2) + \rho^2/4$. Then (Konishi, 1978, pp. 654-655)

$$\Pr\{\sqrt{m}(z - \zeta) < x\} = \Phi(x) - \tfrac{1}{2}(\rho m^{-\tfrac{1}{2}} + x^3 m^{-1}/6)\phi(x) + O(m^{-3/2}).$$

This gives an improvement upon the $N(\zeta, (n-3)^{-1})$ approximation for Fisher's z, and can be used to give an accurate approximation to the distribution of R over its whole domain. Based on calculations for $n = 11, 25$, and 50 and for $\rho = 0.1(0.2)0.9$, Konishi found that the above appears to be more accurate when $\rho \geq 0.3$ than the approximations due to Ruben and Kraemer discussed in [10.3.2.-3]; even when n is as small as 11, the above gives high accuracy over all values of R. For the listed range of values of n and of ρ, Konishi found that, for $\Pr(R \leq x)$,

Max| exact value - approximate value | ≤ 0.00102.

[10.3.6] Samiuddin (1970, pp. 461-464) showed that $\rho \neq 0$ and for moderately large n, the variable

$$t = \frac{(R-\rho)\sqrt{(n-2)}}{\sqrt{\{(1-R^2)(1-\rho^2)\}}}$$

has an approximate t distribution with $n-2$ degrees of freedom, in the set up of *[10.2.4.2]*. Limited calculations for $n = 8$, and for $\rho = 0.3, 0.5, 0.9$, shows the approximation "gives accurate results even for large ρ", with the closest agreement when ρ is small.

[10.3.7] Muddapur (1988, pp. 99-108) shows that a large sample approximation to the distribution of

$$\frac{(1+R)(1-\rho)}{(1-R)(1+\rho)}$$

is that of F with $(n-2, n-2)$ degrees of freedom.

[10.3.8] The *arc sine transformation of R*. The statistic arc sin R is unbiased; E(arc sin R) = arc sin ρ (Harley 1956, pp. 219-2240; Harley, 1957, pp. 273-275; Stuart and Ord, 1991, Sec. 26.16).

Let $X = (R-\rho)/(1-\rho R)$ and $Y = $ arc sin X. Then Y has an approximate normal distribution when n is large, where

$$E(Y) = \frac{\rho}{2(n-1)} + \frac{3\rho + \rho^3}{8(n-1)^2} + \frac{3\rho + 3\rho^5}{16(n-1)^3} + O(n^{-4}),$$

$$\text{Var}(Y) = \frac{1}{n-1} + \frac{2-\rho^2}{2(n-1)^2} + \frac{8 - 3\rho^2 - 6\rho^4}{12(n-1)^3} + O(n^{-4}).$$

This approximation (Sankaran, 1958, pp. 567-571) is accurate to three decimal places in giving probabilities for R; in a simplified version, $Y \sim N(0, (n-2)^{-1})$, approximately, but this is less accurate.

REFERENCES

Abramowitz, M. and Stegun, I. A. (eds.) (1970). *Handbook of Mathematical Functions*, Washington, D.C. : National Bureau of Standards. *[10.2]*

Anderson, T. W. (1984). *An Introduction to Multivariate Statistical Analysis* (2nd ed.), New York: Wiley. *[10.2.1.3; 10.2.3.1; 10.2.5; 10.3.1.1]*

Aroian, L. A. and Öksoy, D. (1982). Garwood's formulas for the distribution of the correlation coefficient revisited, resurrected, revised, clarified and extended, *A.S.A., Proceedings of the Statistical Computing Section*, 306-308. *[10.2.2.2]*

Bhattacharya, P. K. (1984). Induced order statistics: Theory and Applications, *Handbook of Statistics 4: Nonparametric Methods* (Krishnaiah, P. R. and Sen, P. K., eds.), 383-403, New York: North-Holland. *[10.1.5.3]*

Cook, M. B. (1951). Two applications of bivariate k-statistics, *Biometrika* 38, 368-376. *[10.1.6]*

Cramér, H. (1946). *Mathematical Methods of Statistics*, Princeton, N.J.: Princeton University Press. *[10.2.1.3]*

David, F. N. (1938, 1954). *Tables of the Correlation Coefficient*, Cambridge, England: Cambridge University Press. *[10.2.2.1; 10.3.1.1]*

David, H. A. (1973). Concomitants of order statistics, *Bulletin of the Institute of International Statistics* 45, 295-300. *[10.1.5.1]*

David, H. A. (1981). *Order Statistics* (2nd ed.), New York: Wiley. *[10.5.1,2]*

David, H. A. (1982). Concomitants of order statistics: Theory and applications, *Some Recent Advances in Statistics* (J. Tiago de Oliveira, ed.), 89-100, New York: Academic Press. *[10.1.5]*

David, H. A. (1993). Concomitants of order statistics: Review and recent developments, *Multiple Comparisons, Selection, and Applications in Biometry* (F. M. Hoppe, ed.) 507-518, New York: Dekker. *[10.1.5]*

David, H. A. and Galambos, J. (1974). The asymptotic theory of concomitants of order statistics, *Journal of Applied Probability* 11, 762-770. *[10.1.5.2]*

David, H. A., O'Connell, M. J. and Yang, S. S. (1977). Distribution and expected value of the rank of a concomitant of an order statistic, *Annals of Statistics* 5, 216-223. *[10.1.5.2]*

De Lury, D. B. (1938). Note on correlations, *Annals of Mathematical Statistics* 9, 149-151. *[10.1.4]*

Fisher, R. A. (1915). Frequency distribution of the values of the correlation coefficient in samples from an indefinitely large population, *Biometrika* 10, 507-521. *[10.1.2.1; 10.2.1.1]*

Fisher, R. A. (1921). On the probable error of a coefficient of correlation deduced from a small sample, *Metron 1* No. 4, 3-32. *[10.3.1.1]*

Gajjar, A. V. and Subrahmaniam, K. (1978). On the sample correlation coefficient in the truncated bivariate normal population, *Communications in Statistics B7*, 455-478. *[10.1.6]*

Garwood, F. (1933). The probability integral of the correlation coefficient in samples from a normal population, *Biometrika 25*, 71-78. *[10.2.2.2]*

Gayen, A. K. (1951). The frequency distribution of the product-moment correlation coefficient in random samples of any size drawn from non-normal universes, *Biometrika 38*, 219-247. *[10.3.1.2]*

Ghosh, B. K. (1966). Asymptotic expansions for the moments of the distribution of sample correlation coefficient, *Biometrika 53*, 258-262. *[10.2.3.1]*

Guenther, W. C. (1977). Desk calculation of probabilities for the distribution of the sample correlation coefficient, *The American Statistician 31*, #1, 45-48. *[10.2.1.3]*

Hall, I. J. and Sheldon, D. D. (1979). Improved bivariate normal tolerance regions with some applications, *Journal of Quality Technology 11*, 13-19. *[10.1.9]*

Harley, B. I. (1954). A note on the probability integral of the correlation coefficient, *Biometrika 41*, 278-280. *[10.3.1.2]*

Harley, B. I. (1956). Some properties of an angular transformation for the correlation coefficient, *Biometrika 43*, 219-224. *[10.3.8]*

Harley, B. I. (1957). Relation between the distribution of noncentral t and of a transformed correlation coefficient, *Biometrika 44*, 273-275. *[10.3.8]*

Hotelling, H. (1953). New light on the correlation coefficient and its transforms, *Journal of the Royal Statistical Society B 15*, 193-232 (with discussion). *[10.2.1.2; 10.2.3.3; 10.3.1.1]*

Johnson, N. L. and Kotz, S. (1970). *Distributions in Statistics. Continuous Univariate Distributions-2*, New York: Wiley. *[10.6.1; 10.2.1.3]*

Konishi, S. (1978). An approximation to the distribution of the sample correlation coefficient, *Biometrika 65*, 654-656. *[10.3.5]*

Kraemer, H. C. (1973). Improved approximation to the non-null distribution of the correlation coefficient, *Journal of the American Statistical Association 68*, 1004-1008. *[10.2.6.1; 10.2.7.2; 10.3.1.1; 10.3.3, 4]*

Kshirsagar, A. (1972). *Multivariate Analysis*, New York: Dekker. *[10.1.2.1; 10.3.2]*

Mason, R. L., Young, J. C. and Langley, M. P. (1987). Computer program for generating a bivariate normal, *A.S.A., Proceedings of the Statistical Computing Section*, 448-451. *[10.1.8]*

Muddapur, M. V. (1988). A simple test for correlation coefficient in a bivariate normal distribution, *Sankhyā B 50*, 99-108. *[10.2.4.2; 10.3.7]*

Mudholkar, G. S. and Chaubey, Y. P. (1976). On the distribution of Fisher's transformation of the correlation coefficient, *Communications in Statistics B 5*, 163-172. *[10.3.4]*

Niki, N. and Konishi, S. (1984). Higher order asymptotic expansions for the distribution of the sample correlation coefficient, *Communications in Statistics B 13*, 169-182. *[10.3.1.3]*

Odeh, R. E. (1982). Critical values of the sample product-moment correlation coefficient in the bivariate normal distribution, *Communications in Statistics B 11*, 1-26. *[10.2.2.3]*

Öksoy, D. and Aroian, L. A. (1981). Formulas for the rapid calculation of the ordinates, distribution, and percentage points of the coefficient correlation, *A.S.A., Proceedings of the Statistical Computing Section*, 328-332. *[10.2.2.2]*

Öksoy, D. and Aroian, L. A. (1982). Tables of distribution function and of percentage points of the coefficient of correlation, *A.S.A., Proceedings of the Statistical Computing Section*, 302-305. *[10.2.2.2]*

Prabhakaran, V. T., Mahajan, V. K. and Uma (1991). On an identity for finding moments of sample moments of bivariate normal random variables, *Australian Journal of Statistics 33*, 95-102. *[10.1.2.2]*

Rao, B. R., Garg, M. L. and Li, C. C. (1968). Correlation between the sample variances in a singly truncated bivariate normal distribution, *Biometrika 55*, 433-436. *[10.1.6]*

Ruben, H. (1966). Some new results on the distribution of the sample correlation coefficient, *Journal of the Royal Statistical Society B 28*, 513-525. *[10.3.2, 4]*

Samiuddin, M. (1970). On a test for an assigned value of correlation in a bivariate normal distribution, *Biometrika 57*, 461-464. *[10.3.6]*

Sankaran, M. (1958). On Nair's transformation of the correlation coefficient, *Biometrika 45*, 567-571. *[10.3.8]*

Stuart, A. and Ord, J. K. (1987). *Kendall's Advanced Theory of Statistics, Vol. 1* (5th edn.). New York: Oxford. *[10.1.3; 10.2.1.2; 10.2.4.1; 10.3.1.1]*

Stuart, A. and Ord, J. K. (1991). *Kendall's Advanced Theory of Statistics, Vol. 2* (5th edn.). New York: Oxford. *[10.1.3]*

"Student" (1908). On the probable error of a correlation coefficient, *Biometrika 6*, 302-310. *[10.2.4.1]*

Watterson, G. A. (1959). Linear estimation in censored samples from multivariate normal populations, *Annals of Mathematical Statistics 30*, 814-824. *[10.1.5.1]*

Winterbottom, A. (1980). Estimation for the bivariate normal correlation coefficient using asymptotic expansions, *Communications in Statistics B 9*, 599-609. *[10.3.1.1]*

Chapter 11
POINT ESTIMATION

Point estimation is concerned with statistical inference about the unknown parameter(s) of a distribution. It provides a single value for each parameter. In the literature, no unique definition of the optimality of an estimator exists. Hence, various estimators based on different optimality criteria are available. The concepts of sufficiency and completeness play a significant role in the construction of some of these estimators. In the following we first list sufficiency and completeness of some useful statistics and then list some optimal estimators of the parameters and their functions.

More commonly used estimators like Maximum likelihood (ML), Method-of-moments (MM), Mean-squared-error (MSE) consistent, Uniformly minimum variance unbiased (UMVU), Minimax, Bayes, and Best Asymptotically Normal (BAN) are covered in this chapter.

For the following results, let $X_1, X_2, ..., X_n$ represent a single random sample from $N(\mu, \sigma^2)$. For the two sample problem, let $Y_1, Y_2,...,Y_n$ represent another random sample from $N(\theta, \tau^2)$. Both samples are assumed independent of each other. Let

$$\bar{X} = \sum_1^n X_i/n, \quad S^2 = \sum_1^n (X_i - \bar{X})^2 / (n-1), \text{ and } \bar{Y} = \sum_1^n Y_i/n .$$

11.1 SUFFICIENCY AND COMPLETENESS

Sufficient statistics contain all the relevant information in a sample. One useful result is that if a complete sufficient statistic exists then each function of it is a uniformly minimum variance unbiased (UMVU) estimator of the expected value of the function. The property of completeness implies the uniqueness of the estimator.

[11.1.1] Let σ be known. The normal family $f(x, \mu, \sigma^2)$ is complete with respect to μ [Lehmann & Scheffé (1950), p. 313; Stuart and Ord (1991), Sec. 22.10].

[11.1.2] Let μ be known. The normal family $f(x, \mu, \sigma^2)$ is not even boundedly complete with respect to σ^2 [Lehmann & Scheffé (1950), p. 313; Stuart and Ord (1991), Sec. 22.10].

[11.1.3] Let σ be known. The statistic $\sum_1^n X_i$ (or $\bar{X}$) is both complete and minimal sufficient for μ [DeGroot (1975), pp. 303, 310]. If σ is not known, $\sum_1^n X_i$ is not sufficient for μ [Rohatgi (1984), p. 648].

[11.1.4] Let μ be known. The statistic $\sum_1^n (X_i - \mu)^2$ is both complete and minimal sufficient for σ^2 [Roussas (1973), p. 219; Hogg & Craig (1978), p. 211]. If μ is not known, $\sum X_i^2$ (or S^2) is not sufficient for σ^2 [Rohatgi (1984), p. 648].

[11.1.5] Let μ and σ be both unknown. The pair of statistics

$$\left(\sum_1^n X_i, \ \sum_1^n X_i^2 \right) \text{ or } \left(\bar{X}, S^2 \right)$$

are jointly complete and minimal sufficient for (μ, σ^2) [DeGroot (1975), pp. 307, 310; Mood et al. (1974), p. 356].

[11.1.6] Let $X \sim N(\mu, b\mu^2)$, where the coefficient of variation b is assumed known. The minimal sufficient for μ is a pair of statistics

$$\left(\bar{X}, \sqrt{\sum_1^n (X_i - \bar{X})^2 / n} \right),$$

but the pair is not complete or even boundedly complete [Lehmann (1956), pp. 151-152; Gleser & Healy (1976), p. 976].

[11.1.7] Let $X \sim N(\mu, \mu)$. The statistic $\sum_1^n X_i$ (or $\bar{X}$) is both complete and sufficient for μ [Feldman & Fox (1968), p. 154].

[11.1.8] Two-sample problem. Let μ, θ, σ, and τ be unknown. The statistics

$$\left(\bar{X}, \bar{Y}, \sum_1^n (X_i - \bar{X}^2), \sum_1^m (Y_i - \bar{Y})^2 \right)$$

are minimal sufficient and complete for $(\mu, \theta, \sigma^2, \tau^2)$ [Lehmann (1983), p. 88; Zacks (1971), p. 58].

Point Estimation

[11.1.9] Two-sample problem. Let $\mu = \theta$, and θ, σ, τ be unknown. The statistics

$$\left(\overline{X}, \overline{Y}, \sum_1^n (X_i - \overline{X})^2, \sum_1^m (Y_i - \overline{Y})^2\right)$$

are minimal sufficient but not complete for $(\theta, \sigma^2, \tau^2)$. [Lehmann (1983), p. 88; Zacks (1971), p. 58)].

[11.1.10] Two-sample problem. Let $\sigma = \tau$, and $\mu, \theta, \sigma,$ and τ be unknown. The statistics

$$\left(\overline{X}, \overline{Y}, \sum_1^n (X_i - \overline{X})^2, \sum_1^m (Y_i - \overline{Y})^2\right)$$

are minimal sufficient but not complete for (μ, θ, τ^2). [Lehmann (1983), p. 88; Zacks (1971), p. 58)].

[11.1.11] Two-sample problem. Let $\mu = \theta$, and $\sigma^2/\tau^2 = \gamma$ be unknown, and γ be known. The statistics

$$\left(\sum_1^n X_i^2 + \gamma \sum_1^m Y_i^2, \sum_1^n X_i + \gamma \sum_1^m Y_j\right)$$

are jointly minimal sufficient and complete for (θ, τ^2) [Lehmann (1983), p. 88].

11.2 ESTIMATORS OF μ AND ITS FUNCTIONS

Single sample case, σ known.

[11.2.1] The sample mean $\overline{X}$ satisfies $\text{Var}(\overline{X}) = \sigma^2/n$; $\overline{X}$ is
(a) the ML estimator of μ [Mood et al. (1974), p. 281; Roussas (1973), p. 245].
(b) the MM estimator of μ [Mood et al. (1974), p. 275; Roussas (1973), p. 259].
(c) a model unbiased estimator of μ [Wasan (1970), p. 120].
(d) a consistent estimator of μ [Mood et al. (1974), p. 295].
(e) a BAN estimator of μ [Mood et al. (1974), p. 296].
(f) the UMVU estimator of μ [Roussas (1973), p. 233; Lehmann (1983), pp. 84-85].

[11.2.2] (a) Let $a \leq \mu \leq b$, (a and b known constants). If the loss function is squared error, then the sample mean $\overline{X}$ is not minimax and not admissible [Lehmann (1983), pp. 268]. Also, refer Casella & Strawderman (1981).

(b) $\mu > b$, (b a known constant). If the loss function is squared error, then the sample mean $\overline{X}$ is minimax but no longer admissible [Lehmann (1983), p. 267].

[11.2.3] Let the loss function be quadratic. The sample mean $\overline{X}$ is the best invariant estimator of μ and $\overline{X}$ is minimax and admissible [Ferguson (1976), p. 176; Lehmann (1983), pp. 257, 265].

[11.2.4] Suppose μ is known to take one of the values $0, \pm 1, \pm 2, \ldots$. Hammersley (1950), pp. 192, 207-209, has obtained the ML estimator of μ as $[\overline{X}]$, where $[\overline{X}]$ is the integer nearest to $\overline{X}$ with the convention that $[b - \frac{1}{2}] = b$, if b is an integer. The estimator $[\overline{X}]$ is unbiased and consistent. Let

$$\Psi(y) = \sqrt{\tfrac{2}{\pi}} \sum_1^\infty j \int_{jy}^\infty e^{-t^2/2} \, dt \; ;$$

then

$$\text{var}\big([\overline{X}]\big) = \Psi\!\left(\tfrac{\sqrt{n}}{2\sigma}\right) - 2\Psi\!\left(\tfrac{\sqrt{n}}{\sigma}\right) \approx \sqrt{8 \tfrac{\sigma^2}{(n\pi)}} \, \exp\!\left(-\tfrac{n}{(8\sigma^2)}\right)$$

as $n/\sigma^2 \to \infty$. Also,

$$\text{var}\big([\overline{X}]\big) \leq 4 \sum_1^\infty j \left\{ 1 - \Phi\!\left[\tfrac{\sqrt{n}}{\sigma}(j - \tfrac{1}{2})\right] \right\}$$

[Lehmann (1983), p. 143 and Stuart and Ord (1991), p. 701]. Khan (1973), pp. 1039-1041, and Ghosh and Meeden (1978), pp. 1-10, have discussed the admissibility of the estimator under various loss functions. Kojima et al. (1982), pp. 429-440, have obtained the locally best unbiased estimator for μ which is compared with the best invariant estimator of μ.

[11.2.5] Let the prior distribution of μ be $N(\theta, b^2)$ (θ and b^2 both known). Then the Bayes estimator of μ with respect to squared-error loss is

$$(\theta\sigma^2 + n\bar{X}b^2)/(\sigma^2 + nb^2)$$

[DeGroot (1975), pp. 277-278]. Lehmann (1983), pp. 243-244, shows that the above holds for any loss function $\rho(d-\mu)$ for which ρ is convex and even.

[11.2.6] If $\sigma = 1$, the UMVU estimator of μ^k, ($k \geq 1$ known) is

$$T = (-1/\sqrt{n})\, H_k(-\sqrt{n}\,\bar{X}),$$

where $H_k(y)$ are the Hermite polynomials. For $k = 2$, $T = \bar{X}^2 - \frac{1}{n}$; see Zacks (1971), pp. 129-130 and Washio et al. (1956), p. 76, who (p. 79) have also obtained var(T) for $k \geq 1$.

[11.2.7] Let $\sigma = 1$ and $g(\mu) = \frac{1}{\sqrt{2\pi}} e^{-(b-\mu)^2/2}$, (b known). Then the UMVU estimator of $g(\mu)$ is

$$T = \sqrt{\tfrac{n}{n-1}}\; \phi\!\left[\sqrt{\tfrac{n}{n-1}}\,(b-\bar{X})\right],\ n \geq 2\,;$$

$$\mathrm{var}(T) = \phi^2(b-\mu)\left[\tfrac{1}{2}\sqrt{n+1}\,\exp\!\left\{\!\left(1-\tfrac{n}{n^2-1}\right)(b-\mu)^2\right\}-1\right]$$

[Zacks (1971), pp. 132-134; Lehmann (1983), pp. 86-87; Guenther (1978), p. 33].

[11.2.8] Let $X \sim N(\mu, v(\mu))$. The ML estimator of μ is a root of

$$v' = 2(\bar{X}-\mu) + (v'/v)\,\tfrac{1}{n}\sum_1^n (X_i - \mu)^2,\ v' = dv(\mu)/d\mu,$$

and hence if $v(\mu) = \sigma^2 \mu^k$, σ known, $\hat{\mu}$ is a function of both $\bar{X}$ and $\sum_1^n X_i^2$ unless $k = 0$ (when $\hat{\mu} = \bar{X}$) or $k = 1$ (when $\hat{\mu}$ is a function of $\sum_1^n X_i^2$ only; see Stuart and Ord (1991), p. 704.

[11.2.9] Let $p = P(X \leq b)$, (b given). Then the UMVU estimator of p is

$$T_1 = \Phi\!\left[\sqrt{n}\,(b-\bar{X})/(\sigma\sqrt{n-1})\right].$$

For $\sigma = 1$, $\mathrm{var}(T_1) = \Phi_1(b-\mu, b-\mu; 1/n) - \Phi^2(b-\mu)$ where $\Phi_1(a, c, \rho)$ is the bivariate standard normal integral at (a, c), with a correlation coefficient ρ [Zacks (1971), pp. 132-135; Lehmann (1983), pp. 86; Folks et al. (1965), p. 44; Sathe & Varde (1969), p. 712]. The ML estimator of p is

$$T_2 = \Phi\left[(b - \bar{X})/\sigma\right].$$

Lehmann (1983), p. 113, has discussed the performance of T_1 and T_2 based on mean squared error. Also refer Zacks (1966), pp. 1043-1044, 1049. Washio et al. (1956), p. 77, have derived the UMVU estimator of $(1-p)^k$, k a positive integer, k < n.

[11.2.10] Let $Y = f(X)$, where f is a monotone function.

(a) Let $F(y_0) = P[Y \leq y_0]$. Li and Owen (1989), p. 813, have obtained the following UMVU estimator of $F(y_0)$:

$$T = \begin{cases} 1 - T_1 & \text{if f is a decreasing function,} \\ T_1 & \text{if f is an increasing function,} \end{cases}$$

$$T_1 = \Phi\left\{\sqrt{\frac{n}{n-1}} \left[\frac{f^{-1}(y_0) - \bar{X}}{\sigma}\right]\right\}.$$

The UMVU estimator covered in [11.2.9] is a special case of this result.

(b) Let $E(Y) = E[f(X)]$. Li and Owen (1989), p. 813, have also obtained the following UMVU estimator of $E(Y)$:

$$T_2 = \int_{-\infty}^{\infty} f\left(\bar{X} + \sqrt{\tfrac{n-1}{n}}\, \sigma t\right) \frac{1}{\sqrt{2\pi}} e^{-t^2/2}\, dt.$$

They also give the UMVU estimator of var(Y).

[11.2.11] Suppose g(y) has a convergent Taylor's expansion about μ with infinite radius of convergence. Gray et al. (1976), p. 95, have obtained the following UMVU estimator of $g(\mu)$, under certain regularity conditions:

$$g(\bar{X}) + \sum_{1}^{\infty} \frac{(-1)^m g^{(2m)}(\bar{X})}{m!} \left[\sigma^2/(2n)\right]^m,$$

where $g^{(j)}$ denotes the jth derivative of g. Also, refer to Gray et al. (1973), pp. 288-289, 297. Let $\sigma = 1$, and $g(\mu) = \exp(-\mu)$. The UMVU estimator of $g(\mu)$ is $\exp[-\bar{X} - 1/(2n)]$.

[11.2.12] Let X and Y be independent rvs such that $X \sim N(\mu, \sigma^2)$ and $Y \sim N(0, 1)$. Let $p = P(Y<X) = \Phi[\mu/\sqrt{1+\sigma^2}]$.
 (a) The ML estimator of p is $\Phi[\bar{X}/\sqrt{1+\sigma^2}]$ [Mazumdar (1970), p. 161].
 (b) The UMVU estimator of p is $\Phi[\bar{X}/\sqrt{1+(n-1)\sigma^2/n}]$ [Mazumdar (1970), p. 161; Downton (1973), p. 555].

11.3 ESTIMATORS OF σ^2 AND ITS FUNCTIONS

Single sample case, μ known.

[11.3.1] The sample variance $T = \sum_1^n (X_i - \mu^2)/n$ is
(a) the ML estimator of σ^2 [Roussas (1973), pp. 245-246],
(b) the MM estimator of σ^2,
(c) a consistent estimator of σ^2 [Mood et al. (1974), p. 296],
(d) a BAN estimator of σ^2 [Mood et al. (1974), p. 296],
(e) the UMVU estimator of σ^2 [Roussas (1973), p. 241].

[11.3.2] Let $\mu = 0$ and the loss function be squared error. Let $\tau = 1/2\sigma^2$ and let the prior distribution of τ be gamma with the pdf

$$e^{-y\alpha} y^{\alpha-1} g^{\alpha}/\Gamma(g), \quad y > 0, \ g > 0, \ g \text{ constant.}$$

Then the Bayes estimator of σ^2 is

$$T_1 = \left(\alpha + \sum_1^n X_i\right)\Big/\left(n + 2g - 2\right).$$

If the loss function is scale invariant, i.e., $L(d, \sigma^2) = (d - \sigma^2)/\sigma^4$, then the Bayes estimator is

$$T_2 = \left(\alpha + \sum_1^n X_i^2\right)\Big/\left(n + 2g + 2\right)$$

[Lehmann (1983), pp. 246-247].

[11.3.3] Let $\mu = 0$ and the loss function be $L(\sigma^2, \alpha) = (\log a - \log \sigma^2)^2$. Then the best invariant estimator of σ^2 is

$$T_1 = \sum_1^n X_i^2 \Big/ [2 \exp\{\psi(n/2)\}],$$

where $\psi(\alpha)$ is the digamma function (logarithmic derivative of gamma) given by

$$\psi(\alpha) = \frac{d}{d\alpha}\log \Gamma(\alpha) = \frac{1}{\Gamma(\alpha)}\int_0^\infty e^{-x} x^{\alpha-1} \log x\, dx$$

[Ferguson (1967), p. 179].

If the loss function is $L(\sigma^2, a) = (a/\sigma^2 - 1)^2$ then the best invariant estimator of σ^2 is

$$T_2 = \sum_1^n X_i^2/(n+2).$$

It may be of interest to note that $T_1 \approx T_2$. The estimators T_1 and T_2 are also minimax [Ferguson (1967), p. 179].

[11.3.4] The UMVU estimator of $(\sigma)^r$ is

$$\frac{\Gamma(n/2)}{2^{r/2}\Gamma[(n+r)/2]}\left[\sum_1^n (X_i - \mu)^2\right]^{\frac{r}{2}}, \quad n > -r$$

[Lehmann (1983), p. 85].

[11.3.5] The sample standard deviation $\sqrt{\sum_1^n (X_i - \mu)^2/n}$ is

(a) the ML estimator of σ [Mood et al. (1974), p. 284],
(b) the MM estimator of σ.

[11.3.6] Let $\mu = 0$. Then $(\sqrt{\pi/2})\sum_1^n |X_i|/n$ is a consistent estimator of σ [Lehmann (1983), p. 391].

[11.3.7] Let $p = P(X \leq b)$, b known. Folks et al. (1965), pp. 44-45 have obtained the UMVU estimator of p in the following form:

$$T = \begin{cases} 0, & b < \mu - T_1 \\ G[(b-\mu)\sqrt{n-1}/\sqrt{T_1^2 - (b-\mu)^2}], & \mu - T_1 \leq b \leq \mu + T_1 \\ 1, & b > \mu + T_1 \end{cases}$$

where $G(\cdot)$ is the cdf of the t-distribution with $(n-1)$ d.f. and $T_1 =$

$\sum_1^n (X_i - \mu)^2$. Patil & Wani (1966), pp. 41-42 have obtained the UMVU estimator of p, when $\mu = 0$, in a different form:

$$T = T \, V\tfrac{1}{2} \begin{cases} \tfrac{1}{2} I_a(\tfrac{1}{2},(n-1)/2) & \text{if } b \leq 0, \\ & \text{if } b = 0, \\ 1 - \tfrac{1}{2} I_a(\tfrac{1}{2},(n-1)/2) & \text{if } b > 0, \end{cases}$$

where $I_a(c, d)$ is the incomplete beta function ratio, and $a = b^2 / \sum_1^n X_i^2$.

[11.3.8]   Let $Y = f(X)$, where f is a monotone function.

(a) Let $F(y_0) = P(Y \leq y_0)$. Li and Owen (1989), pp. 813-814, have obtained the following UMVU estimator of $F(y_0)$:

$$T = \begin{cases} 1 - T_1 & \text{if f is a decreasing function,} \\ T_1 & \text{if f is an increasing function,} \end{cases}$$

$$T_1 = I_{\max\left\{0, \frac{f^{-1}(y_0) - \mu + \sqrt{n-1}\, S}{2\sqrt{n-1}\, S}\right\}} \left(\tfrac{n-1}{2}, \tfrac{n-1}{2}\right).$$

The UMVU estimator covered in [11.3.7] is a special case of this result.

(b) Let $E(Y) = E[f(X)]$. Li and Owen (1989), p. 814, have also obtained the following UMVU estimator of $E(Y)$:

$$T_2 = \int_{-1}^1 f(\mu + \sqrt{n-1}\, St) \frac{1}{B(\tfrac{1}{2}, \tfrac{n-1}{2})} (1-t^2)^{\frac{n-3}{2}}\, dt\,.$$

They also give the UMVU estimator of var(Y) on p. 814.

[11.3.9]   Suppose g(y) has a convergent Taylor's expansion about σ^2 with infinite radius of convergence. Gray et al. (1976) pp. 95-96, have obtained the following UMVU estimator of $g(\sigma^2)$, under certain regularity conditions:

$$\sum_0^\infty \frac{(-1)^m L_m^{(n_2)}(n_0)(T)^m g^{(m)}(T)}{(n_0)_m},$$

where $L_m^{(\alpha)}$ is the generalized Laguerre polynomial [Abramowitz & Stegun (1970), p. 775], $n_2 = \frac{n-2}{2}$, $n_0 = -\frac{1}{2}$, $g^{(j)}$ denotes the jth derivative of g,

$$(\alpha)_m = \frac{\Gamma(\alpha+m)}{\Gamma(\alpha)}, \quad T = \sum_1^n \frac{(X_i - \mu)^2}{n}.$$

Also, refer Gray et al. (1973), pp. 301-302.

11.4 JOINT ESTIMATORS OF μ AND σ^2, AND THEIR FUNCTIONS

Single sample case, μ and σ^2 unknown.

[11.4.1]   The estimators $\bar{X}$ and S^2 are

(a) jointly ML estimators of μ and σ^2, respectively [Mood et al. (1974), p. 281; Roussas (1973), p. 245]. If $n = 1$, no ML estimator of S^2 exists [Bickel & Doksum (1977), p. 111].

(b) jointly MM estimators of μ and σ^2, respectively [Mood et al. (1974), p. 275; Roussas (1973), p. 259].

(c) the UMVU estimators of μ and σ^2, respectively [Lehmann (1983), p. 85]. Here $\text{var}(\bar{X}) = \sigma^2/n$, $\text{var}(S^2) = 2\sigma^4/(n-1)$.

[11.4.2]

(a) The UMVU estimator of $(\sigma)^r$ is

$$\frac{\Gamma(n-1)/2}{2^{r/2}\Gamma[(n+r-1)/2]} (n-1)^{\frac{r}{s}} S^r, \quad n > -r+1,$$

[Lehmann (1983), p. 85].

(b) The minimum MSE estimators of σ^2 and σ are, respectively, $(n-1)S^2/(n+1)$ and $\sqrt{n-1}\,S/d$, $d = \sqrt{2}\,\Gamma\{\frac{1}{2}(n+1)\}/\Gamma(\frac{1}{2}n)$ [Gurland and Tripathi (1971), p. 26]. For σ^r, such an estimator is

$$\frac{\Gamma[(n-1+r)/2]}{2^{r/2}\Gamma[(n-1+2r)/2]} (n-1)^{r/2} S^r, \quad n > 1 - 2r$$

[D'Agostino (1970), pp. 14-15].

[11.4.3]   Let $x_{(1-\alpha)}$ and $z_{(1-\alpha)}$ be, respectively, the $100(1-\alpha)$th lower percentile of $N(\mu, \sigma^2)$ and $N(0, 1)$. Then $z_{(1-\alpha)} = \mu + \sigma z_{(1-\alpha)}$. The UMVU estimator of $x_{(1-\alpha)}$ is

$$\bar{X} + \frac{\Gamma[(n-1)/2]}{\Gamma(n/2)\sqrt{2}} S\sqrt{n-1}\, z_{(1-\alpha)}$$

[Mood et al. (1974), pp. 357-358; Lehmann (1983), p. 86].

[11.4.4] Suppose $\mu > 0$. Gupta and Rohatgi (1980), pp. 370-371, have considered the following four estimators of μ, where $f(x) = \max(0, x)$, $-\infty < x < \infty$:

$$T_1 = f(\bar{X}), \quad T_2 = \sum_1^n f(X_i)/n, \quad T_3 = |\bar{X}|, \quad T_4 = \sum_1^n |X_i|/n.$$

They have compared these estimators by their biases and MSEs.

[11.4.5] Let $\alpha = \mu/\sigma$, the reciprocal of the coefficient of variation. The UMVU estimator of α is

$$\frac{\bar{X}\sqrt{2}\,\Gamma((n-1)/2)}{S\sqrt{(n-1)}\,\Gamma((n-2)/2)}, \quad n > 2$$

[Lehmann (1983), p. 85; Washio et al. (1956), p. 86].

[11.4.6] Let $\alpha = \mu/\sigma$ be known. Let $p = P(X > L)$. The ML estimator of p is

$$\Phi[(\hat{\mu} - L)/(a\hat{\mu})], \quad \hat{\mu} = \left[-B + \sqrt{B^2 + 4A}\,\right]/2,$$

$$A = \sum_1^n X_i^2/(n\alpha^2), \quad B = \left(\sum_1^n X_i\right)/(\alpha^2 n).$$

Gertsbakh and Winterbottom (1991), p. 1503] discuss this estimator along with two others and compare them va their mean square errors.

[11.4.7] Let $p = P(X < 0) = \Phi(-\mu/\sigma)$. Lieberman and Resnikoff (1955), pp. 457, have obtained the UMVU estimator of p as

$$T = \begin{cases} 0 & \text{if } a < 0, \\ I_a(\tfrac{n}{2}-1, \tfrac{n}{2}-1) & \text{if } 0 \le a \le 1, \\ 1 & \text{if } a > 1, \end{cases} \quad a = \tfrac{1}{2}\left[1 - \frac{\sqrt{n}\,\bar{X}}{(n-1)S}\right],$$

where $I_a(b, c)$ is the incomplete beta function ratio [Zacks (1971), pp. 101, 106-107].

[11.4.8] Let $X \sim N(\mu, a\mu^2)$, $\mu > 0$ and the coefficient of variation $\sqrt{a}$, $a > 0$ be known. Two unbiased estimators of μ are $\bar{X}$ and

$$T = c_n \left[\sum_{i=1}^{n} (X_i - \bar{X})^2 / n \right]^{1/2},$$

$$c_n = \sqrt{n}\, \Gamma\left[(n-1)/2\right] / \left[\sqrt{2a}\, \Gamma(n/2)\right].$$

(a) The ML estimator of μ is

$$\left\{ -\bar{X} + \left[4a(T/c_n)^2 + (1+4a)\bar{X}^2\right] \right\} / (2a).$$

The asymptotic variance of the MLE is $a\mu^2/[n(1+2a)]$, which is the Cramér-Rao lower bound for the variance of unbiased estimators of μ [Khan (1968), p. 1041; Gleser & Healy (1976), p. 978].

(b) Khan (1968), p. 1040, has shown that the estimator

$$\hat{\mu} = (d_n \bar{X} + a\, T/n)/(d_n + a/n), \quad d_n = \left[(n-1)ac_n^2/n - 1\right],$$

has smallest variance, uniformly in μ, among all unbiased estimators that are linear in $\bar{X}$ and T. The efficiency of $\hat{\mu}$ relative to $\bar{X}$ approaches 1 as $a \to 0$ and becomes large for large a, whereas the efficiency of $\hat{\mu}$ relative to T gets large as $a \to 0$ and approaches 1 as $a \to \infty$. Here $\hat{\mu}$ is a BAN estimator and hence the asymptotic efficiency of $\hat{\mu}$ relative to the MLE is 1.

(c) Let C be a class of all estimators of μ which are linear in $\bar{X}$ and T but not necessarily unbiased. Then defining $\hat{\mu}$ as in (b), Gleser & Healy (1976), p. 978, have shown that the estimator

$$\bar{\mu} = \left[(d_n + a/n)/(d_n + a/n + ad_n/n)\right] \cdot \hat{\mu}$$

is the estimator in C with uniformly (in μ) minimum risk under squared-error loss. The estimator $\bar{\mu}$ dominates $\hat{\mu}$ in risk. Gleser and Healy (op. cit.), p. 979, have obtained an admissible, minimum risk, scale equivariant estimator μ that dominates $\hat{\mu}$, $\bar{\mu}$ and the ML estimator in risk. They have

Point Estimation

also constructed a class of Bayes estimators against inverted-gamma priors which includes μ within its closure.

[11.4.9] Let $X \sim N(\mu, a\mu)$ (a known), $w = (\sqrt{nT})/2$, $T = \sum_1^n X_i$.

(a) The ML estimator of μ is $\frac{1}{2}\left[(a^2 + 4T/n)^{1/2} - a\right]$ [Ratani (1977), p. 22; Feldman and Fox (1958), p. 155].

(b) The UMVU estimator of μ is

$$\frac{\frac{aw}{n}\int_{-1}^{1} y(y-y^2)^{\frac{n-3}{2}} \exp(yw)dy}{\int_{-1}^{1} (1-y^2)^{\frac{n-3}{2}} \exp(yw)dy} = \frac{aw}{n}\frac{I_{n/2}(w)}{I_{n/2-1}(w)},$$

where $I_\lambda(w)$ is the modified Bessel function of type I [Ratani (1977), p. 20; Feldman and Fox (1958), p. 154].

[11.4.10] Consider

$$g(\mu) = \frac{1}{\sqrt{2\pi\sigma}} e^{-(b-\mu)^2/2\sigma^2} = f(b, \mu, \sigma)$$

(b known). The UMVU estimator of $g(\mu)$ is given by

$$T = \frac{1}{S} \cdot h\left(\frac{b-\bar{X}}{S}\right),$$

$$h(z) = \frac{\Gamma(\frac{n-1}{2})}{\Gamma(\frac{1}{2})\Gamma(\frac{n-2}{2})} \sqrt{\frac{n}{n-1}} \left(1 - \frac{nz^2}{n-1}\right)^{\frac{n-5}{2}} / \sigma, \quad 0 < |z| < \sqrt{\frac{n-1}{n}}$$

[Lehmann (1983), p. 87].

[11.4.11] Estimation of $p = P(X \leq b)$. We present the ML estimator of p in (a), and various representations of the UMVU estimator of p in (b) – (e). The latter all express the UMVUE in terms of common quantities k and T, where

$$k = \sqrt{n}/(n-1), \quad T = (b - \bar{X})/S.$$

In (c) – (e) Y_1 denotes a variable having a Student t-distribution with $n - 2$ degrees of freedom.

(a) The ML estimator of p is $\Phi[(b-\bar{X})/S^*]$, $S^* = \sqrt{(n-1)/n}\ S$ (Bickel and Doksum, 1977, p. 146). Gertsbakh and Winterbottom (1991, pp. 1497-1514) give a useful review of the comparison of this estimator with that in (b); see also Mee (1988), pp. 1465 – 1479.

(b) Lieberman and Resnikoff (1955), p. 469, have derived the UMVUE of p in the form

$$T_1 = \begin{cases} 0, & kT \leq -1, \\ P[Y < (1+kT)/2], & |kT| < 1, \\ 1, & kT \geq 1, \end{cases}$$

where Y has a beta distribution with parameters $[(n-2)/2, (n-2)/2]$. Bowker and Goode (1952) had earlier given this estimator without derivation. Lieberman and Resnikoff (op. cit.), pp. 474-487, have presented tables giving T_1 for $T < 0$ and $1 - T_1$ for $T > 0$, for $|T| = 0.1(0.1)0.3$ $(0.01)3.90$ and $n = 3, 4, 5, 7, 10(5)40, 50, 75, 100, 150, 200$. For $n=5(1)10$, Gertsbakh and Winterbottom (1991), pp. 1500-1501 give a summary of simple formulas from Rukhin (1986), pp. 91-99, to evaluate $(1-T_1)$.

(c) Based on Barton (1961), p. 228, Basu (1964), pp. 215-219 and Folks et al. (1965), pp. 45, gave the UMVUE of p in the form

$$T_2 = \begin{cases} 0, & kT \leq 1, \\ P[Y_1 < kT\sqrt{(n-2)/(1-k^2T^2)}], & |kT| < 1, \\ 1, & kT \geq 1, \end{cases}$$

Guenther (1971), pp. 18-19, shows that T_1 in (b) and T_2 are equivalent, and discusses the relative merits of these forms.

(d) Sathe and Varde (1969), p. 712 give the UMVU estimator of p in the form

$$T_3 = \begin{cases} \frac{1}{2}\left[1 - I_{a^2}\left(\frac{1}{2}, \frac{n-2}{2}\right)\right] & \text{if } a \leq 0, \\ \frac{1}{2}\left[1 + I_{a^2}\left(\frac{1}{2}, \frac{n-2}{2}\right)\right] & \text{if } a \geq 0, \end{cases}$$

where a = $(b - \bar{X})\sqrt{n} / [S(n-1)] = kT$ in the notation of (b) – (d).

(e) Wheeler (1970), pp. 752-753 gives yet another expression for the UMVU estimator of p when n is even:

$$T_4 = \begin{cases} 0, & kT \leq -1, \\ \frac{1}{2} + \sum_{r=0}^{(n-4)/2} K_2(r)\{\sqrt{n}(\bar{X}-b)/S\}^{2r+1}, & |kT| < 1, \\ 1, & kT \geq 1, \end{cases}$$

in the notation of (b) – (e) for k and T, and where

$$K_2(r) = \left(\frac{n-4}{2}\right) \frac{(-1)^{r+1}(n-1)^{-2r-1}\,\Gamma(\frac{n-1}{2})}{(2r+1)\Gamma(\frac{1}{2})\,\Gamma(\frac{n-2}{2})}.$$

He also derives var(T_5) on p. 753.

(f) Washio et al. (1956), pp. 86-87 have derived the UMVU estimator of $(1-p)^k$, $n > k > 0$.

(g) Angus (1983), pp. 1345-1358 has discussed estimators of the form $\bar{X} + kS$ for estimating the p-quantile, where k is chosen either to satisfy

$$\left[(\bar{X} + kS - \mu)/\sigma\right] = p$$

or to minimize

$$E\{\Phi\left[(\bar{X} + kS - \mu)/\sigma\right] - p\}^2.$$

These estimators are compared with ML, UMVU and other estimators.

[11.4.12] Let $Y = f(X)$, and f be a monotone function.

(a) Let $F(y_0) = P(Y \leq y_0)$. Li and Owen (1989), p. 810 have obtained the UMVUE T of $F(y_0)$ given by

$$T = \begin{cases} 1 - T_1 & \text{if f is a decreasing function,} \\ T_1 & \text{if f is an increasing function,} \end{cases}$$

$$T_1 = I_{\max\{0,\,[f^{-1}(y_0)-b]/a\}}\left(\frac{n-2}{2}, \frac{n-2}{2}\right),$$

$$a = 2(n-1)S/\sqrt{n}, \quad b = \bar{X} - (n-1)S/\sqrt{n}.$$

The UMVU estimator covered in *[11.4.11]* is a special case of this result.

(b) Let $E(Y) = E[f(X)]$. Li and Owen (1989), p. 811 have obtained the UMVU estimator of $E(Y)$

$$T_2 = \frac{1}{B(\frac{1}{2}, \frac{n-1}{2})} \int_{-1}^{1} f\left(\bar{X} + \frac{(n-1)S}{\sqrt{n}} t\right)(1-t^2)^{\frac{n-4}{2}} dt,$$

where $B(\cdot, \cdot)$ is the beta function. Li and Owen have also obtained the UMVU estimator of $var(Y)$ on p. 812.

[11.4.13] Suppose the function f is of recursive type. Shimizu (1983), pp. 975-985 considers UMVU estimators of $E\,f(X)$ and $var\,f(X)$, see also Neyman and Scott (1960), pp. 643-655 and Hoyle (1968), pp. 1125-1143.

[11.4.14] Let $Y = \exp(X)$. Then Y has a lognormal distribution with $E(Y) = \exp(\mu + \sigma^2/2)$ and $var(Y) = \exp(2\mu + \sigma^2)\{\exp(\sigma^2) - 1\}$. Crow and Shimizu (1988) have discussed UMVU and other estimators of several functions of lognormal parameters in Chapter 2 of their book. They also give variances of these estimators. Finney (1941), p. 157 has obtained the UMVU estimators of $E(Y)$ and $var(Y)$.

(a) The ML estimator of $E(Y)$ is

$$T_1 = \exp\{\bar{X} + (n-1)S^2/(2n)\}.$$

It is also a BAN estimator of $E(Y)$ (Zacks, 1971, pp. 247-248; Stuart and Ord, 1991, p. 698).

$$var(T_1) = \exp(2\mu + \sigma^2/n)\left[\exp(\sigma^2/n)(1 - 2\sigma^2/n)^{-(n-1)/2} - (1 - \sigma^2/n)^{-(n-1)}\right]$$

$$\simeq \exp(2\mu + \sigma^2)[\sigma^2 + \sigma^4/2]/n,$$

asymptotically. The asymptotic efficiency of the mean $\bar{Y}$ relative to T_1 is

$$(\sigma^2 + \sigma^4/2)/[\exp(\sigma^2) - 1]$$

(Stuart and Ord, 1991, p. 699). The estimator T_1 is not unbiased and in fact $E(T_1) > E(Y)$.

(b) For a more general result on the UMVU estimator of E(Y), let $\hat{\mu}$ and T be a pair of statistics that are mutually independent and jointly sufficient for μ and σ^2. Suppose also that $\hat{\mu} \sim N(\mu, \lambda^2/\sigma^2)$, where λ^2 is a known positive quantity, and that T/σ^2 has a chi-squared distribution with degrees of freedom ν. Then the UMVU estimator of E(Y) is

$$T_2 = \exp(\hat{\mu}) \, G\left[(1-\lambda^2)S^2, \nu\right],$$

where (Neyman and Scott, 1960, p. 653)

$$G[aS^2, \nu] = \sum_{k=0}^{\infty} \frac{1}{k!} \frac{\Gamma(\nu/2)}{\Gamma(\nu/2 + k)} (aS^2/4)^k \, ;$$

$$\mathrm{var}(T_2) = \exp(2\mu + \sigma^2)\left\{\exp(\lambda^2\sigma^2) \, G\left[(1-\lambda^2)^2\sigma^4, \nu\right] - 1\right\}$$

(Mehran, 1973, p. 726). Bradu and Mundlak (1970), p. 206 have derived $\mathrm{var}(T_2)$ in a different form. Mehran (op. cit.) has compared the efficiency of the sample mean $\overline{Y}$ and T_2 from the same sample. The sample mean performs surprisingly well, especially for small and moderate sample sizes and for coefficients of variation less than unity.

(c) The ML estimator of $\mathrm{var}(Y)$ is

$$T_3 = \exp\left\{2\overline{X} + (n-1)S^2/n\right\}\left[\exp\left\{(n-1)S^2/n - 1\right\}\right]$$

(Zacks, 1971, p. 250; Stuart and Ord, 1991, p. 700). Asymptotically

$$\mathrm{var}(T_3) \simeq \left(2\sigma^2/n\right) \exp\left(4\mu + 2\sigma^2\right)\left[2\left\{\exp(\sigma^2) - 1\right\}^2 + \sigma^2\left\{2\exp(\sigma^2) - 1\right\}^2\right]$$

(Stuart and Ord, 1991, p. 701). The estimator T_3 is not unbiased.

(d) The UMVU estimator of var(Y), as given by Finney (1941), p. 157, can be expressed as

$$\exp(2\overline{X}) \sum_{p=0}^{\infty} \frac{T_0^p}{(n_1)_p \cdot p!} \left[(n-1)^p - (n_2)^p\right],$$

$$T_0 = \sum_{i=1}^{n} (X_i - \overline{X})^2/n, \quad (\alpha)_m = \Gamma(\alpha + m)/\Gamma(\alpha),$$

and $n_i = (n-i)/2$, $i = 1,2,3$ (Gray et al., 1976, p. 96). An alternate form of the UMVU estimator in a more rapidly converging series is given by Gray et al. (1976), p. 96, as

$$\exp(2\bar{X} + T_0) \sum_{p=0}^{\infty} \sum_{q=0}^{\infty} \frac{(-1)^p T_0^p \left[2^q \exp(T) - 1 \right]}{(n_1)_p (p-q)!} L_q^{(p-q+n_3)}(n_0) ,$$

where $L_m^{(\alpha)}(\cdot)$ is the generalized Laguerre polynomial (Abramowitz and Stegun, 1970, p. 775). Also, refer Gray et al. (1973), pp. 305-306.

[11.4.15] Let X and Y be independent rvs such that $X \sim N(\mu, \sigma^2)$ and $Y \sim N(0, 1)$. Then $P(Y < X) = \Phi\left[\mu/\sqrt{1+\sigma^2}\right]$. The estimation of $P(Y < X)$ is of interest in reliability problems (Church and Harris, 1970, pp. 49-54; Downton, 1973, pp. 551-558).

(a) The ML estimator of $P(Y < X)$ is $\Phi\left[\bar{X}/\sqrt{1 + S^2(n-1)/n}\right]$ (Downton, 1973, p. 555; Woodward and Kelley, 1977, p. 96).

(b) The UMVU estimator of $P(Y < X)$ was obtained by Downton (1973), p. 553, to be

$$T = \int_{-1}^{1} \Phi\left[\frac{\bar{X} + u\, S(n-1)}{\sqrt{n}}\right] \frac{\Gamma(\frac{n-1}{2})}{\sqrt{\pi}\,\Gamma(\frac{n-2}{2})} \left(1 - \mu^2\right)^{(n-4)/2} du .$$

Also, see Mazumdar (1970), pp. 161-162. Downton (op. cit.) approached the problem of obtaining approximations to T by considering approximations of the form

$$T_1 = \Phi\left[\bar{X}/\sqrt{1 + c_n S^2}\right].$$

When $c_n = 1$, T_1 coincides with the estimator given by Church and Harris (1970), p. 51; when $c_n = (n-1)/n$, T_1 coincides with the MLE given in (a). For n = 16, Downton (op. cit.), p. 557, has compared these approximations with T; none appear to be superior to others.

(c) An alternate form of the UMVU estimator of $P(Y < X)$ is given by Woodward and Kelley (1977), p. 95-97, as

$$T = \Phi(w) + \phi(w) \sum_{p=1}^{\infty} \sum_{q=0}^{p} \frac{(-1)^{p+1} L_q^{(\alpha)}(n/2) H_{2p-1}(w)}{(p-q)! 2^{2p-q}((n-1)/2)_p} \left(\frac{\hat{\sigma}^2}{1+\hat{\sigma}^2}\right)^p ,$$

Point Estimation

where $w = \bar{X}/\sqrt{1+\hat{\sigma}^2}$, $\hat{\sigma}^2 = (n-1)S^2/n$, $\alpha = p-q+(n-3)/2$, $L_q^{(\alpha)}(\cdot)$ is the generalized Laguerre polynomial, $(a)_b = \Gamma(a+b)/\Gamma(a)$, $H_k(\cdot)$ is the Hermite polynomial. Woodward and Kelley suggest an approximation of T by truncating the series at $p = 3$, which gives

$$T_2 = \Phi(w) + \frac{\phi(w)}{n+1}\left(\frac{\hat{\sigma}^2}{1+\hat{\sigma}^2}\right)^2\left[\frac{1}{4}H_3(w) - \frac{1}{3(n+3)}\left(\frac{\hat{\sigma}^2}{1+\hat{\sigma}^2}\right)H_5(w)\right].$$

The estimator T_2 seems to be superior to those of Downton and of Harris and Church discussed in (a) and (b).

[11.4.16] Let X and Y be independent rvs such that $X \sim N(\mu_1, \sigma_1^2)$ and $Y \sim N(\mu_2, \sigma_2^2)$. Then

$$P[Y < X] = P[Y - X < 0] = \Phi\left[(\mu_1 - \mu_2)/\sqrt{\sigma_1^2 + \sigma_2^2}\right].$$

The UMVU estimator of $P(Y < X)$ is obtained by Downton (1973), p. 556, to be

$$T_2 = c\int_{-1}^{1}\int_{-1}^{g(v)}(1-v^2)^{(n-4)/2}(1-u^2)^{(m-4)/2}\,du\,dv,$$

$$c = \left[\Gamma\left(\frac{n-1}{2}\right)\Gamma\left(\frac{m-1}{2}\right)\right]/\left[\pi\,\Gamma\left(\frac{n-2}{2}\right)\Gamma\left(\frac{m-2}{2}\right)\right],$$

$$g(v) = \frac{(\bar{X}-\bar{Y})\sqrt{m}}{S_2(m-1)} + v\frac{S_1(n-1)}{S_2(m-1)}\sqrt{\frac{m}{n}},$$

and the integral is to be interpreted as zero if $g(v) < -1$ and as unity if $g(v) > +1$ for all $|v| \leq 1$. $\bar{X}$, S_1^2 are respectively, the sample mean and the unbiased sample variance based on n observations from the X-distribution; and $\bar{Y}$, S_2^2 are defined similarly on m observations from the Y-distribution. Downton (op. cit.), p. 558, has proposed an approximation of T_2 in the form

$$\Phi[\bar{X}-\bar{Y}]/\left[c_n S_1^2 + c_m S_2^2\right].$$

[11.4.17] Suppose $g(y)$ has a convergent Taylor's expansion about μ with infinite radius of convergence. Gray et al. (1976), p. 95 have obtained, under certain regularity conditions, the UMVU estimator of $g(\mu)$,

$$g(\overline{X}) + \sum_{1}^{\infty} \frac{(-1)^m g^{(2m)}(\overline{X})}{(n_1)_m m!} (T/4)^m ,$$

where $n_1 = (n-1)/2$, $(\alpha)_m = \Gamma(\alpha + m)/\Gamma(\alpha)$, $T = \sum_1^n (X_i - X)^2/n$, and $g^{(j)}$ denotes the jth derivative of g, . Also, refer to Gray et al. (1973), pp. 291, 295. If $g(\mu) = \exp(-\mu)$, then the UMVU estimator of $g(\mu)$ is $\sum_{m=0}^{\infty} [(-T/4)^m / \{(n_1)_m m!\}]$.

[11.4.18] Suppose g(y) has a convergent Taylor's expansion about σ^2 with infinite radius of convergence. Gray et al. (1976), pp. 95-96, have obtained, under certain regularity conditions, the following UMVU estimator of $g(\sigma^2)$:

$$\sum_{m=0}^{\infty} \frac{(-1)^m L_m^{(n_3)}(n_0) T^m g_{(T)}^{(m)}}{(n_1)_m} ,$$

where $n_3 = \frac{1}{2}(n-3)$, $n_0 = -\frac{1}{2}$, $L_m^{(\alpha)}$ is the generalized Laguerre polynomial (Abramowitz and Stegun, 1970, p. 755) and where $g^{(j)}$, $(\alpha)_m$ and T are defined as in *[11.4.17]*. Also, refer Gray et al. (1973), p. 302.

[11.4.19] Suppose $g(\mu, v)$ has a convergent Taylor's expansion about (μ, σ^2) with infinite radius of convergence. Gray et al. (1976), p. 96, have obtained, under certain regularity conditions, the following UMVU estimator of $g(\mu, \sigma^2)$:

$$\sum_{p=0}^{\infty} \sum_{q=0}^{p} \frac{(-1)^p L_q^{(p-q+n_3)}(n_0)}{(p-q)! 4^{p-q}(n_1)_p} T^p g^{(2p-2q,q)}(\overline{X}, T)$$

where $n_i = (n-i)/2$, $i = 0, 1, 2, 3$, $T = \sum_{i=1}^{n}(X_i - \overline{X})^2/n$, and $g^{(j,k)}$ denotes the mixed (j, k)th partial derivative of g. Also, refer to Gray et al. (1973), p. 304.

11.5 ESTIMATORS OF PARAMETRIC FUNCTIONS: TWO SAMPLES

Here $X_1, \ldots, X_n$ are iid $N(\mu, \sigma^2)$ and $Y_1, \ldots, Y_n$ are iid $N(\theta, \tau^2)$, independently of one another.

[11.5.1] Let $\mu = \theta$ and $\sigma^2/\gamma^2 = \gamma$, ($\gamma$ known). The UMVU estimator of μ is $\alpha \overline{X} + (1-\alpha)\overline{Y}$, where $\alpha = m/(n\gamma + m)$ (Lehmann, 1983, p. 88).

Point Estimation

If γ is unknown, no UMVU estimator of ξ exists (Lehmann, 1983, pp. 88-89). An unbiased estimator (ibid, p. 89) is $\alpha\bar{X} + (1-\alpha)\bar{Y}$, where $\alpha = nS_Y^2 / (mS_X^2 + nS_Y^2)$, $S_X^2 = \sum_1^n (X_i - \bar{X})^2 / (n-1)$, $S_Y^2 = \sum_1^m (Y_i - \bar{Y})^2 / (m-1)$.

[11.5.2] Let $\sigma^2 = \tau^2$. The joint ML estimators of (μ, ξ, τ^2) are

$$\left(\bar{X}, \bar{Y}, \left[\sum_1^n (X_i - \bar{X})^2 + \sum_1^m (Y_i - \bar{Y})^2\right] \Big/ (n+m)\right)$$

(Bickel and Doksum, 1977, p. 112).

[11.5.3] (a) The UMVU estimator of $(\theta - \mu)$ is $(\bar{Y} - \bar{X})$ (Lehmann, 1983, p. 88).

(b) The UMVU estimator of $(\tau/\sigma)^r$ is

$$\frac{k_{m-1,r} \left[\sum_1^m (Y_i - \bar{Y})^2\right]^{r/2}}{k_{n-1,r} \left[\sum_1^n (X_i - \bar{X})\right]^{r/2}}, \quad \min(m, n) > -r + 1,$$

where (ibid, p. 88)

$$k_{p,r} = \Gamma(\tfrac{1}{2}p) \Big/ \left[2^{r/2} \Gamma(\tfrac{1}{2}p + \tfrac{1}{2}r)\right].$$

[11.5.4] Let $\theta = \mu$. If σ^2/n and τ^2/m are known, then $T_1 = (n\bar{X}\tau^2 + m\bar{Y}\sigma^2)/(n\sigma^2 + m\tau^2)$ is the UMVU estimator of μ (Graybill and Deal, 1959, p. 548). If σ^2/n and τ^2/m are unknown, let

$$T_2 = (n\bar{X}S_X^2 + m\bar{X}S_Y^2)/(nS_X^2 + mS_Y^2).$$

Then T_2 is a uniformly better unbiased estimator of μ than $\bar{X}$ or $\bar{Y}$ iff $n > 10$ and $m > 10$ (Graybill and Deal, 1959, p. 548). Nair (1980), pp. 212-226 gives an expression for $\text{var}(T_2)$. See also Cohen and Sackrowitz (1974), pp. 1274-1282.

[11.5.5] Let $p = P(Y < X) = \Phi\left[(\mu - \theta)/\sqrt{\sigma^2 + \tau^2}\right]$. The UMVU estimator of p is

$$\frac{\Gamma\left(\frac{n-1}{2}\right)\Gamma\left(\frac{m-1}{2}\right)}{\pi\,\Gamma\left(\frac{n-2}{2}\right)\Gamma\left(\frac{m-2}{2}\right)} \int_{-1}^{1} \int_{-1}^{g(v)} (1-v^2)^{(n-4)/2}(1-u^2)^{(m-4)/2}\,du\,dv\,,$$

where $g(v) = \dfrac{(\overline{X}-\overline{Y})\sqrt{m}}{S_Y(m-1)} + \dfrac{v\,S_X(u-1)}{S_Y(m-1)}\sqrt{\dfrac{m}{n}}$. This integral is to be taken as zero if $g(v) < -1$ and as unity if $g(v) > 1$. See Downton, 1973, pp. 556-558, who also gives approximations for the UMVU estimator of p.

REFERENCES

Abramowitz, M. and Stegun, I. A. (1970). *Handbook of Mathematical Functions*, Washington, DC: National Bureau of Standards. *[11.3.9; 11.4.14, 18]*

Angus, J. E. (1983). Normal quantile estimation based on minimizing the error in predicted distribution functions, *Communications in Statistics A12*, 1345-1358. *[11.4.11]*

Barton, D. E. (1961). Unbiased estimation of a set of probabilities, *Biometrika 48*, 227-229. *[11.4.11]*

Basu, A. P. (1964). Estimates of reliability for distributions useful in life testing, *Technometrics 6*, 215-219. *[11.4.11]*

Bickel, P. J. and Doksum, K. A. (1977). *Mathematical Statistics: Basic Ideas and Selected Topics*, San Francisco: Holden-Day. *[11.4.1, 11; 11.5.2]*

Bowker, A. H. and Goode, H. P. (1952). *Sampling Inspections by Variables*, New York: McGraw-Hill. *[11.4.11]*

Bradu, D. and Mundlak, Y. (1970). Estimation in lognormal linear models, *Journal of the American Statistical Association 65*, 198-211. *[11.4.14]*

Casella, G. and Strawderman, W. E. (1981). Estimating a bounded normal mean, *Annals of Statistics 9*, 870-878. *[11.2.2]*

Church, J. D. and Harris, B. (1970). The estimation of reliability from stress-strength relationships, *Technometrics 12*, 49-54. *[11.4.15]*

Cohen, A. and Sackrowitz, H. G. (1974). On estimating the common mean of two normal distributions, *Annals of Statistics 2*, 1274-1282. *[11.5.4]*

Crow, E. L. and Shimizu, K. (eds.) (1988). *Lognormal Distributions: Theory and Applications*. Dekker: New York. *[11.4.14]*

D'Agostino, R. B. (1970). Linear estimation of the normal distribution standard deviation, *The American Statistician 14, #3*, 14. *[11.4.12]*

DeGroot, M. H. (1975). *Probability and Statistics* (2nd edn.), New York: Wiley. *[11.1.3, 5; 11.2.5]*

Downton, F. (1973), The estimation of $P(Y < X)$ in the normal case, *Technometrics 15*, 551-558. *[11.2.12; 11.4.15, 16; 11.5.5]*

Feldman, D. and Fox, M. (1968). Estimation of the parameter n in the binomial distribution, *Journal of the American Statistical Association 63*, 150-158. *[11.1.7; 11.4.9]*

Ferguson, T. S. (1967), *Mathematical Statistics*, New York: Academic Press. *[11.2.3; 11.3.3]*

Finney, D. J. (1941). On the distribution of a variate whose logarithm is normally distributed, *Journal of the Royal Statistical Society B 7*, 155-161. *[11.4.14]*

Folks, J. L., Pierce, D. A. and Stewart, C. (1965). Estimating the fraction of acceptable product, *Technometrics 7*, 43-50. *[11.2.9; 11.3.7; 11.4.11]*

Gertsbakh, I. and Winterbottom, A. (1991). Point and interval estimation of normal tail probabilities, *Communications in Statistics - A20(4)*, 1497-1514. *[11.4.6, 11]*

Ghosh, M. and Meeden, G. (1978). Admissibility of the MLE of the normal integer mean, *Sankhya B 40*, 1-10. *[11.2.4]*

Gleser, L. J. and Healy, J. D. (1976). Estimating the mean of a normal distribution with known coefficient of variation, *Journal of the American Statistical Association 71*, 977-981. *[11.1.6; 11.4.8]*

Gray, H. L., Schucany, W. R. and Woodward, W. A. (1976). Best estimates of functions of the Gaussian and gamma distributions, *IEEE Transactions Reliability 25, #2*, 95-99. *[11.2.11; 11.3.9; 11.4.14, 17, 18]*

Gray, H. L., Watkins, T. A. and Schucany, W. R. (1973). On the jackknife statistic and its relation to UMVU estimators in the normal case, *Communications in Statistics 2(4)*, 285-320. *[11.2.11; 11.3.9; 11.4.14, 17, 18]*

Graybill, F. A. and Deal, R. B. (1959). Combining unbiased estimators, *Biometrics 15*, 543-550. *[11.5.4]*

Guenther, W. C. (1971). A note on the minimum variance unbiased estimates of the fraction of a normal distribution below a specification limit, *The American Statistician 25, #2*, 18-20. *[11.4.11]*

Guenther, W. C. (1978). Some easily found minimum variance unbiased estimators, *The American Statistician 32, #1*, 29-34. *[11.2.7]*

Gupta, A. K. and Rohatgi, V. K. (1980). On the estimation of restricted mean, *Journal of Statistical Planning and Inference 4*, 369-379. *[11.4.4]*

Gurland, J. and Tripathi, R. (1971). A simple approximation for unbiased estimation of the standard deviation, *The American Statistician 25, #4*, 30-32. *[11.4.2]*

Hammersley, J. M. (1950). On estimating restricted parameters, *Journal of the Royal Statistical Society B12*, 192-240. *[11.4.4]*

Hogg, R. V. and Craig, A. T. (1978). *Introduction to Mathematical Statistics* (4th ed.), MacMillan: New York. *[11.1.4]*

Hoyle, M. H. (1968). The estimation of variance after using a Gaussianating transformation, *Annals of Mathematical Statistics 39*, 1125-1143. *[11.4.13]*

Khan, R. A. (1968). A note on estimating the mean of a normal distribution with known coefficient of variation, *Journal of the American Statistical Association 63*, 1039-1041. *[11.2.4; 11.4.8]*

Kojima, Y., Morimoto, H. and Takeuchi, K. (1982). Two "best" unbiased estimators of normal integral mean, *Statistics and Probability: Essays in honor of C. R. Rao* (G. Kallianpur, P. R. Krishnaiah, and J. K. Ghosh, eds.) 429-440. *[11.2.4]*

Lehmann, E. L. (1986). *Testing Statistical Hypotheses* (2nd edn.), New York: Wiley. *[11.1.6]*

Lehmann, E. L. (1983). *Theory of Point Estimation*, New York: Wiley. *[11.1.8, 9, 10, 11; 11.2.1,2,3,4,5,7,9; 11.3.2, 4, 6; 11.4.1; 11.5.10, 13]*

Lehmann, E. L. and Scheffé, H. (1950). Completeness, similar regions and unbiased estimation, Part I, *Sankhya 10*, 305-340. *[11.1.1, 2]*

Li, H. and Owen, D. B. (1989). On the UMVU estimators after using a normalizing transformation, *Communications in Statistics - A 18*, 801-816. *[11.2.10; 11.3.8; 11.4.12]*

Lieberman, G. J. and Resnikoff, G. J. (1955). Sampling plans for inspection by variables, *Journal of the American Statistical Association 50*, 457-516. *[11.4.11]*

Mazumdar, M. (1970). Some estimates of reliability using interference theory. *Naval Research Logistics Quarterly*, 17, 159-165. *[11.2.12; 11.4.15]*

Mee, R. W. (1988). Estimation of the percentage of a normal distribution lying outside a specified interval, *Communications in Statistics - A17*, 1465-1479. *[11.4.11]*

Mehran, F. (1973). Variance of the MVUE for the lognormal mean, *Journal of the American Statistical Association* 68, 726-727. *[11.4.14]*

Mood, A. M., Graybill, R. A. and Boes, D. C. (1974). *Introduction to the Theory of Statistics* (3rd edn.), New York: McGraw-Hill. *[11.1.5; 11.2.1; 11.3.1, 5; 11.4.1]*

Nair, K. A. (1980). Variance and distribution of the Graybill-Deal estimator of the common mean of two normal populations, *Annals of Statistics* 8, 212-216. *[11.5.4]*

Neyman, J. and Scott, E. L. (1960). Correction for bias introduced by a transformation of variables, *Annals of Mathematical Statistics 31*, 643-655. *[11.4.13, 14]*

Patil, G. P. and Wani, J. K. (1966). Minimum variance unbiased estimation of the distribution function admitting a sufficient statistic, *Annals of the Institute of Statistical Mathematics* 18, 39-47. *[11.3.7]*

Ratani, R. T. (1977). Estimation of the parameter of a normal distribution $N(\theta, a\theta)$, *Gujarat Statistical Review 4*, 17-24. *[11.4.9]*

Rohatgi, V. K. (1984). *Statistical Inference*, New York: Wiley. *[11.1.3, 4]*

Roussas, G. G. (1973). *A First Course in Mathematical Statistics*, Reading, Mass.: Addison Wesley. *[11.2.1; 11.3.1; 11.4.1]*

Rukhin, A. L. (1986). Estimating normal tail probabilities, *Naval Research Logistics Quarterly 33*, 91-99. *[11.4.11]*

Sathe, Y. S. and Varde, S. D. (1969). On minimum variance unbiased estimation of reliability, *Annals of Mathematical Statistics 40*, 710-714. *[11.2.9; 11.4.11]*

Shimizu, K. (1983). Variances of UMVU estimators for means and variances after using a normalizing transformation, *Communications in Statistics A 12*, 975-985. *[11.4.13]*

Stuart, A. and Ord, J. K. (1991). *Kendall's Advanced Theory of Statistics*, Vol. 2, (5th edn.) New York: Oxford University Press. *[11.1.1, 2; 11.2.4, 8; 11.4.14]*

Wasan, M. T. (1970). *Parametric Estimation*, New York: McGraw-Hill.

[11.2.1]

Washio, Y., Morimoto, H. and Ikeda, N. (1956). Unbiased estimation based on sufficient statistics, *Sankhya*, 69-93. *[11.2.6, 9; 11.4.5, 11]*

Wheeler, D. J. (1970). The variance of an estimator in variable sampling. *Technometrics 12*, 751-755. *[11.4.11]*

Woodward, W. A. and Kelley, G. D. (1977). Minimum variance unbiased estimation of $P(Y < X)$ in the normal case, *Technometrics 19*, 95-98. *[11.4.15]*

Zacks, S. (1966). Unbiased estimation of the common mean of two normal distributions based on small samples of equal size, *Journal of the American Statistical Association 61*, 467-476. *[11.2.9]*

Zacks, S. (1971). *The Theory of Statistical Inference*, New York, Wiley. *[11.2.6, 7; 11.4.7, 14]*

Chapter 12
STATISTICAL INTERVALS

Three types of statistical intervals are commonly in use. These are (i) confidence intervals, (ii) tolerance intervals, and (iii) prediction intervals.

Confidence intervals contain unknown parameters of a statistical population with a specified probability. Tolerance intervals contain a specified proportion of the population values with a specified probability. Prediction intervals use the information from an observed (past) sample to contain some functions of a future independent sample from the same population with a specified probability.

These intervals are available in many forms and hence a large number of results are available in literature. In the following three sections we list several commonly used intervals. Bayesian intervals are not included.

12.1 CONFIDENCE INTERVALS

Confidence interval estimation is another aspect of statistical inference about the unknown parameter(s) of a distribution. It provides some useful information about the precision or possible error of a point estimator. As in point estimation, no unique definition of the optimality of an interval estimator exists. Since confidence intervals can be viewed as equivalent to testing of hypotheses, some of the optimality properties of the tests also carry forward to confidence intervals.

Let $X_1, X_2, ..., X_n$ be a random sample from a $N(\mu, \sigma^2)$ distribution. In the simplest form, a confidence interval (CI) for a parameter can be defined as follows. Consider two statistics $L = L(X_1, X_2, ..., X_n)$ and $U = U(X_1, X_2, ..., X_n)$. Let θ be some parameter that we want to estimate. If

L and U are selected so that

$$P(L \leq \theta \leq U) = 1 - \alpha,$$

then the interval (L, U) is a *two-sided $100(1-\alpha)\%$ CI* for θ. For one-sided intervals, the statistic L (or U) is identified as a *lower confidence limit* (or an *upper confidence limit*). Unless stated otherwise all confidence intervals listed are two-sided. In many cases one-sided confidence intervals can be obtained from two-sided intervals with appropriate changes in the percentile values of the distributions involved.

In the following we include mostly intervals based on one and two complete samples. A few simultaneous confidence intervals are also included. We use the abbreviation CL for confidence limit.

For the following three sections define

$$\bar{X} = \sum_{i=1}^{n} X_i/n, \quad S^2 = \sum_{i=1}^{n}(X_i - \bar{X})^2 \big/ (n-1), \quad T = \sum_{i=1}^{n}(X_i - \mu)^2/n.$$

For the two samples case, let the second sample be $Y_1, Y_2, ..., Y_m$. Then $\bar{X}, \bar{Y}$ denote the sample means and S_1^2, S_2^2 denote sample variances from the X and Y samples, respectively. As before z_α, $t_{(\nu;\alpha)}$, $\chi^2_{(\nu;\alpha)}$ and $f_{(\nu_1,\nu_2;\alpha)}$ represent $100\alpha\%$ upper percentiles of the standard normal, t, chi-square, and F-distributions, respectively.

[12.1.1] Confidence Intervals for μ and Its Functions

[12.1.1.1] A $100(1-\alpha)\%$ CI for μ (σ known) based on $\bar{X}$ and σ has limits given by

$$L = \bar{X} - \frac{\sigma}{\sqrt{n}} z_{\alpha_1}, \quad U = \bar{X} + \frac{\sigma}{\sqrt{n}} z_{\alpha_2},$$

where $0 < \alpha_i < 1$ ($i = 1, 2$), and $\alpha_1 + \alpha_2 = \alpha$. For $\alpha_1 = \alpha_2 = \alpha/2$, this is the "shortest width unbiased" interval (Guenther, 1971, pp. 51-52). Pratt (1961, 1963) has discussed different procedures for obtaining "efficient shorter confidence" intervals. On the basis of a single sample of fixed size there do not exist CIs for μ with a bounded width (Lehmann, 1986, p. 258).

[12.1.1.2] A $100(1-\alpha)\%$ CI for μ (σ not known) based on $\bar{X}$ and S has limits given by

$$L = \bar{X} + \frac{S}{\sqrt{n}} t_{(n-1;\alpha_1)}, \quad U = \bar{X} + \frac{S}{\sqrt{n}} t_{(n-1;\alpha_2)},$$

where $0 < \alpha_i < 1$ (i = 1, 2), and $\alpha_1 + \alpha_2 = \alpha$. For $\alpha_1 = \alpha_2 = \alpha/2$, this is the "shortest width unbiased" interval (Guenther, 1971, p. 52). Also refer Bartoszynski and Chan (1970), pp. 415-417 for a remark on the "shortest" CI for μ.

[12.1.1.3] A $100(1-\alpha)\%$ CI for μ (σ not known) based on $\bar{X}$ and sample mean deviation d has limits given by

$$\bar{X} \pm \frac{d}{\sqrt{n}} h_{(n;\alpha/2)},$$

where $d = \frac{1}{n} \sum_{i=1}^{n} |X_i - \bar{X}|$, and $h_{(n;\alpha/2)}$ is the $100(\alpha/2)$th upper percentile of the probability distribution of the rv $H = \sqrt{n}(\bar{X} - \mu)/d$ (Herrey, 1965, pp. 261-262). Herrey (1971), p. 188 gives $h_{(n;\alpha/2)}$ – values for n = 40, 60, 120, ∞; and $\alpha = 0.10(.10), 0.90, 0.95, 0.98, 0.99; 0.999$. Also, Krutchkoff (1966), p. 669 gives $\frac{1}{\sqrt{n}} h_{(n;\alpha/2)}$ – values for n=3(1)(10), and $\alpha = 0.001$, 0.01, 0.02, 0.05, 0.10, 0.20. Both Herrey (1965) and Krutchkoff have obtained an approximate CI. They have also compared the expected width of this interval with that given in [12.1.1.2]. The results indicate that increase in expected width using d is insignificant whereas there is a decrease in expected width using d when one takes an additional observation.

[12.1.1.4] A $100(1-\alpha)\%$ CI for μ, when the coefficient of variation $\kappa = \sigma/\mu$ is known, and based on $\bar{X}$ and κ, has limits given by

$$L = \bar{X}/(1 + A), \quad U = \bar{X}/(1 - A),$$

where $A = (\kappa/\sqrt{n})z_{\alpha/2}$ (Weiler, 1958, p. 322). Weiler has compared the efficiency of this CI using the expected width of the confidence intervals in [12.1.1.1] and [12.1.1.2].

[12.1.1.5] Let Y be a lognormal rv, i.e., $\ln Y \sim N(\mu, \sigma^2)$. The median of Y is $\theta = \exp(\mu)$.

A $100(1-\alpha)\%$ CI for θ (σ known) based on g and τ has limits given by

$$L = g \cdot \exp(A\tau), \quad L = g \cdot \exp(B\tau),$$

where $g = \left(\prod_{i=1}^{n} Y_i\right)^{\frac{1}{n}}$, and $\tau = \sigma^2/\sqrt{n}$ (Dahiya and Guttman, 1982, p. 282). The constants A and B are selected so as to minimize the width of this CI. They have tabulated on p. 281 A and B values for $\alpha = 0.01, 0.05, .010$, and $\tau = 0.02(.01)\ 1.0, 1.5, 2.0, 3.0, 4.0, 5.0$. On p. 285 they have also obtained a $100(1-\alpha)\%$ Bayesian CI for θ when σ is not known.

[12.1.2] Confidence Intervals for σ^2 and Its Functions

[12.1.2.1] A $100(1-\alpha)\%$ CI for σ^2 (μ known) based on T has limits of the form

$$L = nT/\chi^2_{(n;1-\alpha_1)} = nT/a, \quad U = nT/\chi^2_{(n;\alpha_2)} = nT/b,$$

$0 < \alpha_i < 1$ $(i = 1, 2)$, and $\alpha_1 + \alpha_2 = \alpha$.

This interval has the shortest width (both actual and expected) when the constants a and b are chosen so that

$$\int_a^b g_\nu(t)dt = 1-\alpha \quad \text{and} \quad a^2 g_\nu(t) = b^2 g_\nu(b)$$

where $g_\nu(t)$ is the chi-squared pdf with df $= \nu = n$. Tate and Klett (1959), pp. 678-679 have tabulated a and b values for df $\nu = 2(1)29$; and $(1-\alpha) = 0.90, .095, 0.99, 0.995, 0.999$. For this interval to be 'unbiased'; (i) Tate and Klett (1959), pp. 680-681 give a and b values for ν and $(1-\alpha)$-values listed above, (ii) Pacharaces (1961), p. 86, also gives a, b for $\nu = 1(1)20, 24, 30, 40, 60, 120$; and $(1-\alpha) = 0.90, 0.95, 0.99$ and (iii) Lindley, East, and Hamilton (1960), pp. 436-437 give a, b for $\nu = 1(1)100$ and $(1-\alpha) = 0.95, 0.99, 0.999$. In this case, the unbiased, the shortest width and the equal-tail intervals are all different. Also, refer Pratt (1961), pp. 559-562.

[12.1.2.2] A $100(1-\alpha)\%$ CI for σ^2 (μ not known) based on S^2 has limits of the form

$$L = (n-1)S^2/\chi^2_{(n-1;1-\alpha_1)} = (n-1)S^2/a,$$

$$U = (n-1)S^2/\chi^2_{(n-1;\alpha_2)} = (n-1)S^2/b,$$

where $0 < \alpha_i < 1$ ($i = 1,2$), and $\alpha_1 + \alpha_2 = \alpha$. For this interval to be the shortest (both actual and expected) width interval or the unbiased interval, refer to the interval in *[12.1.2.1]* with df $\nu = n-1$. Also, for this interval to be unbiased, refer *[12.1.2.1]* with df $\nu = n-1$. Shorrock (1990) has constructed a CI for σ^2 based on both S^2 and $\bar{X}$. These intervals have the same width (and depend on S^2) but have higher probability of coverage. Also refer to Cohen (1972) who answers the following two questions, initially posed by Tate and Klett (1959):

(i) Does the interval of shortest width based on $(\bar{X}, S^2)$ depend only on S^2? (NO)

(ii). Among those intervals based only on S^2, is the interval of shortest width necessarily of the form given above? (YES).

[12.1.2.3] A $100(1-\alpha)\%$ CI for σ (μ not known) based on the sample quasi-range W_r has limits give by

$$L = W_r/W_{(r,1-\alpha/2)}, \quad U = W_r/W_{(r,\alpha/2)},$$

where $W_r = X_{(n-r;n)} - X_{(r+1;n)}$ and $W_{(r,p)}$ is the 100 pth upper percentile of the distribution of W_r (Harter, 1969, p. 12). Tables of $W_{(r,p)}$ − values for $r = 0(1)8$; $n = (2r+2)(1)20(2)40(10)100$; and $p = 0.0001, 0.0005, 0.001, 0.005, 0.01, 0.025, 0.05, 0.1(.1)$ 0.9, 0.95, 0.975, 0.99, 0.995, 0.999, 0.9995, 0.9999 are given by Harter on pages 295-319. He has also discussed the efficiency of this interval with that based on sample standard deviation S.

[12.1.2.4] Let $\theta = (\exp(\sigma^2) - 1)^{1/2}$. Then θ is the coefficient of variation of the lognormal distribution of the rv $Y = \ln X$.

A $100(1-\alpha)\%$ CI for θ (μ not known) based on S^2_Y has limits given by

$$L = \exp\left[\{S_Y^2/\chi^2_{(n-1,\alpha_1)}\} - 1\right]^{1/2}, \quad U = \exp\left[\{S_Y^2/\chi^2_{(n-1,1-\alpha_2)}\} - 1\right]^{1/2}$$

where $S_Y^2 = \sum_{i=1}^{n}(Y_i - \bar{Y})^2$, $0 < \alpha_i < 1$ (i = 1, 2), and $\alpha_1 + \alpha_2 = \alpha$ (Koopmans et al. 1964, p. 30). This interval is of finite but unbounded width.

[12.1.3] Confidence Intervals for Functions of μ and σ^2

[12.1.3.1] A minimum $100(1-\alpha)\%$ CI for (σ/μ) ($\mu \geq c > 0$; c known and σ not known) based on $\bar{X}$ and S has limits given by

$$L = \tau_1(t, c/T), \quad U = \tau_2(t, c/T)$$

where for $\lambda > 0$,

$$\tau_1(t,\lambda) = \begin{cases} 1/t_1^{-1}, & t \geq t_1(\lambda), \\ 0, & t < t_1(\lambda), \end{cases} \quad \tau_2(t,\lambda) = \begin{cases} 1/t_2^{-1}, & t \geq t_2(\lambda), \\ 0, & t < t_2(\lambda), \end{cases}$$

$T = \left[(n-1)S^2/\chi^2_{(n-1,\beta_3)}\right]^{1/2}$, $t = \sqrt{n}\,\bar{X}/S$, $G[t_1(\mu/\sigma)] = \beta_1$, $G[t_2(\mu/\sigma)] = 1 - \beta_2$, $0 \leq \beta_1 < \frac{1}{2}$, $0 < \beta_2 < \frac{1}{2}$, $0 < \beta_3 < 1$, and $\beta_1 + \beta_2 + \beta_3 = \alpha$; $G(\cdot)$ is the cdf of the non-central t-distribution with df n and non-centrality parameter $\sqrt{n}\,\mu/\sigma$ (Koopmans et al. 1964, pp. 27-28). This interval has finite width with probability one. Owen (1962) has tabulated $t_1(\lambda)$ and $t_2(\lambda)$ values on pp. 108-112.

[12.1.3.2] Let $X_p = \mu + \sigma Z_{1-p}$ be the pth quantile of X. A $100(1-\alpha)\%$ CI for $X_p = \mu + \sigma Z_p$ based on $\bar{X}$ and S has limits given by

$$L = \bar{X} - k_{(\frac{1+r}{2})} S, \quad U = \bar{X} - k_{(\frac{1-r}{2})} S$$

(Owen 1968, p. 457). Owen has provided tables of k-values.

[12.1.3.3] Let $Y \sim N(\mu, \sigma^2/\gamma^2)$, ($\gamma$ known) and T^2 be distributed, independently of Y, as σ^2/ν times a χ^2 rv with known df ν. A $100(1-\alpha)\%$ CI for $(\mu + \lambda\sigma^2)(\lambda \neq \text{known})$ based on Y and T has limits given by

$$L = Y + \delta\beta T\{\nu^{-\frac{1}{2}} C_{-1}(\beta T; \nu, 1-\alpha) + \beta T/2\},$$

$$U = Y + \delta\beta T\{\nu^{-\frac{1}{2}}C_{-2}(\beta T;\nu, 1-\alpha) + \beta T/2\},$$

where $\delta = (\nu+1)/(2\lambda\gamma^2)$, $\beta = 2|\lambda|\nu/(\nu+1)^{\frac{1}{2}}$ (Land, 1973, p. 963). The critical values $C_{-1}(t;\nu,\alpha)$ and $C_1(t;\nu,\alpha)$ for $\alpha = 0.90, 0.95, .99$; $t = 0.1$, 0.2, 0.5, 1.0, 2.0, 5.0, 10.0; and $\nu = 2, 4, 10, 20$ are tabulated by Land on page 962. He has also tabulated critical values for one-sided CIs on page 961.

[12.1.3.4] Let $Y \sim N(\mu,\sigma^2/\gamma)$, $(\gamma > 0$ known) and T^2 be distributed, independently of Y, as σ^2/ν times a χ^2 rv with known df ν.

A $100(1-\alpha)\%$ CI for $(\mu + k\sigma)$ ($k \neq 0$, known) based on Y and T has limits given by

$$L = Y - \frac{T}{\gamma} t_{(\nu,\eta;1-\alpha_2)}, \quad U = Y - \frac{T}{\gamma} t_{(\nu,\eta;\alpha_1)},$$

where $0 < \alpha_i < 1$, $\alpha_1 + \alpha_2 = \alpha$, and $t_{(\nu,\eta;\alpha)}$ is the $(100)\alpha^{th}$ upper percentile of a non-central t-distribution with df ν, non-centrality parameter $\eta = -\gamma k$ with the symmetry property $t_{(\nu,\eta;\alpha)} = -t_{(\nu,-\eta;1-\alpha)}$ (Land, 1988, p. 101).

[12.1.3.5] Let $\log X \sim N(\mu, \sigma^2)$. Land (1988) gives CIs for several functions of both μ and σ.

[12.1.3.6] Confidence intervals for a proportion of the normal population. Let $\pi_1 = P(X < L^*)$ and $\pi_2 = P(X < U^*)$, where L^* and U^* are specified limits. A $100(1-\alpha)\%$ one-sided lower (upper) CL L(U) for π_1 based on $\overline{X}$ and S is obtained from

$$P\left[T(\sqrt{n}\,Z_L) \leq \sqrt{n}\,k\right] = \alpha, \quad P\left[T(\sqrt{n}\,Z_U) \leq \sqrt{n}\,k\right] = 1-\alpha,$$

where $k = k_i$ (i = 1, 2); $k_1 = \frac{\overline{X} - L^*}{S}$ and $k_2 = \frac{U^* - \overline{X}}{S}$, $T(\lambda)$ has a non-central t-distribution with df $= (n-1)$ and non-centrality parameter λ (Mee, 1988, pp. 1467-1468; Gertsbakh and Winterbottom, 1991, pp. 1505-1506). Odeh and Owen (1980) have tabulated these limits for $k = -3(.20)6.0$; $n = 2(1)18(3)30, 40(20)120, 240, 600, 1000, 1200$; and for $(1-\alpha) = .50, 0.75, 0.90, 0.95, 0.99, 0.995$; see also Kirkpatrick (1970), pp. 150-155. Gertsbakh and Winterbottom (1991) have also obtained a

Bayesian confidence interval for π_2. Resnikoff (1955) has obtained an approximate confidence interval for $\pi = \pi_1 + \pi_2$. Mee (1988) provides a simulation to estimate the actual confidence level for this interval.

[12.1.3.7] A $100(1-\alpha)\%$ joint upper CL for both π_1 and π_2 (refer [12.1.3.6]) based on $\bar{X}$ and S is obtained from

$$P\left[\pi_1 \leq p_1^*, \pi_2 \leq p_2^*\right] = 1 - \alpha$$

(Chou and Owen, 1984, p. 150). Let $k_1 = (\bar{X} - L^*)/S$, $k_2 = (U^* - \bar{X})/S$. When $k_1 = k_2$, take $p_1^* = p_2^* = p^*$. They have tabulated on page 151 p^* value for n = 5(1), 9, 12, 15, 24, 30, 60, 120, 240; k = $-3(.2)6.0$; and $(1-\alpha) = 0.90, 0.95, 0.99$. They have also obtained a joint one-sided lower confidence interval for π_1 and π_2.

[12.1.3.8] Let $Y \sim N(0, 1)$ and $X \sim N(\mu, \sigma^2)$. Suppose X and Y are independent rvs. Then

$$P(Y < X) = \Phi\left(\frac{\mu}{\sqrt{1+\sigma^2}}\right).$$

(a) An approximate $100(1-\alpha)\%$ CI for $\mu/\sqrt{1+\sigma^2}$ based on $\bar{X}$ and S has limits given by $V \pm z_{\alpha/2}\hat{\sigma}_v$ where $V = \bar{X}/\sqrt{1+S^2}$, and

$$\hat{\sigma}_v = \frac{S^2}{1+S^2}\left(\frac{1}{n} + \frac{\bar{X}^2 S^2}{2(n-1)(1+S^2)^2}\right)^{\frac{1}{2}}$$

(Church and Harris, 1970, p. 52).

(b) An approximate $100(1-\alpha)\%$ CI for $P(Y < X)$ based on $\bar{X}$ and S has limits

$$L = \Phi\left[V - z_{\alpha/2}\hat{\sigma}_v\right], \quad U = \Phi\left[V - z_{\alpha/2}\hat{\sigma}_v\right].$$

Church and Harris (1970) have compared confidence intervals given here with one given by Govindarajulu (1967). Refer Mazumdar (1970) and Owen et al. (1964) for one-sided intervals for $P(Y < X)$.

[12.1.3.9] A $100(1-\alpha)\%$ confidence region for (μ and σ^2) based on a single observation x is given by

$$\{x - tk|x| \leq \mu \leq x + tk|x|, \ \sigma \leq k|x|\},$$

where $t = z_{\gamma_1/2}$, $(k)^{-1} = z_{(1-\gamma_2)}$, and $\gamma_1 + \gamma_2 = \alpha$ (Rosenblatt, 1966, p. 367).

[12.1.3.10] A $100(1-\alpha)\%$ confidence region for (μ and σ^2) based on $\overline{X}$ and S^2 is given by

$$\left\{\overline{X} - c\frac{\sigma}{\sqrt{n}} \leq \mu \leq \overline{X} + d\frac{\sigma}{\sqrt{n}}, \ \frac{(n-1)S^2}{b} \leq \sigma^2 \leq \frac{(n-1)S^2}{a}\right\},$$

where the factors a, b, and c, d are determined subject to $\Phi(d) - \Phi(c) = \sqrt{1-\alpha}$, and $G(b) - G(a) = \sqrt{1-\alpha}$, where $G(\cdot)$ is the chi-square cdf with $n-1$ df (Mood et al. 1974, p. 385).

[12.1.4] Confidence Intervals Based on Two Samples

[12.1.4.1] A $100(1-\alpha)\%$ CI for $(\mu_1 - \mu_2)$ (σ_1, σ_2 known) based on $\overline{X}$, $\overline{Y}$, σ_1, σ_2 has limits given by

$$L = (\overline{X} - \overline{Y}) - \left(\sqrt{\frac{\sigma_1^2}{n} + \frac{\sigma_2^2}{m}}\right) z_{\alpha_1}, \quad U = (\overline{X} - \overline{Y}) - \left(\sqrt{\frac{\sigma_1^2}{n} + \frac{\sigma_2^2}{m}}\right) z_{\alpha_2},$$

where $0 < \alpha_i < 1$ (i = 1, 2), and $\alpha_1 + \alpha_2 = \alpha$. The comments in *[12.1.1.1]* also apply here.

[12.1.4.2] A $100(1-\alpha)\%$ CI for $(\mu_1 - \mu_2)$ ($\sigma_1 = \sigma_2$ but not known) based on $\overline{X}$, $\overline{Y}$, S_1^2, S_2^2 has limits given by

$$L = (\overline{X} - \overline{Y}) - S_p\left(\sqrt{\frac{1}{n} + \frac{1}{m}}\right) t_{(n+m-2;\alpha_1)}$$

$$U = (\overline{X} - \overline{Y}) + S_p\left(\sqrt{\frac{1}{n} + \frac{1}{m}}\right) t_{(n+m-2;\alpha_2)}$$

where $S_p^2 = \left[(n-1)S_1^2 + (m-1)S_2^2\right]/(n+m-2)$, $0 < \alpha_i < 1$ (i = 1, 2), and $\alpha_1 + \alpha_2 = \alpha$. The comments in *[12.1.1.2]* should also apply here.

[12.1.4.3] *Paired samples.* Let $d_i = (X_i - Y_i)$, i = 1, 2, ..., n, where $d_1, d_2, ..., d_n$ are iid $N(\mu_d, \sigma_d^2)$ rvs.

A $100(1-\alpha)\%$ CI for μ_d based on $\overline{d}$ and S_d has limits given by

$$\text{LCL} = \overline{d} - \frac{S_d}{\sqrt{n}} \cdot t_{(n\text{-}1;\alpha_1)}, \quad \text{UCL} = \overline{d} + \frac{S_d}{\sqrt{n}} \cdot t_{(n\text{-}1;\alpha_2)},$$

where $\overline{d} = \sum_{i=1}^{n} d_i/n$, $S_d^2 = \sum_{i=1}^{n}(d_i - \overline{d})^2/(n-1)$, $0 < \alpha_i < 1$ ($i = 1, 2$), and $\alpha_1 + \alpha_2 = \alpha$.

[12.1.4.4] Suppose $X \sim N(\mu, \sigma_1^2)$ and $Y \sim N(\mu, \sigma_2^2)$.

A $100(1-\alpha)\%$ CI for the common mean μ based on $\overline{X}$, $\overline{Y}$, $S_X = \sum_{i=1}^{n}(X_i - \overline{X})^2$, $S_Y = \sum_{i=1}^{n}(Y_i - \overline{Y})^2$ has limits given by

$$\overline{X} \pm \frac{(\overline{Y} - \overline{X}) \, an S_X}{n S_X + bm S_Y} - \frac{\sqrt{S_X} \, t_{(m\text{-}1;\alpha/2)}}{\sqrt{m-1}}.$$

(Khatri, 1981, p. 101; Brown and Cohen, 1974, p. 974). Khatri gives a sufficient condition for obtaining constants a and b which makes this confidence interval to have the same width and the probability of coverage greater than the one given in *[12.1.1.2]*. He has tabulated constants a and b on page 105.

[12.1.4.5] A $100(1-\alpha)\%$ CI for (σ_1^2/σ_2^2) based on $\overline{X}$, $\overline{Y}$, S_1, S_2 has limits given by

$$L = S_1^2/(bS_2^2), \quad U = S_1^2/(aS_1^2),$$

where $b > a > 0$ [Levy and Narula (1974), p. 85]. For the equal tail case $a = f_{(n_1\text{-}1, n_2\text{-}1;\alpha/2)}$ and $b = f_{(n_1\text{-}1, n_2\text{-}1;1\text{-}\alpha/2)}$. For this CI to have the shortest (actual and expected) width, they have tabulated a, b values for both $(n_1 - 1)$ and $(n_2 - 1) = 4(1)10, 12, 16, 20, 24, 30, 40, 60$, and for $1 - \alpha = .95$ on page 86. For this CI to be the shortest width unbiased CI, Ramchandran (1958) has tabulated a and b values for both $(n_1 - 1)$ and $(n_2 - 1) = 2, 3, 4, 6, 8, 10, 12, 16, 20, 24, 30, 40, 60$, and for $1 - \alpha = .95$. Also, refer Molinska and Molinski (1988).

[12.1.4.6] A $100(1-\alpha)\%$ CI for σ_1^2/σ_2^2 based on $\overline{X}$, $\overline{Y}$, S_1^2, S_2^2 has limits

$$L = \frac{1}{m_1 c}(N - n_1 c)\left(\frac{S_1^2}{S_p^2}\right), \quad U = \frac{n_1 c}{N - m_1 c}\left(\frac{S_1^2}{S_p^2}\right),$$

where $n_1 = n - 1$, $m_1 = m - 1$, $N = n_1 + n_2$, $c = c(n_1, n_2; \alpha)$. John (1975),

p. 344, calls this interval the uniformly most accurate unbiased interval. He has tabulated on pages 345-346 c-values for $n_1, n_2 = 1(1)20(2)30, 36, 45, 60, 90, 180, \infty$, and for $\alpha = 0.001, 0.01, 0.05, 0.10$.

[12.1.4.7] Suppose $\nu_1 S_1^2/\sigma_1^2 \sim \chi_{\nu_1}^2$, $\nu_2 S_2^2/\sigma_2^2 \sim \chi_{\nu_2}^2$ and that both rvs are independent. An approximate $100(1-\alpha)\%$ one-sided upper CI for $(c\sigma_1^2 + \sigma_2^2)$, $c > 0$, has limit L given by

$$L = (CS_1^2 + S_2^2)/F_{(k,\infty;1-\alpha)}, \quad k = (CS_1^2 + S_2^2)/\left[\frac{C^2 S_1^4}{\nu_1} + \frac{S_2^4}{\nu_2}\right],$$

$\min(\nu_1, \nu_2) \leq k \leq (\nu_1 + \nu_2)$ (Wang, 1988, p. 285). He gives a better approximate interval when ν_1 and ν_2 are large. Also he gives an iterative algorithm which can be used to obtain an exact one-sided interval.

[12.1.5] Simultaneous Confidence Intervals

Suppose samples are independently drawn from k normal populations. Let X_{ij} ($i = 1, 2, ..., k$; $j = 1, 2, ..., n_i$) be a rv from a $N(\mu_i, \sigma^2)$ population. Consider a contrast $\sum_{i=1}^{k} c_i \mu_i$ with known c_i's subject to $\sum_{i=1}^{k} c_i = 0$.

Miller (1981) provides a comprehensive description and comparison of many multiple comparison procedures available in the literature. In the following, we give a few commonly used such procedures. Let

$$\bar{X}_{i.} = \frac{1}{n_i} \sum_{j=1}^{n} X_{ij}, \quad S_*^2 = \frac{1}{\nu} \sum_{k=1}^{k} \sum_{j=1}^{n_i} \left(X_{ij} - \bar{X}_{i.}\right)^2, \quad \text{and} \quad \nu = \sum_{i=1}^{k} (\nu_i - 1).$$

[12.1.5.1] *Tukey's Method.* Suppose $n_i = n$ ($i = 1, 2, ..., k$). A $100(1-\alpha)\%$ simultaneous CI for all contrasts $\sum_{i=1}^{k} c_i \mu_i$ based on the $\bar{X}_{i.}$ and S_*^2 has limits given by

$$\sum_{k=1}^{k} c_i \bar{X}_{i.} \pm q_{(k,k(n-1);\alpha)} \cdot \frac{S_*}{\sqrt{n}} \sum_{i=1}^{k} \frac{|c_i|}{2},$$

where $q_{(k,k(n-1);\alpha)}$ is the $100\alpha\%$ upper percentile of the studentized range distribution for k means and df = $k(n-1)$. (Scheffé, 1959, p. 74; Miller, 1981, p.39). Miller provides on pages 234-237 values of the percentiles in selected cases.

[12.1.5.2] *Scheffé's Method.* A $100(1-\alpha)\%$ simultaneous CI for all contrasts $\sum_{i=1}^{k} c_i \mu_i$ based on the $\overline{X}_{i.}$ and S_*^2 has limits given by

$$\sum_{i=1}^{k} c_i \overline{X}_{i.} \pm \left[(k-1)F_{(k-1,k(n-1);\alpha)}\right]^{\frac{1}{2}} S_* \left(\sum_{i=1}^{k} \frac{c_i^2}{n_i}\right)^{\frac{1}{2}}$$

(Scheffé, 1959, p. 67; Miller, 1981, p. 55).

[12.1.5.3] *Bonferroni Method.* An at least $100(1-\alpha)\%$ simultaneous CI for all differences $(\mu_i - \mu_r)$ ($i \neq r$) based on the $\overline{X}_{i.}$ and S_*^2 has limits given by

$$(\overline{X}_{i.} - \overline{X}_{r.}) \pm t_{(v; \frac{\alpha}{2c})} S_* \left(\frac{1}{n_i} + \frac{1}{n_r}\right)^{\frac{1}{2}}$$

(Miller, 1985, p. 681; Miller, 1981, pp. 67-68). Miller (1981) provides on page 238 values of the t-percentiles in selected cases.

[12.1.5.4] *Studentized Maximum Modulus Method.* Suppose X_i ($i = 1, 2, ..., k$) are independent $N(\mu_i, d_i\sigma^2)$ rvs and that the d_i's are known. Let S^2 be a chi-square rv with df $= \nu$ and independent of rvs X_i. Let $\sum_{i=1}^{k} \ell_i \mu_i$ be a linear combination of the means.

A $100(1-\alpha)\%$ simultaneous CI for all linear combinations $\sum_{i=1}^{k} \ell_i \mu_i$ based on X_i and S^2 has limits given by

$$\sum_{i=1}^{k} \ell_i X_i \pm |m|_{(k,v;\alpha)} \sum_{i=1}^{k} |\ell_i \sqrt{d_i}| \quad ,$$

where $|m|_{(k,v;\alpha)}$ is the $100\alpha^{th}$ upper percentile of the studentized maximum modulus distribution with parameters k and v (Miller, 1981, pp. 71,75). Miller provides on page 239 values of the percentiles in selected cases.

12.2 TOLERANCE INTERVALS

Among various statistical problems arising in the process of controlling quality in mass production, a rather important one appears to be the determination of tolerance limits when the variability of the product is known to be due to random factors. The tolerance limits or specifications could be set up so that any appreciable departure will make the product unusable. The problem then is to produce the product so that an

Statistical Intervals

acceptably high proportion of units will fall within these limits. If the product is made without prior specifications, or if alterations are made, it is desirable to know within what limits the process can hold a quality characteristic reasonably high percentage of the time. The problem of determining tolerance limits was first treated in a pioneer article by Wilks (1941). A review paper on the subject is by Patel (1986). A monograph on the subject is by Guttman (1970).

Let $X_1, X_2, ..., X_n$ be a random sample from a $N(\mu, \sigma^2)$ distribution. Consider two statistics $L = (X_1, X_2, ..., X_n) < U = U(X_1, X_2, ..., X_n)$. Then $P_X(L \leq X \leq U)$ is known as the *coverage* of the random interval (L, U).

Tolerance intervals are mainly of two types: (i) β-content tolerance intervals and (ii) β-expectation tolerance intervals.

Let β and γ be two constants such that $0 < \beta < 1$ and $0 < \gamma < 1$. Now if L and U are determined so that

$$P\big[P_X(L \leq X \leq U) \geq \beta\big] = \gamma ,$$

then (L, U) is a β-*content two-sided tolerance interval* (which controls the center of the distribution) at confidence level γ (Wald and Wolfowitz, 1946, pp. 208-215). If L and U are determined so that

$$P\big[P_X(X \leq L) \leq \beta_1 \text{ and } P_X(X \geq U) \leq \beta_2\big] = \gamma ,$$

then (L, U) is a *two-sided tolerance interval* (which controls both tails of the distribution) at confidence level γ (Owen, 1964, pp. 377-387). It is a common practice to take $\beta_1 = \beta_2 = \beta/2$. In that case we call (L, U) a β-*content tolerance interval* (control both tails).

If L and U are determined so that

$$E\big[P_X(L \leq X \leq U)\big] = \beta$$

for all (μ, σ^2), then (L, U) is a β-*expectation two-sided tolerance interval* (Fraser and Guttman, 1956, pp. 162-179). For one-sided intervals of the form (L, ∞) (or (−∞, U)), we identify L (or U) as a lower (or an upper) tolerance limit.

Construction of tolerance intervals for the normal distribution is indirectly related to construction of *variable sampling plans*. A variable sampling plan provides a procedure for accepting or rejecting a lot of material. A one-sided variable sampling plan, for example, guarantees that at least a proportion β of the population being measured is greater than a value L with a given probability γ, or guarantees that at least a proportion β of the population measured is less than a value U with probability γ. Thus a one-sided plan would accept the lot if $\bar{X} + kS \leq U$. The factor k is the same one given by the one-sided tolerance interval. Sampling plans are available for controlling the center of the distribution as well as for controlling the tails of the distribution.

Tolerance intervals are also related to prediction intervals (see *[12.3]*). A γ-level prediction interval for a single future independent observation X based on a past sample can be interpreted as a lower γ-expectation tolerance limit, via

$$P[L < X] = E[1 - F(L)]$$

(Proschan, 1953, p. 560; Paulson, 1943, p.91).

Tolerance intervals can also be constructed using uniformly most powerful tests. (See Guenther, 1971). Zacks (1971) has discussed how the concept of uniformly most accurate confidence intervals be extended to tolerance intervals.

In the following we use the abbreviation TI for tolerance interval and TL for tolerance limit.

[12.2.1] *β-Content Tolerance Intervals (Control Center)*

[12.2.1.1] A two-sided β-content TI (μ, σ known) has limits given by

$$\mu \pm \sigma z_{(1-\beta)/2}$$

with $\gamma = 1$. One-sided lower and upper β-content TIs are given by

$$L = \mu - \sigma z_{1-\beta}, \quad U = \mu + \sigma z_{1-\beta},$$

respectively, both with $\gamma = 1$.

[12.2.1.2] A two-sided β-content TI (μ known) based on μ and S has limits $\mu \pm kS$. The factor k can be determined from

$$k = z_{(1-\beta)/2} \cdot \left[(n-1)/\chi^2_{(n-1;\gamma)}\right]^{\frac{1}{2}}.$$

(Jilek and Likar, 1960b, p.205). They have tabulated k for $\beta = 0.90, .095, 0.975, 0.99, 0.995, 0.999$; n = 3(1)20(5)50(10)100, 120, 150, 200, 250, 300, 400, 500, 600, 800, 1000, 1200, 1500, 2000, 5000, 10000, ∞; and $\gamma = 0.90, 0.95, 0.99, 0.999$. (Their β is our γ and vice versa).

One-sided lower and upper β-content TLs based on μ and S are given by

$$L = \mu - kS, \quad U = \mu + kS,$$

respectively. In both cases the factor k can be determined from

$$k = z_{1-\beta}\left[(n-1)/\chi^2_{(n-1;\gamma)}\right]^{\frac{1}{2}}$$

(Jilek and Likar, 1960b, p.205). They have tabulated k for $\beta = 0.80, 0.90, 0.95, 0.98, 0.99, 0.998$; and the same n and γ as in two-sided intervals.

[12.2.1.3] A two-sided β-content TI (σ known) based on $\overline{X}$ and σ has limits

$$\overline{X} \pm k\sigma.$$

The factor k can be determined from

$$\Phi\left(\frac{c}{\sqrt{n}} + k\right) - \Phi\left(\frac{c}{\sqrt{n}} - k\right) = \beta,$$

where $c = Z_{(1-\gamma)/2}$ (Proschan, 1953, p. 560). Jilek and Likar (1960a) p. 79 have tabulated k for n = 1(1)20(5)50(10)100(50)300(100)600, 800, 1000, ∞; $\beta = 0.90, 0.95, 0.99$; and $\gamma = 0.95, 0.99, 0.999$.

One-sided lower and upper β-content TLs based on $\overline{X}$ and σ are given by
$$L = \overline{X} - k\sigma, \quad U = \overline{X} + k\sigma.$$

respectively, In both cases the factor k can be determined from

$$k = z_{1-\beta} + \frac{1}{\sqrt{n}} z_{(1-\gamma)/2}.$$

Jilek and Likar (1960a), p. 78, have tabulated k for the same n, β, and γ as in two-sided intervals.

[12.2.1.4] (a) A two-sided β-content TI based on $\overline{X}$ and S has limits

$$\overline{X} \pm kS.$$

Wald and Wolfowitz (1946), pp. 208-209; Weissberg and Beatty (1960), p. 484; Ellison (1964), p. 763; Howe (1969), p. 617 and others have discussed approximations of the factor k. Odeh and Owen (1980), have tabulated k in their table 3 for n = 2(1)100(2)180(5)300(10)400(25)650(50)1000, 1500, 2000, 3000, 5000, 10000, ∞ ; β = 0.75, 0.90, 0.95, 0.975, 0.99, 0.999, 0.999, and γ = 0.50, 0.75, 0.90, 0.95, 0.975, 0.990, 0.995.

One-sided lower and upper β-content TLs based on $\overline{X}$ and S are given by
$$L = \overline{X} - kS, \quad U = \overline{X} + kS,$$

respectiveiy. Odeh and Owen (1980) have tabulated k, in both cases, in their Table 1 for the same n, β, and γ as in two-sided intervals.

(b) A two-sided β-content TI based on $\overline{X}$ and sample range R has limits given by

$$\overline{X} \pm kR,$$

where $R = X_{(n;n)} - X_{(1;n)}$ is obtained from the same sample (Mitra, 1957, p. 89). He has tabulated the factor k for n = 2(1), 20; β = 0.75, 0.90, 0.95,

Statistical Intervals 407

0.99, 0.999; and $\gamma = 0.75, 0.90, 0.95, 0.99$.

(c) A two-sided β-content TI based on grand mean $\overline{\overline{X}}$ and mean range $\overline{R}$ has limits given by

$$L = \overline{\overline{X}} - k\overline{R}, \quad U = \overline{\overline{X}} + k\overline{R},$$

where $\overline{\overline{X}}$ and $\overline{R}$ are obtained from N samples each of size m (Mitra, 1957, p. 89). He has tabulated the factor k for $m = 4.5$; $N = 4(1)20(5)30(10)50(25)(25)$; $\gamma = 0.75, 0.90, 0.95, 0.99$; and $2\beta = 0.75, 0.90, 0.95, 0.99, 0.999$. He also gives a simple formula for approximating k.

(d) Eberhardt et al. (1989), pp. 397-413) provide a self-contained FORTRAN subroutine which computes the factor k.

(e) Canavos and Koutrouvelis (1984) have investigated the robustness of the two-sided β-content tolerance intervals given in (a). They found that these tolerance limits are, for the most part, very sensitive to departure from normality when the confidence coefficient exceeds 0.90.

[12.2.2] β-Content Tolerance Intervals (both tails controlled)

[12.2.2.1] A two-sided β-content TI (μ, σ known) has limits $\mu \pm \sigma z_{\beta/2}$ with $\gamma = 1$.

[12.2.2.2] A two-sided β-content TI (μ known) based on μ and S has limits $\mu \pm kS$. The factor k can be determined from

$$k = z_{\beta/2} \cdot \left[(n-1)/\chi^2_{(n-1;\gamma)}\right]^{\frac{1}{2}}$$

(Owen, 1964, p. 384).

[12.2.2.3] A two-sided β-content TI (σ known) based on $\overline{X}$ and σ has limits $\overline{X} \pm k\sigma$. The factor k can be determined from

$$k = z_{\beta/2} + \frac{1}{\sqrt{n}} z_{(1-\gamma)/2}$$

(Owen 1964, p. 384).

[12.2.2.4] (a) A two-sided β-content TI based on $\overline{X}$ and S has limits

$$L = \overline{X} - k_1 S, \quad U = \overline{X} + k_2 S$$

(Owen, 1964, p. 383). Odeh and Owen (1980) have tabulated k_1 and k_2 in their Table 4 for n = 2(1)100(2)180(5)300(10)400(25)650(50)1000, 1500, 2000, 3000, 5000, 10000, ∞; β_1, β_2 = 0.005, 0.01, 0.025, 0.05, 0.10, 0.125; and γ = 0.50, 0.75, 0.90, 0.95, 0.975, 0.99, 0.995.

If $\beta_1 = \beta_2$ and we set $k_1 = k_2$, values are exact in these tables. If $\beta_1 \neq \beta_2$, the tabulated values of k's are satisfactory for n ≥ 5. See Odeh and Owen (1980, pp. 281-283).

(b) A two-sided β-content TI based on $\overline{\overline{X}}$ and $\overline{R}$ has limits given by $\overline{\overline{X}} \pm k\overline{R}$ where $\overline{\overline{X}}$ and $\overline{R}$ are defined as in *[12.2.1.4(c)]* (Frawley et al. 1971, p. 652). They have tabulated the factor k for N = 1; γ = 0.90, 0.95, 0.99; m = 2(1)15; and $\beta/2$ = 0.1, 0.05, 0.025, 0.01. They also give approximate k for N = 2(1)15; γ = 0.90, 0.95, 0.99; m = 5, 10, and $\beta/2$ = 0.1, 0.05, 0.025, 0.01.

[12.2.3] β-Expectation Tolerance Intervals

[12.2.3.1] A two-sided β-expectation TI (μ, σ known) has limits $\mu \pm \sigma z_{(1-\beta)/2}$. A one-sided lower β-expectation TL is given by $L = \mu - \sigma z_{1-\beta}$. Similarly, a one-sided upper β-expectation TL is $U = \mu + \sigma z_{1-\beta}$.

[12.2.3.2] A two-sided β-expectation TI (μ known) based on μ and $T = \sqrt{\sum(X_i - \mu)^2/n}$ has limits

$$\mu \pm Tt_{(\frac{1-\beta}{2},n)}$$

(Fraser and Guttman, 1956, p. 173). One-sided lower and upper β-expectation TLs based on μ and T are given, respectively, by

$$L = \mu - Tt_{(1-\beta,n)}, \quad U = \mu + Tt_{(1-\beta,n)}.$$

[12.2.3.3] A two-sided β-expectation TI (σ, known) based on $\overline{X}$ and σ has limits

$$\overline{X} \pm \sigma\left(1 + \frac{1}{n}\right)^{\frac{1}{2}} z_{\frac{1-\beta}{2}}$$

(Fraser and Guttman, 1956, p. 173). One-sided lower and upper β-expectation TLs based on $\overline{X}$ and σ are given, respectively, by

$$L = \overline{X} - \sigma(1 + \tfrac{1}{n})^{\tfrac{1}{2}} z_{1-\beta}, \quad U = \overline{X} + \sigma(1 + \tfrac{1}{n})^{\tfrac{1}{2}} z_{1-\beta}.$$

[12.2.3.4] A two-sided β-expectation TI based on $\overline{X}$ and S has limits

$$\overline{X} \pm S(1 + \tfrac{1}{n})^{\tfrac{1}{2}} t_{(n-1;\tfrac{1-\beta}{2})}$$

(Fraser and Guttman, 1956, p. 173; Odeh et al., 1989, p. 461). One-sided lower and upper β-expectation TLs based on $\overline{X}$ and S are given, respectively, by

$$L = \overline{X} - S(1 + \tfrac{1}{n})^{\tfrac{1}{2}} t_{(n-1;1-\beta)}, \quad U = \overline{X} + S(1 + \tfrac{1}{n})^{\tfrac{1}{2}} t_{(n-1;1-\beta)}.$$

Odeh et al. (1989), pp. 461-468, report that these tolerance intervals are sensitive to departure from normality, just as Canavos and Koutrouvelis (1984), pp. 144-149, reported on β-content tolerance intervals.

[12.2.4] Simultaneous Tolerance Intervals

Consider M $N(\mu_i, \sigma^2)$ populations (i = 1, 2, ..., M). Let a random sample of size n_i be drawn from the ith population so that $n = \sum_{i=1}^{m} n_i$.

Consider the sample means $\overline{X}_i$ (i = 1, 2, ..., M) and pooled sample variance $S_p^2 = \sum_{i=1}^{M} n_i S_i^2 / (n - M)$ from these M populations.

[12.2.4.1] A two-sided β-content simultaneous TI (control center) based on $\overline{X}_i$ and S_p has limits $\overline{X}_i \pm k_i S_p$ (i = 1, 2, ..., M). Mee (1990), pp. 87-90, has tabulated the factor $k_i = k$ for $n_i = n = 2(1)10(2)15(5)30(10)$ 100, 300, 500, 1000, ∞; γ, β = 0.90, 0.95, 0.99; and M = 2, 3, 4, 5, 6, 8, 10. Mee also discusses the case of unequal n_i's.

One-sided lower β-content simultaneous TLs based on $\overline{X}_i$ and S_p are given by $L_i = \overline{X}_i - k_i S_p$ (i = 1, 2, ..., M). Similarly, one-sided upper β-content simultaneous TLs are $U_i = \overline{X}_i + k_i S$ (i = 1, 2, ..., M). Mee has tabulated $k_i = k$ on page 85-87 for the same $n_i = n$, γ, β, as in two-sided intervals.

Marcus and Genizi (1991), pp. 1861-1870, have considered the problem of constructing simultaneous tolerance intervals for sets of contrasts of normal variables.

[12.2.5] Miscellaneous Results

[12.2.5.1] Hall and Sampson (1973a), pp. 317-324, have given a procedure for constructing one-sided tolerance limits based on censored samples. A lower β-content limit based on estimators $\hat{\mu}$ and σ using m smallest observations is $L = \hat{\mu} - k_1\sigma$. Similarly $U = \hat{\mu} + k_2\hat{\sigma}$. They have tabulated factors k_1 and k_2 for n = 5, 7, 10, 15, 20; β = 0.90, 0.95; γ = 0.90; and m = 2(1)n.

[12.2.5.2] Consider a censored sample of size n = 2r where the smallest r and the largest r observations are censored. Let $\hat{\mu}$ and $\hat{\sigma}$ be the modified maximum likelihood estimators (Tiku, 1967) of μ and σ, respectively. Balakrishnan and Kocherlakota (1985) p. 177, have obtained a two-sided β-content tolerance interval (control both tails) based on $\hat{\mu}$ and $\hat{\sigma}$ with limits $\hat{\mu} \pm k^*\hat{\sigma}$. They have tabulated on page 180, k^* for γ = 0.90; β = 0.02, 0.05, 0.10, 0.20; and n = 5(1)10(5)50(10)100. Fung and Balakrishnan (1989), p. 1136, give correct Monte Carlo simulated values of γ. They report that this interval is more sensitive than the classical procedure of *[12.2.1.4]* to departure from normality.

[12.2.5.3] Consider a one-sided β-content tolerance limit given by L when both μ and σ are not known. Suppose it is desired to specify $L = L(S; \beta, \gamma) = B$ and solve it for γ. Then γ is a rv. Folks and Browne (1975), p. 288, have obtained the probability distribution of the rv γ.

[12.2.5.4] Consider the coverage

$$c = \Phi\left(\frac{\overline{X} - \mu}{\sigma} + k_1\frac{S}{\sigma}\right) - \Phi\left(\frac{\overline{X} - \mu}{\sigma} - k_1\frac{S}{\sigma}\right)$$

of the tolerance interval $(\overline{X} - k_1 S, \overline{X} + k_2 S)$. Papp (1992), pp. 1309-1318, has obtained the probability distribution, $P(c \geq q)$, when $q > .5$. This should be useful for finding two-sided tolerance intervals and calculating lower confidence bounds for the fraction of products conforming to two-sided specifications.

[12.2.5.5] Mee (1989) pp. 99-105, provides methods for computing tolerance limits from a stratified random sample. He assumes that the characteristic of interest is normally distributed within each stratum and that the within-stratum variances are equal. He has obtained one-sided approximate β-content and β-expectation tolerance limits, and he has examined the adequacy of the proposed limits using simulation.

[12.2.5.6] *Tolerance limits for a population of sample variances when μ and σ are unknown.* Consider a process which generates a population of sample variances. Let each S_i^2 be based on a sample of size n and based on a set of N such variances, define a pooled estimator $S_p^2 = \sum_{i=1}^{N} S_i^2 / N$.

A one-sided upper β-content TL based on S_p^2 is given by

$$U = \frac{N \chi^2_{(n-1;\beta)}}{\chi^2_{(N(n-1),\gamma)}} \cdot S_p^2$$

(Tietjen and Johnson, 1979, p. 109). If the individual observations corresponding to each S_i^2 are available, then they give on page 110 a better tolerance interval.

[12.2.5.7] *Tolerance limits for the distribution of the product and quotient of normal rvs.* Hall and Sampson (1973b), pp. 109-119, give a procedure to obtain tolerance limits under the assumption that the distributions have positive means with small coefficients of variations. Such intervals would be useful in the pharmaceutical industry, where, for example, the dose of medication in a tablet is the product of the weight of the tablet and the proportion of medication in the tablet.

[12.2.6] Sample Size Determination

Suppose μ and σ are known. Let $c(\overline{X}, S, k) = F(\overline{X} + kS) - F(\overline{X} - kS)$ denote the coverage of the tolerance interval $(\overline{X} - kS, \overline{X} + kS)$.

[12.2.6.1] A two-sided β-content tolerance interval (control center) satisfies

$$P\left[c(\overline{X}, S, k) \geq \beta\right] = \gamma .$$

Faulkenberry and Weeks (1968), pp. 343-348, considered finding the

smallest n which for given β, γ, γ^1, and d also satisfy
$$P\big[c(\overline{X}, S, k) \geq \beta + d\big] \leq 1 - \gamma^1.$$

This additional requirement would make the probability of the interval covering too large a proportion of the sampled population to be small. Odeh et al. (1987) pp. 980-982, have tabulated minimum n for $\beta = 0.75$, 0.90, 0.95; γ, γ^1, 0.95, 0.975, 0.99; and various values of d which depend on β. They have also tabulated on pages 972-979 values of d for n = 25(25)500(50)700(100)1000, 1500, 10000; $\beta = 0.75, 0.90, 0.95, 0.975$; γ, $\gamma^1 = 0.90, 0.95, 0.975, 0.99$.

Also refer Faulkenberry and Daly (1970), pp. 812-821, and Guenther (1972), for finding n using some approximations.

[12.2.6.2] Odeh et al. (1989), pp. 461-468, have also modified Faulkenberry and Weeks' criterion for determining n for a two-sided β-expectation tolerance interval. This is to find the smallest n for given β, γ, δ which satisfies the additional requirement

$$P\big[c(\overline{X}, S, k) \leq \beta + \delta\big] \geq \gamma.$$

Odeh et al. (1989), p. 467, have tabulated n for $\gamma = 0.90, 0.95, 0.975, 0.99$; $\beta = 0.75, 0.90$, and various values of δ which depend on β. They have also tabulated δ on pp. 465-466 for n =25(25)500(50)700(100)1000,1500, 200(1000)10000; $\beta = 0.75, 0.90, 0.95, 0.975$; and $\gamma = 0.90, 0.95, 0.975, 0.99$.

[12.2.6.3] A two-sided β-expectation tolerance interval satisfies

$$E\big[c(\overline{X}, S, k)\big] = \beta.$$

Wilks (1941), pp. 91-96, considered finding the smallest n which for given γ, $\delta_1 > 0$, $\delta_2 > 0$ satisfies

$$P\big[\beta - \delta_1 \leq c(\overline{X}, S, k) \leq \beta + \delta_2\big] \geq \gamma.$$

This additional requirement controls the variation of the coverage around its expected value so that the coverage has a specified degree of stability.

Odeh et al. (1989), p. 464, have tabulated minimum n for $\gamma = 0.90, 0.95,$ 0.975, 0.99; $\beta = 0.75, 0.90$ and various values of $\delta_1 = \delta_2 = \delta$ that depend on β. They have also tabulated on pp. 462-463 $\delta_1 = \delta_2 = \delta$ for n = 25(25)500(50)700(100)1000, 1500, 10000; and $\beta = 0.75, 0.90, 0.975, 0.99$. Also, refer Guenther (1970) for finding n.

12.3 PREDICTION INTERVALS

Prediction intervals use results from a past sample to contain the results of a future sample from the same population with a specified probability. Prediction intervals are available in several forms and can be useful in solving statistical problems arising in quality control and business. One of the earliest work on prediction intervals is that by Baker (1935), pp. 197-201. Hickman (1963), pp. 1104-1112, has called such intervals *forecast intervals*. A review paper on the subject is by Patel (1989), pp. 2393-2466.

Let $X_1, X_2, ..., X_n$ be an observed (past) random sample from a $N(\mu, \sigma^2)$ distribution. Consider two statistics $L_X = L(X_1, X_2, ..., X_n) < U_X = U(X_1, X_2, ..., X_n)$. Let $Y_1, Y_2, ..., Y_m$ be a future random sample from the same distribution, drawn independently of the past sample. Consider a statistic $T_Y = T(Y_1, Y_2, ..., Y_m)$.

Let $0 < \gamma < 1$. If L_X and U_X are determined so that

$$P(L_X \leq T_Y \leq U_X) = \gamma,$$

then (L_X, U_X) is a *two-sided prediction interval* for T_Y. For one-sided intervals of the form (L_X, ∞), (or $(-\infty, U_X)$), we identify L_X (or U_X) as a one-sided lower (or a one-sided upper) prediction limit. Prediction intervals are also related to tolerance intervals (see *[12.2]*). We have not covered Bayesian prediction intervals. Unless otherwise stated, all listed intervals are two-sided. We use PI to stand for 'prediction interval', PL to stand for 'prediction limit', and $L = L_X$, $U = U_X$. Also $\overline{X}$ and S^2 are defined as in *[12.2]* using the observed sample.

Not much work has been done on sample size determination. Meeker and Hahn (1982), pp. 201-206, have considered sample size requirements for obtaining a prediction interval to contain the mean of a future sample. The paper deals with the effect of the initial sample size on the length of the interval.

[12.3.1] Prediction Intervals to Contain All Future Observations

[12.3.1.1] A $100\gamma\%$ PI for all future observations (μ, σ known) has limits

$$\mu \pm \sigma z_\alpha, \quad \alpha = \tfrac{1}{2}\{1 - (1-\gamma)^{1/m}\}$$

(Chew, 1968, pp. 323-324).

[12.3.1.2] A $100\gamma\%$ PI for all future observations (μ known) based on μ and $T = \sum_{i=1}^{n}(X_i - \mu)/n$ has limits $\mu \pm r_{(m,n;\gamma)} \cdot T$ (Chew, 1968, p. 328), where

$$r_{(1,n;\gamma)} = t_{(n-1;\,\frac{1-\gamma}{2})}$$

(Whitmore, 1986, p. 142). The factor $r_{(m,n;\gamma)}$ has not been tabulated for $m > 1$. Chew, p. 328 gives the following two conservative approximations to compute r:

$$r_{(m,n;\gamma)} \approx \sqrt{mf_{(m,n;1-\gamma)}} \quad , \quad r_{(m,n;\gamma)} \approx t_{(\frac{1-\gamma}{2m};n)}$$

[12.3.1.3] A $100\gamma\%$ PI for all future observations (σ known) based on $\overline{X}$ and σ has limits $\overline{X} \pm \sigma r_{(m,n;\gamma)}$ (Chew, 1968, p. 329). For $m = 1$,

$$r_{(m,n;\gamma)} = z_{(1-\gamma)/2} \cdot \sqrt{1 + \tfrac{1}{n}}$$

(Whitmore, 1986, p. 142). The factor $r_{(m,n;\gamma)}$ has not been tabulated for $m > 1$. Chew, p. 329, gives the following two conservative approximations to compute r:

$$r_{(m,n;\gamma)} = \sqrt{(1 + \tfrac{1}{n})\,\chi^2_{(m;1-\gamma)}} \quad , \quad r_{(m,n;\gamma)} = z_{(1-\gamma)/(2m)} \cdot \sqrt{1 + \tfrac{1}{n}}$$

[12.3.1.4] Let μ and σ both be not known.
For two-sided intervals, refer section [12.3.2] with $k = m$. Similarly for

one-sided intervals refer section *[12.3.3]*.

[12.3.2] Prediction Intervals to Contain at Least k of m Future Observations

[12.3.2.1] A $100\gamma\%$ PI to contain k or more of m future observations based on $\overline{X}$ and S has limits

$$\overline{X} \pm Sr_{(k,m,n;\gamma)}.$$

Odeh (1990), pp. 204-213, has tabulated values of the factor r for $\gamma = 0.90$, 0.95, 0.99; n = 8(1)12, 15, 20, 25, 30, 40, 60, 120, 240, 480, ∞; m = 1(1)9 for k = m; m = 10, 15, 80, 100, for k = m − j where j = 0(1)8. When k = m = 1,

$$r_{(k,m,n;\gamma)} = t_{(n-1;\frac{1-\gamma}{2})} \cdot \sqrt{1 + \frac{1}{n}}$$

(Whitmore, 1986, p. 142). For m = k, Hahn (1969, 1970) previously had tabulated this factor in selected cases and had given some approximations of these factors. Also refer Chew (1968), pp. 326-327, and Lieberman (1961), pp. 23-24, for these approximations. Angers and McLaughlin (1979), p. 385, gave an inequality to obtain less conservative factors than Chew's approximation.

Odeh reports that these prediction intervals are extremely sensitive to the assumption of normality.

[12.3.2.2] A $100\gamma\%$ one-sided lower PL for $(m - \ell)$ or more of m future observations based on $\overline{X}$ and S has limit

$$L = \overline{X} - Sr_{(\ell,m,n;\gamma)},$$

(Fertig and Mann, 1977, pp. 167, 176). They have tabulated on pages 168-175 values of r for ℓ = 1(1)8; n = 2(1)15(5)30(10)50, ∞; m = 20, 25, 30(10)80; and γ = 0.90, .095, .0.99. Previously Fertig and Mann (1975), pp. 533-556, had tabulated r for the case $\ell = 0$.

[12.3.3] Simultaneous Prediction Intervals

[12.3.3.1] A $100\gamma\%$ one-sided lower PL to contain all of k future sample means (using samples of size m each) based on $\overline{X}$ and S has limit

$$L = \overline{X} - Sr_{(k,m,n;\gamma)}.$$

Similarly, a one-sided upper PL is $\overline{X} + Sr_{(\ell,m,n;\gamma)}$ (Odeh, 1989, p. 1557). He has tabulated on pages 1559-1581 values of factor r for $\gamma = 0.90, .095, .0.99$; n = 8(2)12, 15(5)30, 40(20)100, 250, 500, 1000, ∞; k = 1(1)14(2)20, 25, 30(10)60, 80, 100; and m = 1, 5(5)20. For m = 1, previously Hahn (1970), p. 1670, Fertig and Mann (1975), p. 538, and Hall and Prairie (1973), pp. 898-905, had tabulated (or drawn graphs) to find the factor r in selected cases. Fertig and Mann (1975), p. 548, also give an approximation to the factor r.

[12.3.3.2] A $100\gamma\%$ PI to contain ℓ or more of k future sample means (using samples of size m each) based on $\overline{X}$ and S has limits

$$\overline{X} \pm Sr_{(\ell,k,m,n;\gamma)}$$

(Odeh, 1989, p. 429). He has tabulated on pages 431-453 values of the factor r for $\gamma = 0.90, .095, .0.99$; n = 8(1)12, 15, 20, 25, 30, 40, 60, 120, 240, 480, ∞; m = 1, 5(5)20; k = 1(1)9 and ℓ = k; k = 10, 15, , 80, 100 and $\ell = k - j$, for j = 0, 1. Hahn and Hendrickson (1971), pp. 323-332, have also considered this problem when ℓ = k and have tabulated appropriate factors in selected cases. They have obtained a conservative prediction interval when sample sizes for future samples are unequal.

[12.3.3.3] Simultaneous prediction intervals to contain the sample standard deviations or sample ranges of future samples.

Refer Hahn (1972), pp. 938-942, for these intervals. These are mostly one-sided exact or conservative ones.

[12.3.3.4] Simultaneous prediction intervals to contain at least $(m_i - k_i + 1)$, (i = 1, 2, ..., ℓ) out of m_i observations in ℓ future samples.

A $100\gamma\%$ one-sided lower simultaneous PI to contain at least $(m_i - k_i + 1)$ observations in ℓ future samples based on $\overline{X}$ and S has the limit L given by

$$L = \overline{X} - Sr_{(n,\ell,m_i,k_i;\gamma)}$$

Statistical Intervals

(Chou and Owen, 1986, p.248). They have tabulated on pages 249-250 the values of factor r for $\ell = 2$; $m_1 = m_2 = m = 20(10)80$; $k_1 = k_2 = k = 1, 2, 3$; $n = 2(1)15(5)30(10)50, \infty$; and $\gamma = 0.90, .095$. These intervals, which extend intervals given by Fertig and Mann (1977), pp. 167-177, are useful when a manufacturer wishes to assure the acceptance of future shipments of a product.

[12.3.4] Miscellaneous Prediction Intervals

[12.3.4.1] A $100\gamma\%$ PI based on S, to contain the future sample standard deviation has limits

$$L = S\sqrt{f_{(m-1,\ n-1;\ \frac{1+\gamma}{2})}}, \quad U = S\sqrt{f_{(m-1,\ n-1;\ \frac{1+\gamma}{2})}}$$

(Hahn, 1972, p. 939).

[12.3.4.2] A $100\gamma\%$ PI, based on $X_{(n)}$ and σ^*, to contain the largest future observation has limits

$$L = Y_n + \sigma^* \cdot u_{(r,n;\frac{1+\gamma}{2})}, \quad U = Y_n + \sigma^* \cdot u_{(r,n;\frac{1+\gamma}{2})},$$

where it is assumed that only the first r ($1 \le r < n$) ordered observations $X_{(1)}, X_{(2)}, ..., X_{(r)}$ are available and $\sigma^* = \sum_{i=1}^{n} b(n,r;i) X_{(i)}$ is the best linear unbiased estimator of σ; the coefficients $b(n,r;i)$ are tabulated by Sarhan and Greenberg (1962). Y_n is computed using

$$Y_n = X_{(n)} + \left[E(Z_{(n)}) - E(Z_{(r)}) \right] \sigma^*,$$

where $Z_{(i)}$ is the ith order statistic from a standard normal distribution (Nelson and Schmee, 1981, p. 462). They have tabulated on p. 463 the factor $u_{(r,n;\delta)}$ for $n = 3(1)10$; $r = 2(1)(n-1)$; $(1-\delta)$ and $\delta = 0.005, 0.01, 0.025, 0.05, 0.10(.10)0.50$.

[12.3.4.3] *Prediction intervals to contain the sample mean and sample variance using a subset of the past sample.*

Suppose that only the first $k (2 \le k < n)$ ordered observations $X_{(1)}$,

$X_{(2)}, \ldots, X_{(k)}$ are available. Let $S_1^2 = \sum_{i=1}^{k}(X_{(i)} - \overline{\overline{X}}_{(i)})^2/(k-1)$, where $\overline{\overline{X}}_1 = \sum_{i=1}^{k} X_{(i)}/k$. Similarly, define $\overline{X}$ and S^2 for the complete sample.

(a) A $100\gamma\%$ PI based on $\overline{\overline{X}}_1$ and S_1^2, and to contain $\overline{X}$ when μ and σ are both unknown, has limits

$$\overline{\overline{X}}_1 \pm \sqrt{\frac{S_1^2}{k-1}\left(\frac{n-k}{n}\right)} \cdot t_{(k-1;\frac{\gamma}{2})}$$

(Hickman, 1963, p. 1107). He also gives on page 1107 the following prediction interval.

(b) A $100\gamma\%$ PI based on S_1^2, and to contain S^2, has limits

$$L = \frac{kS_1^2}{n}\left[1 + \frac{n-k}{k-1} f_{(n-k,k-1;1-\frac{\gamma}{2})}\right], \quad U = \frac{kS_1^2}{n}\left[1 + \frac{n-k}{k-1} f_{(n-k,k-1;1-\frac{\gamma}{2})}\right].$$

[12.3.4.4] Suppose the past and future samples came from a lognormal distribution. Let

$$T = \left(\prod_{i=1}^{n} X_i\right)^{1/n}$$

be the geometric mean using the past sample. A $100\gamma\%$ 'shortest' PI to contain a single future observation based on T (σ known) has limits

$$L = T \exp(A_c), \quad U = T \exp(B_c),$$

where $c^2 = \sigma^2(1 + 1/n)$. Dahiya and Guttman (1982), p. 281, have tabulated factors A_c and B_c for $c = 0.02(.01)0.80(.05)1, 1.5, 1, 3, 4, 5$; and 0.95, 0.99. If σ is not known, they use a Bayesian argument to obtain a prediction interval.

[12.3.4.5] Prediction intervals using two samples. Let $\overline{X}_i$ ($i = 1, 2$) be the sample mean from a past sample of size n_i from a $N(\mu_i, \sigma_i^2)$ population and S_i^2 ($i = 1, 2$) be the corresponding sample variance with $(n_i - 1)$ as a denominator. Let S_p be the pooled sample standard deviation obtained using S_1^2 and S_2^2. Let $\overline{Y}_i$ ($i = 1, 2$) be defined similarly using future samples of size m_i from the same populations, $i = 1, 2$.

(a) A $100\gamma\%$ PI to contain $(\overline{Y}_1 - \overline{Y}_2)$ based on $(\overline{X}_1 - \overline{X}_2)$ and S_p when $\sigma_1 = \sigma_2$ but all parameters unknown has limits

$$(\overline{X}_1 - \overline{X}_2) \pm S_p \left[\frac{1}{n_1} + \frac{1}{n_2} + \frac{1}{m_1} + \frac{1}{m_2}\right]^{\frac{1}{2}} \cdot t_{(n_1 + n_1 - 2; \frac{1+\gamma}{2})}.$$

(Hahn, 1977, p. 132). This prediction interval could be useful in situations where a producer has sampled the performance of both his/her product and that of a competitor, and wishes to determine the difference in future mean performance. Hahn (1977), p. 133, gives an approximate interval when $\sigma_1 \neq \sigma_2$.

(b) Let S_1^2 and S_2^2 be sample variances with common df ν_1 obtained from the past sample. Similarly, let T_1^2 and T_2^2 be sample variances defined with common df ν_2 from future samples from the same population. Let $R_1 = S_1^2/S_2^2$ and $R_2 = T_1^2/T_2^2$. A $100\gamma\%$ PI for R_2 based on R_1 with μ_1 and σ_2 known or not has limits

$$L = R_1/g_{(\nu_1,\nu_2;\frac{1+\gamma}{2})}, \quad U = R_1/g_{(\nu_1,\nu_2;\frac{1+\gamma}{2})}$$

(Meeker and Hahn, 1980, p. 358). They have tabulated on pages 359-363 the values of the factor g for all combinations of $\nu_2 \leq \nu_1 = 1(1)12, 15, 20, 24, 30, 60, 120, \infty$; and $\gamma = 0.75, 0.90, 0.95, 0.975, 0.99, 0.995$. They also discuss the case of unequal sample sizes.

REFERENCES

Angers, C. and McLaughlin, G. (1979). A note on prediction intervals for large future samples, *Technometrics 21*, 383-385. *[12.3.2.1]*

Baker, G. A. (1935). The probability that the mean of a second sample differ from the mean of a first sample by less than a certain multiple of the standard deviation of the first sample, *Annals of Mathematical Statistics 6*, 197-201. *[12.2]*

Balakrishnan, N. and Kocherlakota, S. (1985). Robust two sided tolerance limits based on MML estimators, *Communication in Statistics - A14(1)*, 175-184. *[12.2.5.2]*

Bartoszynski, R. and Chan, W. (1990). A remark on the shortest confidence interval of a normal mean, *The American Mathematical Monthly 97*, 415-417. *[12.1.1.2]*

Brown, L. D. and Cohen, A. (1974). Point and confidence estimation of a common mean and recovery of interblock information, *The Annals of Statistics 2*, 963-976. *[12.1.4.4]*

Canavos, G. S. and Koutrouvelis, I. A. (1984). The robustness of two-sided tolerance limits for normal distributions, *Journal of Quality Technology 16*, 144-149. *[12.2.1.4; 12.2.3.4]*

Chew, V. (1968). Simultaneous confidence regions for predictions, *Technometrics 10*, 323-330. *[12.3.1.1, 2, 3; 12.3.2.1]*

Chou, Y. M. and Owen, D. B. (1984). One-sided confidence regions on the upper and lower tail areas of the normal distribution, *Journal of Quality Technology 16*, 150-158. *[12.1.3.7]]*

Chou, Y. M. and Owen, D. B. (1984). One-sided simultaneous lower prediction intervals for ℓ future samples from a normal distribution, *Technometrics 28*, 247-251. *[12.3.3.4]*

Church, J. D. and Harris, B. (1970). The estimation of reliability from stress-strength relationships, *Technometrics 12*, 49-54. *[12.1.3.8]*

Cohen, A. (1972). Improved confidence intervals for the variance of a normal distribution, *Journal of the American Statistical Association 67, 382-287*. *[12.1.2.2]*

Dahiya, R. C. and Guttman, I. (1982). Shortest confidence and prediction intervals for the log-normal, *The Canadian Journal of Statistics 10*, 277-291. *[12.1.1.5; 12.3.4.4]*

Eberhardt, K. R., Mee, R. W. and Reeve, C. P. (1989). Computing factors for exact two-sided tolerance limits for a normal distribution, *Communications in Statistics B18*, 397-404. *[12.2.1.4]*

Ellison, B. E. (1964). On two-sided tolerance intervals for a normal distribution, *Annals of Mathematical Statistics 35*, 762-772. *[12.2.1.4]*

Faulkenberry, G. D. and Daly, J. C. (1970). Sample size for tolerance limits on a normal distribution, *Technometrics 12*, 812-821. *[12.2.6.1]*

Faulkenberry, G. D. and Weeks, D. L. (1968). Sample size determination for tolerance limits, *Technometrics 10*, 343-348. *[12.2.6.1]*

Fertig, K. W. and Mann, N. R. (1975). A new approach to the determination of exact and approximate one-sided prediction intervals for normal and lognormal distributions, with tables, *Reliability and*

Fault Tree Analysis (R. E. Barlow, J. B. Fussell and N. D. Sinpurwalla, eds.), 533-556, *SIAM*. *[12.3.2.2; 12.2.3.1]*

Fertig, K. W. and Mann, N. R. (1977). One-sided prediction intervals for at least p out of m future observations from a normal population, *Technometrics 19*, 167-177. *[12.3.2.2; 12.3.3.4]*

Folks, J. L. and Browne, R. (1975). On the interpretation of the observed confidence in certain reliability assessments, *Technometrics 17*, 287-290. *[12.2.5.3]*

Fraser, D. A. S. and Guttman, I. (1956). Tolerance regions, *Annals of Mathematical Statistics 27*, 162-179. *[12.2; 12.2.3.2, 3, 4]*

Frawley, W. H., Kapadia, C. H., Rao, J. N. K. and Owen, D. B. (1971). Tolerance limits based on range and mean range, *Technometrics 12*, 651-656. *[12.2.2.4]*

Fung, K. Y. and Balakrishnan, N. (1989). On the robust two-sided tolerance limits based on MML estimators, *Communications in Statistics - A18(3)*, 1125-1128. *[12.2.5.2]*

Gertsbakh, I. and Winterbottom, A. (1991). Point and interval estimation of normal tail probabilities, *Communications in Statistics - A20(4)*, 1497-1514. *[12.1.3.6]*

Govindarajulu, Z. (1967). Two-sided confidence limits for $P(X < Y)$ based on normal samples of X and Y, *Sankhya B, 29*, 35-40. *[12.1.3.8]*

Guenther, W. C. (1971). Unbiased confidence intervals, *The American Statistician 25, #1*, 51-53. *[12.1.1.1, 2]*

Guenther, W. C. (1971). On the use of best tests to obtain best b-content tolerance intervals, *Statistica Neerlandica 25*, 191-202. *[12.2]*

Guenther, W. C. (1972). Tolerance intervals for univariate distributions, *Naval Research Logistics Quarterly 19*, 309-333. *[12.2.6.1, 3]*

Guttman, I. (1970). *Statistical Tolerance Regions-Classical and Bayesian.* London: Griffin. *[12.1]*

Hahn, G. J. (1969). Factors for calculating two-sided prediction intervals for samples from a normal distribution, *Journal of the American Statistical Association 64*, 878-888. *[12.3.2.1]*

Hahn, G. J. (1972). Simultaneous prediction intervals to contain the standard deviations or ranges of future samples from a normal distribution, *Journal of the American Statistical Association 67*, 938-942. *[12.3.3.3; 12.3.4.1]*

Hahn, G. J. (1977). A prediction interval on the difference between two future sample means and its application to a claim of product superiority, *Technometrics 19*, 121-134. *[12.3.4.5]*

Hahn, G. J. and Hendrickson, R. W. (1971). A table of percentage points of the distribution of the largest absolute value of k student t variates and its applications, *Biometrika 58*, 323-332. *[12.3.3.2]*

Hall, I. J. and Prairie, R. R. (1973). One-sided prediction intervals to contain at least m out of k future observations, *Technometrics 15*, 897-914. *[12.3.3.1]*

Hall, I. J. and Sampson, C. B. (1973a). One-sided tolerance limits for a normal population based on censored samples, *Journal of Statistical Computation and Simulation 2*, 317-324. *[12.2.5.1]*

Hall, I. J. and Sampson, C. B. (1973b). Tolerance limits for the distribution of the product and quotient of normal variates, *Biometrika 29*, 109-119. *[12.2.5.7]*

Harter, H. L. (1969). *Order Statistics and Their Use in Testing and Estimation*, Vol. 2, Aerospace Research Lab., USAF. [12.1.2.3]

Herrey, E. M. J. (1965). Confidence intervals based on the mean absolute deviation of a normal sample, *Journal of the American Statistical Association 60*, 257-259. *[12.1.1.3]*

Herrey, E. M. J. (1971). H-distribution for commuting confidence limits or performing t-tests by way of the mean absolute deviation, *Journal of the American Statistical Association 66*, 187-188. *[12.1.1.3]*

Hickman, J. G. (1963). Preliminary regional forecasts for the outcome of an estimation problem, *Journal of the American Statistical Association 58*, 1104-1112. *[12.3.4.3]*

Howe, W. G. (1969). Two-sided tolerance limits for normal populations - some improvements, *Journal of the American Statistical Association 64*, 610-620. *[12.2.1.4]*

Jilek, M. and Likar, O. (1960a). Tolerance limits of the normal distribution with known variance and unknown mean, *Australian Journal of Statistics 2*, 78-83. *[12.2.1.3]*

Jilek, M. and Likar, O. (1960b). Tolerance regions of the normal distribution with known μ, unknown σ^2, *Biometrische Zeitschrift 2*, 204-209. *[12.2.1.2]*

John, S. (1975). Tables for comparing two normal variances or two gamma means, *Journal of the American Statistical Association 70*, 344-347. *[12.1.4.6]*

Khatri, C. G. and Shah, K. R. (1981). Interval estimation of the common mean, *Communications in Statistics - B10(2)*, 99-107. *[12.1.4.4]*

Kirkpatrick, R. L. (1970). Confidence limits on a percent defective characterized by two specification limits, *Journal of Quality Technology 2*, 150-155. *[12.1.3.6]*

Koopmans, L. H., Owen, D. B. and Rosenblatt, J. I. (1964). Confidence interval for the coefficient of variation for normal and lognormal distributions, *Biometrika 51*, 25-32. *[12.1.2.4, 3, 4]*

Krutchkoff, R. G. (1966). The correct use of the sample mean absolute deviation in confidence intervals for a normal variate, *Technometrics 8*, 663-674. *[12.1.1.3]*

Land, C. E. (1973). Standard confidence limits for linear functions of the normal mean and variance, *Journal of the American Statistical Association 68*, 960-963. *[12.1.3.3]*

Land, C. E. (1988). *Hypothesis tests and interval estimates, Lognormal Distributions - Theory and Applications*, E. L. Crow and K. Shimizu, eds., 37-86, New York: Marcel Dekker. *[12.1.3.2, 4, 5]*

Lehmann, E. L. (1986). *Testing Statistical Hypotheses*, (2nd ed.), New York: Wiley. *[12.1.1.1]*

Levy, K. J. and Narula, S. C. (1974). Shortest confidence intervals for the ratio of two normal variances, *The Canadian Journal of Statistics 2*, 83-87. *[12.1.4.5]*

Lieberman, G. J. (1961). Prediction regions for several predictions from a single regression line, *Technometrics 3*, 21-27. *[12.3.2.1]*

Lindley, D. V., East, D. A. and Hamilton, P. A. (1960). Tables of making inferences about the variance of a normal distribution, *Biometrika 47*, 433-437. *[12.1.2.1]*

Marcus, R. and Genizi, A. (1991). Simultaneous one-sided tolerance intervals for sets of contrasts from normal populations, *Communications in Statistics - A20(5-6)*, 1861-1870. *[12.2.4.1]*

Mazumdar, M. (1970). Some estimates of reliability using interference theory, *Naval Research Logistics Quarterly 17*, 159-165. *[12.1.3.8]*

Mee, R. W. (1988). Estimation of the percentage of a normal distribution lying outside a specified interval, *Communications in Statistics - A17(5)*, 1465-1479. *[12.1.3.6]*

Mee, R. W. (1989). Normal distribution tolerance limits for stratified random samples, *Technometrics 31*, 99-105. *[12.2.5.5]*

Mee, R. W. (1990). Simultaneous tolerance intervals for normal populations with common variance, *Technometrics* 32, 83-92. *[12.2.4.1]*

Meeker, W. Q. and Hahn, G. J. (1980). Prediction intervals for the ratio of normal distribution sample variances and exponential distribution sample means, *Technometrics* 22, 357-366. *[12.3.4.5]*

Meeker, W. Q. and Hahn, G. J. (1982). Sample sizes for prediction intervals, *Journal of Quality Technology* 14, 201-206. *[12.2]*

Miller, R. G. (1981). *Simultaneous Statistical Inference* (2nd Ed.). New York: Springer. *[12.1.5; 12.1.5.1, 2, 3,4]*

Miller, R. G. (1985). Multiple comparisons, *Encyclopedia of Statistical Science*, N. L. Johnson, S. Kotz, and C. B. Read, eds., 678-689. New York: Wiley. *[12.1.5.3]*

Mitra, S. K. (1957). Tables for tolerance limits for a normal population based on sample mean and range or mean range, *Journal of the American Statistical Association* 5, 88-94. *[12.2.1.4]*

Molinska, A. and Molinski, K. (1988). Relationship between confidence intervals for the ratio of two variances of independent normal normal distributions, *Communications in Statistics - A17(3)*, 925-933. *[12.1.4.5]*

Mood, A. M., Graybill, R. A., and Boes, D. C. (1974). *Introduction to the Theory of Statistics*, 3rd ed., New York: McGraw-Hill. *[12.1.3.10]*

Nelson, W. and Schmee, J. (1981). Prediction limits for the last failure time of a (log) normal sample from early failures, *IEEE Transactions on Reliability* 30, 461-463. *[12.3.4.2]*

Odeh, R. E. (1989). Simultaneous one-sided prediction intervals to contain all of k future means from a normal distribution, *Communications in Statistics - B 18(4)*, 1557-1585. *[12.3.3.1, 2]*

Odeh, R. E. (1990). Two-sided prediction intervals to contain at least k out of n future observations from a normal distribution, *Technometrics* 32, 203-216. *[12.3.2.1]*

Odeh, R. E., and Owen, D. B. (1980). *Tables for Normal Tolerance Limits, Sampling Plans, and Screening*, New York: Marcel Dekker. *[12.1.3.6; 12.2.1.4; 12.2.2.4]*

Odeh, R. E., Chou, Y. M. and Owen, D. B. (1987). The precision for coverages and sample size requirements for normal tolerance intervals, *Communications in Statistics - B16*, 969-985. *[12.2.6.1]*

Odeh, R. E., Chou, Y. M. and Owen, D. B. (1989). Sample-size determination for two-sided b-expectation tolerance intervals for a normal distribution, *Technometrics 31*, 461-468. *[12.2.3.4; 12.2.6.2, 3]*

Owen, D. B. (1962). *Handbook of Statistical Tables*. Reading: Addison-Wesley. *[12.1.3.1]*

Owen, D. B. (1964). Control of percentage in both tails of the normal distribution, *Technometrics 6*, 377-387; errata 8, 570 (1966). *[12.2; 12.2.2.2, 3, 4]*

Owen, D. B. (1968). A survey of properties and applications of the noncentral t-distribution, *Technometrics 10*, 445-478. *[12.1.3.2]*

Owen, D. B., Craswell, K. J. and Hanson, D. L. (1964). Nonparametric upper confidence bounds for $P(Y < X)$ and confidence limits for $P(Y < X)$ when X and Y are normal, *Journal of the American Statistical Association 59*, 906-924. *[12.1.3.8]*

Pacharaces, J. (1961). Tables for unbiased tests on the variance of a normal population, *The Annals of Mathematical Statistics 32*, 84-87. *[12.1.2.1]*

Papp, Z. (1992). Two-sided tolerance limits and confidence bounds for normal populations, *Communications in Statistics - A21(5)*, 1309-1318. *[12.2.5.3]*

Patel, J. K. (1986). Tolerance limits - A review, *Communications in Statistics - A15(9)*, 2719-2762. *[12.2]*

Patel, J. K. (1989). Prediction intervals - A review, *Communications in Statistics - A18(7)*, 2393-2466. *[12.3]*

Paulson, E. (1943). A note on tolerance limits, *Annals of Mathematical Statistics 14*, 90-93. *[12.2]*

Pratt, J. W. (1961). Length of confidence intervals, *Journal of the American Statistical Association 56*, 549-567. *[12.1.1.1, 12.1.2.1]*

Pratt, J. W. (1963). Shorter confidence intervals for the mean of a normal distribution with known variance, *The Annals of Mathematical Statistics 34*, 574-586. *[12.1.1.1]*

Proschan, F. (1953). Confidence and tolerance intervals for the normal distribution, *Journal of the American Statistical Association 48*, 550-564. *[12.2; 12.2.1.3]*

Ramchandran, K. V. (1958). A test of variances, *Journal of the American Statistical Association 53*, 741-747. *[12.1.4.5]*

Resnikoff, G. J. (1955). *Interval estimation of proportion defective in sampling inspection by variables.* Tech. Report 21, Applied Math. & Statistics Lab., Stanford Univ. Stanford, CA. *[12.1.3.6]*

Rosenblatt, J. (1966). Confidence interval for standard deviation from a single observation, *Technometrics* 8, 367-368. *[12.1.3.9]*

Sarhan, A. E. and Greenberg, B. G. (1962). *Contributions to Order Statistics.* NY: Wiley. *[12.3.4.2]*

Scheffé, H. (1959). *The Analysis of Variance.* New York: Wiley. *[12.1.5.1, 2]*

Shorrock, G. (1990). Improved confidence intervals for a normal variance. *The Annals of Statistics* 18, 972-980. *[12.1.2.2]*

Tate, R. F. and Klett, G. W. (1959). Optimal confidence intervals for the variance of a normal distribution, *Journal of the American Statistical Association,* 54, 674-682. *[12.1.2.1, 2]*

Tietjen, G. L. and Johnson, M. E. (1979). Exact statistical tolerance limits for sample variances, *Technometrics* 21, 110-110. *[12.2.5.6]*

Tiku, M. L. (1967). Estimating the mean and standard deviation from a censored normal sample, *Biometrika* 54, 155-165. *[12.2.5.2]*

Wald, A. and Wolfowitz, J. (1946). Tolerance limits for a normal distribution, *Annals of Mathematical Statistics* 17, 208-215. *[12.2; 12.2.1.4]*

Wang, C. M. (1988). One-sided confidence intervals for the positive linear combination of two variances, *Communications in Statistics - B17(1),* 283-292. *[12.2.4.7]*

Weiler, H. (1958). Confidence limits for the mean of a normal distribution with known coefficient of variation, *Australian Journal of Applied Sciences* 9, 321-325. *[12.1.1.4]*

Weissberg, A. and Beatty, G. H. (1960). Tables of tolerance limit factors for normal distribution, *Technometrics* 4, 483-500. *[12.2.1.4]*

Whitmore, G. A. (1986). Prediction limits for a univariate normal observation, *The American Statistician* 40, 141-143. *[12.3.1.2, 3; 12.3.2.1]*

Wilks, S. S. (1941). Determination of sample sizes for setting tolerance limits, *Annals of Mathematical Statistics* 12, 91-96. *[12.2; 12.2.6.3]*

Zacks, S. (1971). *The Theory of Statistical Inference.* New York: Wiley. *[12.2]*

INDEX

Abbe, Ernst, 7
Absorption distributions,
　approximation, 198
Adrain, Robert, 3
Angular transformations, 183
Asymptotic normality, 153-159, 227

Bernoulli, Jacob, 1
Berry-Esseen Theorem, 51, 160-165
　for order statistics, 278
Bessel, Friedrich Wilhelm, 4
Beta distribution, approximations,
　Camp-Paulson, 200-201
　from binomial, 199-200
　Molenaar, 200-202
　Peizer-Pratt, 201
Bienaymi
Binomial distribution, 1-2, 180-228
　approximations, 180-187
　　Borges, 185-186
　　classical, 180-181
　　Ghosh, 182-183
　　Gram-Charlier, 181
　　inverse sine, 183-184
　　Molenaar, 181-182, 185-187
　　Peizer-Pratt, 186
　bounds, 187-188
Birnbaum-Saunders distributions, 115
　approximation, 225
Bivariate Cauchy distribution, 104
Bivariate normal distribution, 295-359,
　313-316
　algorithms, 307-308, 310-311
　approximations, 308-313
　circular coverage function, 324-329
　conditional distributions, 298-299,
　313-315
　　correlation, 296
　　cumulants, 299

Bivariate normal distribution (con't)
　density function, 295
　marginal distributions, 295-296
　moment-generating function, 299
　moments, 299-301
　products, 318
　quotients, 316-317
　regression, 298-299, 313-314
　standardized, 296-297, 319
　tables, 304-306
　truncated, 319-324
Bivariate normal probabilities, 114,
　301-304
Bivariate normal samples, 341-359
　concomitants, 344-346
　correlation coefficient, 348-359
　joint density, 341-342
　means, 341-342
　regression coefficient, 343
　variances, 341-343
Box-Cox family, 205
Box-Muller transformation, 231
Burr distributions, 71

Cauchy distribution, 113, 316
Central Limit Theorems,
　classical, 146-151
　for order statistics, 277-278
　general, 151-153
　rapidity of convergence, 159-164
Characteristic functions,
　convergence of, 159
Characterizations, 81-106
　and admissibility, 102-103
　and identifiability, 100
　and independence, 92-93
　and sufficiency, 100-101
　and transformations, 96-99
　by conditional distributions and
　　regression, 88-92

Characterizations, (continued)
  by characteristic functions, 94-96
  by linear statistics, 82-85
  by moments, 94-96
  via order statistics, 103-104
Chi-squared distribution, 7-8, 32, 95, 98 116-118, 203, 286, 332, 381
  approximations, 204-210
    Fisher, 204
    fourth-root, 205-206
    Peizer-Pratt, 206-207
    quantiles, 207-209
    Wilson-Hilferty, 204-205
Circular coverage function, 324-329
Circular normal distribution, 297, 317, 325
Cochran's Theorem, 120-121
Coefficient of variation, 124
Concomitants, 344-346
Confidence intervals, 391-402
  based on two samples, 399-401
  for a proportion, 397-398
  for m, 392-394
  for s^2, s, 394
  for functions, 396-399
  k groups, 401-402
Continued fractions, 48, 55
  Laplace's, 50, 55-57, 64
  Patry and Keller, 61-62
  Shenton's, 56-57, 65
Convergence to normality,
  rapidity of, 159-165
Cornish-Fisher expansions, 172-174, 209, 215, 218
Correlation coefficient, 10, 296
Correlation coefficient, sample,
  approximations, 353-359
  arc sine transformation, 359
  density function, 349-350, 352-353
  distribution function, 350
  Fisher's z, 353-358
  moments, 351-352
  percentage points, 350-351
  transformations, 353-359
Covariance, 299

Darmois-Skitovich Theorem, 84
De Moivre, Abraham, 1-3, 10
De Moivre-Laplace Limit Theorem, 146-147, 149, 180-188
Densities,
  limit theorems for, 153

Distance distributions,
  approximation, 225
Distribution function,
  expansions for, 167-172

Edgeworth expansions, 168-170, 282
Edgeworth, Francis, 7, 10
Elementary Errors, Hypothesis of, 5
Elliptical normal distribution, 297
  offset circles, ellipses, 329-332
Entropy, 99-100
Error function, 20, 50
Estimation,
  of normal probability, 377-379, 382-383
Estimators,
  of m, 361-371, 374-377
  of s^2, 371-374
  two samples, 384-386
Exponential distributions, 27
Exponential families, 26
  normal a member of, 26-28

F distribution, 126-127, 212-213, 218-222
  approximations, 218-222
    Fisher, 219
    Paulson, 219
    Peizer-Pratt, 220
    percentiles, 220-222
    Wilson-Hilferty, 219
Farlie-Gumbel-Morgenstern distribution 297
Feller's Theorem, 149-150, 164
Fisher distribution,
  approximation, 226
Fisher information, 99
Fisher's z-transformation, 353-358
Fisher, R. A., 10

Galileo, 1
Galton, Sir Francis, 1, 4, 6-7, 10-11
Gamma distribution, 203-204, 228, 281
Gauss, Carl Friedrich, 3-5, 7, 10
Generalized variance,
  approximation, 225-226
Geometric distribution, 191
Gini's mean difference, 280-281
Gosset, W. S., 8-9
Gram-Charlier expansions, 170-172, 18

Index 429

Helmert, 7
Hermite polynomials, 21, 49-50, 168-171, 174, 297, 355-356, 369, 383
Hypergeometric distribution, 195
    approximations, 195-198, 348-349
        classical, 195
        Ling-Pratt, 196-197
        Molenaar, 195-196
    asymptotic normality, 158
Hypergeometric function, 348-349
Incomplete beta function, 243
Incomplete gamma function, 20, 328
Independence, 114-116
Infinitely divisible distributions, 29
Inverse transformations, 228

Johnson system, 230-231

k-statistics, 132
kurtosis, 134
Laguerre polynomials, 138, 374, 382-384
Laplace distribution, 114
Laplace, Pierre Simon de, 2-5, 7, 10
Large Numbers, Weak Law of, 1
Law of Errors, 4
Least squares, 2-4, 6
Legendre, 3
Levy's theorem, 149, 154, 156, 164
Limit Theorem, Central, 2, 9-10
Lindeberg condition, 148-149
Lindeberg's Theorem, 148-150, 154, 156, 164
Linear and quadratic characterizations, 85-88
Location statistics, 280
Location-free statistics, 280
Lognormal distribution, 32, 229, 380, 395, 418
Lyapunov's Theorem, 148-149
Lyapunov, Alexander M., 9

m-dependent variables, 151-152
Markov, Andrei A., 9
Martingales, 151
Maxwell, James Clerk, 5-6, 10, 82
Mean deviation, 129-130
Mean squared successive difference, 138
Mean, deviates from, 242
Mean, ordered deviates, 258-260
    studentized, 246-258, 260-263
Median, 269-272, 275
Mehler's identity, 296-297

Mellin transform, 139
Midrange, 269, 272-273, 275
Mills' ratio, 48
    approximations to, 52-53, 55-63
    bounds, 56-60
    continued fractions, 55-62
    expansions, 55-62
    Laplace's expansions, 57-58, 65-66
Moment ratios, 132-136
Monotone likelihood ratio, 28

Near-characterizations, 105
Negative binomial distribution, 191-194, 228
    approximations to, 191-194
        Camp-Paulson, 193
        from binomial, 192
        Molenaar, 192-193
        Peizer-Pratt, 194
Neyman's Type A distribution,
    approximation to, 158
Noncentral beta distribution, 128, 137
Noncentral chi-square, 119-121, 137, 156, 210, 326
    approximations, 156, 210-212
        Sankaran, 210-211
        to percentiles, 212
        Wilson-Hilferty, 210
Noncentral F distribution, 127-128, 222-223, 229
    approximations, 222-223
Noncentral t distribution, 137, 216-218, 228-229, 396-397
    approximations, 216-218
        based on c, 216-217
        to percentiles, 217-218
Normal density function, 47-48
    approximations to
Normal distribution, 37-40
    algorithms, 46
    characterization, 81
    cumulants, 24-25
    density function, 19
    Geary's a, 130-132
    genesis, 1-11
    kurtosis, 24-25
    incomplete moments, 25
    moment ratios, 132-136
    moments, 24-25
    nomogram, 46
    order statistics, 241-287
    quantiles, 20, 66-70

Normal Distribution, (continued)
  repeated derivatives, 21
  repeated integrals, 22-23
  sample mean, 115-116, 122, 124, 138
  sample mean deviation, 129-130
  sample quantiles, 278-280
  sample standard deviation, 118-119
  sample variance, 115-118, 122,
      124, 131, 138
  truncated, 33
Normal distribution function,
  rational functional approximations,
      51-53
  series expansions, 49-50
  simple approximations, 53-55
Normal integrals, 34-37
Normal quantiles,
  approximations to, 66-70
Normal samples,
  completeness, 365-367
  sufficient statistics, 365-367
Normal scores, 256-257, 279

Order statistics, 103, 241-287
  asymptotic properties, 273-278, 286
  concomitants, 344-346
  central limit theorem, 277-278
  distributions, 243-245
  linear trend, 287
  moments, 250-258
  percent points, 243-246
  sample range, 116
  tables, 245-259

Pearson system, 28
Pearson, Karl, 6-8, 11
Pierce, Charles S., 11
Point estimation, 365-386
Poisson distribution, approximations,
    188-191
  classical, 188
  Molenaar, 189-191
  Peizer-Pratt, 190, 194
Polar coordinates, 96, 306-307
Polya-type distributions, 29
Poplar coordinates, 325
Power transformations, 229
Prediction intervals, 404, 413-419
  simultaneous, 415-417
  some future observations, 415
  two samples, 418-419

Probability integral transformation, 2!
Probit transformation, 231

Quadratic forms, 120-121, 326
  approximations, 223-225
Quantiles, 231-232
  expansions for, 172-174
Quasi-medians, 272-273
Quasi-ranges, 242, 248, 268-269
Quetelet, Adolphe, 4, 6, 10, 11

Random normal deviates, 46
Range, 242, 246, 248-249
Rayleigh distribution, 317
Reach statistic, 242-243
Records, 199
Regression, 298-299
Runs, distribution of, 199

Sample coefficient of variation, 124-12
  asymptotic normality, 156-157
Sample correlation coefficient, 348-359
Sample fractiles,
  limit theorems for, 165-167
Sample mean, 115-116, 122, 124, 138
    242-243, 276, 281-282
  as estimator, 367-371, 374-377
Sample mean deviation, 129-130
Sample median, 269-272, 275-276
Sample midrange, 269, 272-273, 275
Sample quantiles, 278-280
Sample range, 263-268, 276, 284-285
  approximations, 265-266, 276
  distribution, 263-264, 276
  moments, 264-265
  studentized, 266-268
Sample variance, 115-118, 122, 124, 13
    138, 242-243, 285-286
  as estimator, 374
Sampling distributions, 113-139
Sargan distributions, 73
Selection differential, 242-243
Series expansions, 49-50
  normal cdf
Skew-normal distribution, 31
Skewness, 132
Square root transformations, 228
Stable distributions, 28
Student t distribution, 121-123, 137
Student's t distribution, 212
Studentized maximum modulus
    distribution, 402-413

Studentized ordered deviates, 246, 258, 260-263
Studentized range, 266-268, 401
Symmetric power distributions, 30

t distribution, 8, 121-123, 137
    bounds, 213-214
    inequalities, 123
t distribution, approximations, 213-216
    Fisher expansions, 214-215
    Mill, 215
    percentiles, 215-216
    Peizer-Pratt, 214
    Wallace, 213-214
Tchebyshev, Pafnuti L., 9-10
Tchebyshev-Hermite polynomials
  see Hermite polynomials
Testing normality, 97
Tolerance intervals,
    beta-content, 403-408, 409-412
    beta-expectation, 403-404, 408-409, 412-413

Tolerance intervals, (continued)
    simultaneous, 409-410
Tolerance regions, 347-348
    bivariate normal
Transformations, 226-232
    to normality, 226-227, 229-231
    variance stabilizing, 226-229
Trimmed mean, 282-284

U-statistics,
    asymptotic normality, 157-158
Uniform distribution, 32, 51, 99
Unimodal Distributions, 29

Van der Waerden statistic, 279-280
Variable sampling plans, 404
Von Mises distribution, approximations, 202-203

Weibull distributions, 72
Weldon, W. F. R., 6-8
Wilson-Hilferty approximation, 219